Lecture Notes in Mechanical Engineering

For further volumes:
http://www.springer.com/series/11236

S. Sathiyamoorthy · B. Elizabeth Caroline
J. Gnana Jayanthi
Editors

Emerging Trends in Science, Engineering and Technology

Proceedings of International Conference, INCOSET 2012

Editors
S. Sathiyamoorthy
J.J. College of Engineering
  and Technology
Tiruchirappalli, India

J. Gnana Jayanthi
Department of Computer Applications
J.J. College of Engineering
  and Technology
Tiruchirappalli, India

B. Elizabeth Caroline
Department of Electronics
  and Communication Engineering
J.J. College of Engineering
  and Technology
Tiruchirappalli, India

ISSN 2195-4356
ISBN 978-81-322-1006-1
DOI 10.1007/978-81-322-1007-8

ISSN 2195-4364   (electronic)
ISBN 978-81-322-1007-8   (eBook)

Springer New Delhi Heidelberg New York Dordrecht London

Library of Congress Control Number: 2012954493

© Springer India 2012

This work is subject to copyright. All rights are reserved by the Publisher, whether the whole or part of the material is concerned, specifically the rights of translation, reprinting, reuse of illustrations, recitation, broadcasting, reproduction on microfilms or in any other physical way, and transmission or information storage and retrieval, electronic adaptation, computer software, or by similar or dissimilar methodology now known or hereafter developed. Exempted from this legal reservation are brief excerpts in connection with reviews or scholarly analysis or material supplied specifically for the purpose of being entered and executed on a computer system, for exclusive use by the purchaser of the work. Duplication of this publication or parts thereof is permitted only under the provisions of the Copyright Law of the Publisher's location, in its current version, and permission for use must always be obtained from Springer. Permissions for use may be obtained through RightsLink at the Copyright Clearance Center. Violations are liable to prosecution under the respective Copyright Law.

The use of general descriptive names, registered names, trademarks, service marks, etc. in this publication does not imply, even in the absence of a specific statement, that such names are exempt from the relevant protective laws and regulations and therefore free for general use.

While the advice and information in this book are believed to be true and accurate at the date of publication, neither the authors nor the editors nor the publisher can accept any legal responsibility for any errors or omissions that may be made. The publisher makes no warranty, express or implied, with respect to the material contained herein.

Printed on acid-free paper

Springer is part of Springer Science+Business Media (www.springer.com)

Preface

Dear Distinguished Delegates, Guests, and Readers,

J.J. College of Engineering and Technology (J.J.C.E.T.), one of the premier educational institutions, located in Tiruchirappalli, Tamilnadu, India amid pleasant surroundings is a delightful place to hold an international conference (www.jjcet.ac.in). The management, staff, and students together are highly interested in promoting India as a developed country. As a stepping-stone in their efforts, they hosted this first International Conference on Emerging Trends in Science, Engineering and Technology (INCOSET-2012), getting technical sponsors from the leading and most prestigious publications, IEEE and Springer. The Conference INCOSET-2012 was held on 13 and 14 December 2012.

In the INCOSET-2012 conference, there were five themes covering different disciplines. Theme 1 Integrated Information, Communication, and Computing Technology. Theme 2 comprised Engineering Disciplines of Production, Aeronautical, Mechanical, Material Science, and Energy. Theme 3 consisted of Environmental and Civil Engineering. Theme 4 covered Electrical Power Drives and Controls, Instrumentation, Automation, and Robotics. Theme 5 included Science and Management. The INCOSET-2012 Conference published the proceedings in two volumes: Volume-I comprised Theme 1 and Theme 4 which is technically sponsored by IEEE. Volume-II comprised Themes 2, 3, and 5 which are published by Springer.

Both the proceedings volumes I and II of the conference contain the written versions of most of the contributions presented by the researchers and academicians who came from all over the world to present and discuss topics in their respective research areas. Around 250 scientific researchers and academicians had a good opportunity for having many fruitful discussions and exchanges that contributed to the success of the conference. The contributions of their research that were presented on the first day formed the heart of the conference to provide ample opportunity for discussion.

This conference attracted more than 700 papers from authors. All these papers were sent to the members of the Peer Reviewing Technical Committee. The reviews were based on the papers' relevance to the theme, originality and presentation of the work, importance and contribution of work towards current research. The selected, accepted, and registered papers 160 in number were split almost equally between the five main themes and were distributed across the days of the conference so that approximately equal numbers of papers in the different areas were scheduled on both days for oral presentation. One paper was selected from each theme and was awarded as the best paper in the respective theme. There were nearly 160 oral presentations during the conference's 20 pre-planned technical sessions. The Conference program was made available on the college and conference web site (www. jjcet.ac.in and www. incoset.jjcet.ac.in).

The inclusion of video conferencing technology in the conference INCOSET-2012 facilitated the keynote speaker's efforts to present the ideas and methods in a lively and accessible way. The authors of accepted and registered papers who were unable to participate in the conference INCOSET-2012 were also permitted to present from their own place. This adoption of Video Conferencing technology in the conference INCOSET-2012 was remarkable in the history of J.J.C.E.T and was highly appreciated.

There were five Keynote Speakers: In Theme 1, Dr. Mohd Zaid Bin Abdullah (USM's School of Electrical and Electronic Engineering, Malaysia) gave a very enlightening speech on *"Microwave Technology"*. In Theme 2, Dr. S. G. Ponnambalam, (Monash University, Malaysia) talked on *"Evolutionary Algorithm for Flexible Assembly Line Scheduling"* and Dr. Yukio Tamura, Director, Global Center of Excellence Program, Tokyo Polytechnic University, Japan, enlightened with the topic on *"Recent topics in wind engineering and building aerodynamics special reference to 634 m high Tokyo Sky Tree"*. In Theme 3, **Dr. Shen-En-Chen**, Associate Professor, University of North Carolina, Charlotte, USA, spoke on the topic **"Network-Wide Transportation Infrastructure Monitoring Strategies and Enabling Technologies"**. In Theme 4, Dr. Samavedham Lakshmi Narayanan (National University of Singapore, Singapore) enlightened with his talk on *Mathematical Modeling and its Relevance to the Management of Epidemics/ Infectious Disease*. In Theme 5, Dr. S. R. S. Prabaharan (Manipal International University, Malaysia) conversed on the topic *"Will NANO Technology Dominate 21Century and Beyond?"*.

All in all, the INCOSET-2012 in J.J.C.E.T., Tiruchirappalli was successful. The keynote speeches and plenary sessions gained insight into new areas. Also, included among the speakers were several young scientists, academicians, industrialists, Ph.D. research scholars and PG students, who brought new perspectives to their fields.

Given the rapidity with which engineering and science is advancing in all of the areas covered by INCOSET-2012, we expect that the future INCOSET conferences will be more stimulating than this most recent one, as indicated by the contributions presented in this proceedings volume.

Dr. S. Sathiyamoorthy
Dr. B. Elizabeth Caroline
Dr. J. Gnana Jayanthi

Acknowledgments

At the outset we gratefully acknowledge the moral support, permission, and guidance offered by Government of India—Ministry of Human Resource Development (MHRD), Ministry of External Affairs (MEA), Ministry of Home Affairs (MHA), the Government of Tamilnadu and Anna University to hold the conference INCOSET-2012 at our campus J.J. College of Engineering and Technology (J.J.C.E.T.), Tiruchirappalli, Tamilnadu, India.

We whole heartedly thank All India Council for Technical Education (AICTE), Department of Science and Technology (DST) India, Council of Scientific and Industrial Research (CSIR) India, and Tamil Nadu State Council for Science and Technology (TNSCST), all other sponsors and advertisers for their financial support to this conference.

We place on record our deep indebtedness and appreciation for the yeoman service rendered by members of the Advisory Committees, Keynote Speakers, Invited Resource Persons, and Session Chairs for sharing their expertise. We would like to acknowledge the services rendered by the reviewers who had spared considerable time in their busy schedule. Their valuable support helped in improving the quality of revised submissions of many of the papers submitted to the Conference.

We profusely thank our chief patrons Prof. K. Ponnusamy, Mrs. Pon. Priya, Mr. G. Venkatesan for their whole-hearted support and encouragement right from the beginning. Without lending their hands it would not have been possible to hold this conference.

We thank the patrons, our advisor Dr. V. Shanmuganathan from the deepest corner of our heart for his colossal support to the conference and offering valuable suggestions and our executive director Dr. T. Sivasankaran for extending his help in organizing this conference.

We place on record our deep indebtedness and appreciation for the yeoman service rendered by the theme Organizers and Conveners from all spheres in making this conference a grand success.

We express our sense of gratitude to all the authors of presented papers and participants without their presence, this conference could not have been successful.

Last but not the least; our thanks are due to Faculty Members, Staff and Students of J.J.C.E.T., who held the flag of the Conference very high among the scientific community of India and abroad.

Dr. S. Sathiyamoorthy, Conference Chairman
Dr. B. Elizabeth Caroline, Organizing Secretary
Dr. J. Gnana Jayanthi, Organizing Secretary

Contents

Part I   Thermal Engineering

A Review and New Design Options to Minimize the Capital and Operational Cost of Single Effect Solar Absorption Cooling System for Residential Use 3
V. Boopathi Raja and V. Shanmugam

Ant Colony Optimization for the Minimization of Internal Combustion Engine Forces and Displacements 19
T. Ramachandran, K. P. Padmanaban and J. Vinayagamoorthy

Augmentation of Heat Transfer and Pressure Drop Characteristics Inside a Double Pipe U-Tube Heat Exchanger by using Twisted Tape Inserts 33
P. V. Durga Prasad and A. V. S. S. K. S. Gupta

Design of Integrated R134a Vapor Compression Heating and Cooling Cycle 47
Priyank Agarwal, R. Shankar and T. Srinivas

Effect of Thermal Barrier Coating on DI Diesel Engine Fuelled with P20 Biodiesel 55
A. R. Pradeep Kumar, K. Annamalai and S. R. Premkartikkumar

Effects of Compression Ratio on the Performance and Emission Characteristics of Diesel Engine Fuelled with Ethanol Blended Diesel Fuel 63
M. Santhosh and K. P. Padmanaban

Green Carrier 81
Vikrant Goyal and Pankhuri Arora

Implementation of Ansys for Analyzing Windmill Tower
with AISI 302 Stainless Steel.................................... 93
K. Vinoth Raj, N. Shankar Ganesh and T. Elamaran

Integration of LiBr-H$_2$O Vapor Absorption Refrigeration
Cycle and Power Cycle 103
R. Shankar and T. Srinivas

Implementation of Methanol as Fuel in SI Engine
for the Measurement of Formaldehyde Emission.................... 111
T. Elamaran, N. Shankar Ganesh and K. Vinoth Raj

Performance Characteristics of Diesel Engine Operating
with Preheated Palm Biodiesel.................................. 123
Sumedh Ingle, Vilas Nandedkar and Madhva Nagarhalli

Recent Trends in Development of Hydraulic Press
for Damper Flap Notching 131
K. Vinoth Raj

Theoretical Analysis of Overall Heat Loss Coefficient
in a Flat Plate Solar Collector with an In-Built Energy Storage
Using a Phase Change Material 145
R. Sivakumar and V. Sivaramakrishnan

Part II Design and Manufacturing Techniques

Design and Development of Nonautomatic Tabletop Mini Lathe 157
T. T. M. Kannan and P. Marimuthu

Experimental Analysis of the Pressure Distribution
on Different Models of Carbody in the Windtunnel................. 165
K. Selvakumar and K. M. Parammasivam

Neuro-Genetic Hybrid System for 2D Fixture Layout............... 185
M. Vasundara, K. P. Padamanaban, M. Sabareeswaran and K. Kalaivanan

Part III Materials and Machining Processes

Basic Properties and Performance of Vegetable Oil-Based
Boric Acid Nanofluids in Machining 197
P. Vamsi Krishna, R. R. Srikant, R. Padmini and Bharat Parakh

Cuckoo Search Algorithm for Optimization of Sequence in PCB Holes Drilling Process 207

Wei Chen Esmonde Lim, G. Kanagaraj and S. G. Ponnambalam

Efficient Cooling of Building Using Phase Change Materials Along with Coolant 217

S. Mathana Krishnan, M. Joseph Stalin and P. Barath

Fatigue Damage Mechanisms in Fiber Reinforced with Al_2O_3 Composites Under Cyclic Reversed Loading 231

K. Mohamed bak and K. Kalaichelvan

Investigation of Chip Morphology and Tool Wear in Precision Turning Process 241

R. Vinayagamoorthy and M. Anthony Xavior

Investigation of Micromachining on CNC 251

D. Rajkumar, P. Ranjithkumar and C. Sathiyanarayanan

Machining Performance of TiCN/Al_2O_3 Multilayer and B-TiC Nano Multilayer Coated Inserts on Martensitic Stainless Steel in CNC Turning 261

Kamaraj Chandrasekaran, P. Marimuthu and K. Raja

Metallurgical Test on 20MnCr5 Steel: To Suggest as a Suitable Crankshaft Material 273

M. Umashankaran, J. Hari Vignesh, A. Saravana Kumar, R. Shyamaprasad and R. Saravana Prabhu

Multi-Objective Optimization and Empirical Modeling of Centerless Grinding Parameters 285

N. Senthil Kumar, C. K. Dhinakarraj, B. Deepanraj and G. Sankaranarayanan

Nanomaterials in Gas Sensors: A Big Leap in Technology! 297

G. Paulraj, A. Evangeline and G. Arun Prakash

Statistical Analysis of Tool Wear Using RSM and ANN 305

A. Arun Premnath, T. Alwarsamy and T. Abhinav

Surface Integrity of Ti-6Al-4V Precision Machining Using Coated Carbide Tools Under Dry Cutting Condition 317

R. Vinayagamoorthy and M. Anthony Xavior

Surface Roughness Evaluation in Drilling Hybrid
Metal Matrix Composites 325
T. Rajmohan, K. Palanikumar and G. Harish

Thermomechanical Behavior of Commercially Pure Titanium
(CP-Ti) During Isothermal Compressive Deformation 333
M. Vetrivel, T. Senthilvelan and G. Sriram

Part IV Intelligent Manufacturing

A Heuristic toward Minimizing Waiting Time of Critical
Jobs in a Flow Shop 343
R. Pugazhenthi and M. Anthony Xavior

A Study on Development of Knee Simulator for Testing
Artificial Knee Prosthesis 351
R. B. Durairaj, J. Shanker, P. Vinoth Kumar and M. Sivasankar

An Integrated Methodology for Geometric Tolerance
Analysis and Value Specification Based on Arithmetical,
Graphical and Analytical Methods................................ 361
R. Panneer and V. Sivaramakrishnan

Complexity on Parallel Machine Scheduling: A Review 373
D. K. Behera

Design of Optimal Maintenance Strategy 383
A. Sarkar and D. K. Behera

Effect of Dummy Machines in the Gupta's Heuristics
for Permutation Flow Shop Problems 391
A. Baskar and M. Anthony Xavior

Productivity Improvement in Railway Reservation Process:
A Case Study 399
Krishna Kumar Singh Sengar and Apratul Chandra Shukla

Simulation Study of Servo Control of Pneumatic Positioning
System Using Fuzzy PD Controller 409
D. Saravanakumar and B. Mohan

Part V Propulsion Systems

Studies on Starting Thrust Oscillations in Dual-Thrust Solid Propellant Rocket Motors 421
S. Deepthi, S. K. Kumaresh, D. Aravind Kumar,
J. Darshan Kumar and M. Arun

Vibration Analysis of a Constant Speed and Constant Pitch Wind Turbine 429
T. Sunder Selwyn and R. Kesavan

Part VI Aircraft Design and Manufacturing

Comparative Study Between Experimental Work and CFD Analysis in a Square Convergent Ribbed Duct 447
K. Sivakumar, E. Natarajan and N. Kulasekharan

Optimization of Rounded Spike on Hypersonic Forebody Reattachment 459
A. Sureshkumar, S. Nadaraja Pillai and P. Manikandan

Thermo Mechanical Analysis and Design of Rotating Disks 473
Priyambada Nayak and Kashi Nath Saha

Part VII Recent Trends in Civil Engineering

A Study on Radioactivity Levels in Sedimentary and Igneous Rocks Used as Building Materials in and Around Tiruchirappalli District, Tamil Nadu, India 487
P. Shahul Hameed, G. Sankaran Pillai, G. Satheeshkumar and K. Jeevarenuka

An Experimental Approach on Enhancing the Concrete Strength and its Resisting Property by Using Nano Silica Fly Ash 497
Yuvaraj Shanmuga Sundaram, Dinesh Nagarajan and Suji Mohankumar

Numerical Analysis and Study of Nanomaterials for Repair in Distressed Irrigation Structures 505
J. Nirmala, G. Dhanalakshmi and A. Rajaraman

Part VIII Geotechnical Engineering

Russell Projection and Mercator Projection by Harmonic Equations Comparison Study 517
Mohammed S. Akresh and Ali E. Said

Stable Isotopic Analysis Using Mass Spectrometry and Laser Based Techniques: A Review 523
M. Someshwar Rao

Part IX Waste Management

An Analytical Study of Physicochemical Characteristics of Groundwater in Agra Region 541
Chandel Prerna, Piyus Kumar Pathak, Deepak Singh and Amitabh Kumar Srivastava

Experimental Studies on Decolourisation of Textile Effluent by Using Bioremediation Technique 551
S. Sathish and D. Joshua Amarnath

Shade Improvement in Textiles Through Use of Water From RO-Treated Dye Effluent 561
S. Karthikeyan and R. I. Sathya

Part X Physics

Fabrication and Electrochemical Properties of All Solid State Lithium Battery Based on PVA–PVP Polymer Blend Electrolyte 571
N. Rajeswari, S. Selvasekarapandian, J. Kawamura and S. R. S. Prabaharan

Modeling Thermal Asymmetries in Honeycomb Double Exposure Solar Still 579
K. Shanmugasundaram and B. Janarthanan

Optical Band Gap Studies on Dy^{3+} Doped Boro-Tellurite Glasses 595
K. Maheshvaran and K. Marimuthu

Preparation and Characterization of Cobalt Doped Mn-Zn Ferrites 603
M. Bhuvaneswari and S. Sendhilnathan

Contents

Resonant Frequency Analysis of Nanowire-Based Sensor and Its Applications Toward Ethanol Sensing 611
D. Parthiban, M. Alagappan and A. Kandaswamy

Spectroscopic Analysis on Sm^{3+} Doped Fluoroborate Glasses 619
S. Arunkumar and K. Marimuthu

Synthesis and Spectroscopic Characterization of Pure and l-Arginine Doped KDP Crystals 627
K. Indira and T. Chitravel

Structural and Dielectric Studies on Dy^{3+} Doped Alkaliborate Glasses 637
S. Arunkumar and K. Marimuthu

Part XI Chemistry

Crystal Growth and Characterization of Biologically Essential Drug Materials 647
K. Bhavani, K. Sankaranarayanan and S. Jerome Das

Indigo Dye Removal by Using Coconut Shell Adsorbent and Performance Evaluation by Artificial Neural Network 655
S. K. Deshmukh

Structural and Electrical Conductivity of ZnO Nano Bi-pyramids-like Structure 665
Subbaiyan Sugapriya, Rangarajalu Sriram and Sriram Lakshmi

Part XII Mathematics

An Ordering Policy for Deteriorating Items with Quadratic Demand, Permissible Delay, and Partial Backlogging 673
K. F. Mary Latha and R. Uthayakumar

Complex Dynamics of BRD Sets 683
Bhagwati Prasad and Kuldip Katiyar

Coordinating Supply Chain Inventories with Shortages 689
R. Uthayakumar and M. Rameswari

Corrosion Inhibition of Mild Steel in Hydrochloric Acid Medium by 1-Methyl-3-Ethyl-2, 6-Diphenyl Piperidin-4-One Oxime 697
K. Tharini, K. Raja and A. N. Senthilkumar

New Mutation Embedded Generalized Binary PSO 705
Yograj Singh and Pinkey Chauhan

On Fuzzy Fractal Transforms 717
R. Uthayakumar and M. Rajkumar

Part XIII   Management Studies

An Application of Graph Theory Towards Portfolio Judgement 729
Tuhin Mukherjee and Arnab Kumar Ghoshal

A Study on CRM Influence in Small and Medium Retail B2B Industries in India 737
T. Narayana Reddy and G. Silpa

A Study on the Effectiveness of Campus Recruitment and Selection Process in IT Industries 745
Geeta Kesavaraj and Manjula Pattnaik

Financial Transactions Over Wired and Wireless Networks: Technical Perspective 759
J. Gnana Jayanthi, J. Felicita, S. Albert Rabara, A. Arun Gnanaraj and P. Manimozhi

Job Satisfaction as Predictor of Organizational Commitment 767
P. Na. Kanchana

Knowledge Management Key to Competitive Advantage 775
V. Maria Tresita Paul and G. Prithiviraj

Part XIV   Digital Libraries: Information Management for Global Access

Knowledge Management and Academic Libraries 785
K. Mahalakshmi and S. Ally Sornam

Author Index 791

About the Editors

Dr. S. Sathiyamoorthy, B.E., M.E., Ph.D., is the Principal of J.J. College of Engineering and Technology (J.J.C.E.T.), Tiruchirappalli, in Tamilnadu, India with over 17 years of experience in teaching, research, and industry. He has completed the B.E. degree in Electrical and Electronics Engineering (EEE) from Regional Engineering College, Tiruchirappalli in 1995; M.E. Control and Instrumentation (CIE) degree from Guindy Engineering College, Anna University, Chennai in 2002. He has received his Doctorate degree from the National Institute of Technology, Tiruchirappalli in 2008. He has attended several National and International conferences and workshops. He has published more than 15 papers in leading international and national journals. He has given numerous expert lectures in various forums and colleges in the area of Neural Networks and Fuzzy Logic Control. His research areas include Process Control and Industry Automation, Multiphase Flow Identification using Electrical Capacitance Tomography. He is a recognized Ph.D. Research Supervisor by the Anna University, Chennai, Tamilnadu, India since 2009 and St. Peter's University, Chennai, Tamilnadu, India since 2010. He is guiding ten Ph.D. research scholars. He is acting as Doctoral Committee member for five research scholars. He is an active member of the Institute of Electrical and Electronics Engineers (IEEE), and a life member of the Indian Society for Technical Education (ISTE), and the International Society of Automation (ISA).

Dr. B. Elizabeth Caroline, B.E., M.E., Ph.D., is an Electronics and Communication Engineering graduate of the 1992 Batch from Karunya Institute of Technology, Coimbatore in Tamilnadu, India. She obtained her Master's degree in Communication System in the year 2000 from Regional Engineering College (NIT), Trichy. She was awarded the Ph.D. (Optical Signal Processing) by Anna University, Chennai, Tamil Nadu, India in 2010. She has concurrent Teaching, Research, and Industrial experience of two decades. She is working as a Professor and Head of the Department of Electronics and Communication Engineering J.J.C.E.T., affiliated to Anna University, Chennai, Tamilnadu, India. She has published research papers in international journals, national and international Conferences. She traveled to Victoria University, Australia in the year 2005 to present her research paper in an international conference. Her areas of interest include digital signal processing, image processing, optical signal processing, and optical computing. She is an active member of IEEE and a life member of ISTE and Broadcast Engineering Society (BES). She is a recognized supervisor under the faculty of Information and Communication Engineering, Anna University, Tiruchirappalli, Tamilnadu, India since 2011. She is guiding two research scholars in the area of image processing and wireless sensor network. She is acting as Doctoral Committee member for three research scholars. She is a reviewer for SPIE (Photonic Society of Optical Engineering) journal, Elsevier International Journal, Academic Journal of Engineering and Computer Innovation, Journal of Scientific Research and Reviews (JSRR). She has served as Chair-Person, Resource Person, and Keynote Speaker for National and International Conferences.

Dr. J. Gnana Jayanthi, M.C.A., M.Phil., Ph.D., is an Associate Professor in the Department of Computer Applications, J.J.C.E.T., affiliated to Anna University, Chennai, Tamilnadu, India. She received the M.C.A. degree (1996) in Computer Applications from St. Joseph's College (Autonomous), affiliated to Bharathidasan University, India; M.Phil. (2003) in Computer Science from Mother Teresa Women's University, India; and Ph.D. (2012) in Computer Science from Bharathidasan University, India. She has more than 17 years of service experience in educational institutions to promote Research and Teaching-Learning processes. She traveled to Cambridge University, U.K. during Feb 2009 to present her paper on Mobile Commerce. Her paper was awarded as one of the best papers among 627 papers. She has published more than 25 research papers in International and National Conferences and Technical Journals. She has also been invited to chair the technical sessions sponsored technically by Springer in international conferences. She is a life member of the Computer Society of India (CSI), member of the World Scientific and Engineering Academy and

Society (WSEAS), International Association of Computer Science and Information Technology (IACSIT) and member of International Association of Engineers (IAENG). Her research interests include network-related fields such as the transition of IPv4/IPv6, mobile IP (v4/v6) networks, security, and mobile commerce. She has rich experience in research on Internet Protocol and Mobile Ad hoc NETworks. Currently, she focuses on issues related to inter-mobility of IPv4 and IPv6 nodes which are the immediate requirement in the Internet world.

Conference Details

INCOSET

The INternational Conference On Emerging Trends in Science, Engineering and Technology (INCOSET), largely deliberates on recent advances and innovations in Science, Engineering, Technology and Management Studies broadly classified under five themes. INCOSET-2012 had paper presentations which provided a platform for interaction and exchange of knowledge expertise and experience between the participants drawn from academia, industries as well as research laboratories and invited speakers from India and abroad. It greatly helped in innovative invention and modern product design for the budding entrepreneurs to be successful entrepreneurial endeavors. The young Scientists and Researchers got a fillip to proceed further with confidence in their work. This offered profitable solutions to the industrialists to face their problems with courage. The research community was benefitted by this conference to broaden their knowledge-base and hone their skill sets.

About Institution

J.J. College of Engineering and Technology (J.J.C.E.T.), was established in the year 1994–1995 by a munificent trust "*J.J. Educational Health and Charitable Trust*"

formed by benevolent and philanthropic educationists of Tiruchirappalli. The College is situated in Ammapettai Village, 18 km from the city abutting the NH 45 in a green eco friendly and serene area sprawling over more than 100 acres. It is approved by AICTE, New Delhi and affiliated to Anna University. Over the years, the college has grown by leaps and bounds in every aspect and has become one of the pioneering technical institutions in this educationally, economically, and socially backward rural area. The college has become the cynosure of craving young minds for excellence in knowledge for the best all round exposure since the day of its inception.

The college is offering nine Under Graduate and nine Post Graduate courses in all disciplines of prime and paramount importance. The unique character of this institution is its constant and continuous perseverance for perfection in all disciplines, both at UG and PG levels in teaching, learning, and research. The college has excellent infra structural facilities, well equipped and modern state-of-the-art laboratories, enriched and endowed library and to crown above all a team of well qualified, experienced and motivated faculty to mould the career of young students.

Tiruchirappalli city is centrally located on the bank of Holy river Cauvery and is known for historical places like Rockfort Temple, St. Lourde's Church and Natharsha Mosque. It is well connected to the rest of the country by rail, road, and air.

Organizing Committee

Chief Patrons

Prof. K. Ponnusamy, Chairman, J.J. Group of Institutions.
Mrs. Pon. Priya, Vice-Chairperson, J.J. Group of Institutions.
Mr. G. Venkatesan, Secretary, J.J. Group of Institutions.

Patrons

Dr. V. Shanmuganathan, Director, J.J. Group of Institutions.
Dr. T. Sivasankaran, Executive Director, J.J. Group of Institutions.

Conference Chairman

Dr. S. Sathiyamoorthy, Principal, J.J.C.E.T.

Conference Co-Chairman

Dr. P. Shahul Hameed, Research Advisor, J.J.C.E.T.

Secretaries

Dr. B. Elizabeth Caroline, HOD, Department of Electronics and Communication Engineering, J.J.C.E.T.
Dr. J. Gnana Jayanthi, Associate Professor, Department of Computer Applications, J.J.C.E.T.

Theme Organizers

Dr. S. Thayumanavan, HOD, Department of Civil Engineering, J.J.C.E.T.
Dr. P. D. Sheba Kezia Malarchelvi, HOD, Department of Computer Science and Engineering, J.J.C.E.T.
Dr. A. Josephine Amala, HOD, Department of Electrical and Electronics Engineering, J.J.C.E.T.
Dr. Sw. Rajamanoharane, HOD, Department of Management Studies, J.J.C.E.T.
Dr. G. Paulraj, HOD, Department of Production Engineering, J.J.C.E.T.

Conveners

Dr. M. Sujaritha, Department of Computer Science and Engineering, J.J.C.E.T.
Dr. T. Ayyamperummal, Department of Civil Engineering, J.J.C.E.T.
Dr. R. Pavendan, HOD, Department of Mechanical Engineering, J.J.C.E.T.
Dr. M. Sabibullah, HOD, Department of Computer Applications, J.J.C.E.T.
Dr. Kavitha Shanmugam, Department of Management Studies, J.J.C.E.T.
Dr. R. Nagarajan, HOD, Department of Mathematics, J.J.C.E.T.
Dr. P. K. Manimozhi, HOD, Department of Physics, J.J.C.E.T.
Dr. A. Prabhaharan, HOD, Department of Chemistry, J.J.C.E.T.
Dr. G. Sathurappa Samy, HOD, Department of English, J.J.C.E.T.
Dr. T. Veema Raj, Department of Chemistry, J.J.C.E.T.
Dr. S. Selvandan, Department of Physics, J.J.C.E.T.
Dr. S. Boobalan, Department of Chemistry, J.J.C.E.T.
Dr. M. Ramesh, Department of Chemistry, J.J.C.E.T.
Dr. R. Surendrakumar, Department of Chemistry, J.J.C.E.T.
Dr. S. Kolangikannan, Department of Physics, J.J.C.E.T.
Dr. C. Hariharan, Department of Physics, J.J.C.E.T.
Dr. K. Jayaraja, Department of Physical Education, J.J.C.E.T.
Cdr. I. Sharfudeen, HOD, Department of Aeronautical Engineering, J.J.C.E.T.
Prof. A. Saravanan, HOD, Department of Information and Technology, J.J.C.E.T.
Prof. K. Dhayalini, Department of Electrical and Electronics Engineering, J.J.C.E.T.
Prof. R. Sumathi, Department of Computer Science and Engineering, J.J.C.E.T.
Prof. M. Maheswari, Department of Electronics and Communication Engineering, J.J.C.E.T.
Prof. M. Banu Sundareswari, HOD, Department of Electronics and Instrumentation Engineering, J.J.C.E.T.
Prof. K. Thamodaran, Department of Computer Applications, J.J.C.E.T.
Prof. V. Balasubramaniam, Department of Mechanical Engineering, J.J.C.E.T.
Prof. S. Sankaran Pillai, HOD, Department of Library, J.J.C.E.T.
Prof. M. Mahendran, HOD, Department of Training and Placement, J.J.C.E.T.
Prof. A. Noble Jeba Kumar, Department of English, J.J.C.E.T.
Prof. K. Soundaranayaki, Department of Civil Engineering, J.J.C.E.T.

Advisory Committee

Dr. M. Ponnavaikko, SRM University, Chennai, India.
Dr. S. Sundarrajan, NIT, Tiruchirappalli, India.
Dr. Profulla Agnihotri, IIM, Tiruchirappalli, India.
Dr. M. Chithambaram, IIT, Chennai, India.
Dr. R. Krishnamoorthi, AUT, Tiruchirappalli, India.
Dr. A. P. Kabilan, CCET, Karur, India.
Dr. S. Albert Rabara, SJC, Tiruchirappalli, India.
Dr. S. Annadurai, SKCET, Coimbatore, India.
Dr. N. Mathivanan, MKU, Madurai, India.
Dr. S. Sundaramoorthy, PEC, Pondicherry, India
Dr. B. Vijaya Kumar, BITS-Pilani, Dubai Campus, UAE.
Dr. A. Alphones, Nanyang Technological University, Singapore.
Dr. Sitti Mariyam Shamsudeen, Universiti Teknologi, Malaysia.
Dr. Vijay K. Gurbani, Bell Laboratories, Alcatel-Lucent, USA.
Dr. R. Janadhanam, University of North Carolina, Charlotte, USA.
Dr. Mohd Zaid Abdullah, University Sains Malaysia, Malaysia.
Dr. Shen-En-Chen, University of North Carolina, Charlotte, USA.
Dr. K. Ravichandran, New York Institute of Technology, Abudhabi.
Dr. R. Ramanathan, University of Bedfordshire Business School, UK.
Dr. Ganesh D. Sockalingum, Universite de Reims Champagne-Ardenne, France.
Dr. Samavedham Lakshmi Narayanan, NUS, Singapore.
Dr. M. D. Jose, IGCAR, Kalpakkam, India.
Dr. P. M. Ravi, BARC, Mumbai, India.
Dr. Vijayamohanan K. Pillai, CSIR-CERI, Karaikudi, India.
Dr. A. Sundaramurthy, ISRO, Bangalore, India.
Dr. M. Sithartha Muthu Vijayan, C-MMACS, Bangalore, India.
Dr. N. Rajendran, NPCL, Mumbai, India.
Dr. C. Narendiran, Samsung, Bangalore, India.
Mr. T. S. Rengarajan, IEEE Madras Section, Chennai, India.
Dr. R. Jeyapal, Cethar Vessels Ltd, Tiruchirappalli, India.
Dr. C. Ramesh, Tata Elxsi Ltd, Chennai, India.

Reviewers

Dr. C. Sathiyanarayanan, National Institute of Technology, Tiruchirappalli, Tamilnadu, India.

Dr. N. Sivasankaran, Indian Institute of Management, Shillong, India.

Dr. M. Kannadasan, Indian Institute of Management, Raipur, India.

Dr. N. Nandagopal, PSG Institute of Management, Coimbatore, Tamilnadu, India.

Dr. R. Krishnamoorthi, Anna University of Technology, Tiruchrappalli, Tamilnadu, India.

Dr. S. Albert Rabara, St.Joseph's College, Tiruchirappalli, Tamilnadu, India.

Dr. L. Arockiam, St. Joseph's College, Tiruchirappalli, Tamilnadu, India.

Dr. V. Arul Mozhi Selvan, National Institute of Technology, Tiruchirappalli, Tamilnadu, India.

Dr. S. Suresh, National Institute of Technology, Tiruchirappalli, Tamilnadu, India.

Dr. R. Ramanathan, University of Bedfordshire Business School, UK.

Dr. R. Jeyapaul, National Institute of Technology, Tiruchirappalli, Tamilnadu, India.

Dr. N. Rajendran, NPCL, Mumbai, India.

Dr. V. Anandakrishnan, National Institute of Technology, Tiruchirappalli, Tamilnadu, India.

Dr. S. Vinoth, National Institute of Technology, Tiruchirappalli, Tamilnadu, India.

Dr. R. Balu, NICHE, Kumarakovil, Tamilnadu, India.

Dr. S. G. Ponnambalam, Monas University, Malaysia.

Dr. Rohit Nongthobam, Asia Institute of Technology, Bangkok, Thailand.

Dr. P. Pradeep Majuwdar, Indian Institute of Science, Bangalore, India.

Dr. Indu Uprety, Gautam Buddha University, Nepal, India.

Dr. Elizabeth Kay Kay, GTZ, Frankfurt, Germany.

Dr. Shen-en Chen, University of North Carolina, USA.

Shri. R. Mathiyarasu, IGCAR, Kalpakkam, India.

Dr. K. Shivanna, BARC, Mumbai, India.

Dr. Ganesh D. Sockalingum, Universite de Reims Champagne-Ardenne, France.

Dr. G. Muralidharan, Gandhigram University, Gandhigram, Tamilnadu, India.

Dr. J. P. Malathy, Kalasalingam University, Srivilliputtur, Tamilnadu, India.

Dr. S. Ravi, Mepco Schelnik Engineering College, Sivakasi, Tamilnadu, India.

Dr. B. Natarajan, BARC, Mumbai, India.

Dr. K. G. Sekar, National College, Tiruchirappalli, Tamilnadu, India.

Dr. N. Kannan, ANJA College, Sivakasi, Tamilnadu, India.

Dr. M. Selvakumar, MIT, Manipal University, Karnataka, India.

Dr. B. Vijaya Kumar, BITS-Pilani, Dubai Campus, UAE.

Dr. M. Madurai, Bharathidasan University, Tiruchirappalli, Tamilnadu, India.

Dr. L. Rajendran, Madura College, Madurai, Tamilnadu, India.

Dr. A. Solairaju, Jamal Mohamed College, Tiruchirappalli, Tamilnadu, India.

Dr. K. Kannan, SASTRA University, Thanjavur, Tamilnadu, India.

Dr. R. Subramanian, BIM, Tiruchirappalli, Tamilnadu, India.

Dr. K. Karthikeyan, Saranathan College of Engineering, Tiruchirappalli, Tamilnadu, India.

Dr. S. Arunkumar, Saranathan College of Engineering, Tiruchirappalli, Tamilnadu, India.

Dr. V. R. Rajan, Pondicherry University, Puducherry, India.

Dr. C. Narendiran, Samsung, Bangalore, India.

Part I
Thermal Engineering

A Review and New Design Options to Minimize the Capital and Operational Cost of Single Effect Solar Absorption Cooling System for Residential Use

V. Boopathi Raja and V. Shanmugam

Abstract Solar energy is an alternative energy source for cooling systems where electricity is demand or expensive. Many solar assisted cooling systems have been installed in different countries for domestic purpose. Many researches are going on to achieve economical and efficient thermal systems when compared with conventional systems. This paper reviews the past efforts of solar assisted-single effect vapor absorption cooling system using LiBr-H_2O mixture for residential buildings. Solar assisted single-effect absorption cooling systems were capable of working in the driving temperature range of 70–100 °C. In this system LiBr-H_2O are the major working pairs and has a higher COP than any other working fluids. Besides the review of the past theoretical and experimental investigations of solar single effect absorption cooling systems, some new ideas were introduced to minimize the capital and operational cost, to reduce heat loss from generator and thus to increase COP to get effective cooling.

Keywords Solar absorption cooling system • New design options • Minimize cost • Suggestions

1 Introduction

Cold production through absorption cycles has been traditionally considered one of the most desirable applications for solar thermal energy. So, the most commercially developed solar cooling technologies are the absorption systems [1]. Conventional cold producing machines that are based on vapor compression

---

V. Boopathi Raja (✉)
Department of Mechanical Engineering, Narasu's Sarathy Institute of Technology,
Salem, Tamilnadu, India
e-mail: boops.raja@gmail.com

V. Shanmugam
Sri Shanmugha College of Engineering and Technology, Sankari, Tamilnadu, India

S. Sathiyamoorthy et al. (eds.), *Emerging Trends in Science,*
Engineering and Technology, Lecture Notes in Mechanical Engineering,
DOI: 10.1007/978-81-322-1007-8_1, © Springer India 2012

principle are primary electricity consumers and their working fluids are being banned by international legislation. Solar powered cooling systems as a green cold production technology are the best alternative. Absorption refrigeration is a mature technology that has proved its applicability with the possibility to be driven by low grade solar and waste heat [2]. The CFCs and HCFCs gases, which are used by the conventional vapor compression cooling system, have high global warming potential and also high ozone depleting potential. These effects can be remedied by choosing the environmental friendly cooling system like $H_2O/LiBr$ absorption system. An advantage of this absorption system is that it can be driven by low grade thermal energy like solar energy [3]. An absorbent and a refrigerant in an absorption cycle form a working pair. The most common working pairs in absorption refrigeration are lithium bromide-water and water-ammonia. [4] For cooling purposes, the $LiBr-H_2O$ system is the cheapest, while the cost of H_2O-NH_3 is high. The optimum generator temperatures in the $LiBr-H_2O$ system for cooling purposes are around 75–92 °C. The lithium bromide-water pair is widely used for cooling applications, with evaporation temperatures about 5–10 °C.

1.1 Working Principle of Solar Assisted Lithium Bromide-Water Absorption Cycle

The basic working principle of solar assisted lithium bromide-water absorption cycle is depicted in Fig. 1, which is the simplest and most commonly used design. In the absorption cycle, compressing refrigerant vapor is achieved by the absorber, the solution pump and the generator. Water evaporated from evaporator (which outputs a cooling effect) is absorbed into a strong lithium bromide solution in the absorber, and the absorption process need to release heat of absorption to the ambient. After absorbing the water vapor, the lithium bromide solution becomes a weak solution, which is then pumped to the generator. As heat is added to the generator, water will be desorbed from the solution in a vapor form. The vapor then flows to the condenser, where it is condensed and condensing heat is rejected to the ambient. The condensed water flows through an expansion device, where the pressure is reduced. The strong solution from the generator flows back to the absorber to absorb water vapor again, a heat exchanger could be used between the strong solution and weak solution lines. The entire cycle operates below atmospheric pressure, since water is used as the refrigerant. The advantages of absorption method over conventional method is that they consume little electricity, they can be run by low thermal energy source, they have very few moving parts leading to low noise and vibration levels, and they do not emit ozone depleting substances [5].

Jaruwongwittaya et al. [6] pointed that the absorption cooling technology using lithium bromide/water was the most appropriate for the solar applications in Thailand. Fong et al. [7] compared five types of solar cooling systems for Hong Kong, the solar electric compression refrigeration, solar mechanical compression refrigeration, solar absorption refrigeration, solar adsorption refrigeration, and

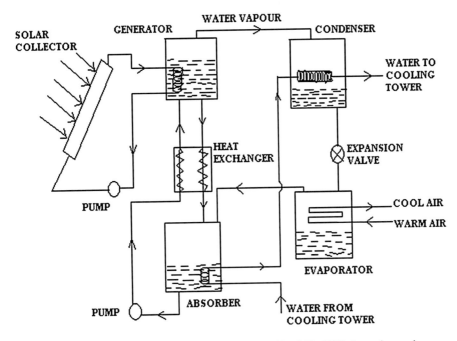

Fig. 1 Schematic representation of solar assisted single effect LiBr-H2O absorption cycle

solar solid desiccant cooling. Through this comparative study, it was found that solar electric compression refrigeration and solar absorption refrigeration had the highest energy saving potential in Hong Kong. Commercially available absorption chillers for air conditioning applications usually operate with solution of lithium bromide in water and use steam or hot water as the heat source [8]. According to the operating temperature range of driving thermal source, single-effect LiBr/H_2O absorption chillers have the advantage of being powered by ordinary flat-plate or evacuated tubular solar collectors. Under normal operation conditions, such machines need typically temperatures of the driving heat of 80–100 °C and achieve a COP of about 0.7. The hot water storage tank is used in the system as a heat reservoir. [9] For solar thermal cooling the cost of solar collection is much lower as a percentage of the overall cost, but the cost of the refrigeration system often represents a larger percentage of the cost. If the costs of refrigeration were to come down as well as thermal refrigeration COP increases, especially to values greater than 1, it could be expected that solar thermal cooling costs would be competitive with electric cooling costs. Solar thermal systems will be cost-competitive by 2030, if COP improvements and/or thermal collector costs comes down considerably. Dieter Boer et al. [10] states that the problem of satisfying a given cooling demand at minimum cost and environmental impact is formulated as a bi-criterion non-linear optimization problem that seeks to minimize the total cost of the cooling application and its contribution to global warming. The results obtained by them shows that with the current energy price and without considering government subsides on solar technologies, the use of solar energy in cooling applications is not economically appealing.

In this paper, the research works of solar-powered single effect absorption cooling systems were reviewed. Some new options were given for solar-powered single-effect absorption cooling systems for domestic purpose, economical comparison with vapor compression system was made and suggestions were given to reduce capital and operational cost.

2 Experimental Investigations

There were some solar absorption cooling systems in large capacities up to 100 kW, the experimental investigations were mainly based upon medium and small-sized solar cooling systems. Usually, the cooling capacity and COP of solar cooling systems were tested under practical operating conditions. Rosiek and Batlles [11] reported the solar-powered single-effect absorption cooling system installed in the Solar Energy Research Center of Spain. According to the calculation, the cooling demand during the whole year was 13,255 kWh. The flat-plate solar collectors with the area of 160 m^2 and a single-effect absorption chiller with the cooling capacity of 70 kW were used to meet the energy demands for cooling in summer. The performance of the solar-powered cooling system was monitored and controlled by a control and data acquisition system. During one year of operation, it could be seen that the solar collectors were able to provide sufficient energy to supply the absorption chiller during the summer. The average values of COP and the cooling capacity were calculated for summer months, obtaining values of the order of 0.6 and 40 kW, respectively. Ortiz et al. [12] and Mammoli et al. [13] carried out the experiments of a solar cooling system for a 7,000 m^2 educational building situated in a high desert climate. There were two kinds of solar collectors in this system, which included 124 m^2 of flat-plate collectors and 108 m^2 of vacuum tubular collectors. A water–glycol mixture was pumped through the arrays and a heat exchanger, which was connected to a hot water storage tank with approximate volume of 34 m^3. The absorption chiller was a Yazaki single-effect LiBr/H_2O water fired chiller. This 70 kW absorption chiller was designed to work with hot water supply temperatures in the range from 70 to 95 °C. The cold water produced by the absorption chiller is stored in seven 50 m^3 cold water tanks and supplied to the cooling coils. Large chilled water storage tanks were charged off-peak and discharged during the day, cooling the building in parallel with the chiller. According to the experimental results, in the peak of summer, the solar cooling system could supply approximately 18 % of the total cooling load. This percentage could be increased to 36 % by tuning the air handler operation and by improving the insulation in the storage tank.

Syed et al. [14] investigated a solar cooling system consisting of a 35 kW LiBr/ H_2O absorption machine energized by 49.9 m^2 of flat-plate collectors. Thermal energy was stored in a 2 m^3 stratified hot water storage tank during hours of bright sunshine. The generator design of the machine allowed the use of hot water in the temperature range of 65–90 °C. The maximum measured instantaneous daily average and period average COP were 0.60 (at maximum capacity), 0.42, and 0.34, respectively. The daily average collector efficiency (without considering pipe and plate heat exchanger losses) was 50 %. Through the analysis of energy flows in the system, it was demonstrated that the technology worked best in dry and hot climatic conditions where daily large variations in relative humidity and dry bulb temperature prevailed. Praene et al. [15] presented a solar-powered 30 kW LiBr/H$_2$O single-effect absorption cooling system which was designed and installed at Institute Universitaire Technologique of Saint Pierre. It was reported that the solar loop could produce hot water to fire the absorption chiller from 8:00 a.m to 5:00 p.m. According to the first field test, the system was sufficient to obtain thermal comfort with the mean air temperature inside the classrooms of about 25 °C. Li and Sumathy [16] studied the performance of a solar-powered absorption air conditioning system with a partitioned hot water storage tank. The system employed a flat-plate collector array with the surface area of 38 m^2 to drive a LiBr/H$_2$O absorption chiller of 4.7 kW cooling capacity. The system was provided with a storage tank (2.75 m^3) which was partitioned into two parts. The upper part had a volume of about one-fourth of the entire tank. The study revealed that the solar cooling effect could be realized nearly 2 h earlier to the system operating in partitioned mode. In this system a solar COP of about 0.07, which was about 15 % higher than traditional whole-tank mode, was attained. Experimental results also showed that during cloudy days, the system could not provide a cooling effect, when operated conventionally, however in the partitioned mode-driven system the chiller could be energized, using solar energy as the only heat source.

Agyenim et al. [17] developed a domestic-scale prototype experimental solar cooling system, which consisted of a 12 m^2 vacuum tubular solar collector, a 4.5 kW LiBr/H$_2$O absorption chiller, a 1,000 l cold storage tank, and a 6 kW fan coil. The average COP of the system was 0.58. Experimental results proved the feasibility of the concept of cold store at this scale, with chilled water temperatures as low as 7.4 °C, demonstrating its potential use in cooling domestic scale buildings. The existing experimental results showed that solar-powered single-effect absorption cooling systems were capable of working in the driving temperature range of 70–100 °C. Generally, the system COP of about 0.6 could be obtained under the design condition. However, when a chiller worked at partial load, a low efficiency from solar to cooling would be obtained. Rodríguez Hidalgo et al. [18] developed an experimental facility with 50 m^2 flat-plate solar collectors. It fed a 35 kW single-effect LiBr/H$_2$O absorption machine. The chiller worked at partial load, during 2004 summer season for COP of 0.33. Izquierdo [19] conducted trials to determine the performance of a commercial 4.5-kW air-cooled, single effect LiBr/H$_2$O absorption chiller for residential use. Measurements were recorded over a 20-day period. The hot water inlet temperature in the generator varied throughout the day from 80 to 107 °C. The results for the period as a whole showed that cooling power tended to decline with rising outdoor dry bulb temperatures. At outdoor temperatures from 35 to 41.3 °C the chilled water outlet temperature in the evaporator climbed to over 15 °C. The total energy supplied to the generator came to 1085.5 kWh and the heat removed in the evaporator to 534.5 kWh. The average COP for the period as a whole was 0.49. Al-Dadah [20] conducted Propane miscibility tests in various lubricating oils and concluded that Propane is most miscible in Alkylated Benzene AB300 as compared to AB150 and shell Clavus oils 32 and 64. They used Propane as refrigerant and Alkylated Benzene as absorbent and its performance was evaluated. The absorption system gave an effective cooling capacity of 1.3 kW. The COP increased with increasing generator temperature and by decreasing the absorber temperature. A COP was found up to 1.

3 Theoretical Analysis and Simulation

The main components of a solar absorption cooling system are the solar field, the absorption cooling system and the heat storage water tank. The overall system performances depend on the coupling of these three components. Such research works were carried out mainly by theoretical analysis and simulation with the aid of Simulation softwares like TRNSYS [21]. Balghouthi et al. [22] presented a research project aiming at assessing the feasibility of solar-powered absorption cooling technology under Tunisian conditions. The system was modeled using the TRNSYS and EES programs with a meteorological year data file containing the weather parameters of Tunis, the capital of Tunisia. The optimized system for a typical building of 150 m^2 was composed of a LiBr/H_2O absorption chiller of a capacity of 11 kW, a 30 m^2 flat-plate solar collector area tilted 35° from the horizontal and a 0.8 m^3 hot water storage tank. Florides et al. [23] designed a LiBr/H_2O absorption unit with the cooling capacity of 11 kW, which could cover the cooling load of a typical model house in Cyprus. The optimum system as obtained from the complete system simulations consisted of 15 m^2 compound parabolic collectors tilted at 30° from horizontal and a 600 l hot water storage tank. Assilzadeh et al. [24] reported a solar cooling system that had been designed for Malaysia and similar tropical regions using evacuated tubular solar collectors and a LiBr/H_2O absorption unit. It was shown that a 0.8 m^3 hot water storage tank was essential in order to achieve continuous operation and increase the reliability of the system. The optimum system for Malaysia's climate for a 3.5 kW system consisted of 35 m^2 evacuated tubular solar collector sloped at 20°. Atmaca and Yigit [25] simulated a solar cooling system based on a 10.5 kW constant cooling load. A modular computer program was developed for the absorption system to simulate various cycle configurations and solar energy parameters for Antalya, Turkey. It was shown that the solar collector area of 50 m^2, a 3,750 kg storage tank mass seemed to be the best choice. Joudi and Abdul-Ghafour [26] developed an integrated program for the complete simulation of a solar cooling system with a LiBr/H_2O absorption chiller. The results obtained from the simulation were used to develop a general design procedure for solar cooling systems, presented in a graphical form called the cooling chart. Using this design chart could simplify the designer's task for predicting the long term cooling energy supplied from a solar collector array serving an absorption chilled water system. Besides, a correlation was developed from the simulation results for estimating the hot water storage size necessary for the solar cooling system. The coupling of the main components of a solar cooling system is determined by the cooling demand, solar resource availability, climatic conditions, component cost and component performance characteristics. The specific cooling capacity was observed to be 0.1–0.7 kW/m^2. Jian Sun [27] made a mathematical model of a single effect, LiBr-H2O absorption heat pump operated at steady conditions. He took into consideration of crosscurrent flow of fluids for heat and mass exchangers, two-dimensional distribution of temperature and concentration fields, local values of heat and mass transfer coefficients, thermal parameter dependent physical properties of working fluids and operation limits due to the danger of the LiBr aqueous solution hydrates and crystallization. It was found that the mass flux of vapor increased with the increase of absorber pressure, coolant flow rate, spray density of LiBr solution and decrease of coolant and input temperature of solution. And the vapor mass flux increased almost linearly with the increase of absorber pressure. Results derived from this model show agreement within 7 % with experimental values. Xiaohong Liao [28] focused on the crystallization issues and control strategies in LiBr-H_2O air-cooled absorption chillers. He specified six causes which may trigger crystallization: (1) high ambient temperature; (2) low ambient temperature with full load; (3) air leak into the machine or non-absorbable gases produced during corrosion; (4) too much heat input to the desorbed; (5) failed dilution after shutdown; and (6) chilled water supply temperature is set too low when the weather and/or exhaust are too hot. Bahador Bakhtiari [29] conducted an experimental and simulation analysis of a laboratory single-stage H_2O–LiBr absorption heat pump with a cooling capacity of 14 kW. Design characteristics of the machine was given and the temperature of chilled, cooling and hot water and, the flow rate of cooling water and hot water were found to be the most influential operating parameters. A design and dimensioning model of H_2O–LiBr absorption heat pumps was developed. The steady-state simulation results of the model were compared with experimental measurements.

4 New Design Options of Solar-Powered Single-Effect Absorption Cooling System

The schematic diagram of the proposed new design is shown in Fig. 2. As per the new design, three major electrical components are required to run the system. A fan in the air cooled condenser, a blower in the evaporator and a solution pump to pump weak solution from the absorber to the generator. In this new design the generator is placed inside the insulated hot water storage tank.

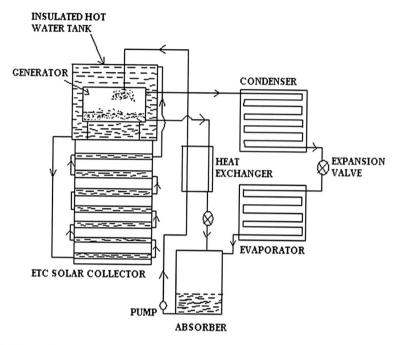

Fig. 2 New design with generator placed inside insulated hot water tank

This prevents heat loss from the generator to the surrounding. Also the insulation cost for the generator gets minimized. If the storage tank and generator are separate units then the hot water has to be circulated from the tank. As per the new design the generator is placed inside the tank, the heat transfer loss from the tank to generator is minimized. The hot water tank is located at the top of the collector, thermo siphon principle is used to transfer heat from the solar collector to the storage tank. This avoids the usage of pump to circulate the water from collector to tank. Thus the initial and operating cost of the VACS can be reduced. The two main circuits: (1). Hot water circuit (solar collector—hot water storage tank—solar collector). (2). Refrigerant circuit (generator—condenser—evaporator—absorber—solution pump—generator) is evacuated and the vacuum pressure can be raised to reduce the boiling point of water. This helps to run the absorption cooling system even in low solar thermal intensity. The refrigerant vapor and the strong solution from the generator are passed to the absorber by siphon principle and the components of the cooling system are placed at proper altitudes so that both fluids will flow in the required pressure. Zhaolin Gu [30] applied vacuum membrane distillation process in Li-Br absorption system. Hollow fiber module with Polyvinylidene fluoride (PVDF) membrane material is used. This process has the advantages of lower temperature driving heater and enormous contact area per unit volume. This membrane can be used in the generator to operate at low temperature than traditional generator. The vapor absorption cooling system can also be integrated with components used in conventional cooling system to get better performance and reduce cost.

4.1 Solar Collectors

Nearly all solar cooling systems were driven by ordinary flat plate or evacuated tubular solar collectors, which are available in the market. However, some new types of solar collectors have also been taken into account. Mazloumi et al. [31] simulated a single-effect absorption cooling system designed to supply the cooling load of a typical house in Ahwaz where the cooling load peak was about 17.5 kW. Solar energy was absorbed by a horizontal N–S parabolic trough collector and stored in an insulated thermal storage tank. It was concluded that the minimum required collector area was about 57.6 m^2, which could supply the cooling loads for the sunshine hours of the design day. The parabolic trough collectors obtained more solar heat energy in the areas with suitable direct radiation, which caused the cooling systems to operate earlier. Mittelman et al. [32] investigated the performance and cost of a CPVT system with single-effect absorption cooling. Concentrating photovoltaic (CPV) systems can operate at higher temperatures than flat-plate collectors. Collecting the rejected heat from a CPV system leads to a CPV/thermal (CPVT) system, providing both electricity and heat at medium rather than low temperatures. CPVT collectors may operate at temperatures above 100 °C, and the thermal energy can drive processes such as refrigeration, desalination and steam production. In the CPVT system, the thermal energy is a low cost byproduct and, therefore, could lead to a much more competitive solar cooling solution. The results showed that under a wide range of economic conditions, the combined solar cooling and power generation plant could be comparable to, and sometimes even significantly better than, the conventional alternative.

4.2 Auxiliary Energy Systems

For the purpose of all-weather operation, it is necessary to install auxiliary energy systems to supplement solar-powered cooling systems. Apart from electric heaters and oil boilers, almost all the auxiliary energy-to be used in case of scarce solar irradiation is supplied by a gas fired auxiliary boiler. The use of a gas-fired heater can be acceptable only when the auxiliary energy to be supplied is low [33]. Such a conclusion was also drawn by Calise et al. [34]. Compared with the auxiliary energy system of a gas-fired heater, the layout of an electric water-cooled chiller showed better energetic performance. It was showed that the primary energy saving of such a system vs. a traditional electric heat pump was close to 37 %. In addition, the auxiliary energy systems for solar cooling systems could be used with other options of clean energy or renewable energy. Pongtornkulpanich et al. [35]

designed a solar-driven 10- ton LiBr/H_2O single-effect absorption cooling system. It was shown that the 72 m^2 evacuated tube solar collector delivered a yearly average 81 % of the thermal energy required by the chiller, with the remaining 19 % generated by a LPG-fired backup heating unit. Prasartkaew and Kumar [36] presented a solar-biomass hybrid absorption cooling system which was suitable for residential applications. It consisted of three main parts: solar water heating with a storage tank, biomass gasifier fired hot water boiler, and single effect absorption chiller. The biomass gasifier hot water boiler was located between the hot water storage tank and absorption chiller machine. This insulated boiler had two functions: it worked as an auxiliary boiler when solar energy was not enough and worked as main heat source when the solar radiation was not available. Based upon the Bangkok meteorological data, the COP of the chiller and the overall system was found to be 0.7 and 0.55, respectively. Besides, the biomass (charcoal) consumption for 24 h operation was 24.44 kg/day. Ahmed Hamza et al. [37] reported the performance of an integrated cooling plant including both free cooling system and solar powered single-effect LiBr/H_2O absorption chiller, which had been in operation since August 2002 in Oberhausen, Germany. A floor space of 270 m^2 was air-conditioned by the plant. The plant included 35.17 kW cooling absorption chiller, vacuum tube collectors' aperture area of 108 m^2, hot water storage capacity of 6.8 m^3, cold water storage capacity of 1.5 m^3 and a 134 kW cooling tower. It was shown that free cooling in some cooling months could be up to 70 % while it was about 25 % during the 5 years period of the plant operation. For sunny clear sky days, collector's field efficiency ranged from 0.352 to 0.492 and chiller COP varied from 0.37 to 0.81, respectively. Li et al. [38] designed a 200 kW solar absorption cooling system assisted by a ground source heat pump (GSHP) with a rated cooling capacity of 391 kW. The chilled water produced by the solar cooling system was stored in a cold storage water tank. In order to maintain the setting temperature inside the cold storage water tank, the GSHP was turned on either when the water temperature was higher than 18 °C or during the period from 22:00 to 7:00 with cheaper electricity tariff. As for a solar cooling system without any backup system, Marc et al. [39] indicated that it was very difficult to design this kind of installation and particularly to define the appropriate refrigerating capacity of the chiller. In a case where the chiller is undersized and runs in nominal conditions with good performances, thermal comfort inside the building will not be achieved in some critical periods of the year. In a second case where the chiller is oversized and does not run in nominal conditions with low performances, thermal comfort inside the building is achieved.

4.3 Cooling Modes

Although the COP would be higher with a wet cooling tower, a dry cooling tower could be selected in order to avoid the usual problem of the legionella of the wet cooling towers. Monné et al. [40] reported a solar cooling system which consisted of 37.5 m^2 of flat-plate collectors, a 4.5 kW single-effect LiBr/H$_2$O absorption chiller and a dry cooling tower. The performance analysis of the solar driven chiller showed the average values of COP close to 0.6 in 2007 and between 0.46 and 0.56 in 2008. Concerning to the average cooling power, the chiller reached values between 4.0 and 5.6 kW in 2007 and between 3.6 and 5.3 kW in 2008. The studies indicated the great influence of the temperature of the heat rejection sink on the machine performance. Helm et al. [41] suggested that a low temperature latent heat storage together with a dry air cooler in solar-driven absorption cooling systems was a promising alternative to a conventional wet cooling tower. The reject heat of the absorption chiller was buffered by the heat storage and transferred to the ambient during periods of low ambient temperatures, e.g. night time or off-peak situations. An analysis of the thermal design of the different system components showed that latent heat storage allowed for moderate temperatures of the driving heat and thus substantially reduced the over-sizing of the solar collector system. In this study, the phase-change material (PCM) calcium chloride hexahydrate with phase transition, i.e., melting and solidification, in the temperature range of 27–29 °C was applied. The latent heat storage provided a 10 times higher volumetric storage density in comparison to a conventional water heat storage. By means of a latent heat storage integrated into the heat rejection loop of the chiller, a part of the auxiliary power demand can be shifted to off-peak hours with only a marginal increase of the overall electric consumption of the solar cooling system. As a consequence, a reduction of the operating cost is accomplished due to the reduced night tariff for electricity. And for the operation of the electric grid, a more even load profile with reduced daytime peaks is achieved, allowing for increased efficiency and reduced cost in power generation [42].

4.4 Formulae for the Calculation of Various Parameters

The instantaneous values of solar thermal plant during the day were found from Eqs. (1–3).

$$Q_t = A_{col}\, G_t \tag{1}$$

$$Q_{col} = m_{fcol} C p_{fcol} \left(t_{ocol} - t_{icol}\right) \tag{2}$$

$$Q_{itank} = m\, Cp\, \left(t_{otank} - t_{itank}\right) \tag{3}$$

The collector efficiency (η_{col}) calculated from Eq. (4).

$$\eta_{col} = Q_{col}/Q_t \tag{4}$$

The heat transfer rate of generator (Q_g), evaporator (Q_e), absorber and condenser (Q_{a+c}) are calculated from Eqs. (5–7).

$$Q_g = m\, Cp\, \left(t_{ig} - t_{og}\right) \tag{5}$$

$$Q_e = m_{\text{chill}} Cp_w \left(t_{ie} - t_{oe} \right) \tag{6}$$

$$Q_{a+c} = m_{\text{air}} Cp_{\text{air}} \left(t_{oair} - t_{odb} \right) \tag{7}$$

The thermal, electric and primary energy COPs, calculated from Eqs.(8–10).

$$COP_{th} = Q_e / Q_{,} \tag{8}$$

$$COP_{elec} = Q_e / W_{elec-prot} \tag{9}$$

$$COP_{\text{prim energy}} = Q_e / Q_g + \left(W_{elec-prot} / \eta_{conv} \right) \tag{10}$$

The solar facility performance (SFP) and solar coefficient of performance (SCOP) were calculated from Eqs. (11) and (12).

$$SFP = Q_g / Q_t \tag{11}$$

$$SCOP = Q_e / Q_t \tag{12}$$

5 Economical Comparison

Analysis of overall initial and operating costs and comparisons of alternatives require an understanding of the cost, the comfort demands and the environmental impacts of the system. According to Daini [43] there are two main methods to compare the cost of any two or more systems. (1) *First Cost Comparison*. It reflects only the initial price, installed and ready to operate, and ignores such factors as expected life, ease of maintenance ,and even to some extent, efficiency. (2) *Life-cycle cost* (LCC), which includes all cost factors (first cost, operating cost, maintenance, replacement, and estimated energy use) and can be used to evaluate the total cost of the system over the complete life of the system.

A comparison of general cost associated with single effect vapor absorption and vapor compression air-conditioning system is made. The cost analysis covers the initial costs and the operating costs of each of the systems. The selection depends on system which requires the minimum LCC and can perform the intended function for its life span.

5.1 Initial Costs

The initial costs for the single- effect solar vapor absorption systems include the absorption machine, heat rejection equipment, and solar energy collection system. The physical size of the absorption system is larger than the size of the vapor compression system; this increase in size requires a larger building, moving equipment, and support systems. This results in a higher installation cost for the vapor absorption system. The initial cost therefore should include, in addition to the purchase and installation of the systems, the various subsystems necessary for effective operation. This includes piping, wiring, and specific structures. The solar energy collection system plays the most significant role in the initial cost of the vapor absorption system.

5.2 Operating Costs

Operating costs, which include the costs of electricity, wages of employees, supplies, water, and materials, are those incurred by the actual operation of the system. With regard to the vapor compression system, the operating costs are dominated by the electricity required to drive the compressor. Additional electricity is used to drive the condenser water pump and the cooling tower fans.

5.3 Maintenance Costs

There are various levels of maintenance that may be applied to building air conditioning services. The three most common levels are run-to-failure, preventive, and finally predictive maintenance. The maintenance cost is difficult to quantify because it depends on a large number of variables such as local labor rates, their experience, the age of the system, length of time of operation, etc. The maintenance cost for the heat rejection subsystem tends to be higher for the VAS due to more rapid scaling; however, this could be offset by the maintenance cost of the VCS because it is a work-operated cycle. Maintenance costs cited in various studies show that the vapor absorption system's maintenance costs range from 0.6 to 1.25 times the maintenance costs of the vapor compression system.

6 Conclusion and Suggestions

It is concluded that single effect absorption cooling method using LiBr-H$_2$O as working fluid pair is more suitable for domestic purpose. Flat plate and evacuated tube solar collectors are more reliable and economical for this system. It is found that two important parameters determine the most economical solar cooling system. They are: (1) the cost of the solar collection and storage technologies (2) the performance of the cooling technologies. By considering the two parameters some suggestions were given as follows: (1) By placing hot water storage tank above the solar collector and thermo siphon principle can be used for transferring the heat from collector to tank (2) Heat loss due to transfer of hot water from storage tank to generator can be avoided by placing the generator inside the insulated storage tank. Also the insulation cost for the generator can be minimized (3) The vacuum pressure of the cooling circuit can be increased to enhance the boiling of water inside the generator (4) According to the new idea given in this paper only three major electrical equipments are used (condenser fan, cooling coil fan, and a pump). Hence, the operational cost is very much minimized when compared with compression system. Though the initial cost is more, there are many possibilities for reducing the operational cost of the solar cooling system. Finally it is indicated that solar assisted single effect absorption cooling system would be competitive with compression cooling system when compared for long-term operation.

References

1. Henning H-M (2007) Solar assisted air conditioning of buildings—an overview. Appl Therm Eng 27(10):1734–1749
2. Hassan HZ, Mohamad AA (2012) A review on solar cold production through absorption technology. Renew Sustain Energy Rev 16:5331–5348
3. Jaruwongwittaya T, Chen G (2010) A review: renewable energy with absorption chillers in Thailand. Renew Sustain Energy Rev 14:1437–1444
4. Altamush Siddiqui M (1997) Economic analyses of absorption systems: part-B-optimization of operating parameters. Energy Convers Mgmt 38(9):905–918
5. Deng J, Wang RZ, Han GY (2011) A review of thermally activated cooling technologies for combined cooling, heating and power systems. Progress Energy Combust Sci 37:172–203
6. Jaruwongwittaya T, Chen G (2010) A review: renewable energy with absorption chillers in Thailand. Renew Sustain Energy Rev 14(5):1437–1444
7. Fong KF, Chow TT, Lee CK, Lin Z, Chan LS (2010) Comparative study of different solar cooling systems for buildings in subtropical city. Sol Energy 84(2):227–244
8. Gomri R (2010) Investigation of the potential of application of single effect and multiple effect absorption cooling systems. Energy Convers Manage 51(8):1629–1636
9. Otanicar T, Robert A, Taylor B, Patrick EP (2012) Prospects for solar cooling—an economic and environmental assessment. Solar Energy 86:1287–1299
10. Berhane HG, Gonzalo GG, Laureano J, Dieter B (2012) Solar assisted absorption cooling cycles for reduction of global warming: a multi-objective optimization approach. Solar Energy 86:2083–2094
11. Rosiek S, Batlles FJ (2009) Integration of the solar thermal energy in the construction: analysis of the solar-assisted air-conditioning system installed in CIESOL building. Renew Energy 34(6):1423–1431
12. Ortiz M, Barsun H, He H, Vorobieff P, Mammoli A (2010) Modeling of a solar-assisted HVAC system with thermal storage. Energy Build 42(4):500–509
13. Mammoli A, Vorobieff P, Barsun H, Burnett R, Fisher D (2010) Energetic economic and environmental performance of a solar-thermal-assisted HVAC system. Energy Build 42(9):524–1535
14. Syed A, Izquierdo M, Rodríguez P, Maidment G, Missenden J, Lecuona A et al (2005) A novel experimental investigation of a solar cooling system in Madrid. Int J Refrig 28(6):859–871
15. Praene JP, Marc O, Lucas F, Miranville F (2011) Simulation and experimental investigation of solar absorption cooling system in Reunion Island. Appl Energy 88(3):831–839
16. Li ZF, Sumathy K (2001) Experimental studies on a solar powered air conditioning system with partitioned hot water storage tank. Sol Energy 71(5):285–297
17. Agyenim F, Knight I, Rhodes M (2010) Design and experimental testing of the performance of an outdoor LiBr/H2O solar thermal absorption cooling system with a cold store. Sol Energy 84(5):735–744

18. Rodríguez Hidalgo MC, Rodríguez Aumente P, Izquierdo Millán M, Lecuona Neumann A, Salgado RM (2008) Energy and carbon emission savings in Spanish housing air-conditioning using solar driven absorption system. Appl Therm Eng 28(14–15):1734–1744
19. Izquierdo M, Lizarte R, Marcos JD, Gutierrez G (2008) Air conditioning using an air cooled single effect lithium bromide absorption chiller: results of a trial conducted in Madrid in August 2005. Applied Thermal Engg 28:1074–1081
20. Al-Dadah RK, Jackson G, Ahmed R (2011) Solar powered vapor absorption system using propane and alkylated benzene AB300 oil. Appl Thermal Eng 31:1936–1942
21. Zhai XQ, Qu M, Li Y, Wang RZ (2011) A review for research and new design options of solar absorption cooling systems. Renew Sustain Energy Rev 15:4416–4423
22. Balghouthi M, Chahbani MH, Guizani A (2008) Feasibility of solar absorption air conditioning in Tunisia. Build Environ 43(9):1459–1470
23. Florides GA, Kalogirou SA, Tassou SA, Wrobel LC (2002) Modelling, simulation and warming impact assessment of a domestic-size absorption solar cooling system. Appl Therm Eng 22(12):1313–1325
24. Assilzadeh F, Kalogirou SA, Ali Y, Sopian K (2005) Simulation and optimization of a LiBr solar absorption cooling system with evacuated tube collectors. Renew Energy 30(8):1143–1159
25. Atmaca I, Yigit A (2003) Simulation of solar-powered absorption cooling system. Renew Energy 28(8):1277–1293
26. Joudi KA, Abdul-Ghafour QJ (2003) Development of design charts for solar cooling systems. Part I: computer simulation for a solar cooling system and development of solar cooling design charts. Energy Convers Manage 44(2):313–339
27. Sun J, Lin F, Zhang Shigang, Hou W (2010) A mathematical model with experiments of single effect absorption heat pump using LiBr-H_2O. Appl Therm Eng 30:2753–2762
28. Liao X, Radermacher R (2007) Absorption chiller crystallization control strategies for integrated cooling heating and power systems. Int J Refrig 30:904–911
29. Bakhtiari B, Fradette L, Legros R, paris J (2011) A model for analysis and design of H_2O-LiBr absorption heat pumps. Energy Convers Manage 52:1439–1448
30. Wang Z, Zhaolin G, Feng S, Li Y (2009) Application of vacuum membrane distillation to Lithium Bromide absorption refrigeration system. Intl J Refrig 32:1587–1596
31. Mazloumi M, Naghashzadegan M, Javaherdeh K (2008) Simulation of solar lithium bromide–water absorption cooling system with parabolic trough collector. Energy Convers Manage 49(10):2820–2832
32. Mittelman G, Kribus A, Dayan A (2007) Solar cooling with concentrating photovoltaic/thermal (CPVT) systems. Energy Convers Manage 48(9):2481–2490
33. Calise F (2010) Thermoeconomic analysis and optimization of high efficiency solar heating and cooling systems for different Italian school buildings and climates. Energy Build 42(7):992–1003
34. Calise F, Palombo A, Vanoli L (2010) Maximization of primary energy savings of solar heating and cooling systems by transient simulations and computer design of experiments. Appl Energ 87(2):524–540
35. Pongtornkulpanich A, Thepa S, Amornkitbamrung M, Butcher C (2008) Experience with fully operational solar-driven 10-ton LiBr/H2O single-effect absorption cooling system in Thailand. Renew Energy 33(5):943–949
36. Prasartkaew B, Kumar S (2010) A low carbon cooling system using renewable energy resources and technologies. Energy Build 42(9):1453–1462
37. Ahmed Hamza HA, Noeres P, Pollerberg C (2008) Performance assessment of an integrated free cooling and solar powered single-effect lithium bromide-water absorption chiller. Solar Energy 82(11):1021–1030
38. Li J, Bai N, Ma W (2006) Large solar powered air conditioning-heat pump system. Acta Energiae Solaris Sinica 27(2):152–8 (in Chinese)
39. Marc O, Lucas F, Sinama F, Monceyron E (2010) Experimental investigation of a solar cooling absorption system operating without any backup system under tropical climate. Energy Build 42(6):774–782

40. Monné C, Alonso S, Palacín F, Serra L (2011) Monitoring and simulation of an existing solar powered absorption cooling system in Zaragoza (Spain). Appl Therm Eng 31(1):28–35
41. Helm M, Keil C, Hiebler S, Mehling H, Schweigler C (2009) Solar heating and cooling system with absorption chiller and low temperature latent heat storage: energetic performance and operational experience. Int J Refrig 32(4):596–606
42. Qu M, Yin H, David HA (2010) A solar thermal cooling and heating system for a building: experimental and model based performance analysis and design. Sol Energy 84(2):166–182
43. Elsafty A, Al-Daini AJ (2002) Economical comparison between a solar powered vapor absorption air-conditioning system and a vapor compression system in the middle east. Renew Energy 25:569–583

Ant Colony Optimization for the Minimization of Internal Combustion Engine Forces and Displacements

T. Ramachandran, K. P. Padmanaban and J. Vinayagamoorthy

Abstract The unbalanced forces from the engine reciprocating and rotating component are the major contributors for the internal combustion engine vibration. The reciprocating and rotational inertial forces are modeled to determine the total forces excited and displacements caused at the engine mounts. A heuristic optimization (ACA) is used to optimize the design variables to reduce the forces and displacements at the supports.

Keywords IC engine vibration · Inertial force · ACA

1 Introduction

The reasons for the internal combustion (IC) vibration are fluctuating forces at the skirt of the piston and at the reciprocating components [1]. Impact forces at piston, due to the unbalanced forces of the rotating and reciprocating parts, are the common impact phenomenon existing in the reciprocating IC engine. It also plays an important role in the transverse vibration of the engine [2]. In the crank and slider mechanism of the IC engine, the piston is driven by the cylinder gas pressure and moves inside the cylinder along with the connecting rod and the crankshaft [3]. At the same time the periodical changing of direction of the connecting rod leads to the variation of inertial force in the cylinder wall and crank shaft. The unbalanced forces from the piston and connecting rod setup create not only the variational

---

T. Ramachandran (✉)
PSNA College of Engineering and Technology, Dindigul, Tamilnadu, India
e-mail: ramji_kkp@yahoo.com

K. P. Padmanaban
SBM College of Engineering & Technology, Dindigul, Tamilnadu, India
e-mail: padmarubhan@yahoo.co.in

J. Vinayagamoorthy
Mechanical Engineering, PSNA College of Engineering and Technology, Dindigul, Tamilnadu, India
e-mail: Vinamech0909@gmail.com

S. Sathiyamoorthy et al. (eds.), *Emerging Trends in Science,*
Engineering and Technology, Lecture Notes in Mechanical Engineering,
DOI: 10.1007/978-81-322-1007-8_2, © Springer India 2012

impact force on the cylinder walls but also the variational speed fluctuation on the crank shaft. So, the cylinder block experiences the vibrational force due to the impact force by slap as well as the speed variation in the crank shaft.

The solutions attempted to reduce the engine forces and displacements due to the diesel excitations are usually to isolate the engine from the vibratory forces. In this chapter, a 4-cylinder IC engine is considered as rigid body and the mathematical modeling is done to determine the forces and displacements exerted by the engine during the operation. Attempts are made to reduce the engine forces and displacements by optimizing the crank angles of the counter weights. Ant colony optimization [4] technique is used to predict the optimum positions of counter weights under minimum displacements at the engine supports.

2 Rigid Body Modeling of the Engine

The engine vibration can be classified into two types: vibration of engine components with respect to each other, named as internal vibration, and the vibration of engine block as a total, named as external vibration. The unbalanced forces from the engine components lead to the external vibration [1]. To analyse the engine vibration the engine will be considered as rigid body with 6° of freedom in x, y and z axis about its centre of gravity [5].

Consider an engine as a rigid body of mass M connected to a rigid chassis by three rubber engine mounts, as shown in Fig. 1. The origin of the fixed global coordinate system is located at the center of mass of engine. The z-axis is parallel to the crank shaft, and the x-axis is in the vertical direction. The general translation [6] is in the y-direction.

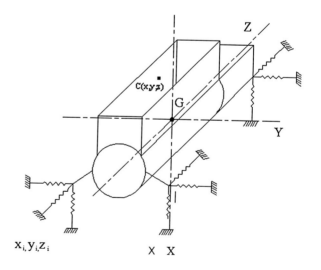

Fig. 1 Rigid body model of the engine

F_x, F_y and F_z are the sum of forces acting on the engine block in the x, y, and z direction, respectively and x, y, and z denote position of center of gravity of engine block. The engine's rotational equations of motion become principal axes of inertia; therefore, the products of inertia are eliminated. Three kinds of forces and moments act on the engine block: (1) shaking forces (inertia forces) and moments generated by moving links of the slider-crank mechanism, (2) forces and moments associated with the balancing system, and (3) reaction forces exerted to the engine block by three engine mounts.

2.1 Forces and Displacements on Engine

The engine is supported by three mounts positioned between the engine and the supporting structures. To achieve decoupling of the roll mode, the engine mounting system is designed to allow the elastic axis of the isolators to coincide with the engine roll inertia axis. This technique is achieved by inclining the mounts. In order to achieve the inclination of the engine mounts to be effective, the compression rate of the mounts must be significantly higher than the shear rate. The relation between the inclination angle of an isolator and the angle of elastic axis is kept equal. The mounts considered in this model are standard viscoelastic material with the stiffness of the mounts [7] are assumed to constant in the three axis and for the entire analysis. The components of the engine assumed as rigid bodies and the force and displacements of the components during the cyclic motion of the engine was calculated by the kinematic and dynamic analysis of the crank-slider arrangement.

In this work a three dimensional model is proposed for the 4-cylinder diesel engine as shown in Fig. 1. This shows the piston motion in the directions along x, y, and z axis. The forces developed by the engine components in the cylinder are transferred to the engine supports as three dimensional forces and displacements. The translation forces from the cylinder are given by

$$\vec{F} = \begin{bmatrix} \sum F_x \\ \sum F_y \\ \sum F_z \end{bmatrix} \tag{1}$$

From the piston and connecting rod arrangement as shown in Fig. 2, the mass of the crank and the mass of the connecting rod are assumed to be concentrated at the mass center of the connecting rod. The masses of rotating and reciprocating components are calculated [8] as

$$m_{rot} = \left(^c/_l\right) m_{crk} + \left(\frac{l-b}{l}\right) m_{con} \tag{2}$$

Fig. 2 Crank and slider arrangement

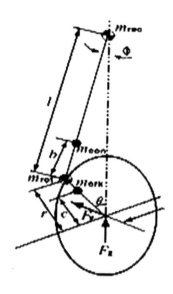

$$m_{\text{rec}} = \left(b/l\right) m_{\text{con}} + m_p \tag{3}$$

The forces from the engine cylinder [9] are calculated by adding the gas force and piston force from the reciprocating components and resolved into the vertical and horizontal [10] (radial and tangential) components.

x-directional force exerted at the ith cylinder

$$\vec{F}_{x,i} = (F_{\text{rot}} + F_{\text{pr}} + F_{\text{rd}})_x \vec{e}_x + (F_{\text{rt}} + F_{\text{rec}})_y \vec{e}_y; \tag{4}$$

y-directional force exerted at the ith cylinder

$$\vec{F}_{y,i} = (F_{\text{rot}} + F_{\text{rec}})_y \vec{e}_y \tag{5}$$

Gas pressure exerted at the ith cylinder

$$F_{\text{pr},i} = \pi/4 d^2 (a_1 \cos k\theta + a_2 \sin k\theta) \vec{e}_y \tag{6}$$

Unbalanced force exerted at the counter weight disc in the x direction

$$F_{\text{rd},x} = m_{\text{rd}} r_{\text{rd}} \omega^2 \sin(\theta + \psi); \tag{7}$$

y-directional force due to rotary components

$$\vec{F}_{\text{rot},y} = m_{\text{rot}} r \omega^2 \cos \theta; \tag{8}$$

Unbalanced force exerted at the counter weight disc in the y direction

$$F_{\text{rd},y} = m_{\text{rd}} r_{\text{rd}} \omega^2 \cos(\theta + \psi); \qquad (9)$$

y-directional force due to reciprocating components Table 1

$$F_{\text{rec}} = m_{\text{rec}} r \omega^2 \left(\cos \theta + 4A_2 \cos 2\theta + 16A_4 \cos 4\theta \right) \qquad (10)$$

where,

$$A_2 = \frac{l}{r} \left[\frac{1}{4} (l/r)^2 + \frac{1}{16} (l/r)^4 \right];$$

$$A_4 = -\frac{l}{r} \left[\frac{1}{64} (l/r)^4 + \frac{1}{256} (l/r)^6 \right];$$

In finalising the model, the model allowances must be included which are must for the determination of dissipation of energy due to damping forces at the supports. Substituting the equation for the inertial forces and reaction forces [11] unbalance forces associated with the engine in terms of Newton law motion gives the following translational and rotational equations of motion of the system under the idling condition. The set of differential equations are solved numerically [12] over any respective time interval [0, T] to determine the displacements at the supports.

$$M\ddot{x} = \sum F_x$$

$$M\ddot{y} = \sum F_y$$

$$M\ddot{z} = \sum F_z$$

3 Ant Colony Algorithm for Optimization

Colonies of social insects can exhibit an amazing variety of complex behaviors and have always captured the interest of biologists and entomologists. The study of ant colonies behavior turned out to be very fruitful, giving rise to a completely novel field of research, now known as ant algorithms. In ant algorithms, a colony

Table 1 Force components

$F_{x,i}, F_{y,,i}, F_{z,i}$	x, y and z axis forces of the ith cylinder
F_{rot}	Forces from the rotational parts
F_{pr}	Forces due to the gas pressure
F_{rd}	Reaction forces at the rotating disc
F_{rec}	Forces from the reciprocating parts
θ	Crank angle
ψ	Connecting rod angle at the piston
a_1, a_2	Fourier constants

of relatively simple agents called as ants, efficiently carries out complex tasks such as resource optimization and control [13]. Ants deposit pheromones while carrying out their own tasks. These modifications change the way sensed by the other ants in the colony and implicitly act as a signal triggering other ants' behaviors that again generate new modifications that will simulate other ants and so on.

3.1 ACA-Based Engine Vibration Optimization Method

To apply the ant colony algorithm for the engine vibration optimization problems, randomly 'R' solutions are selected from different possible solutions. A critical value is fixed about which number of superior and inferior solutions are defined. Global search is carried for inferior solutions, whereas the local search is carried out for the superior solutions. In this chapter, the ant colony [4] algorithm is used in continuous engine vibration displacement optimization. The various processes involved in ant colony algorithm-based vibration displacement optimization method are as follows: (1) initialization, (2) global search, and (3) local search. Figure 2 explains the distribution of ants for local search and global search.

3.2 Initialization

1. 20 vibration displacements are randomly generated from the range defined for each vibration displacement model.
2. Then, the vibration displacement models are sorted according to ascending order of the solutions.
3. The solutions from 1 to 12 are named as superior solutions and from 13 to 20 are named as inferior solutions.

3.3 Global Search

The global search is done to improve the inferior solutions. This search includes crossover or random walk, mutation and trail diffusion.

Crossover or random walk In this process, the inferior solutions from 13 to 18 are replaced by the superior solutions. This process includes the following steps:

(a) Replacement of each inferior solution by a superior solution is decided based on the crossover probability.
(b) To replace 13th solution, a random number between 1 and 12 is generated. Then, the corresponding solution in the superior region replaces the 13th inferior solution.

Ant Colony Optimization for the Minimization of Internal Combustion Engine

(c) The selected solutions in the superior region should be excluded, so that it is not selected again for replacement.
(d) The above procedure is repeated up to the 18th solution.

Mutation The mutation process further improves the replaced solutions. This process includes the following steps:

(a) The crank angle of each model in the replaced 13th model modified by adding or subtracting with mutation step size (Δ).

The mutation step size (Δ) is obtained as

$$\Delta = R(1 - r^{(1-T)^b}) \tag{11}$$

where,
$$R = (X_{j}\text{max} - X_i)$$

X_{max} maximum range of crank angle defined for the cylinders
X_i angle of the respective cylinder
R random number
T ratio of current iteration to the total no of iteration
B constant (obtained by trial)

(b) The mutation probability (Pm) is set. Then a random number is generated between 0 and 1. If the random number generated is less than Pm, the mutation step size (Δ) is subtracted to the node number of the respective displacement configuration or else it is added to the node number of the respective crank angle configuration element. The same procedure is repeated up to the 18th solution.

Trail diffusion The trail diffusion improves the 19 and 20th solutions. This process includes the following steps.

(a) Two displacements are randomly selected from the superior solutions, and they are named as parent-1 and parent-2. The crank angle of each configuration in parent-1 and parent-2 is termed as XP1 and XP2, respectively. The new set angles obtained from parent-1 and parent-2 is termed as Child. The angle of each configuration in the Child configuration is termed as XC.
(b) One more random number (α) is generated between 0 and 1 for the angle of each crank angle configuration.
(c) If α is between 0 and 0.5, then the new crank angle position of each configuration element of the new layout is obtained by XC $= (\alpha)$XP$_1$ $+(1-\alpha)$XP$_2$
(d) If α is between 0.5 and 0.75, then the new position of each cylinder of the configuration is obtained by XC $=$ XP$_1$

(e) If α is between 0.75 and 1, then the new angle position of each cylinder of the new configuration is obtained by $XC = XP_2$

(f) The above procedure is repeated for the 20th solution also. After the crossover or random walk, mutation, and trail diffusion processes, the solutions for the modified configurations from 13 to 20th are found using calculations.

3.4 Local Search

The local search is done to improve the superior solutions from 1 to 12.

This process includes the following steps.

The average pheromone value is calculated by $P_{avg} = \sum P / N_s$

where,

P pheromone value of each solution (Initial value is assumed to be 1.0)

N_S number of superior solutions

(a) A random number is generated between 0 and 1. If the number generated is less than average pheromone value (P_{avg}), the search is further pursued or else the ant quits, and then leaves the solution without any alteration.

(b) A limiting step value LS, which is added to the number of the respective cylinder when the random number generated is greater than 0.5 and subtracted to the number of the respective cylinder when the random number generated is less than 0.5, is calculated as follows: $L_S = K_1 - A \times K_2$) where, K_1 and K_2 are the values chosen such that $K1 > K2$. 'A' is the age, which is assumed to be 10 for all the solutions in the first iteration.

(c) All the crank angles corresponding to the superior solutions are modified by local search and solutions for the modified angles from 1 to 12 are found using model calculations.

(d) The new age for each solution for the next generation is calculated as follows:

If the current solution is less than the previous solution, the age for the new solution is calculated as follows:

$$A = A_{i-1} + 1$$

If the new solution is greater than the previous solution, the age for the new solution is calculated as follows:

$$A = A_{i-1} - 1$$

where,

A_i is the age for the new iteration, A_{i-1} is the age for the previous iteration.

(e) The new pheromone value of the ant in the next iteration is also calculated as follows: $P_i = \frac{S_i - S_{i-1}}{S_{i-1}} + P_{i-1}$

where,

P_i—pheromone value for the new solution, S_i—the value of current solution, P_{i-1}—pheromone value for the old solution, S_{i-1}—the value of old solution.

The above steps, i.e., local and global searches are performed in all the iterations to improve the solutions.

4 Convergence of ACA

For each crank angle considered in the particular ACA iteration, the forces are applied sequentially for the crank and slider arrangement of all cylinders. The maximum deformation (Δ_{max}) among the maximums for force application is found out. The same procedure is repeated for all the crank position for all the iteration. Then using ACA, the minimum deformation (Δ_{min}) among maximums for each crank angle is found. The same procedure is repeated for all the iterations (N_G). The algorithms converge if either the number of iterations over which no change in the objective function value is obtained, N_{chg}, or if the number of iterations, N, reaches the defined maximum number of iterations, N_G whichever is earlier. The optimization of the crank position is carried out for the whole process in a single step. The different runs are performed in ACA-based continuous optimization method until the minimum engine displacement corresponding to optimal crank angle is determined.

5 Numerical Simulations and Results

The calculations of the mathematical model are performed using Matlab to determine the forces and displacements of the engine. To determine the engine forces the simulation of motion of engine over a given time interval for specific speed is made. The calculated time history of the behavior of the system enables the determinations of the objective function of the engine model (Table 2). In the simulation the initialization of crank angle of the any one cylinder is set to zero. If the first cylinder angle is set to zero, then the subsequent angle are assigned as design variables. The engine is simulated at a speed of 1,500 rpm for specified time duration of 0.05 s to obtain the forces and displacements exerted by the engine at the supports. In the AC optimization process the objective function is defined with the convergence parameters and then the optimization of the objective function is made after the predefined iterations. The results of every set of generations of each run are determined and the optimum values of the each design variables (Tables 3 and 4) with respect to the minimum forces and displacements (Figs. 3 and 4) are revealed. The minimum of minimum force and displacement of 5 runs is considered to be the optimum conditions of the engine variables. The force and displacement results of the model are compared with and without optimization. The results obtained in the ACA shows good convergence (Fig. 5).

6 Conclusion

In this research work, the forces and displacements caused in a 4-cylinder diesel engine are determined by considering the engine as the rigid model. To determine the forces and displacements at the engine supports the forces from the various engine components (rotating and reciprocating components) are modeled through the Newton law of motion by considering the different constants of the engine specifications. The force models of the all components are coupled together and made into the single force system by considering the force system as one way coupling model. By summing up the force components of the all cylinders into single system the final model is developed (Tables 2, 3, 4 and 5, and Figs. 3, 4 and 5).

The model developed is simulated with a help of Matlab such that the design variables are specified with a range and the response is the displacements at the engine mounts. Finally, the optimization of the design variables is made using the

Table 2 Engine parameters

Abbreviation	Parameter	Value
M	Mass of the engine	250 kg
R	Radius of the crank	0.02 m
L	Length of the connecting rod	0.11 m
m_{con}	Mass of the connecting rod	1.2 kg
m_{crk}	Mass of the crank shaft	2.5 kg
m_p	Mass of the piston	2.65 kg
N	Engine rotational speed	1500 rpm
K_x	Elastomeric mount stiffness in x axis	3.5E5 N/m
K_y	Elastomeric mount Stiffness in y axis	3.9E5 N/m
K_z	Elastomeric mount Stiffness in z axis	2.3E5 N/m

Table 3 Optimised crank angles for the counter weights

Cylinder	Optimum crank angle	
	Before using ACA	After using ACA
1.	$0°$	$4°$
2.	**$255°$**	**$257.5°$**
3.	$105.3°$	$100.1°$
4.	$150°$	$156°$

Table 4 Optimised transmitted forces

Force components	Force transmitted (N)	
	Before using ACA	After using ACA
F_x	421.5	400.22
F_y	2500.8	1800.6
F_z	130.13	90.64

Table 5 Optimised transmitted forces

Force components	Displacement (mm)	
	Before using ACA	After using ACA
X	0.00534	0.0031
Y	−0.0296	−0.0283
Z	0.001051	0.00104

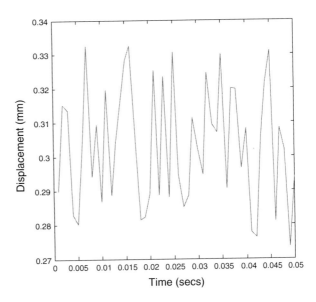

Fig. 3 Engine displacement variation without optimized values

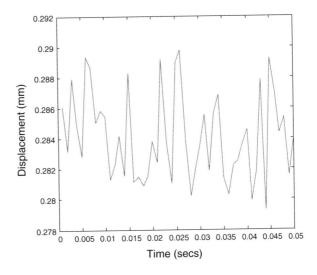

Fig. 4 Engine Displacement variation with optimised value the minimum displacement

Fig. 5 Convergence of ACA towards

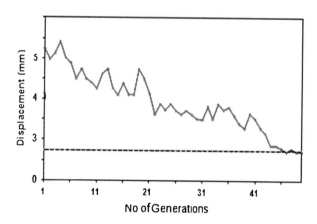

Ant Colony Optimization method to determine the minimum displacements and corresponding forces at the engine supports. In the AC optimization the convergence is obtained after 40 generations. The optimized parameters are once again simulated for the time duration of 0.05 s at a speed of 1500 rpm and the results of the optimized and non-optimized values of displacements are compared. The comparison shows the improvement in the reduction of the displacements and forces at the mounts when AC optimizations is used.

References

1. Snyman JA, Heyns PS (1995) Vermeulen PJ Vibration isolation of a mounted engine through optimization. Mech Mach Theory 30(1):109–118
2. Ma Z-D, Perkins NC (2003) An efficient multibody dynamics model for internal combustion engine systems. Multibody SysDyn 10:363–391
3. Cho S-H, Ahn S-T, Kim Y-H (2002) A simple model to estimate the impact force induced by piston slap. J Sound Vib 255(2):229–242
4. Padmanaban KP, Arulshri KP, Prabhakaran G (2009) Machining fixture layout design using ant colony algorithm based continuous optimization method. Int J Adv Manuf Technol 45:922–934
5. García de Jalón J, Callejo A (2011) A straight methodology to include multibody dynamics in graduate and undergraduate subjects. Mech Mach Theory 46 168–182
6. Aslanov V, Kruglov G, Yudintsev V (2011) Newton–Euler equations of multibody systems with changing structures for space applications. Acta Astronautica 68:2080–2087
7. Ali A, Hosseini M, Sahari BB (2010) A review of constitutive models for rubber-like materials. Am J Eng Appl Sci 3(1):232–239
8. Ohadi AR, Maghsoodi G (2007) Simulation of engine vibration on non-linear hydraulic engine mounts. ASME J Vib Acoust 129:417–424
9. Zheng H, Liu GR, Tao JS, Lam KY (2001) FEM/BEM analysis of diesel piston-slap induced ship hull vibration and underwater noise. Appl Acoust 62:341–358
10. Ebrahimi S, Kövecses J (2010) Unit homogenization for estimation of inertial parameters of multibody mechanical systems. Mech Mach Theory 45:438–445

11. Hoffman DMW, Dowling DR (2001) Fully coupled internal combustion engine dynamics and vibration—part I: model development. ASME J Eng Gas Turbine Power 123:677–684
12. MbonoSamba YC, Pascal M (2001) Dynamic analysis and numerical simulation of flexible multibody systems. Mech Struct Mach 29(3):295–316
13. Prabhaharan G, Padmanaban KP (2007) Machining fixture layout optimization using FEM and evolutionary techniques. Int J Adv Manuf Technol 32(11):1090–1103

Augmentation of Heat Transfer and Pressure Drop Characteristics Inside a Double Pipe U-Tube Heat Exchanger by using Twisted Tape Inserts

P. V. Durga Prasad and A. V. S. S. K. S. Gupta

Abstract A twisted tape is inserted into a double pipe u-tube heat exchanger with a view to generate swirl flow, thereby enhancing the heat transfer rate of the fluids flowing in it. Experimental investigations of heat transfer enhancement and variation of friction factor have been presented for a typical test section of a u-tube double pipe heat exchanger involving a circular tube filled with a full length insert. The flow is under forced convection with Reynolds Number varying from 3,000 to 31,000. From the experimental data, heat transfer coefficient, friction factor, and thermal performance with the twisted tape inserts were calculated and compared with the plain tube data. The results show that there were a significant increase in heat transfer coefficient, friction factor, and it was found that the thermal Performance of smooth tube is better than the full length twisted tape by 2.0–2.2 times.

Keywords Heat transfer enhancement • Heat exchanger • Friction factor • Performance ratio • Twist ratio

Nomenclature

A	Area of heat transfer m^2
d_i	Inside tube diameter, m
d_0	Outside tube diameter, m
D	Twist diameter of inserted tape, m
H/D	Twist ratio = Pitch/twist diameter

P. V. Durga Prasad (✉)
Department of Mechanical Engineering, Narsimha Reddy Engineering College,
Maisammaguda, Hyderabad, AP, India
e-mail: pvdurgap@gmail.com

A. V. S. S. K. S. Gupta
Department of Mechanical Engineering, JNTUH College of Engineering, Hyderabad, AP, India

S. Sathiyamoorthy et al. (eds.), *Emerging Trends in Science,*
Engineering and Technology, Lecture Notes in Mechanical Engineering,
DOI: 10.1007/978-81-322-1007-8_3, © Springer India 2012

L_1	Length of heat transfer test section, m
L_2	Length of pressure drop test section, m
Δp	Pressure drop, kg/m^2
m	Mass flow rate, kg/sec
Q	Rate of heat transfer, W
C_p	Specific heat of fluid, KJ/kg-K
T_{hi}, T_{ho}	Inlet and outlet temperature of hot fluid, °C
T_{ci}, T_{co}	Inlet and outlet temperature of cold fluid, °C
h_i	Inside heat transfer coefficient, W/m2-k
h_o	Out side heat transfer coefficient, W/m2-k
ΔT_{LMTD}	Log mean temperature difference, °C
Re	Reynolds Number
Pr	Prandtl Number
Nu	Nusselt Number
K	Thermal conductivity of fluid, W/m$-$k
f	Friction factor
U	Overall heat transfer coefficient, W/m$^2-$k

1 Introduction

In the convective mode of heat transfer involving heat exchangers used in industrial applications, double pipe U-tube heat exchangers find wide usage. Typically, it involves exchange of heat between two fluids separated by a layer of pipes, with one fluid flowing inside the tube and the other fluid flowing outside the tube (called the annulus). As such, a good design of the heat exchanger is to extract maximum possible amount of heat during the heat exchange process. Based on Newton's law of cooling, the heat transfer coefficient is one parameter which can be enhanced to increase the rate of heat transfer. As the usual economic considerations require a compact design, enhancement of area of heat exchanger serves no purpose. In forced convection under higher Reynolds number, higher heat transfer rates are possible when turbulence or swirl is created in the flow. Hence to achieve this turbulence in the flow, a twisted tape is used as an insert in the pipe with an objective to enhance the heat transfer rates. As the heat exchangers of double pipe U-tube type are frequently used in many applications involving heat transfer between two or more fluids, enhancement of heat transfer rate between the fluids is of great significance for the industry. Besides saving of primary energy, it also leads to reduction in the size and weight of the heat exchanger, thereby saving costs to the users. After a survey of techniques used to enhance the heat transfer rate, an attempt is made in this work to investigate the same using twisted tape inserts.

The various techniques of augment convective heat transfer are given by Hong et al. [1] found that in the case of high Prandtl Number fluid in laminar flow, the heat transfer rate increases considerably for a moderate increase in pressure drop. Eiamsa-ard et al. [2] for experimentally investigating the effect of twisted tape insert on heat transfer in the tube side of a double pipe heat exchanger measured the heat transfer without twisted tape [3]. Twisted tape techniques have been used to augment heat transfer in double pipe heat exchanger. Ahmed et al. [4] measured the tube side heat transfer coefficient for turbulent flow of air in a smooth tube for the purpose of investigating the heat transfer enhancement using twisted tape inserts. An overview of hundred research works of enhancement heat transfer rate by using twisted tape was reported by Dewan et al. [5]. Anil Singh Yadav [6] investigated heat transfer and the pressure drop characteristics in a double pipe heat exchanger with full length twisted tape inserts. Naphon [7] reported heat transfer results from the plain tube for comparing with those from coil-wire inserted tube [8]. Heat transfer and pressure drop characteristics of Laminar flow through a circular tube fitted with regularly spaced twisted tape elements with multiple twists.

The objective of this paper is to study heat transfer and pressure drop characteristics of the U-tube double pipe heat exchanger with and without twisted tape inserts. The effect of various relevant parameters on heat transfer characteristics and pressure drop are also investigated. The experiments on heat transfer and pressure drop characteristics for the U-tube double pipe heat exchanger with full length twisted taped insert have been analysed for their suitability to practical applications in industries.

2 Experimental Equipment and Procedure

2.1 Description of Experimental Equipment

The schematic representation of experimental set up is shown in Fig. 1a. It consists of, Water tank with heater of 0.64 m^3 capacity placed on floor, Double pipe u bend heat exchanger, measuring devices like totalizer, temperatures, Water pumps for pumping hot water, and cold water (0.25 HP). The heat exchanger consists of 2 m long tubes of stainless steel (304 SS), 0.019 m inner diameter, and 0.025 m outer diameter. The heat transfer test section is fitted with jacket (annulus) for admitting hot water. Over the jacket asbestos rope is wound to a thickness of 1 inch to minimize heat loss.

Two RTD PT 100 type temperature sensors one just before the test section and other after the test section are placed to measure the inlet and outlet of fluid. A hot water tank (insulated with asbestos rope) with the in built PID controller and heater (1 kW) is provided to supply hot water at constant temperature. The temperature of hot water is maintained at 75 °C with the deviation of plus or minus 2 °C by thyristor controlled heating. Also two more RTD PT 100 type temperature sensors, one at the inlet and another at the outlet of jacket are placed to measure the hot fluid temperatures.

The cold water at room temperature is being taken at different Reynolds numbers ranging from 3,000 to 33,000 with the aid of a centrifugal pump. The by-pass valve

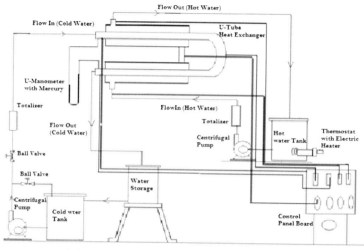

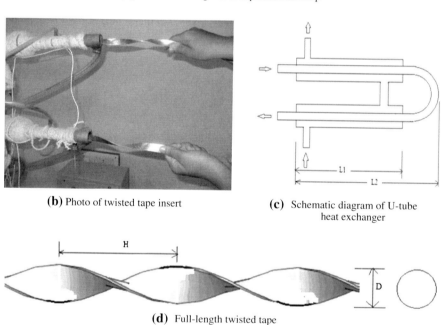

Fig. 1 a Schematic diagram of experimental setup b Photo of twisted tape insert c Schematic diagram of U-tube heat exchanger d Full-length twisted tape

attached to totalizer is used to regulate the flow rate to the test section. The two ends of the U-tube manometer pressure tapes, one just before the test section and the other just after test section are provided for pressure drop measurement. Mercury is used as the manometric liquid for determining the pressure drop at different flow rates.

The photo of twisted tape inserted into the tube and schematic representation of diagram of U-tube heat exchanger is shown in the Fig. 1b, c. The twisted tapes were made of 1.00 mm thick aluminum strips, the width of the strip being 1 mm less than the inside diameter of the test section tube and the dimensions are shown in Table 1. The strips were twisted on a lathe by manual rotation of the chuck. Twisted strips were heated periodically by flame to relax the stresses to prevent them from buckling and untwisting when they are removed from the lathe. The schematic of this twisted tape used in the section is shown in Fig. 1d

2.2 Experimental Procedure

In this study, the centrifugal pump was switched on, and the water flow rate to the test section was adjusted using the by-pass valve. The hot water at a particular flow rate is allowed to flow through jacket side to maintain constant wall temperature until steady state is attained. First the plain tube double pipe heat exchanger (i.e., without turbulator) was tested. At the beginning of series of tests, the cold water was circulated through inner tube and hot water through annulus tube in parallel flow and counter-flow configuration. The flow rate of hot water is maintained constant. The cooling water flowing in the heat exchanger was at room temperature. First the cold water flow rate was fixed to 2 l/min (0.033 kg/sec). A prescribed heat input was given to the water in hot water tank in sufficient state. Usually, 30 min was required for the attainment of steady state for a run. Once the steady state was reached readings were taken for the flow rates of hot and cold water, temperatures at inlet and outlet section of hot and cold fluid and U-tube manometer. The flow rate of hot water was kept constant, and the above procedure was repeated for different flow rates of cold fluid.

Experiments were first conducted for the plain tube heat exchanger and subsequently with twisted tape inserts by inserting the straight full twist insert with H/D ratios of 6, 10, 15, and 20. Full-length twisted tape was inserted into the both straight legs (2.3 m each) of the u-tube. The two tapes were inserted into the tube by pushing from both ends of the tube. Cold water was circulated inside the tube and hot water through annulus in counter flow arrangement and then the data was observed and readings were tabulated.

Table 1 Dimensions of the twisted tape inserts

S. No.	Parameter	Twist ratio, H/D			
		6	10	15	20
1	H (Pitch)	0.108	0.18	0.27	0.36
2	D (Diameter)	0.018	0.018	0.018	0.018

3 Data Reduction Equations

The data reduction of the measured results is summarized in the following procedures. The mean Nusselt number and friction factor are based on the inside diameter of the plain tube. Heat transfer to the cold fluid in the tube, Q_c in the test tube can be determined using

$$Q_c = m_c C_{pc}(T_{co} - T_{ci}) \tag{1}$$

where m_c is the mass flow rate of cold water C_{pc} is the specific heat of water, T_{ci} and T_{co} are the cold water inlet and outlet temperatures respectively.

The heat transfer rate from the hot water in the annulus side, Q_h can be written as

$$Q_h = m_h C_{ph}(T_{hi} - T_{ho}) \tag{2}$$

where m_h is the mass flow rate of hot water, C_{ph} is the specific heat of water, T_{hi} and T_{ho} are the hot water inlet and outlet temperatures respectively.

The average heat transfer rate Q_{ave} used in the calculation is estimated from the hot water and cold water sides as follows:

$$Q_{ave} = \frac{(Q_c + Q_h)}{2} \tag{3}$$

For fluid flows in a U-tube heat exchanger, the heat transfer coefficient, h_i is calculated from

$$Q_{ave} = U A_i \Delta T_{LMTD} \tag{4}$$

where $A_i = \pi d_i (2L_1)$

The heat transfer coefficient h_i for the inner side of the tube is determined by using

$$\frac{1}{U} = \frac{1}{h_i} + \frac{d_c}{k} \ln \frac{d_c}{d_i} + \frac{d_i}{d_c} \frac{1}{h_c} \tag{5}$$

The thermal conductivity k of the fluid is calculated from the fluid properties at the mean fluid temperature.

The annulus side heat transfer coefficient h_o is calculated by using

$$Q_h = m_h C_{ph}(T_{hi} - T_{ho}) = h_o A_o \Delta T$$

From which we get

$$h_o = {Q_h}/{A_o \Delta T} \tag{6}$$

Then substitute values of U from Eq. (4) and h_o from Eq. (6) in Eq. (5) and calculate h_i, thus

$$Nu = {h_i d_i}/{k} \tag{7}$$

The Reynolds number is based on the different flow rates at the inlet of the test section.

$$\mathrm{Re} = 4^r h / \pi d_i \mu \tag{8}$$

where μ is the dynamic viscosity of the working fluid.
Friction factor, f can be calculated from

$$f = 2\Delta p d_i / \rho u^2 L_2 \tag{9}$$

where Δp is the pressure drop across the test section, ρ is the density of water, d_i is the inner diameter of tube, u is the velocity of water, and L_2 is the length of tube.

4 Results and Discussion

4.1 Validation of Plain Tube Data

The results obtained during experimental investigations are presented and discussed in this section. The obtained experimental Nusselt number from Eq. (7) is compared with the correlations available in the literature.

(a) Dittus and Boelter [9]correlation for Nusselt number

$$Nu = 0.023\, \mathrm{Re}^{0.8}\, \mathrm{Pr}^{0.4} \tag{10}$$

(b) Petukhov [10] correlations for Nusselt number

$$Nu = \frac{(f/g)\mathrm{Re}\,\mathrm{Pr}}{1.07 + 12.7(f/g)^{0.5}(\mathrm{Pr}^{2/3-1})} \tag{11}$$

A comparison of Nusselt number was done for the theoretical data and experimental data. A graph was plotted for the results obtained from the data collected during the experimental investigation. Figure. 2 shows variation of Nusselt number with Reynolds number for plain tube for turbulent flow.

It has been observed that the Nusselt number increases with the increase of Reynolds number. Experimental results were compared with the Dittus-Boelter [9] Eq. (10) and Petukhov [10] Eq. (11) with deviations of ± 8 % and ± 12 % respectively.

Friction Factor:
The friction factor using Eq. (9) as obtained during the experiment trial is compared with the correlations available in the literature.

Fig. 2 Comparison of present experimental data with the existing literature values for plain U-tube double pipe heat exchanger with water as the fluid

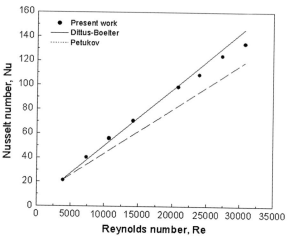

(c) Blasius [11] correlation for friction factor

$$f = 0.3614/Re^{0.25} \qquad (12)$$

(d) Petukhov [10] correlation for friction factor

$$f = (0.790 \, In \, Re - 1.64)^{-2} \text{ for } 3000 \leq Re \leq 5 \times 10^6 \qquad (13)$$

It has been observed that from Fig. 3, as the Reynolds number increases, the pressure drop correspondingly increases along the length of the tube which in turn decreases the friction factor. The friction factor obtained by the experiment is compared with the Petukhov [10] Eq. (13) and Blasius [11] Eq. (12), for the validation of Reynolds number. The experimental results have better agreement with the above equations.

Fig. 3 Variation of friction factor with Reynolds number in a plain tube

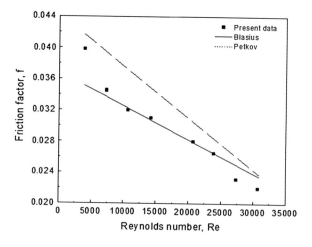

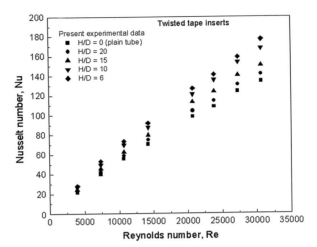

Fig. 4 Effect of H/D ratio on Nusselt number of water in plain tube with and without twisted tape inserts

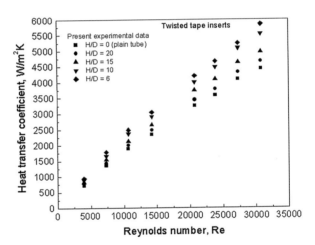

Fig. 5 Validation of heat transfer coefficient of water in plain tube with and without twisted tape inserts

4.2 Effect of Twist Ratio on Heat Transfer Augmentation

The experimental Nusselt number in plain tube with different twist ratios of twisted tape inserts are shown in Fig. 4 and it is found that higher Nusselt numbers are obtained for twist ratio of H/D = 6 compared to other twist ratios of twisted tape inserts. The Nusselt number increases with increase of Reynolds number and decrease of twist ratio.

The heat transfer coefficient as obtained from the experiment at various Reynolds number is shown in Fig. 5

It has been observed that the heat transfer coefficient increases with increase of Reynolds number and decrease of twist ratio.

4.3 Effect of Twist Ratio on Friction Factor

The experimental friction factor of different twisted tape inserts is shown in Fig. 6 and it is found that friction factor increases with increase of Reynolds number and decrease of twist ratios of twisted tape inserts.

4.4 Analysis of Performance Evaluation

The variation of performance ratio with equal mass flow rate basis, for full length twisted tapes is plotted in Fig. 7. On comparing the performance curves of smooth tube with twist ratios of 6, 10, 15, and 20, it has been observed that from the figure, the average heat transfer coefficient for the inside of the tube increases with increase in the flow rate of fluid in each case and heat transfer performance is maximum at H/D = 6. The heat transfer coefficient for a full length twisted tape is increased by approximately 32 % on an average compared to that of a smooth tube.

Pressure Drop Variation

On unit pressure drop basis, the variation of pressure drop with equal mass flow rate for full length twisted tapes is shown in Fig. 8.

It has been observed that, as the pressure drop increases with the increase of mass flow rate, the pressure drop is more in case of the smooth tube than the tube with turbulators. The increase of pressure drop is certainly a disadvantage resulting out of the use a full length twisted tapes inside the tube (turbulators) compared to the advantage gained in terms of increase of average heat transfer coefficient by

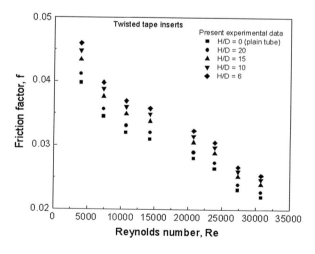

Fig. 6 Comparison of experimental friction factor of water in plain tube with and without twisted tape inserts

Fig. 7 Mass flow rate versus heat transfer coefficient

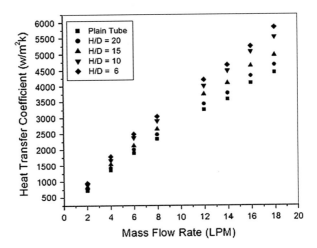

Fig. 8 Variation of pressure drop (Δp) with mass flow rate (LPM)

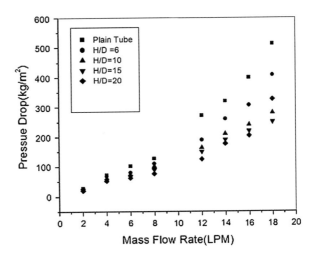

using turbulators. In order to study the advantage of use of turbulators, study of the parameter heat transfer coefficient per unit pressure drop is considered.

The thermal performance ratio of a heat exchanger is the ratio of heat transfer coefficient (h_i) to pressure drop (Δp). It has been observed from Fig. 9 that the heat transfer performance of smooth tube is maximum followed by full length twisted tape with different twist ratios by 2.0–2.2. Thermal performance decreases with use of turbulators because of increase in pressure drop is more than in increase in heat transfer coefficient.

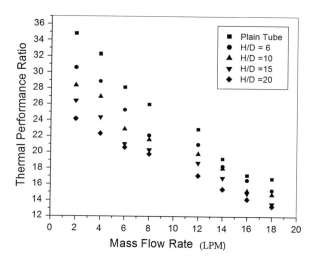

Fig. 9 Variation of thermal performance ratio ($hi/\Delta p$) with mass flow (LPM)

5 Conclusions

Experimental investigation of heat transfer and friction factor characteristics of circular tube fitted with twisted tape inserts of different twist ratios has been studied in a double pipe U-tube heat exchanger using cold water and hot water as the test fluids. From the experimental results, twisted tapes could be inserted inside flow of double pipe U-tube heat exchanger for enhancing heat transfer rate. Decreasing values of the twist-pitch to the diameter ratio lead to increasing values of heat transfer rate and the pressure drop as well. The enhancement in heat transfer coefficient with H/D = 6 is 29.98 % at Reynolds number 3,000 and 32.03 % at the Reynolds number 31,000. The enhancement in friction factor is 17.94 % at Reynolds number 3,000 and 20.93 % at Reynolds number 31,000. On equal mass flow rate basis, the heat transfer performance of full length twisted tape is maximum when compared with smooth tube. On unit pressure drop basis, the heat transfer performance of smooth tube is maximum when compared with full length twisted tapes with tube. It has been observed that thermal performance of smooth tube is better than full length twisted tape by 2.0–2.2 times.

References

1. Hong SW, Bergles AE (1976) Augmentation of laminar flow heat transfer in tube by means of twist tape insert. J Heat Transf, Trans ASME 98(2):251–256
2. Eiamsa-ard S, Thianpong C, Promvonge P (2006) Experimental Investigation of heat transfer and flow friction in a circular tube fitted with regularly spaced twisted tape elements. Intl Commun Heat Mass Transf 33:1225–1233
3. Bergles AE (1995) Techniques to augment heat transfer readings. Hand book of heat transfer applications,Chap 3. McGraw Hill, NewYork

4. Ahmed M, Deju L, Sarkar MAR, Islam SMN (2005) Heat transfer in turbulent flow through a circular tube with twisted tape inserts. Proc Intl Conf on Mech Eng (ICME2005)
5. Dewan A, Mahanta P, Sumithra Raju K, Suresh Kumar P (2004) Review of passive heat transfer augmentation techniques. J Power Energy 218:509–525
6. Yadav AS (2008) Experimental investigation of heat transfer performance of double pipe U-bend heat exchanger using full length twisted tape. Int J Appl Eng Res (IJAER) 3(3):399–407
7. Naphon P (2006) Heat transfer and pressure drop in the horizontal double pipes with and without twisted tape insert, Intl Commun Heat Mass Transf 33:166–175
8. Saha SK, Gaitonde UN, Date AW (1989) Heat transfer and pressure drop characteristics of laminar flow through a circular tube fitted with regularly spaced twisted tape elements with multiple twists. Exp Therm Fluid Sci 2(3):310–322
9. Dittus FW, Boelter LMK (1930) Univ Calif Pub Eng, Berkely, vol 2 p 443
10. Petukhov BS (1970) Heat transfer and friction in turbulent pipe flow with variable physical properties. In: Hartnett JP, Irvine TF (eds) Advances in heat transfer, Academic Press, New York, pp 504–564
11. Blasius H (1908) Grenzschichten in Flussigkeiten mit kleiner Reibung (German). Z Math Phys 56:1–37

Design of Integrated R134a Vapor Compression Heating and Cooling Cycle

Priyank Agarwal, R. Shankar and T. Srinivas

Abstract Although the world is aware of the hiking level of global warming, still numerous conventional air-conditioners are being put up for application in buildings adding contribution to global warming. To resolve this issue, the combined heating and cooling cycle is proposed by providing an additional heat rejecter next to compressor in a traditional vapor compression refrigeration cycle. Design and analysis for the proposed cycle is done with R134a as working fluid. The optimum operating conditions are predicted, delivering maximum COP of 3.36 at 3 bar at mass flow rate of 1 kg/s. Approximately, 31 kW of process heat has been gained with 45.5 tons of refrigeration. This work infers that with small amendment in traditional refrigeration cycle results in an increased efficiency and converts from only cooling to cooling cogeneration plant.

Keywords R134a • Heating • Cooling • Vapor compression • Integration

1 Introduction

The refrigeration and heat pump cycles have a widespread usage for their individual applications. But, individual cycle means more heat loss and lesser efficiency. Thus, researchers moved to integration of systems to utilize waste heat through processes categorized under waste heat recovery. Adding a theme to the current trend, the proposal presented aims at integration of refrigeration cycle for combined heating and cooling cycle using R134a as the working fluid. R134a has

---

P. Agarwal (✉) · R. Shankar · T. Srinivas
CO_2 Research and Green Energy Technologies Centre, School of Mechanical and Building Sciences, VIT University, Vellore, India
e-mail: priyank077@gmail.com

R. Shankar
e-mail: gentlewise26@yahoo.com

T. Srinivas
e-mail: srinivastpalli@yahoo.co.in

S. Sathiyamoorthy et al. (eds.), *Emerging Trends in Science, Engineering and Technology*, Lecture Notes in Mechanical Engineering, DOI: 10.1007/978-81-322-1007-8_4, © Springer India 2012

zero ozone depletion potential and moderate global warming potential of 1,300, as stated by Karagoz et al. [1] and can be used as common working fluid for both refrigeration and heat pump cycles. Havelsky [2] illustrates that R134a possess least total equivalent warming impact (TEWI) and it is much efficient than R12. Tchanche et al. [3] adding support to the Havelsky, states that R134a is better compared to hydrocarbons like butane and propane i.e., they have been categorized in security group A3 while R134a in security group A1 with water.

The effort to improve efficiency by Yamaguchi et al. [4] of heat pump running with CO_2 as working fluid at the refrigerant mass flow rate of about 0.10 kg/s and compressor inlet and outlet pressures as 4 and 12 MPa respectively, they achieved net heat capacity of 22.3 kW with comparatively lesser coefficient of performance. Also maintenance of heat pump with CO_2 as working fluid at such high pressure is quite challenging. El-Meniawy et al. [5] studied the heat pump with R22, a hazardous refrigerant, as working fluid and compression pressure ranging from 1 to 21.7 bar, delivering maximum of 12.002 kW.

The theoretical performance of study on a vapor compression refrigeration system done by Dalkilic and Wongwises [6] shows that the power required for per ton of refrigeration for their cycle at pressure ratio of 6.57 is 3.88 kW with R134a as working fluid. On the other hand, experimental investigation on R134a vapor ejector refrigeration system with rated cooling capacity of 0.5 kW done by Selvaraju and Mani [7] achieved maximum COP of about 0.475 only with refrigerant flow rate varying from 0 to 256.03 lph and generator temperature ranging from 338 to 363 K. Hence, a successful attempt has been made to optimize the performance of these systems.

2 Energy Model for Combined Heating and Cooling Cycle

With intention to impose combined heating and cooling cycle, increment in efficiency of conventional cycle was achieved by reducing the input electrical power required and also procuring a net heating effect through process heat. The design and modeling is done for proposed combined heating and cooling cycle that can be used in various application sectors requiring heating process and refrigeration simultaneously, ultimately saving huge amount of capital investment. There were few assumptions considered which are included in Sect. 2.1 with description of cycle in Sect. 2.2.

2.1 Assumptions

- Compressor inlet: 1.013 bar and 5 °C in dry saturated vapor state.
- Condenser inlet is at 40 and exit at 30 °C
- Isentropic and mechanical efficiencies of the compressor are 75 and 95 % respectively.

2.2 Cycle Description and Working

The schematic diagram of the proposed combined heating and cooling cycle is shown in Fig. 1. R134a is used as the working fluid and the required properties were calculated. For thermodynamic equations used to calculate the properties of R134a can be referred from Appendix.

The refrigeration cycle has been modified with appendage of heat recovery mechanism next to the compressor reducing the condenser load and incrementing refrigeration effect. The refrigerant is compressed from 1.013 bar to varying output pressure ranging from 2 to 15 bar. R134a in dry saturated vapor phase transforms to superheated state after compression and due to high compression ratio, temperature and enthalpy of fluid increases. The heat content of the fluid is utilized in the heat rejecter, providing net process heat with 40 °C as constant outlet temperature. Fixing temperature constraints for the condenser, as mentioned in Sect. 2.1, reduces its load and consequently liquid state is attained at the exit. Liquid refrigerant expands in the expansion valve and moves to evaporator, providing refrigeration effect along with fluid for compressor inlet. The formulae used for calculation are specified in Eqs. (1–7) where $\dot{m}$ symbolizes mass flow rate of refrigerant and h_x symbolizes enthalpy of refrigerant for particular phase at x position in the cycle (x varies from 1 to 5).

Heat input from compressor

$$(Q_{\text{Compressor}}) = \dot{m}\,(h_2 - h_1) \tag{1}$$

Heat utilised in heat rejecter

$$(Q_{\text{Extractor}}) = \dot{m}\,(h_2 - h_3) \tag{2}$$

Heat rejected in condenser

$$(Q_{\text{Condenser}}) = \dot{m}\,(h_3 - h_4) \tag{3}$$

Refrigeration effect from evaporator

$$(Q_{\text{Evaporator}}) = \dot{m}\,(h_1 - h_5) \tag{4}$$

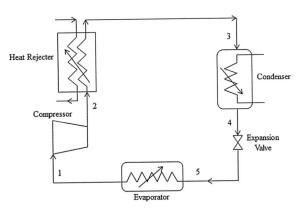

Fig. 1 Combined heating and cooling cycle

Coefficient of Performance for heating process

$$(COP_H) = \frac{h_2 - h_3}{h_2 - h_1} \quad (5)$$

Coefficient of Performance for refrigeration process

$$(COP_R) = \frac{h_1 - h_5}{h_2 - h_1} \quad (6)$$

Coefficient of Performance for integration cycle

$$(COP) = \frac{h_2 - h_3 + h_1 - h_5}{h_2 - h_1} \quad (7)$$

3 Results and Discussions

Figure 2 provides variation of load on evaporator, heat rejecter, and compressor with varying outlet pressure from compressor. It is observable that both compressor and heat rejecter load increases continuously with increase in pressure, but the evaporator load remains constant throughout. This is because of fixed constraints across evaporator. Although only evaporator outlet (or compressor inlet) was kept constant as 1 atm and 5 °C, the inlet conditions also become constant because of predetermined conditions of R134a across condenser. Thus, a constant refrigeration load of about 160 kW (45.5 tons of refrigeration) was obtained.

With variation in the compressor and heat rejecter load, coefficient of performance for both process heat gain part and refrigeration part continuously varies. With their variation, COP for overall cycle also varies. For a heat pump cycle, COP can reach maximum up to the value of 1, indicating ideal working as per Carnot, but, as the

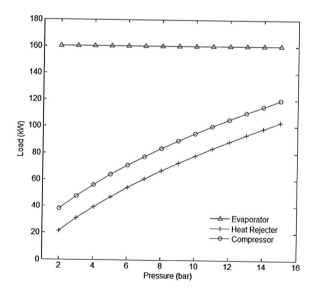

Fig. 2 Variation of evaporator load, heat rejecter load and compressor load with compressor outlet pressure

cycle primarily focuses upon refrigeration effect, we can compromise with COP of heat rejecter to attain higher COP for refrigeration effect. COP for cooling process decreases and for heating process COP increases, but decrease in COP for cooling being dominant over increasing COP of heating, overall coefficient of performance decreases. This variation of COP can be illustrated through Fig. 3.

In spite of these performance characteristics, choosing of optimum working pressure seems to be difficult as one can choose the optimum pressure as per the need. Being primarily a refrigeration cycle, least pressure should be preferred to reduce compressor load achieving higher COP for cooling and moderate COP for heating effects. To aid choosing of optimum pressure, Fig. 4

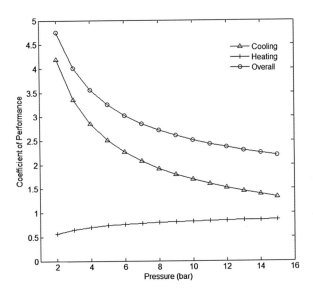

Fig. 3 Variation of COP with compressor outlet pressure

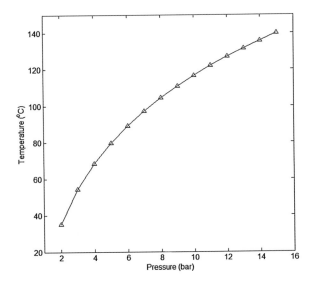

Fig. 4 Variation of compressor outlet temperature with pressure

delivers wavering outlet temperature of refrigerant as per exit temperature from compressor. Although low pressure is efficient, pressure of 2 bar cannot be chosen because of low delivery temperature than the assumed value of 40 °C. Hence, we opt 3 bar as optimum pressure for the working of the proposed cycle.

Figure 5 presents P-h diagram for both conventional and modified cycles at the chosen optimum pressure of 3 bar. Cycle 1-2-4-5-1 represents conventional cycle for which condenser load is $\dot{m} \times (h_4-h_2)$ whereas cycle 1-2-3-4-5-1 represents modified cycle with reduced condenser load as $\dot{m} \times (h_4-h_3)$. The difference of $\dot{m} \times (h_3-h_2)$ contributes to useful heat gain and resembles simple but basic illustration of integrated energy systems. With consideration of compressor efficiencies, we obtain a negatively sloped curve for the heat gain from compressor. The refrigerant first gets compressed in process 1-2, increasing pressure of refrigerant from 1.013 to 3 bar with transformation of phase from saturated vapor state to superheated vapor state and hence there is increase in its heat content. Isobaric processes 2-3 and 3-4 are heat rejection processes with 2–3 being heat rejecter process and 3-4 being condensation process. The superheated state of fluid changes again to saturated vapor state at 40 °C from the exit of heat rejecter. Process 4-5 is constant enthalpy process of expansion which reduces the pressure of the refrigerant to 1.013 bar and the process 5-1 is the evaporation state in which the refrigerant provides refrigeration effect by gaining heat and it achieves vapor state at the exit of evaporator.

Also comparing to the referred works presented earlier, the presented integrated cycle dispensing better performance and results, thus resulting into better efficiency. The work presented by the coauthors Shankar and Srinivas [8, 9] provides another good example for cogeneration but with integration of cooling and power cycles.

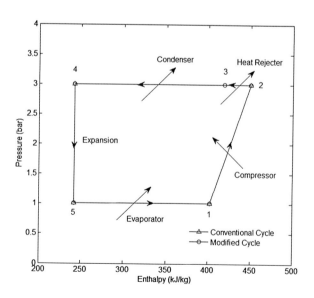

Fig. 5 P-h diagram for conventional and modified cycles at 3 bar of condenser pressure

4 Conclusions

The integration cycle of vapor compression heating and cooling cycle has been proposed and evaluated. The results showed that it gives a higher performance compared to the regular design. The cycle delivers about 45.5 tons of refrigeration and 31 kW of process heat at 1 kg/s of mass flow rate and 3 bar compressor outlet pressure for R134a with maximum COP of 3.36. Moreover, the cycle can further be improved using different fluids and modifying the cycle. Hence, integration poses a good scope for future technology with better performance and reduced investment.

Appendix

Thermophysical properties of R134a (1,1,1,2-tetrafluoroethane)

- Vapor Pressure: [10]

$$P = P_c \times \exp\left(3.946984 \cdot x^{1.66} - \frac{11.313271 \cdot x}{1 - x} + 3.693108 \cdot x + 5.566337 \cdot x^3\right)$$

$$x = 1 - \frac{T}{T_c}; \quad T_c = 374.179 \text{ K}; \quad P_c = 40.56 \text{ bar}$$

- Liquid Enthalpy: [11]

$$h_l = -6.702179 + 0.1675422 \cdot T + 0.2154294 \times 10^{-2} T^2$$

- Vapor Enthalpy: [11]

$$h_v = 83.23572 + 1.742258 \cdot T - 0.2140479 \times 10^{-2} T^2$$

- Superheated Enthalpy: [11]

$$h_{\text{sup}} = B_0 + B_1 T + B_2 T^2 + \omega \cdot (B_3 + B_4 T + B_5 T^2)$$
$$\omega = C_0 + C_1 T + C_2 T^2 + C_3 T^3$$

$B_0 = 155.1313$; $B_1 = 0.8471667$; $B_2 = 0.209139$ E-3; $B_3 = 34.7401$

$B_4 = -0.3860322$; $B_5 = 0.672008$ E-3; $C_0 = -35.94481$; $C_1 = 0.265213$

$C_2 = -0.6782399$ E-3; $C_3 = 0.6323821$ E-6

References

1. Karagoz S, Yilmaz M, Comakli O, Ozyurt O (2004) R134a and various mixtures of R22/R134a as an alternative to R22 in vapor compression heat pumps. Energy Convers Manage 45:181–196
2. Havelský V (2000) Investigation of refrigerating system with R12 refrigerant replacements. Appl Therm Eng 20:133–140
3. Tchanche BF, Papadakis G, Lambrinos G, Frangoudakis A (2009) Fluid selection for a low-temperature solar organic rankine cycle. Appl Therm Eng 29:2468–2476
4. Yamaguchi S, Kato D, Saito K, Kawai S (2011) Development and validation of static simulation model for CO2 heat pump. Int J Heat Mass Transf 54:1896–1906
5. El-Meniawy SAK, Watson FA, Holland FA (1981) A study of the operating characteristics of a water-to-water heat pump system using R22. Heat Recovery Syst 1(3):209–217
6. Dalkilic AS, Wongwises S (2010) A performance comparison of vapor-compression refrigeration system using various alternative refrigerants. Int Commun Heat Mass Transfer 37:1340–1349
7. Selvaraju A, Mani A (2006) Experimental investigation on R134a vapor ejector refrigeration system. Int J Refrig 29:1160–1166
8. Shankar R, Srinivas T (2012) Modelling of energy extraction in vapor absorption refrigeration system. Procedia Eng 38:98–104
9. Shankar R, Srinivas T (2012) Solar thermal based power and vapor refrigeration absorption system. Procedia Eng 38:730–736
10. Huber ML, Ely JF (1992) An equation of state formulation of the thermodynamic properties of R134a (1,1,1,2-tetrafluoroethane). Int J Refrig 15(6):393–400
11. Abou-Ziyan HZ, Ahmed MF, Metwally MN, El-Hameed HMA (1997) Solar-assisted R22 and R134a heat pump systems for low-temperature applications. Appl Therm Eng 17(5):455–469

Effect of Thermal Barrier Coating on DI Diesel Engine Fuelled with P20 Biodiesel

A. R. Pradeep Kumar, K. Annamalai and S. R. Premkartikkumar

Abstract The faster depletion of fossil fuels and rapid increasing cost of the petroleum fuel are the main reasons for this experimental investigation to arrive at an alternative fuel for the present diesel engines. Many researchers have submitted their work and succeeded with 20 % volume of biodiesel fuel with diesel at various ambient conditions. The alphabet "P" refers to Pongamia, which is best suited for the Indian climate and can be cultivated even on a nonagricultural land. The other biodiesels under research are Jatropha, Karanja, Neem, Mahua, Karanja, cotton seed oil, etc. In India, the edible oil is used for cooking; it is decided to use nonedible oils for the research. Out of the total energy produced in compression ignition (CI) engines, one third of heat energy alone is used for useful work as the remaining energy is rejected to coolant and surrounding. To retain the heat energy in the combustion chamber a thermal barrier coating for a thickness of 0.5 mm is applied on the top surface of piston and bottom surface of cylinder to retain the heat energy. The ceramic material preferred for coating is partially stabilized zirconia (PSZ). The tests were conducted with Diesel as the reference fuel and 20 % Pongamia biodiesel blended with diesel (P20) as the test fuel. The results were compared and a discussion has been made with suitable justifications. Compromising results were obtained for brake thermal efficiency (BTE), specific energy consumption (SEC), hydrocarbon emission (HC), and carbon monoxide emission (CO). Nitric oxide (NOx) emission is higher in the experimental work, for which the suitable means have been discussed in the scope for the future work.

Keywords Biodiesel • Thermal barrier coating • PSZ

A. R. Pradeep Kumar (✉)
Department of Mechanical, Dhanalakshmi College of Engineering, Chennai, India
e-mail: Dearpradeepkumar@gmail.com

K. Annamalai
Automobile Department, Anna University, Chennai, India

S. R. Premkartikkumar
SMK Fomra Institute of Technology, Chennai, India

S. Sathiyamoorthy et al. (eds.), *Emerging Trends in Science,*
Engineering and Technology, Lecture Notes in Mechanical Engineering,
DOI: 10.1007/978-81-322-1007-8_5, © Springer India 2012

1 Introduction

The intensive research is an urgent requirement for the replacement of conventional petroleum diesel as the depletion rate is more day by day. Use of vegetable oil is already demonstrated in the 1900 in an exhibition at Paris by Rudolf Diesel [1]. He demonstrated the engine with peanut oil. Since the vegetable oils have higher viscosity, it leads to a poor combustion and further exhibits poor performance characteristics. The emission characteristics of the biodiesel blends indicate that the blends provided a good alternative to conventional diesel. Experimental investigation with Waste cooking oil study showed that Waste Cooking Oil methyl esters have similar properties with diesel fuel [2]. Transesterified vegetable oil, which is known as biodiesel has lesser viscosity and can be comfortably used in compression engines as it has almost compromising level of Calorific value as Petroleum diesel. But calorific value alone is not the property to compare with, but other properties such as Fire point, kinematic viscosity, flash point, and fire point. Unlike fossil fuels, the use of biodiesel does not contribute to global warming as CO_2 emitted is once again absorbed by the plants grown for vegetable oil/biodiesel production. Thus, CO_2 balance is maintained [3]. Another factor to be considered in compression ignition (CI) engines is the heat loss to surroundings. Out of total energy developed, only one third of energy is used for the useful work and the remaining energy is rejected to coolant and to the surroundings. Hence, in this experimental work it is planned to retain the heat energy and also the heat can be used to reduce the viscosity of biodiesel. A thermal barrier coating is applied on the bottom surface of cylinder head and top of the piston surface for a thickness of 0.5 mm by Plasma arc spraying.

2 Biodiesel and its Production

Biodiesel is defined as mono-alkyl esters of long chain fatty acids derived from vegetable oils or animal fats, which conform to ASTM D6751 specifications for use in diesel engines. Biodiesel contains no petroleum, but it can be blended at any level with petroleum diesel to create a biodiesel blend [4]. Biodiesel is the final product obtained by the process called Transesterification [5]. The source for the production of biodiesel is vegetable oil. The block diagram for the transesterification is shown in Fig. 1.

The Pongamia oil is mixed with a catalyst sodium hydroxide (NaOH) pallets and Methanol. The stirring has been done for proper mixing of vegetable oil, methanol, and the catalyst. The mixture is heated up to a temperature of 60–65 °C with continuous stirring. The solution is poured in a container and a settling time of 9 h is allowed. The glycerin is settled at the bottom and methyl esters of vegetable oil are formed at the top. The methyl esters of vegetable is further heated to the boiling point of water (100 °C) and maintained for 10–15 min to remove the unused methanol. It is further washed with water to remove dissolved NaOH.

Effect of Thermal Barrier Coating

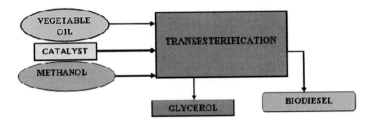

Fig. 1 Transesterification process

Table 1 Comparison of properties of reference fuel and test fuel

S.No	Property	Diesel	Methyl esters of pongamia
1	Caloric value, kJ/kg	43,100	36,409
2	Kinematic viscosity at 40 °C, cSt	3.1	4.8
3	Density at 15 °C, kg/mm^3	830	899
4	Flash point, °C	56	86
5	Fine point, °C	64	95

The Table 1 compares the properties of the diesel and esterified pongamia biodiesel which shows significantly lower calorific value, higher viscosity, density, flash point, and fire point. Use of thermal barrier coating will play a vital role in resolving lack in few properties. At higher combustion zone temperature the viscosity may fall down and there is a possibility of higher efficiency due to utilization of more heat energy.

3 Thermal Barrier Coating

The diesel engine with thermal barrier coating by the ceramic material is referred to as a semi adiabatic engine (SA Engine) or a low heat-rejection (LHR) engine. Partially stabilized zirconia (PSZ) have the highest maximum service temperatures (~2,000 °C) among all the ceramics and they retain some of their mechanical strength close to their melting point (2,750 °C). They have a low thermal conductivity of about 2 W/m K, which make them a good thermal insulator and thus, able to reduce the heat lost to the surroundings [6].

4 Experimental Setup

The Fig. 2 shows the experimental setup used for the experimental investigation [7]. The engine used was a Kirloskar SV 1 model, single cylinder, direct injection (DI) Diesel Engine with Eddy Current dynamometer. AVL Smoke meter

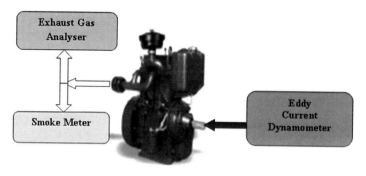

Fig. 2 Experimental setup

is used for testing smoke emission five gas analyser is used for hydrocarbon emission (HC), Carbon monoxide (CO), and Nitric Oxide (NOx) emission. The engine is a water cooled engine with a compression ratio of 17.5:1, injection timing of 27° before Top Dead Center (bTDC) and with a constant injection pressure of 200 bar. All the tests were carried out at standard conditions.

5 Results and Discussion

5.1 Brake Thermal Efficiency

Figure 3 shows the characteristic curve between Brake power and brake thermal efficiency (BTE). The engine efficiency for the petroleum diesel attains a maximum value of 32.4 % and for the P20 biodiesel fuel is 31.5 % which is almost close to the efficiency of Diesel [8, 9].

The utilization of most of the heat during combustion may be the reason for the improvement of BTE though P20 has a lower calorific value. The complete combustion due to the higher combustion zone temperature and oxygen content biodiesel may be the other reasons can be addressed for the better efficiency.

Fig. 3 Brake power versus BTE

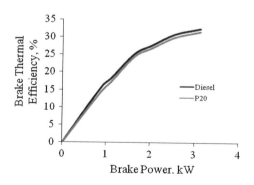

5.2 Brake Power Versus Specific Energy Consumption

In this experimental work specific energy consumption (SEC) has been taken for considered instead of specific fuel consumption because, diesel and Pongamia biodiesel have different calorific values. Figure. 4 illustrates that the SEC for P20 at all loads is higher compared to diesel. This may be due to the lower energy content in the test fuel [10].

5.3 Brake Power Versus Hydrocarbon Emission

The HC with P20, is lesser compared to diesel because of more oxygen content which enhances the combustion reaction (Fig. 5).

Another reason for the complete combustion may be the higher combustion temperature due to the ceramic coating which retains the temperature produced during combustion. There is a gradual decrease of HC emission, in the beginning and the an increase of HC emission at higher loads, which may be due to the rich fuel intake due to the higher operating loads [11].

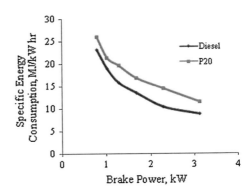

Fig. 4 Brake power versus SEC

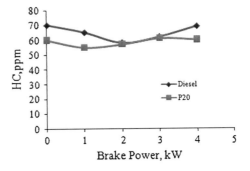

Fig. 5 Brake power versus hydrocarbon combustion

5.4 Brake Power Versus CO Emission

There is lesser CO emission with P20 compared to diesel because of higher oxygen content which enables all the carbon to be converted into CO and further presence of oxygen converts all the CO into CO_2. The higher combustion zone temperature can also be a reason for the complete combustion (Fig. 6).

5.5 Brake Power Versus Nitric Oxide Emission

NOx emission is due to the oxygen content and elevated temperature during combustion. Since the oxygen content is more in the biodiesel and the temperature is much higher in the coated engine, NOx emission is more. The higher NOx emission is a major draw back with the biodiesel fuel both in conventional and coated engines. NOx emission is still higher in coated engine compared to uncoated engine (Fig. 7).

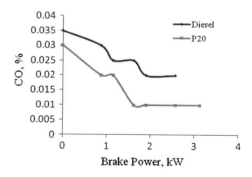

Fig. 6 Brake power versus CO emission

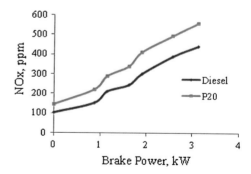

Fig. 7 Brake power versus NOx emission

6 Conclusion

The following results were obtained during this investigation.

1. BTE of P20, is much closer to the petroleum fuel, which shows a positive sign of using Pongamia biodiesel in the automotives.
2. SEC is more in the biodiesel fuel due to its lesser energy content. The cost would be the only compromising factor for this limitation, which may come down when mass production of biodiesel is being done.
3. Both HC and CO emission are lesser which is a positive result on the emission parameter.
4. NOx emission is more when biodiesel is used in the coated engines due to the elevated temperature and presence of oxygen in the biodiesel.

Majority of the results show a good sign of using P20 biodiesel in future. The experimental work may be continued to reduce the NOx emission.

7 Scope for Future Work

The following methods can be adopted to reduce the NOx emission when biodiesel is being used as the fuel.

1. Retarding the injection timing without affecting the BTE to a greater extent.
2. Internal exhaust gas recirculation (EGR).

Acknowledgments Author A. R. Pradeep Kumar, sincerely thank Dr. K. Annamalai, Ph.D., Assistant Professor, Department of Automobile Engineering, Anna University (MIT Campus) for his valuable guidance throughout the experimental investigation. He also thank Dr. V·P. Ramamoorthy, Ph.D., chairman, Dhanalakshmi College of Engineering, Chennai for his motivation at every stage of the experimental work.

References

1. Clean Alternative Fuels (2002) Biodiesel, EPA420-F-00-032. United States Environmental Protection Agency, Mar 2002
2. Arslan R (2011) Emission characteristics of a diesel engine using waste cooking oil as biodiesel fuel. Afr J Biotechnol 10(19):3790–3794
3. Jaichandar S, Annamalai K (2011) The status of biodiesel as an alternative fuel for diesel engine: an overview. J Sustain Energy Environ 2:71–75
4. Kapilan N, Ashok Babu TP, Reddy RP (2009) Technical aspects of biodiesel and its oxidation stability. Int J Chem Tech Res 1(2):278–282
5. Meher LC, Vidyasagar D, Naik SN (2004) Technical aspects of biodiesel production by transesterification-a review. Renew Sustain Energy Rev XXL, pp 1–21
6. Pradeep Kumar AR, Annamalai K, Prabhakar S, Banugopan VN (2010) Analytical investigations on heat transfer in a low heat rejection DI diesel engine, Frontiers in Automobile and Mechanical Engineering (FAME). IEEE Explore Digital Libr, pp 25–27, 189–191

7. Widyan MI Al, Shyoukh AOAL (2002) Experimental evaluation of the transesterification of waste palm oil into biodiesel. Bio Resour Technol 85:253–256
8. Canakei M, Gerpen JV (2003) Comparison of engine performance and e missions for petroleum diesel fuel, yellow grease biodiesel and soybean oil biodiesel. Am Soc Agric Eng 46(4):937–944
9. Mazel MA, Summers JD, Batchelder DG (1985) Peanut, soybean and cottonseed oil as diesel fuels. Trans ASA E 28(5):1375–1377
10. Bora DK, Das LN, Gajendra Babu MK (2008) Performance of a mixed biodiesel fueled DI diesel engine. J Sci Ind Res 67:73–76
11. Agarwal AK (2005) Experimental investigation of the effect of bio diesel utilization on lubricating oil tribology in diesel engines. J Automobile Engg 219:703–713

Effects of Compression Ratio on the Performance and Emission Characteristics of Diesel Engine Fuelled with Ethanol Blended Diesel Fuel

M. Santhosh and K. P. Padmanaban

Abstract The study of effects of compression ratio (CR) in a diesel engine fuelled with ethanol–diesel blends and diesel is important as they significantly affect its performance and emissions. The present paper focuses on the experimental investigation of the influence of CR of the engine on the performance and emission characteristics of ethanol– diesel blended fuel and the results compared with diesel. Tests were carried out using three different compression ratios (CR 16.5:1, 14.3:1, and 12.6:1) at 40 and 60 % of maximum engine load. At lower CR the carbon monoxide (CO), carbon dioxide (CO_2), hydro carbon emission (HC) increases and nitrogen oxides emissions (NO_X) decreases. At higher CR the brake thermal efficiency (BTE) increases. It has been observed that the increase in BTE and brake specific fuel consumption (BSFC) was obtained with the increase of ethanol blend proportions with diesel. The emissions of NO_X and CO_2 decreases and the emission of CO and HC increases with increase of ethanol blend proportions with diesel.

Keywords Compression ratio • Brake thermal efficiency • Emissions

---

M. Santhosh (✉)
P.S.N.A College of Engineering and Technology, Dindigul 624 622, India
e-mail: srinithims78@gmail.com

K. P. Padmanaban
S.B.M College of Engineering and Technology, Dindigul, India
e-mail: padmarubhan@yahoo.co.in

S. Sathiyamoorthy et al. (eds.), *Emerging Trends in Science,*
Engineering and Technology, Lecture Notes in Mechanical Engineering,
DOI: 10.1007/978-81-322-1007-8_6, © Springer India 2012

1 Introduction

In recent studies on the reduction of exhaust emissions from diesel engines into the atmosphere have focused mainly on alternative fuels with reduced environmental impact. The investigations on diesel engines have expanded in the area of alternative fuels, among which ethanol–diesel blend represents a very promising fuel. Ethanol can be produced from biomass by fermentation of sugar, by converting the starch content of biomass feed stocks into alcohol (bio ethanol) or by hydration of ethylene which is obtained from petroleum and other sources. Ethanol usually replaces gasoline in petrol engines and bio diesel makes the same for diesel engines, but diesel fuel blended with low concentrations of ethanol can also run a diesel engine. Diesel engines have better engine performance, low fuel consumption, and greater power output compared to those of gasoline-fueled spark ignition engines. Many investigations are related to the influence of blending ethanol with mineral diesel fuel on engine performance and exhaust emissions, showing an increase in brake thermal efficiency (BTE) and in specific fuel consumption [4], a slight decrease in engine power and a more significant decrease in exhaust emissions compared to the use of diesel fuel. It is reported that the ethanol have been considered as alternative fuels for diesel engines [1–3]. Raheman and Ghadge [12] researched the performance of diesel engine using biodiesel and its blends with diesel at varying compression ratios (18, 19, and 20:1). Results showed that BTE increased with the increase in the compression ratio (CR). Laguitton et al. [10] studied the effect of CR on the emissions of a diesel engine. The results of effect of reduction in CR from 18.4 to 16.0 were presented. It was found that carbon monoxide (CO) and hydro carbon emission (HC) increases and nitrogen oxides emissions (NO_X) decreases.

In this paper, impact of CR of ethanol–diesel fuel blends are analyzed and compared to those of standard diesel. Among the various proportions only up to 15 % of ethanol–diesel fuel blend is tested since the higher bio ethanol concentrations would show ignition problems when used to run a diesel engine, furthermore, the heating value would significantly differ from that of a diesel fuel causing important reduction in engine power. As seen, ethanol– diesel blends and operating parameters influence the diesel engine performance [such as brake specific fuel consumption (BSFC) and BTE] and exhaust emissions (such as HC, CO and NOx. Therefore, this study focused on the effects of CR on ethanol-blended diesel fuel on the engine performance and exhaust emissions of the same DI diesel engine and the results were compared with diesel.

2 Experimental Setup

The experiments were conducted on a four-stroke, naturally aspirated, single cylinder DI diesel engine. Details of the engine specification are shown in Table 1 and set-up of the test bench is shown in Fig. 1. An electrical

Table 1 Technical features of the test engine

Parameter	Specification
Engine type	DI, naturally aspirated, air cooled
Number of cylinders	1
Bore (mm)	80
Stroke (mm)	110
Displacement (cm)	395
Compression ratio (CR)	16.5:1
Maximum power (kW) at rated rpm	5.59
Specific fuel consumption	238–860 g/Kwhr@full load/nominal
Rated rpm	1,500
Power	3.7 KW or 5HP@RTP conditions

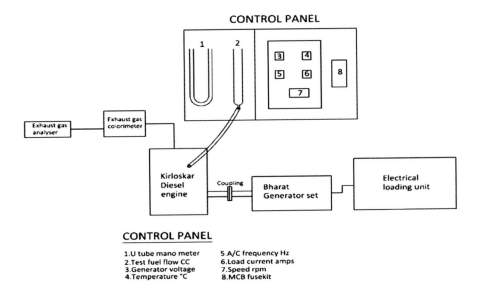

Fig. 1 Schematic diagram of engine set-up

dynamometer has been directly coupled to the engine output shaft. CO, Carbon Dioxide (CO_2), NO_X and HC emissions were measured with a Multi gas analyzer (MARS TECH) with an accuracy of ±0.02, ±0.5 % and ±15, ±10 ppm respectively. The accuracies of measurements are shown in Table 2. The original injection pressure (IP), injection timing (IT) and CR of the engine are 210 bar, 20.9°CA BTDC, 16.5 and 3 respectively. The experiments were carried out at different modes of the compression ratios of the direct injection diesel engine fuelled with ethanol–diesel blends (b5, b10, and b15) and the results were compared with diesel. The properties of diesel and ethanol are shown in Table 3. The different modes of the compression ratios are 16.5, 14.3, and 12.6 respectively.

Table 2 Parameters range and accuracy of the measurements

MARS five gas analyzer	Measuring range	Accuracy
Carbon monoxide (CO)	0–15 %vol	0.01 %
Carbon dioxide (CO_2)	0–20 %vol	0.01 %
Hydrocarbon (HC)	0–30,000 ppm	1 ppm
Oxygen (O2)	0–25 %vol	0.01 %
Nitric oxide (NO)	0–5,000 ppm	±15 ppm
Exhaust gas temperature (EGT)	0–700 °C	±1 °C
Digital stop-watch	–	±0.2 s
Manometer	300 mm	±1 mm
Specific fuel consumption	238–860 g/KWhr	±0.005 kg/kW h
Calculated results	Uncertainty	–

Table 3 The Properties of diesel and ethanol are given below

Properties	Diesel	Ethanol
Density (kg/m^3) at 15 °C	837.8	799.4
Kinematic viscolity at 40 °C (mm^2/s)	2.649	1.1
Cetane index	54	08
Calorific value (KJ/kg)	44893	28180
Flash point (°C)	50	12
Oxygen content (% by weight)	0	34.73
Carbon content (% by weight)	86.14	52.14
Hydrogen content (% by weight)	13.86	13.13
Stoichiometric A/F ratio	14.75	9.06

By increasing the piston to head clearance length the CR decreased. The CR of the piston to head clearance length of 7.096, 8.276, and 9.456 mm are 16.5:1, 14.3:1, and 12.6:1 respectively.

3 Results and Discussion

3.1 Brake Thermal Efficiency

The BTE for different compression ratios and for different proportions of blend and for the standard diesel is predicted and shown in Figs. 2, 3, 4, 5, 6, 7, and 8. The BTE increases with increase in the load for different blends and different compression ratios. The figure is evident that the increase in the proportions of blends very slightly increases the BTE when compared to standard diesel. This is due to the improved mixing of ethanol blends during ignition delay and oxygen enrichment of blends [5, 19]. It is found that the BTE increases with increase of CR for different blend proportions. This is due to the oxygen

Effects of Compression Ratio on the Performance and Emission

Fig. 2 The change in the BTE with the blends compared to diesel at CR 16.5

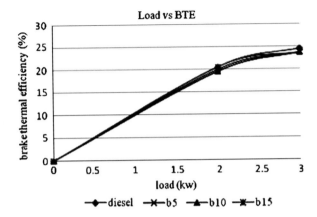

Fig. 3 The change in the BTE with the blends compared to diesel at CR 14.3

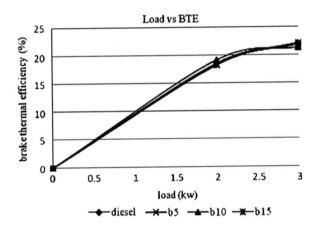

Fig. 4 The change in the BTE with the blends compared to diesel at CR 12.6

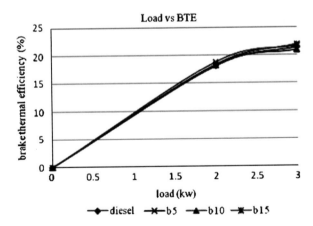

Fig. 5 The change in the BTE for diesel at different CR

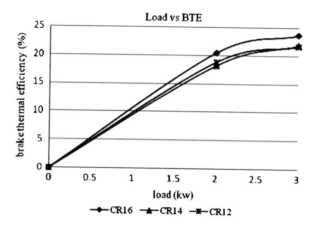

Fig. 6 The change in the BTE for B5 at different CR

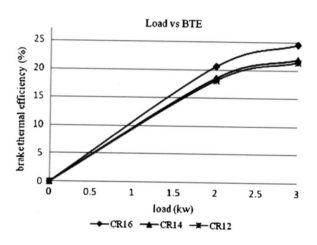

Fig. 7 The change in the BTE for B10 at different CR

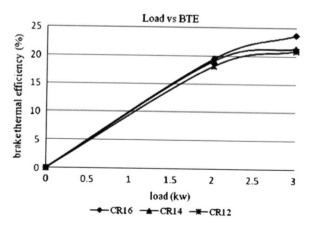

Fig. 8 The change in the BTE for B15 at different CR

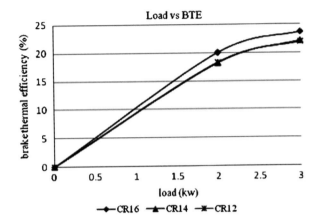

enrichment of ethanol–diesel blend leads to better combustion. BTE is directly proportionate to the CR [8].

3.2 Brake Specific Fuel Consumption

The BSFC for different compression ratios and for different proportions of blend and for the standard diesel is predicted and shown in Figs. 9, 10, and 11. The figure is evident that the BSFC slightly increases with the increase of blend proportions compared to standard diesel. This is due to the lower calorific value of the ethanol compared to standard diesel which causes the amount of fuel injected into the cylinder is to be greater for ethanol–diesel blends [7, 15]. Also, it has been observed that at higher loads the BSFC decreases due to the increase in

Fig. 9 The change in the BSFC with the blends compared to diesel at CR 16.5

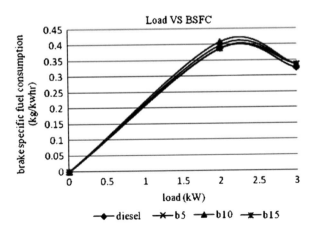

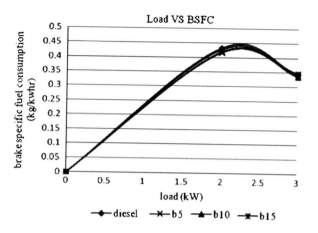

Fig. 10 The change in the BSFC with the blends compared to diesel at CR 14.3

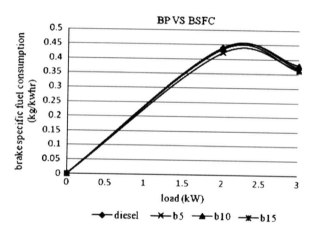

Fig. 11 The change in the BSFC with the blends compared to diesel at CR 12.6

combustion temperature which in turn increases the conversion of heat energy to mechanical work and thereby the BSFC decreased.

3.3 Nitrogen Oxides Emissions

The emission of NOx for different compression ratios and for different proportions of blend and for the standard diesel is predicted and shown in Figs. 12, 13, and 14. It is found that the emission of NOx decreases with decrease in CR for all blend proportions. The formation of NOx is highly dependent on the in-cylinder temperature which is reduced at lower CR and thereby the emission of NOx decreases [11, 12]. The emission of NOx increases with increase in the engine load due to the more amount of fuel injected at higher load which causes the higher cylinder

Effects of Compression Ratio on the Performance and Emission 71

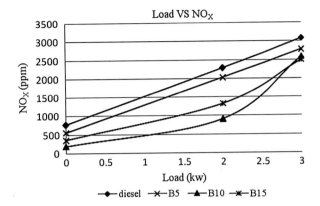

Fig. 12 The emission of NO_X for different blends and compared to diesel at CR 16.5

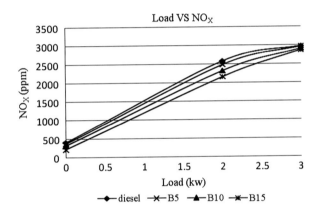

Fig. 13 The emission of NO_X for different blends and compared to diesel at CR 14.3

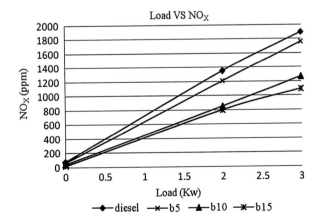

Fig. 14 The emission of NO_X for different blends and compared to diesel at CR 12.6

pressure and increase in combustion cylinder temperature. The emission of NOx decreases with increase in blend proportions of ethanol in diesel due to the lower heating value and the higher latent heat of vaporization of ethanol fuel [20, 21]. The high latent heat of vaporization of ethanol resulted in a decrease in the combustion cylinder temperature, which in turn induced a decrease in NOx emissions [13].

3.4 Carbon Monoxide

The emission of CO for different compression ratios and for different proportions of blend and for the standard diesel is predicted and shown in Figs. 15, 16, and 17. It is found that the emission of CO increases at reduced compression ratios for all blend proportions and standard diesel. At lower CR, insufficient heat of

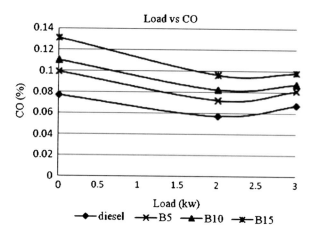

Fig. 15 The emission of CO for different blends and compared to diesel at CR 16.5

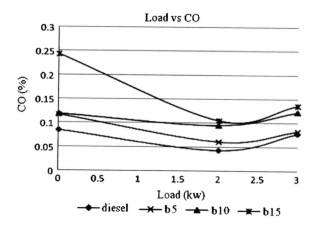

Fig. 16 The emission of CO for different blends and compared to diesel at CR 14.3

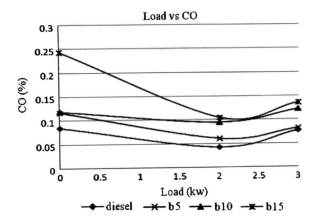

Fig. 17 The emission of CO for different blends and compared to diesel at CR 12.6

compression delays ignition and so the emission of CO increases. The possible reason for this trend could be that the increased CR actually increases the air temperature inside the cylinder therefore reducing the ignition lag causes better and more complete burning of the fuel [12]. The emission of CO decreases with increase in the engine load because of the increased air–fuel ratio and more complete combustion [14–17]. The emission of CO increases with an increase in the proportions of blend with diesel. This is due to the ethanol blending supplied more molecular oxygen to the combustion chamber, the high heat of evaporation of ethanol fuel leads to a lower temperature in the combustion cylinder.

3.5 Carbon Dioxide

The emission of CO_2 for different compression ratios and for different proportions of blend and for the standard diesel is predicted and shown in Figs. 18, 19, and 20. The CO_2 emissions of blends at lower compression ratio (CR 12.6) are higher

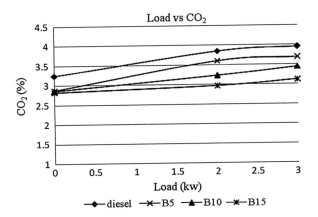

Fig. 18 The emission of CO_2 for different blends and compared to diesel at CR 16.5

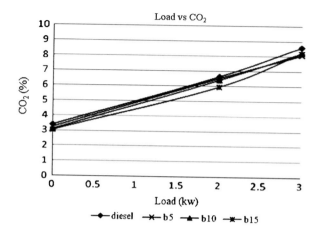

Fig. 19 The emission of CO_2 for different blends and compared to diesel at CR 14.3

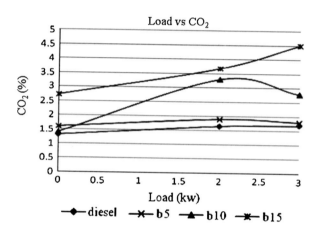

Fig. 20 The emission of CO_2 for different blends and compared to diesel at CR 12.6

than that of diesel fuel. This is due to complete combustion of fuel in combustion chamber. The emission of CO_2 decreases with increase in blend proportions when compared to diesel at CR 16.5. This is due to insufficient oxygen supply during combustion. The CO_2 emissions are not affected for blends compared to diesel fuel at CR 14.3. Also, the emission of CO_2 increases with increase in the load.

3.6 Hydro Carbon Emission

The emission of HC for different compression ratios and for different proportions of blend and for the standard diesel is predicted and shown in Figs. 21, 22, and 23. The figure is evident that the emission of HC increases with increase in ethanol blend proportions in diesel. This is due to the higher heat of evaporation of the ethanol

Effects of Compression Ratio on the Performance and Emission

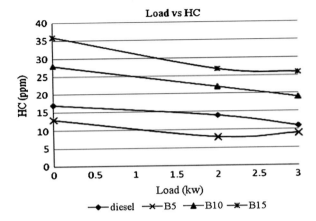

Fig. 21 The emission of HC for different blends and compared to diesel at CR 16.5

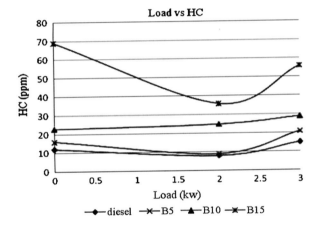

Fig. 22 The emission of HC for different blends and compared to diesel at CR 14.3

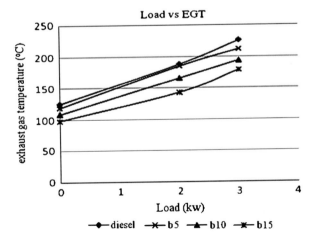

Fig. 23 The emission of HC for different blends and compared to diesel at CR 12.6

blends causing slower evaporation and poor air–fuel mixture [2, 3, 20]. The emission of HC increases for different ethanol–diesel blends due to the low cetane number and a long ignition delay [6, 18]. In general, HC emissions arise due to wall wetting, insufficient oxygen, and residual fuel in cylinder. At lower CR due to the insufficient heat of compression delays ignition and so the emission of HC increases [9].

3.7 Exhaust Gas Temperature

The exhaust gas temperature (EGT) for different loads and different compression ratios are shown in Figs. 24, 25, 26, 27, and 28. The figure is evident that the EGT decreases for all blends with increase of blend proportions of ethanol with diesel. Also, it is found that the EGT increases with increase of load. The EGT increases with increase of CR for all blend proportions.

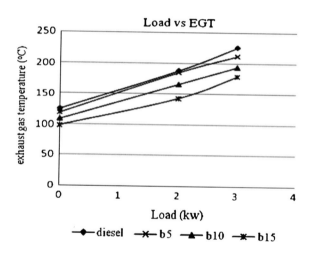

Fig. 24 The EGT for different blends and diesel at CR 16.5

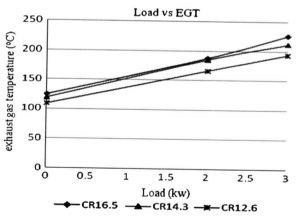

Fig. 25 The EGT for diesel at different CR

Effects of Compression Ratio on the Performance and Emission   77

Fig. 26 The EGT for blend b5 at different CR

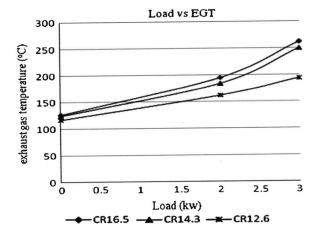

Fig. 27 The EGT for blend b10 at different CR

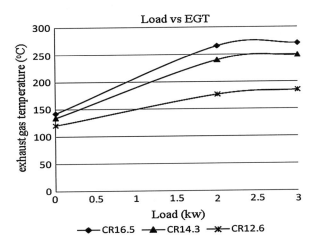

Fig. 28 The EGT for blend b15 at different CR

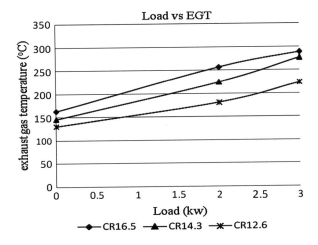

4 Conclusion

The engine performances and emission of exhaust gases of a kirloskar engine fuelled with different proportions of ethanol blended with standard diesel are experimentally investigated for different compression ratios ranging from 12.6, 14.3, and 16.5.

The engine performances and emission of exhaust gases of a kirloskar engine is also experimentally predicted using standard diesel and compared with that of different proportions of ethanol blended with standard diesel.

The investigation concludes as follows:

1. **Influences on engine performances**
   - The brake thermal efficiency increases with increase in load for different blends and CR.
   - The brake thermal efficiency slightly increases with increase in blend proportions.
   - The brake thermal efficiency increases with increase in CR.
   - The BSFC and brake power decrease with increase in CR
   - The BSFC slightly with increase of blend proportions than that of standard diesel and also, BSFC decreases at higher loads.

2. **Influences on emission of exhaust gases**
   - The emission of hydrocarbon increases with increase in blend proportions than that of standard diesel.
   - The emission of CO increases with increase in blend proportions.
   - The emission of CO increases at lower CR for ethanol blends and standard diesel.
   - The emission of CO decreases with increase in engine load.
   - The emission of CO_2 higher at lower CR than that of standard diesel.
   - At CR 16.5 the emission of CO_2 decreases with increase in blend proportions than that of diesel.
   - The emission of nitrogen oxides increases with increase in load and CR for all blend proportions.
   - The emission of NOx decreases with increase in blend proportions with standard diesel.

References

1. Likos B, Callahan TJ, Moses CA (1982) Performance and emissions of ethanol and ethanol–diesel blends in direct-injected and pre-chamber diesel engines. SAE Paper no 821039
2. Ecklund EE, Bechtold RL, Timbario TJ, McCallum PW (1984) State-of-the-art report on the use of alcohols in diesel engines. SAE Paper no 840118
3. Corkwell KC, Jackson MM, Daly DT (2003) Review of exhaust emissions of compression ignition engines operating on E diesel fuel blends, SAE Paper no. 2003-01-3283

4. Buyukkaya E (2010) Effects of biodiesel on a DI diesel engine performance, emission and combustion characteristics. Fuel 89:3099–3105
5. Hansen AC, Taylor AB, Lyne PWL, Meiring P (1989) Heat release in the compression ignition combustion of ethanol. Int J Am Soc Agric Biol Engrs Trans ASAE 32(5):1507–1511
6. Rakopoulos CD, Antonopoulos KA, Rakopoulos DC (2007) Experimental heat release analysis and emissions of a HSDI diesel engine fueled with ethanol–diesel fuel blends. Energy 32:1791–1808
7. Abdel-Rahman AA (1998) On the emissions from internal-combustion engines. Int J Energy Res 22:483–513
8. Ramadhas AS, Jayaraj S, Muraleedharan C (2006) Theoretical modeling and experimental studies on biodiesel–fueled engine. Renew Energy 31:1813–1826
9. Jindal S, Nandwana BP, Rathore NS, Vashistha V (2010) Experimental investigation of the effect of compression ratio and injection pressure in a DI diesel engine running on Jatropha methyl ester. Appl Therm Eng 30:442–448
10. Laguitton O, Crua C, Cowell T, Heikal MR, Gold MR (2007) The effect of compression ratio on exhaust emissions from a PCCI diesel engine. Energy Convers Manage 48:2918–2924
11. Heywood JB (1988) Internal combustion engine fundamentals. McGraw- Hill, NewYork
12. Raheman H, Ghadge SV (2008) Performance of diesel engine with biodiesel at varying compression ratio and ignition timing. Fuel 87:2659–2666
13. Xingcai L, Zhen H, Wugao Z, Degang L (2004) The influence of ethanol additives on the performance and combustion characteristics of diesel engines. Combust Sci Technol 176:1309–1329
14. Eugene EE, Bechtold RL, Timbario TJ, McCallum PW (1984) State-of-the-art report on the use of alcohols in diesel engines, SAE Paper no 840118
15. Ajav EA, Singh B, Bhattacharya TK (1999) Experimental study of some performance parameters of a constant speed stationary Diesel engine using ethanol–diesel blends as fuel. Biomass Bio Energy 17(4):357–365
16. Likos B, Callahan TJ, Moses CA (1982) Performance and emissions of ethanol and ethanol–Diesel blends in direct injected and pre-chamber diesel engines, SAE Paper no 821039
17. Choi CY, Reitz RD (1999) An experimental study on the effects of oxygenated fuel blends and multiple injection strategies on DI Diesel engine emissions. Fuel 78(11):1303–1317
18. Park SH, Youn IM, Lee CS (2011) Influence of ethanol blends on the combustion performance and exhaust emission characteristics of a four-cylinder diesel engine at various engine loads and injection timings. Fuel 90:748–755
19. Hulwan DB, Joshi SV (2011) Performance, emission and combustion characteristic of a multi cylinder DI diesel engine running on diesel–ethanol–biodiesel blends of high ethanol content, Appl Energy
20. Rakopoulos DC, Rakopoulos CD, Kakaras EC, Giakoumis EG (2008) Effects of ethanol–diesel fuel blends on the performance and exhaust emissions of heavy duty DI diesel engine. Energy Convers Manage 49:3155–3162
21. Weidmann K, Menrad H (1984) Fleet test, performance and emissions of diesel engines using different alcohol–diesel fuel blends, SAE Paper no 841331

Green Carrier

Vikrant Goyal and Pankhuri Arora

Abstract The evolution in aircraft industry has brought to us many new aircraft designs. Each and every new design is a step toward a greener tomorrow. Design plays a vital role in deciding the flight characteristics and determining its efficiency. The proposed design has been designed keeping 100,000 lbs as the payload. Raked wingtips, canards, and elliptical shaped fuselage are the highlighting features of the proposed design. Reduction of drag will also be observed due to delay in separation point. Expected outcome of proposed design is less amount of fuel burn because of reduction in drag.

Keywords Design • Drag • Fuselage • Raked wing tips

1 Introduction

With the advent of globalization, air freighters are influencing the various global aspects like environment and economy. Due to the rapid pace of deterioration of the environment, the eye of aviation industry is now focused on sustaining the balance of the environment. The proposed freighter closely addresses the need of greener tomorrow keeping in mind the today's constraints.

Calculations of the individual parts are shown below:

1.1 Fuselage Geometry

Fuselage is a long hollow tube which holds all the pieces of an airplane together. Fuselage contributes a significant portion of the weight of an aircraft. Fuselage of EMERALD D-7 is flat and elliptical as per calculations.

V. Goyal (✉) · P. Arora
S.R.M University, Chennai, India
e-mail: vikrant.goyal1991@gmail.com

P. Arora
e-mail: pankhuriarora@hotmail.com

S. Sathiyamoorthy et al. (eds.), *Emerging Trends in Science,*
Engineering and Technology, Lecture Notes in Mechanical Engineering,
DOI: 10.1007/978-81-322-1007-8_7, © Springer India 2012

1.1.1 Calculations

Cargo container used: **LD3** (IATA type 8)
Length of container: 200 cm
Height of container: 153 cm
Depth of container: 162 cm
Volume of one container: 4.3 m^3
Weight of 1LD3 container: 1,588 kg
(Reference no. 3)
Given Payload: 45,372.05 kg
Number of containers = Payload/weight of one container
= 45,372.05/1588 = 28.571819
= 28 containers + 0.571819 bulk cargo
Bulk cargo = 0.571819 × 1,588 = 908.05 kg
Now as per cargo arrangement, there are two columns of 14 container each, thus length of fuselage,
Length of fuselage = (Length of one LD3 container × 14) + (bulk cargo) + clearance
= (1.53 × 14) + 10.72 + 3.98 = 36.13 m
Height of fuselage = (Height of a container + clearance)
= (1.62 + 1.22) = 2.85 m
Breadth of fuselage = (Breadth of a container × 2) + (clearance)
= (2.0066 × 2) + 1.3868 = 5.4 m

Clearance kept in fuselage is calculated on basis of reference cargo aircrafts i.e., Galaxy C-5 and Beluga (Fig. 1).

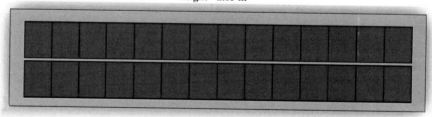

Fig. 1 Cargo arrangement in fuselage

1.2 Airfoil Selection

An airplane wing is a 3-D model of an airfoil. The main purpose of airfoil is the generation of lift force. Using "Design-foil R6 demo", six series airfoil NACA 67−217 a = 0.6 is selected, where "a" is mean-line perimeter (Fig. 2).

Various factors are considered such as:

- High lift coefficient.
- Low drag coefficient.
- High lift to drag ratio.
- Low pitching moment coefficient.
- High Stalling Angle.

Values from graph (Figs. 3, 4):
Stall angle = 15.24, C_{dmin} = 0.0032, C_M = −0.049, C_L = 1.332
Images obtained by substituting data in Design Foil R6 DEMO.

Observations:

1. The airfoil with the highest maximum lift coefficient, max C_L
2. The airfoil with the lowest minimum drag coefficient, min C_D
3. The airfoil with the highest lift-to-drag ratio ((C_L/C_D) max).
4. The airfoil with the highest lift curve slope.
5. The airfoil with the lowest pitching moment coefficient (C_M).
6. The proper stall quality in the stall region (the variation must be gentile, not sharp).

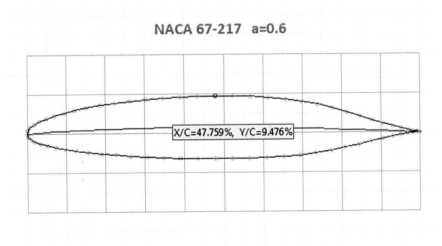

Fig. 2 NACA 67−217 a = 0.6, image obtained from design foil R6 DEMO

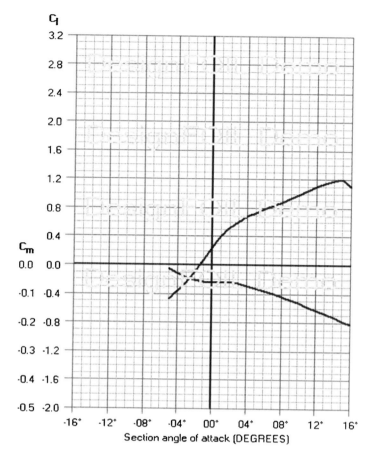

Fig. 3 C_L versus α, C_M versus α

1.3 Wing Geometry

Wings are considered to be lift producing devices. The designed aircraft has high wing. selections of high wings are on the following basis [2]:

1. Eases and facilitate the loading and unloading of loads and cargo into and out of aircraft.
2. High wing will increase the dihedral effect (ω). It makes the aircraft laterally more stable. The reason lies in the higher contribution of the fuselage to the wing dihedral effect (ω).
3. The aerodynamic shape of the fuselage lower section can be smoother.
4. There is more space inside fuselage for cargo, luggage, or passenger.
5. The wing drag is producing a nose-down pitching moment, so it is longitudinally stabilizing. This is due to the higher location of wing drag line relative to the aircraft centre of gravity ($M_{Dcg} < 0$).

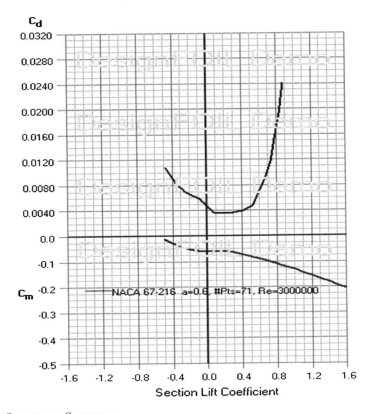

Fig. 4 C_D versus α, C_M versus α

1.4 Raked Wingtips

One of the new features introduced in the wings is the raked wingtip. Raked wingtips are the most recent winglet variants (they are probably better classified as special wings, though), where the tip of the wing has a higher degree of sweep than the rest of the wing. They are widely referred to as winglets, but they are better described as **integrated wingtip extensions** as they are (horizontal) additions to the existing wing, rather than the previously described (near) vertical solutions. Key features of raked wingtip are:

1. Improved fuel economy. An approximate 2 % increase in fuel efficiency.
2. Improved climb performance. Faster climb performance can mean quieter neighbourhoods.
3. Shortened take off field length.
4. Reduction of drag by 5.5 % (Fig. 5).

Fig. 5 Raked wing

1.4.1 Calculations

(1) **Wing loading**: 750 kg/m² (assumed value from reference cargo aircrafts) [1]
MTOW: 213918.6479 kg (as calculated in weight estimation)
Wing area = MTOW/Wing loading
=213918.6479/750 = 326.67 m²

(2) **Aspect Ratio**: 9.89
AR = b2/S
B = √AR × S
=√9.89 × 326.67 = 56.68 m
And also, AR = b/C, 9.89 = 56.86/C, C = 5.74 m

(3) λ = Ct/Cr = 0.1906

$$C^- = \frac{2(CR\ (1+\lambda+\lambda 2))}{3(1+\lambda)} = 5.94\ m$$

CR = 8.23 m and CT = 1.57 m
All calculation formulae are from (Ref. no. 2).

1.5 Horizontal Stabilizer

The stabilizer is a fixed wing section whose job is to provide stability for the aircraft, to keep it flying straight. The horizontal stabilizer prevents up-and-down or pitching motion of the aircraft nose.

$V_{HT} = (S_{HT} \times l_{HT})/(S \times C)$
(Assuming V_{HT} = 0.50)
$\underline{S_{HT}}$ = 0.125, S_{HT} = 40.81 m²
 S
$\underline{L_{HT}}$ = 4, L_{HT} = 19.524 m
 C

Green Carrier

$$C = \frac{C_T + C_R}{2} = 4.888 \text{ m}$$

$AR_{HT} = 0.6 \times AR_{WING}, AR_{HT} = 5.934$
$AR_{HT} = b_{ht}^2 / S_{HT}$
$b_{ht} = \sqrt{(40.81 \times 5.934)} = 14.23 \text{ m}$
Assuming, $\lambda = 0.32$

$$C_R = \frac{2 \times S}{b_{ht} (1 + \lambda) \, 14.23 \, (1.32)} = 2 \times 40.81 = 4.34 \text{ m}$$

$C_T = 0.32 \times 4.34 = 1.28 \text{ m}$

1.6 Vertical Stabilizer

The vertical stabilizer keeps the nose of the plane from swinging from side to side, which is called yaw.

$$V_{VT} = \frac{S_{VT} \times l_{VT}}{S \times C}$$

We assume, $V_{VT} = 0.075$

$$\frac{S_{VT}}{S} = 0.020, S_{VT} = 6.53 \, m^2$$

$L_{VT} = 6.75 \text{ m}$

$$\frac{L_{VT}}{C} = 3.75$$

$AR_{VT} = 1.2$

$$\frac{AR_{vt}}{S_{VT}} = b_{2vt}$$

bvt $= 3.13 \text{ m}$
Assuming $\lambda = 0.6$

$$CR = \frac{2S_{VT}}{b (1 + \lambda)} = 4.569 \text{ m}$$

$CT = 2.2032 \text{ m}$
Back—Deflection $= 18.78$ degree from horizontal axis.
All calculation formulae are from (Ref. no. 3) (Figs. 6, 7, 8).

1.6.1 Analysis of Aircraft Model

1. Meshing of 2-D airfoil (Fig. 9).
2. Flow past airfoil (Fig. 10).

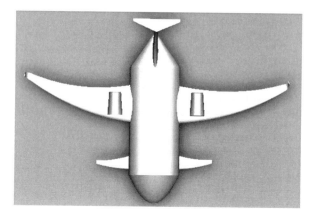

Fig. 6 Top view

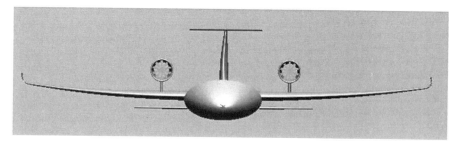

Fig. 7 Front view

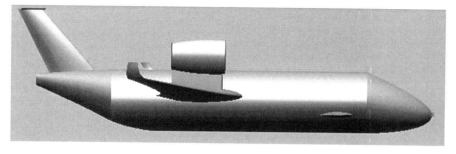

Fig. 8 Side view

1.6.2 Airfoil

CFD analysis of NACA 67−217 a = 0.6 is done and results are obtained as follows after 2,000 iterations. Low pressure on upper surface and high pressure on lower surface, thus airfoil selected is generating required lift force [3].

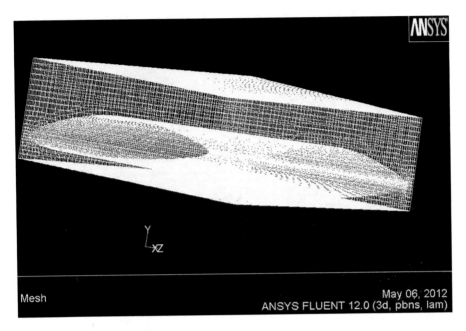

Fig. 9 2-D mesh of airfoil

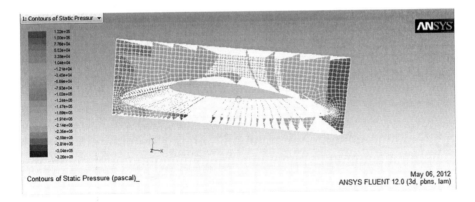

Fig. 10 2-D flow past airfoil

3. Elliptical fuselage: Figure 11.

Computational fluid analysis (CFD) has been done for elliptical fuselage with marine shaped nose. 5,000 iterations has been done and pressure distribution obtained is as desired.

CFD analysis of raked wings and complete design is the scope of future research.

Complete Vehicle Design: Figure 12

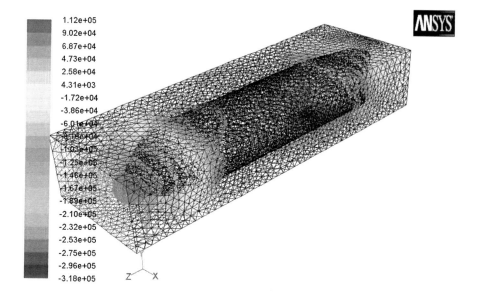

Fig. 11 Pressure distribution along fuselage

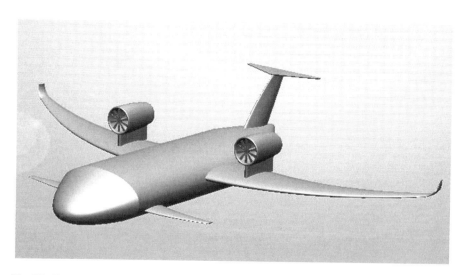

Fig. 12 Complete aircraft design

2 Conclusion

To sum up, a greener and more efficient aircraft model has been designed using SOLID WORKS and analyzed using ANSYS 12.

Acknowledgments We are greatly indebted to our esteemed institution SRM UNIVERSITY. We extend our gratitude for the assistance provided by our department faculty whenever required. Provision of CAD lab and its guidance by the associated faculty is highly appreciated.

References

1. Air freight containers datasheet, Air freight container specifications, IATA and ATA guidelines
2. Mohammad H, Sadraey (2012) Chapter-5 Wing design in aircraft design. Syst Eng Approach, ISBN 9781119953401, Wiley
3. Mohammad H. Sadraey (2010) Chapter-6 Tail design in aircraft design. Syst Eng Approach

Implementation of Ansys for Analyzing Windmill Tower with AISI 302 Stainless Steel

K. Vinoth Raj, N. Shankar Ganesh and T. Elamaran

Abstract Energy is one of the crucial inputs for social-economy and human lives. The global population is increasing day by day with the development of activities resulting in increased energy demand. The sources of energy are mainly from fossil fuels like oil, coal, etc., and also from renewable energy like wind hydro, geothermal energy etc. Wind energy is converted to electrical energy by means of wind turbines which are installed in regions where the wind speed is more and these are mounted on steel structured towers. The tower of a windmill is mainly affected by various loads acting on it, such as air forces, rotating rotor forces, blade weights, and atmospheric temperatures. Therefore the tower will soon be meeting with failure. Practically, it is impossible to check the failures of the components in running condition. Also, it is very difficult and makes a lot of losses (material, cost, time, etc.). Hence with the help of FEA SOFTWARE different materials have been analyzed and it was found that the AISI 302 stainless steel material deflection was low.

Keywords Windmill tower • AISI 302 stainless steel • Finite element analysis • IS 1239 MTD Steel • Deflection

1 Introduction

Generation of power other than conventional sources of energy was developed in the eighteenth century. Two great pioneers that led to the advent of windmills were Charles Brush and Poul La Cour. Charles Brush invented an efficient DC dynamo used in the public electrical grid for commercial electric lights and for an efficient method for manufacturing lead acid batteries. His company, Brush electric in Cleveland Ohio, was sold in 1889 and in 1892 it was merged with Edison General

---

K. Vinoth Raj (✉) · N. Shankar Ganesh · T. Elamaran
Department of Mechanical Engineering, Kingston Engineering College,
Vellore 632002, India
e-mail: nivi619@gmail.com

S. Sathiyamoorthy et al. (eds.), *Emerging Trends in Science, Engineering and Technology*, Lecture Notes in Mechanical Engineering, DOI: 10.1007/978-81-322-1007-8_8, © Springer India 2012

Electric Company under the name General Electric Company (GE). During the winter of 1887–1988 Brush built what is today believed to be the first automatically operating wind turbine for electricity generation.

It was a giant—the world's largest with a rotor diameter of 17 m (50 ft) and 144 rotor blades made of cedar wood. Note that the person mowing the lawn to right of the wind turbine the turbine ran for 20 years and charged the batteries in the cellar of this mansion. Despite the size of the turbine, the generator was only a 12 kW model. This is due to fact that slowly rotating wind turbines of the American wind rose type do not have particularly high average efficiency. La Cour was one of the pioneers of modern aerodynamics and built his own wind tunnel for experiments. He was concerned with storage of energy, and used the electricity from his wind turbines for electrolysis in order to produce hydrogen for gaslight in his school. One basic drawback of this scheme was the fact that he had to replace the windows of several school buildings several times as the hydrogen exploded due to small amounts of oxygen in the hydrogen. The same fate was suffered by the larger turbines used abroad. The turbines became extremely expensive, and the high energy price subsequently became a key argument against wind energy.

2 Megawatt-Sized Wind Turbines

The prototype of the NEG Micon 1500 kW Turbine was commissioned in September 1995. The original model has a 60 m rotor diameter and two 7,509 KW generators operating in parallel. The most recent version is a 1,500/750 kW model (with two 750 kW generators) with a 64 m rotor diameter. The prototype of the Vistas 1,500 kW turbine was commissioned in 1996. The original model has a 63 m rotor diameter and a 1,500 kW generator. The most recent version has a 68 m rotor diameter and a dual 1,650/300 kW generator.

Design should not be confused with discovery. We can discover what has already existed but not what has not been known before. But a design is the product of planning and work. Good design requires both analysis and synthesis. The challenges presented by the design environment are to think of the C's of design, Creativity, Complexity, Choice, and Compromise. In this work, the optimized material has been suggested using Ansys.

2.1 Terminology

2.1.1 The Energy in the Wind: Air Density and Rotor Area

The energy in the wind: Air Density and Rotor Area A wind turbine obtains its power input by converting the force of the wind into torque (turning force) acting on the rotor blades. The amount of energy which the winds transfer to the rotor depends on the density of the air, the rotor area, and the wind speed.

2.1.2 Air Density

The kinetic energy of a moving body is proportional to its mass (or weight). The kinetic energy in the wind thus depends on the density of the air, i.e., its mass per unit of volume. In other words, the "heavier" the air, the more energy is received by the turbine. At normal atmospheric pressure and at 15 °C air weighs some 1.225 kg/m^3, but the density decreases slightly with increasing humidity. Also, the air is denser when it is cold than when it is warm. At high altitudes, (in mountains) the air pressure is lower and the air is less dense.

2.1.3 Betz' Law

The Ideal Braking of the wind is the more kinetic energy a wind turbine pulls out of the wind, the more the wind will be slowed down as it leaves the left side of the turbine in the picture. If we tried to extract all the energy from the wind, the air would move away with the speed zero, i.e., the air cannot leave the turbine. In that case we would not extract any energy at all, since all of the air would obviously also be prevented from entering the rotor of the turbine.

In the other extreme case, the wind could pass through our tube above without being hindered at all. In this case we would likewise not have extracted any energy from the wind. We can therefore assume that there must be some way of braking the wind which is between these two extremes and is more efficient in converting the energy in the wind into useful mechanical energy. It turns out that there is a surprisingly simple answer to this: an ideal wind turbine would slow down the wind by 2/3 of its original speed.

2.1.4 The Cut in Wind Speed

Generally, wind turbines are designed to start running at wind speeds somewhere around 3–5 m/s. This is called the cut in wind speed. The blue area to the left shows the small amount of power we lose due to the fact that the turbine only cuts in after, say 5 m/s.

2.1.5 The Cut Out Wind Speed

The wind turbine will be programmed to stop at high wind speeds above, say 25 m/s, in order to avoid damaging the turbine or its surroundings. The stop wind speed is called the cut out wind speed. The tiny blue areas to the right represent that loss of power.

2.2 Design Calculation of Windmill Tower

Design Input for Tower

Rated power of wind turbine	$= 350$ MW
Load due to blade	$= 1,100$ kg
Number of blades	$= 3$
Total load due to blades W_1	$= 3,300$ kg
Load due to gear box and Shaft W_2	$= 5,500$ kg
Load due to generator W_3	$= 1,030$ kg
Total load on the tower W_4	$= W_1 + W_2 + W_3$
	$= 3,300 + 5,500 + 1,030$
	$= 9,830$ kg
Factor of safety F.S	$= 1.25$
Working load on the tower	$= \text{F.S} \times (W_4)$
	$= 1.25 \times 9830$
	$= 12,287.5$ kg
Basic wind speed	$= 55$ m/sec
Height of the tower	$= 50$ m

2.3 The Following Materials are Selected for Construction of Tower

IS 1161 YST 315 steel
Steel alloy 1,040 cold drawn
AISI 302 stainless steel
IS 1161 YST 310 steel
IS 1239 MTD steel

2.4 First Material IS 1161 Yst 315 Steel Material Properties

1. Material $=$ as per IS 1161
2. Young's modulus E $= 1.99 \times 10^5$ N/mm^2
3. Yield strength $= 315$ N/mm^2
4. Tensile strength $= 500$ N/mm^2
5. Poisson Ratio $= 0.3$

2.4.1 Dimension of Tower

Base diameter of tower	$= 3,500$ mm
Top end diameter	$= 2,275$ mm
Thickness of the tower shell	$= 25$ mm

The tower cross section is conical so the standard moment of inertia is calculated by taking six sections and the average of six sections.

Standard formulae for Moment of inertia of Round hollow section

$$I = (\pi/64) \times (D^4 - d^4) \text{ mm}^4$$

Area A $= (\pi/4) \times (D^2 - d^2) \text{ mm}^2$

4 At Section 1-1

Moment of inertia at Section 1-1

$$I_1 = (\pi/64) \times (D_1^4 - d_1^4)$$
$$= 4.119 \times 10^{11} \text{ mm}^4$$

Area a_1 $= (\pi/4) \times (D^2 - d^2)$
$= 272.925 \times 10^3 \text{ mm}^2$

At Section 2-2

Moment of inertia at Section 2-2

$$I_2 = (\pi/64) \times (D_2^4 - d2^4)$$
$$= 3.308 \times 10^{11} \text{ mm4}$$

Area a_2 $= (\pi/4) \times (D_2^2 - d_2^2)$
$= 253.68 \times 103 \text{ mm}^2$

At Section 3-3

Moment of inertia at Section 3-3

$$I_3 = (\pi/64) \times (D_3^4 - d_3^4)$$
$$= 2.611 \times 10^{11} \text{ mm}^4$$

Area a_3 $= (\pi/4) \times (D_3^2 - d_3^2)$
$= 234.441 \times 10^3 \text{ mm}^2$

At Section 4-4

Moment of inertia at Section 4-4

$$I_4 = (\pi/64) \times (D_4^4 - d_4^4)$$
$$= 2.0197 \times 10^{11} \text{ mm}^4$$

Area a_4 $= (\pi/4) \times (D_4^2 - d_4^2)$
$= 215.199 \times 10^3 \text{ mm}^2$

At Section 5-5

Moment of inertia at Section 5-5

$$I_5 = (\pi/64) \times (D_5^4 - d_5^4)$$
$$= 1.524 \times 10^{11} \text{ mm}$$

Area a_5 $= (\pi/4) \times (D_5^2 - d_5^2)$
a_5 $= 195.956 \times 10^3 \text{ mm}^2$

At Section 6-6

Moment of inertia at Section 6-6

$$I_6 = (\pi/64) \times (D_6^4 - d_2^4)$$
$$= 1.118 \times 10^{11} \text{ mm}^4$$

Area a_6 $= (\pi/4) \times (D_6^2 - d_6^2)$
$= 176.71 \times 10^3 \text{ mm}^2$

Average moment of inertia

I

$= (I_1 + I_2 + I_3 + I_4 + I_5 + I_6)/6$
$= 2.449 \times 10^{11}$ mm^4

Average area, a

a

$= (a1 + a2 + a3 + a4 + a5 + a6)/6$
$= 224.819 \times 10^3$ mm^2

3 Crippling Check

The tower has a fixed foundation at the bottom and the entire load acts on the top of the tower; hence the tower may buckle due to this static load, so check for buckling is required according to Euler's theory.

$$\text{Crippling load } P = \left(\pi^2 \times E \times I\right) / (L)^2$$

where

P = crippling load N
E = young modulus of the material N/mm^2
I = moment of inertia mm^4
L = effective length

One end is fixed and the other end is loaded.

Crippling load $P = (\pi^2 \times E \times I)/(L)^2$
$P = 48.099 \times 10^6$ N

4 Check for Deflection at the End of the Tower

The total wind load acts uniformly on the tower and also the tower is rigidly reinforced internally up to 15 m to control the deflection at the end. The entire tower is assumed to be cantilever with uniformly distributed load for this condition.

The formulae are as follows

Deflection at the top end of tower

$$\delta = \left(w \times L^4\right) / (8 \times E \times I)$$

where

δ = Deflection in mm
w = Load per unit length in N/mm
L = length of the tower in mm

Implementation of Ansys for Analyzing Windmill Tower

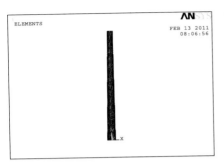

Fig. 1 Shows the mesh model and the load applied on the windmill tower

Total load on tower Due to wind $W = 226.372 \times 10^3$ N
W $\quad = 4.527$ N/mm
$\delta = (w \times L^4)/(8 \times E \times I)$
$\delta = 17.5085$ mm

4.1 FEA Model of Windmill Tower

Define the structural problem:
 Meshed model of tower
 Figure 1 shows the meshed model of tower. This step is after loads applying and checking the results.
 The figure shows the loads are applied on tower after checking the results.

5 Results and Discussion

Vector for the AISI 302 Stainless Steel material is low.

Comparison of AISI 302 Stainless Steel IS 1161 YST 310 Steel and IS 1239
Figure 2 shows the vonmises stress of AISI 302 Stainless Steel, IS 1161 YST 310 Steel, and IS 1239 MTD Steel are 0–25.548 N/mm^2, 0–25.767 N/mm^2 and 0–25.425 N/mm^2 respectively.

Deflections of Five Materials

Figure 3 represents the theoretical and analysis graphical deflections of different materials.

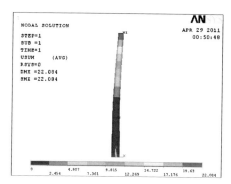
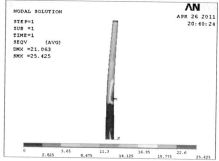

Fig. 2 Shows Vonmises stress for IS 1239 MTD Steel Materials

The above graph in x-axis represents the number of materials and the y-axis represents the deflections in mm.

The AISI 302 Stainless Steel and 1239 MTD Steel graph represents the minimum and maximum deflections respectively.

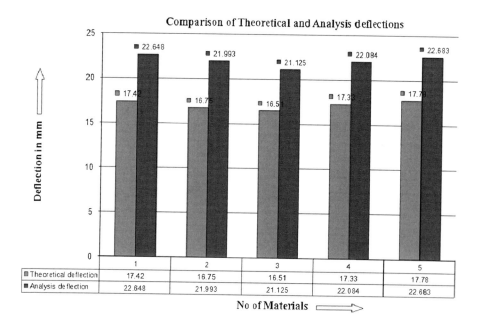

Fig. 3 Deflections of five materials

6 Conclusion

I have drawn the conclusion that **AISI 302 stainless steel Material** is the best material due to its **lower deflection** properties when compared to other materials. The obtained theoretical, analysis deflection valves of **AISI 302 stainless steel Material** and **IS 1239 MTD Steel** are 16.51, 21.125, and 17.78, 22.683 mm respectively. The **Mn** percentage in **AISI 302 stainless steel** material slightly increases the strength of ferrite and the hardness penetration of steel. The **Cr** percentage in the material increases the toughness and wear resistance of steel. Due to the above chemical composition the strength of the material increases which directly reflects the reduction in deflection. From the above properties and results we can confirm that the **AISI 302 stainless steel Material** is the best for construction of the windmill tower.

References

1. Aravinthan T, Omar T (2008) Fibre composite windmill structure: challenges in the design and development. (cice2008) 22–24 July 2008, Zurich, Switzerland no. 1
2. Chien C-W, Jang J-J (2009) a study of wind resistant safety designs of wind turbines tower system. The seventh Asia-pacific conference on wind engineering, Nov 8–12, Taipei, Taiwan
3. Jasbir S Arora (2008) Formulations for the optimal design of RC wind turbine towers. EngOpt 2008: international conference on engineering optimization Rio de Janeiro, Brazil, 01–05 June 2008
4. Leithead Analysis of tower/blade interaction in the ancellation of the tower fore. Aft mode via control w.e. no. 1
5. Peeters J structural analysis of a wind turbine and its drive train using the flexible multimode simulation technique
6. Abdel AF (1995) Dynamics of vertical axis wind turbines (darrieus type). Int J Rotating Mach 2(1): 33–41
7. Scott Larwood Mike ZuteckUC Davis MDZ Consulting Davis, CA Clear Lake Shores TX. Swept wind turbine blade aero elastic modeling for loads and Dynamic Behavior
8. Yildirim S, ozkol I (2010) Wind turbine tower optimization under various requirements by using genetic algorithm. Eng 2: 641–647
9. Singh AN (2007) Concrete construction for wind energy towers. Indian Concr J 43
10. Gutierrez E, Primi S, Taucer F, Caperan P, Tirelli D (ELSA-JRC), Mieres J, Calvo I, Rodriguez J, Vallano F (NECSO), Galiotis C, Mouzakis D (FORTH) A Wind turbine tower design based on the use of fibre-reinforced composites
11. Design of steel structures by P. Dayaratnam
12. Strength of material by Ramamrutham
13. Indian standards (IS 875) for wind loads

Integration of LiBr-H₂O Vapor Absorption Refrigeration Cycle and Power Cycle

R. Shankar and T. Srinivas

Abstract In this proposed integrated power and cooling plant, water/steam is the working fluid and it offers a cost-effective solution compared to aqua-ammonia-based plant. In only cooling system, i.e., without steam turbine, a more amount of heat (sensible heat of vapor and latent heat) is rejected at the condenser. So, this integration minimizes the condenser load by limiting to latent heat rejection only. The plant results 70 kW of power and 433 kW of cooling at the optimized conditions of strong solution (1 kg/s and 0.48 concentrations) and the supply temperature of 250 °C.

Keywords Cooling • Integration • LiBr-water • Power

1 Introduction

The LiBr-water vapor absorption refrigeration (VAR) system is suitable at the cooling requirement above 0 °C. The latent heat of water is high compared to other working fluids and so, it absorbs more heat at the evaporator and results high coefficient of performance (COP). Florides et al. [1] designed and constructed a single effect VAR plant and recommended 0.58 strong solution concentrations at high sink temperature of 35 °C. An experiment has been carried out by Asdrubali and Grignaffini [2] on single effect LiBr-water VAR system and used an electrical heater in a boiler. They did experiment at different mass flow rates and various conditions of condenser, absorber, and evaporator.

To avoid the crystallization formation, James et al. [3] tested the refrigeration cycle by adding various additives and provided suitable additives for high

---

R. Shankar (✉) · T. Srinivas
CO₂ Research and Green Energy Technologies, School of Mechanical and Building Sciences, VIT University, Vellore 632014, TamilNadu, India
e-mail: gentlewise26@yahoo.com

T. Srinivas
e-mail: srinivastpalli@yahoo.co.in

S. Sathiyamoorthy et al. (eds.), *Emerging Trends in Science, Engineering and Technology*, Lecture Notes in Mechanical Engineering, DOI: 10.1007/978-81-322-1007-8_9, © Springer India 2012

temperature application. Similarly the water is mostly used for power production but the cost for making the water into steam is high so the researcher and scientists tried, direct steam solar thermal power plant. [4]. Yongping Yang et al. [5] explained the low heat solar thermal power plant and oil as thermic fluid and produces maximum power. The design and cost analysis is done for solar thermal power plant is connected to the grid by Montes [5]. Whatever the solar thermal power plant the separate power should be used for refrigeration cycle. To avoid the power to the refrigeration cycle the integration is needed with common working fluid. The integration of power and refrigeration cycle and helps to reduce the power used for refrigeration cycle, hence the vapor compression cycle still needs power for compressor. The LiBr-Water vapor refrigeration is good for cooling, but maintaining the vacuum is needed. The water is common for both power and cooling cycle, for the power it requires high pressure but for cooling less than atmospheric pressure. For both industrial and domestic needs power and cooling at a time, hence integration of power and cooling cycle helps to satisfy the need of power and cooling and reduces the power used for the cooling cycle. The main objective of this is paper is integration of solar power and cooling cycle and analysis are made to find the suitable working condition and parameters. The same author's presented a paper about power and cooling run on aqua-ammonia cycle and explained more energy is available at the exit of the generator [6]. The thermo physical properties of LiBr-water are carried out by the equation proposed by Kaita [7] and ASHRAE [8].

1.1 Assumptions and Thermodynamic Equations

- The absorption and condenser exit temperature is maintained as 30 °C.
- The pinch point variation is 10 °C.
- The isentropic efficiency of the pump and turbine is 75 %.
- The mechanical efficiency of pump and turbine is 96 %.

Thermodynamic equations:

$$m_4 = (x_{11} - x_3)/(x_{11}) \tag{1}$$

$$m_{11} = (x_3)/(x_{11}) \tag{2}$$

For power

$$W = m_5(h_5 - h_6) \tag{3}$$

For cooling

$$E = m_9(h_9 - h_8) \tag{4}$$

For generator

$$Q_G = m_4h_4 + m_{11}h_{11} - m_3h_3 \tag{5}$$

For re-heater

$$Q_{SH} = m_5(h_5 - h_4) \tag{6}$$

For efficiency

$$\eta = \frac{m_5(h_5 - h_6) + m_9(h_9 - h_8) - m_2(h_2 - h_1)}{m_4 h_4 + m_{11} h_{11} - m_3 h_3 + m_5(h_5 - h_4)} \tag{7}$$

2 Principle and Working

The proposed cycle has a power cycle followed by the refrigeration cycle and interlinked as shown in the Fig. 1. Similar to the refrigeration cycle the absorber and pump is used as heat absorption and increasing pressure respectively. At the exit of the turbine the high pressure and temperature is there to run the turbine, so turbine is placed. The Reheater is provided next to the generator to make the water into superheated state. At the exit of the turbine the pressure is maintained as atmospheric pressure and the condenser is used to condense the water vapor.

The expansion valve used to expand up to absorber pressure, evaporator is placed as same as in the refrigeration cycle. The liquid mixture of LiBr-water from absorber is pumped into the generator via heat exchanger. As difference in the boiling point the water evaporates and pure water vapor moves to the super heater. The advantage of LiBr—water VAR is no need of reflux condenser as in aqua ammonia cycle. The water vapor is from generator is superheated before the turbine to get more power output. The proposed cycle has a power cycle followed by the refrigeration cycle and interlinked as shown in the Fig. 1. Similar to the refrigeration cycle the absorber and pump is used as heat absorption and increasing pressure respectively. At the exit of the turbine the high pressure and temperature is there to run the turbine, so turbine is placed. The Reheater is provided next to the generator to make the water into superheated state. At the exit of the turbine the pressure is maintained as atmospheric pressure and the condenser is used to condense the water vapor. The expansion valve used to expand up to absorber pressure, evaporator is placed as same as in the refrigeration cycle.

The liquid mixture of LiBr-water from absorber is pumped into the generator via heat exchanger. As difference in the boiling point the water evaporates and pure water vapor moves to the super heater. The advantage of LiBr—water VAR is no need of reflux condenser as in aqua ammonia cycle. The water vapor is from generator is superheated before the turbine to run the turbine more efficient. After work done by the turbine the exit is saturated water vapor with low pressure and low temperature. Because of the low pressure and temperature the condenser load is reduced for refrigeration cycle. Low pressure is needed for the water to complete refrigeration cycle, so the exit pressure of the turbine is fixed as 1.013 bar (atmospheric pressure). The liquid water is expanded from atmospheric pressure to absorber pressure that is the useful cooling is used in the evaporator.

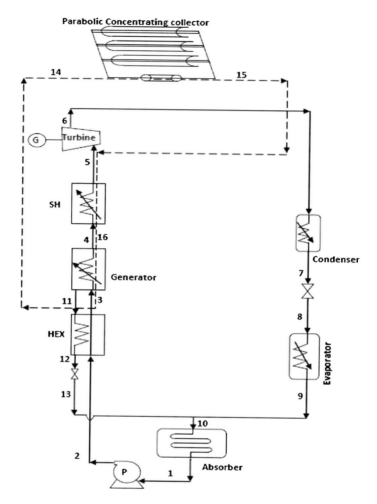

Fig. 1 Energy flow diagram of the proposed combined power and cooling cycle

3 Results and Discussion

The power and cooling output will vary as shown in the Fig. 2 and it depends on the working fluids. The aqua-ammonia as working fluids shows maximum power and cooling output of 18.6 kW and 107 kW, respectively [6]. The enthalpy of vaporization for ammonia is less compared to the LiBr-H$_2$O pair and produces less power and cooling. The LiBr-H$_2$O pair produces more power and cooling but the cooling temperature is more than 0 °C and aqua–ammonia pair is best suitable for less than 0 °C. Absorber concentration of 0.45 and 30 °C produces cooling temperature of −12 °C and reflux condenser exit concentration of 0.99 [9].

Integration of LiBr-H₂O Vapor Absorption Refrigeration

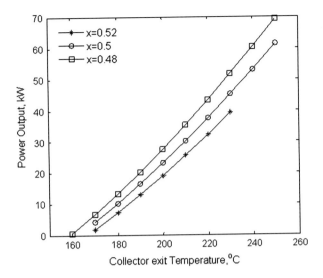

Fig. 2 Changes in specific power output with change in supply temperature and strong solution concentration

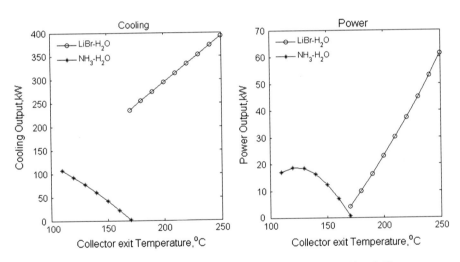

Fig. 3 Comparison of outputs (power and cooling) at two options in working fluids

Figure 3 shows the power produced by turbine in the interlinked VAR cycle. Increase in the collector exit temperature increase input load to the generator, so more water vapor will be produced by the generator. More water vapor results increase in concentration of LiBr into the water and exit concentration is limited to $x = 0.6$ to avoid the blockage in the heat exchanger. The absorber concentration plays a major role in production of power because of the mass variation. At the low concentration of $x = 0.48$, it gives maximum power output of 69.28 kW.

Whereas high concentration of $x = 0.52$ it gives maximum power output of 39.22 kW and the operating temperature is low.

As similar to the power the cooling also gets increased as the same mass as shown in the Fig. 4. The latent heat of water is high and gives more cooling load output compared to the ammonia cooling load. In this proposed cycle it gives maximum cooling load of 433.56 kW at $x = 0.48$ and temperature of 250 °C. The cooling load is increases with respect to solar exit temperature that is directly proportional to the solar trough collector input load the generator. Water is used as thermic fluid for parabolic trough collector.

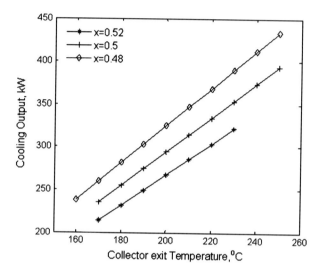

Fig. 4 Variations in cooling output with a change in source temperature and strong solution concentration

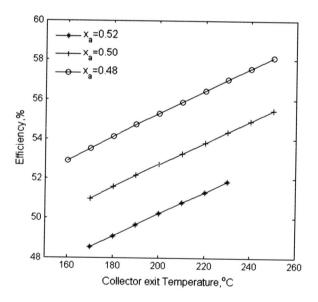

Fig. 5 Energy efficiency of integrated power and cooling cycle with changes in strong solution concentration and supply temperature at the sing temperature of 30 °C

The addition of turbine for producing the power in the refrigeration cycle does not affect the cooling performance. The absorber concentration of $x = 0.48$ has lowest and highest possible working temperature range of 170–230 °C whereas the $x = 0.52$ limited working temperature range of 170–200 °C. The return solution concentration from the generator is limited to $x = 0.6$, so there is variation in temperature working range.

The efficiency of integrated cycle is increased and also depends on the absorber concentration and atmospheric conditions. The proposed cycle produces maximum efficiency of 58 % as shown in the Fig. 5 and it is high compared to aqua-ammonia pair at 30 °C atmospheric temperature [6].

4 Conclusion

The integration of power and VAR cycle improves the energy utilization and also reduces the condenser load. The strong solution concentration, 0.48 with the combination of 250 °C source temperature results 9.6 bar pressure at the turbine inlet. At these conditions, the maximum specific power is 70 kW and maximum cooling is 433 kW obtained at optimum conditions.

References

1. Florides GA, Kalogirou SA, Tassou SA, Wrobel LC (2003) Design and construction of a LiBr–water absorption machine. Energ Convers Manage 44:2483–2508
2. Asdrubali F, Grignaffini S (2005) Experimental evaluation of the performances of a H_2O–LiBr absorption refrigerator under different service conditions. Int J Refrig 28:489–497
3. Dirksen JA, Ring TA et al (2001) Testing of crystallization inhibitors in industrial LiBr solutios, Int J Refrig, 24:856–859
4. Nezammahalleh H, Farhadi F, Tanhaemami M (2010) Conceptual design and techno-economic assessment of integrated solar combined cycle system with DSG technology, Sol Energ 84:1696–1705
5. Yongping Yang et al (2011) An efficient way to use medium-or-low temperature solar heat for power generation e integration into conventional power plant. Appl Therm Eng 31:157–162
6. Shankar R, Srinivas T (2012) Modeling of energy in vapor absorption refrigeration system. Procedia Eng 38:98–104
7. KaitaY (2001) Thermodynamic properties of Lithium bromide-water solutions at high temperatures. Int J Refrig 24(5):374–390
8. ASHRAE (1991) Thermo physical property data for lithium bromide/water solutions at elevated temperature, Final Rep
9. Shankar R, Srinivas T (2012) Solar thermal based power and vapor absorption refrigeration system, Procedia Eng 38:730–736

Implementation of Methanol as Fuel in SI Engine for the Measurement of Formaldehyde Emission

T. Elamaran, N. Shankar Ganesh and K. Vinoth Raj

Abstract The Methanol-gasoline blend is utilized in this work to measure the formaldehyde (HCHO) coming out from the engine exhaust using the alcohol/gasoline blends by the following method and to Characterize the aldehyde emission, study the performance characteristics, CO, HC, emission Levels in the Various blends. Formaldehyde can be sampled with the use of a standard miniature glass fiber filter coated with 3-methyl-2-benzothiazolinone hydrazone hydrochloride (MBTH).The absorption measurement is done immediately by using UV–visible spectrophotometer. Upon increasing the methanol blend over petrol the readings were identified. On increasing methanol blend the PPM has been found to decrease at 100 % methanol at 100 % load, 3.5 PPM has been noticed.

Keywords Alchols • MBTH • UV–visible spectrophotometer

1 Introduction

1.1 Alcohol Fuels for S.I. Engines

In this century, it is believed that petroleum products will become very scarce and costly to find and produce. Increase in number of automobiles will lead to great demand for fuel in the near future. Gasoline and diesel will become scarce and more costly. Alternative fuel technology, availability, and use will become more common in the coming decades and considerable attention was focused on the development of alternative fuels, with particular reference to Alcohols. Among the various Alcohols, methanol (CH_3OH) and ethanol (C_2H_5OH) is considered to be the best blends with petroleum fuels or neat use in SI Engine. The use of alcohol

---

T. Elamaran (✉) · N. Shankar Ganesh · K. Vinoth Raj
Department of Mechanical Engineering, Kingston Engineering College,
Vellore 632002, India
e-mail: elamarant@yahoo.co.in

S. Sathiyamoorthy et al. (eds.), *Emerging Trends in Science,
Engineering and Technology*, Lecture Notes in Mechanical Engineering,
DOI: 10.1007/978-81-322-1007-8_10, © Springer India 2012

fuels has demonstrated a path toward a cost-effective alternative to conventional gasoline engines and economical means of utilizing methanol or ethanol fuels. These alcohols can be readily made from a number of nonpetroleum sources. Methanol or methyl alcohol (CH_3OH) can be produced from coal, a relatively abundant fossil fuel. Ethanol or ethyl alcohol (C_2H_5OH) can be produced by fermentation of carbohydrates which occur naturally and abundantly in some plants like corn and potatoes. Hence these fuels can be produced from highly reliable and long lasting raw material sources.

Methods

Solution or blend-mixture of gasoline and alcohol. Neat fuel mode-100 % ethanol or methanol Feedstock for ethers.

Solution

M85-85 % methanol and 15 % gasoline
Flexible fuel vehicle-M0–M85
Gasohol-10 % ethanol by volume with gasoline.

Need for the Project:

Aldehyde emissions that are coming out of the Engine exhaust, which are harmful gases to human health causes irritation of eyes and may cause headaches, general discomfort, and irritability. There are reports of asthma caused by respiratory tract irritation due to formaldehyde exposure. Additionally, they can cause potential harm to flora, including potherbs, and fauna, especially, unicellular organisms that are relatively sensitive to formaldehydes. It can be involved in chemical reactions in the atmosphere, generating other compounds. Among them the photochemical Smog formation that mostly produces oxidizing gases, especially gas ozone. Besides causing respiratory conditions to human beings, oxidizing gases can also damage materials, mainly rubbers.

Thus it is very necessary to know the measurement level of Aldehyde and to reduce Aldehyde emissions in the exhaust of alcohol fuelled S.I. Engine.

Overview of the project:

This project deals with the characterization of formaldehyde and acetaldehyde emissions in the alcohol fuelled S.I. Engine. The variable compression ratio petrol engine with a capacity of 2.5 kw @ 3,000 rpm. The Engine testing with various alcohol blends with gasoline for various loads (0–100 %) was carried out on an Eddy current dynamometer. High volume sampler were used for collecting the Aldehyde from the exhaust by placing the MBTH Coated filter paper inside the sampler and the absorbance of the sample was immediately measured in the UV Spectrophotometer at 628 nm for formaldehyde and 612 nm for acetaldehyde. Horiba Automotive

Implementation of Methanol as Fuel in SI Engine

Table 1 Important properties of gasoline and methanol

Sl.No.	Property	Gasoline	Methanol
1	Chemical formula	MC_nH_{2n}	CH_3OH
2	Molecular weight	112	32
3	Specific gravity at 15.5 °C	0.7–0.75	0.796
4	Boiling point or range °C	30	65
5	Latent heat of vaporization Kcal/kg	70–100	264
6	Vapor pressure at 58 °C	0.8	0.32
7	Calorific value Kcal/kg	10,500	4,700
8	Stoichiometric air/fuel ratio	14.7:1	6.4:1
9	Self ignition temperature °C	300–450	478
10	Octane numbers	91	114
	(a) Research	82	94
	(b) Motor		
11	Cetane number	8.14	3

Emission Gas Analyzer MEXA 554 J Series were used for measuring HC, CO and CO_2 emissions. NO_X Analyzers were used for measuring NO_X emissions Table 1.

Octane number (ON): Octane number of a fuel can be defined as the percentage by volume of iso-octane in a mixture of iso-octane and n-heptane.

Cetane number (CN): Cetane number of a fuel is defined as the percentage by volume of cetane in a mixture of cetane and a-methylnapthalene ($C_{10}H_7CH_3$).

$$CN = (104 - ON)/2.75$$

2 Experimental Programme

2.1 Preparation of Filter Paper

A 50 μl aliquot of 0.5 % MBTH solution was transferred to a Gellman type AE 13 mm diameter glass fibre filter. The filters were then allowed to dry under vacuum for 30 min and stored in a closed brown bottle, placed in a desiccators.

3 Preparation of Reagents

3.1 Preparation of MBTH Solution

MBTH-3-Methyl-2-Benzothiazolinone Hydrozone Hydrochloride Absorbing Solution (0.05 %).

Dissolve 0.050 g of MBTH in distilled water and dilute to 100 ml. It is stable for at least one week after which it becomes pale yellow.

3.2 Preparation of Oxidizing Reagent

Dissolve 1.6 g of sulphamic acid (NH_2SO_3H) and 1.0 g of ferric chloride ($Fecl_3.6H_2O$) in distilled water and dilute to 100 ml.

3.3 Instrumentation

Absorption measurements were performed on a UV–visible Spectrophotometer. The basic methodology for sampling formaldehyde/acetaldehyde produced from the combustion of methanol/ethanol in S.I. engine exhaust includes the use of a standard miniature glass fiber filter coated with 3-methyl-2-benzothiazolinone hydrazone hydrochloride (MBTH). The formaldehyde hydrazone formed (i.e., the adduct of formaldehyde ($HCHO/CH_3CHO$) and MBTH) was desorbed from the filter with distilled water and then oxidized by ferric chloride-sulfamic acid solution to form a blue cationic dye in acidic medium which can be subsequently determined by visible absorption at 628 nm. The amount of formaldehyde/ Acetaldehyde in the engine exhaust is measured at a flow rate of 15 L/min.

The oxidizing reagent solution is prepared by adding 0.5 gm of ferric chloride hexahydrate and 0.8 gm of sulfamic acid in a 50 ml standard flask and diluted to 50 ml with distilled water. All the chemicals used are of analytical reagent (AR) grade. The formaldehyde from the exhaust that is adsorbed on the MBTH coated glass fiber filter is desorbed from it by adding 2 ml of distilled water and then kept for 20 min and after that 2 ml of oxidizing reagent is added and again kept for 20 min and then the bluish colour is developed and then the absorption measurement is done immediately by UV–visible spectrophotometer. The reaction scheme of the proposed method is shown in Fig. 1.

This method provides a simple and reliable way of determining Aldehyde.

The boiling and condensation process for pure ammonia and ammonia-water mixture was shown in Fig. 1. For pure ammonia the space between the source and working component is wide as ammonia boils and condenses at constant temperature. The gap between the working component and the source is reduced as compared with the pure ammonia. Due to varying boiling and condensation process this is achieved which makes it possible to utilize the energy efficiently form the source.

3.4 Experimental Setup

The experimental set up for this Aldehyde emission measurement purpose using methanol/ethanol-gasoline blends and 100 % methanol consists of a high volume sampler that is placed near the engine exhaust. Above the high volume sampler

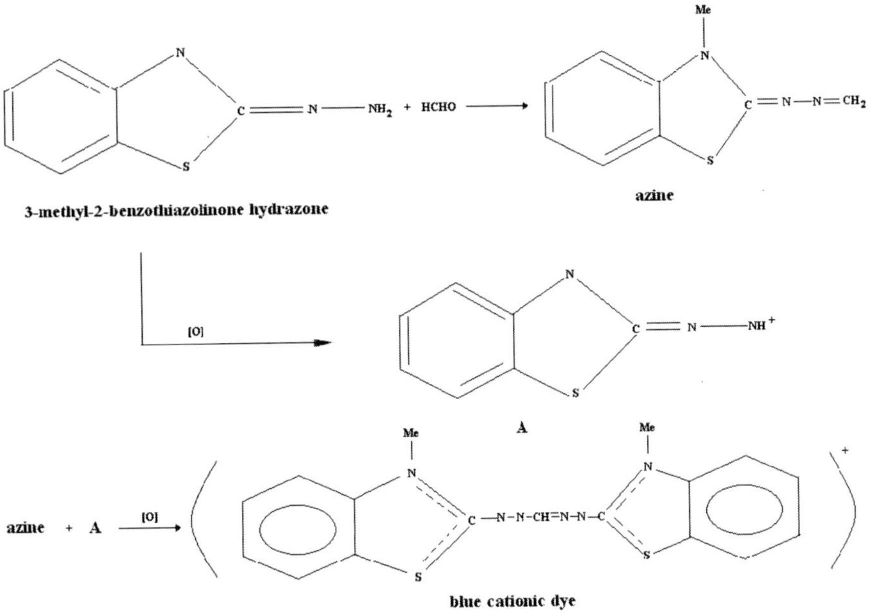

Implementation of Methanol as Fuel in SI Engine

3-methyl-2-benzothiazolinone hydrazone

azine

blue cationic dye

Fig. 1 The reaction scheme

a conical shaped apparatus made of CR sheet is prepared with a hole at the top. A hose pipe is connected between the exhaust tail-pipe and the top hole of the conical shaped apparatus. By this process the exhaust gas passes through the hose pipe and gets adsorbed on the MBTH coated filter that is placed on the high volume sampler. All other sides around the filter is covered by a rubber sheet so that the exhaust gas gets coated on the filter only. The exhaust gas adsorbed filter is then immediately taken for measuring the absorption in UV–visible spectrophotometer. Figure 2 shows the schematic of the Engine test setup.

4 Engine Specifications

See Figure 3, Table 2

4.1 Load Cell for Torque Measurement

A load cell is mounted on the loading end of the dynamometer, as the loading arm hits the load cell which sense the load and read out will be indicated in terms of torque (Nm) by the digital torque indicator. (Load cell capacity = 20 kg).

Fig. 2 Load cell for torque measurement

Fig. 3 Schematic of the engine test setup

4.2 Load Cell Arrangement for Fuel Flow Measurement

A fuel tank having a capacity of 4.5 L is mounted on a load cell (10 kg) which helps to measure the fuel rate. The loss in weight of the fuel is converted into flow rate in kg/hr and is directly indicated on the digital fuel rate indicator and also a burette is connected parallelly as stand-by option.

4.3 Air-Rate Measuring System

An air tank is mounted below the control panel is connected to the air-rate indicator and also to the manometer across an orifice of 20 mm diameter.

The manometric height is converted into actual volume in m^3/hr and the same will be directly indicated on the air-rate indicator.

Table 2 Variable compression ratio petrol engine test rig specifications

Engine	4-Stroke, 1-cylinder
Make	Greaves
BHP	2.5 KW
RPM	3,000 RPM
Fuel	1. **Petrol**
Bore	70 mm
Stroke length	66.7 mm
Starting	Self starter
Method of ignition	Spark ignition
Orifice DIA.	20 mm
Compression ratio	2.5–8
Cooling-bore-head	Air cooled, water cooled
Load	Eddy current dynamometer
Capacity	2.5 kW
Speed	3,000 RPM

4.4 Self Start Arrangement

A Self starter is mounted perpendicular to the gear wheel, a self starter switch is mounted on the control panel which is connected across the battery is used for cranking the engine.

All the various readings of speed, torque, air flow rate, fuel mass flow rate, and exhaust gas temperature are noted down from electronic meters fitted on the stand near the engine.

A HORIBA Exhaust gas analyser is also placed near the exhaust tail-pipe that is used to measure the HC, CO, CO_2 and lambda values. Electronic meters on a stand near the engine are fitted that gives the air flow rate, fuel flow rate, speed, and torque values. A burette is also connected on the meter board that gives the rise and fall of fuel with respect to time. Figure 4a shows the horiba exhaust gas analyzer and Fig. 4b shows the Electronic meters on a stand.

Fig. 4 a HORIBA Exhaust gas Analyzer b heating element arrangement

For running the engine with alcohol-gasoline blends the jet diameter of the carburetor is increased so that more amount of fuel can pass through it. The calorific value of alcohol is lower than gasoline, so to match the performance of engine running like gasoline, the supply of fuel has to be increased. Moreover, another set up is fabricated for heating the air before supplying to the carburetor. This includes a pipe that is clamped near the choke and around the pipe an electric heater is fitted that will heat the air. Figure 4 shows the heating element arrangement fitted before the inlet of the carburetor.

4.5 Preparation of Stock Solution for Formaldehyde

The standard solution is prepared by taking 0.27 ml of 37 % formaldehyde solution and then diluted to 100 ml having a concentration of 1,000 ppm. From this solution 10 ml is taken in another 100 ml standard flask and diluted with distilled water to 100 ml mark. This made the concentration of formaldehyde solution to 100 ppm. This is the stock solution and to this solution 2 ml of oxidizing solution (0.5 gm ferric chloride hexahydrate and 0.8 gm sulfamic acid is added together and diluted to 100 ml with distilled water in a 100 ml standard flask) and 50 µL of 0.5 % MBTH solution are added and this turned the stock solution to bluish colour. The absorption measurement of this stock solution is done immediately at 628 nm using UV–visible spectrophotometer.

4.6 Preparation of Standard Solution for Formaldehyde

Three standard solutions of 20, 30 and 40 ppm are prepared from the stock solution by appropriate quantity in a 25 ml standard flask and diluted to the mark with distilled water. The calculations for that are as follows:

$$N_1V_1 = N_2V_2$$
$$20 \times 25 = 100 \times V_2$$
$$V_2 = 5 \text{ ml}$$
$$N_1V_1 = N_2V_2$$
$$30 \times 25 = 100 \times V_2$$
$$V_2 = 7.5 \text{ ml}$$
$$N_1V_1 = N_2V_2$$
$$40 \times 25 = 100 \times V_2$$
$$V_2 = 10 \text{ ml}$$

So by taking 5, 7.5 and 10 mL from the stock solution the required standard solutions of 20, 30 and 40 are prepared. By using the standard solutions the standard Fig. 5, 6.

Implementation of Methanol as Fuel in SI Engine

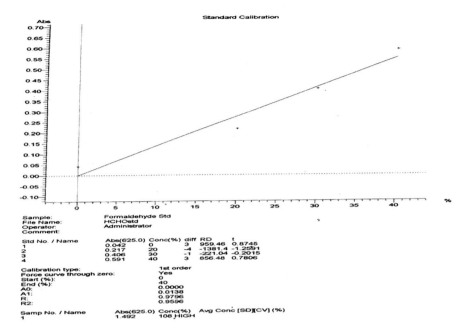

Fig. 5 Formaldehyde standard calibration

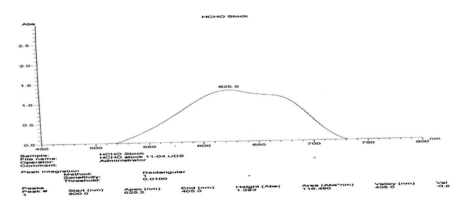

Fig. 6 HCHO stock

4.7 Analysis in UV-Spectrophotometer

The analysis in UV-Spectrophotometer is done by using sample cuvettes.

1. The first step in this analysis work involves the measurement of the absorbance of the stock solution (100 ppm). The Wavelength at which the stock solution gives maximum absorbance is noted.

2. Then the standard solutions of 20, 30, and 40 ppm are prepared from the stock solution and their absorbance is noted at that particular wavelength of the maximum absorbance of the stock solution.
3. From this absorbance of the three standard solutions, a standard curve is prepared. The curve is a linearly increasing straight line with absorbance along the Y-axis and concentration (ppm) along the X-axis.
4. Then the blank solution (i.e., 50 µL of 0.5 % MBTH solution +2 ml of oxidizing reagent solution +2 ml of distilled water) is prepared and their absorbance is measured in the wavelength range of 400–800 nm. This is called as base line correction.
5. After this the sample solution of the engine exhaust is kept in the cuvette and the maximum absorbance at the particular wavelength is measured. The maximum absorbance is found to be around 628 nm. since both the standard, stock and the sample solutions are having bluish color.
6. From that maximum absorbance of the sample solution the corresponding concentration is measured manually from the standard curve.
7. The software used for this measurement work is UV Solutions Figs. 7, 8, Tables 3, 4.

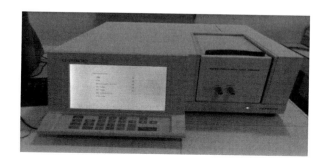

Fig. 7 Schematic of UV-spectrophotometer

Fig. 8 Shows UV-spectrophotometer with cuvettes

Table 3 Readings with M50. Fuel: 50 % Methanol + 50 % Gasoline

S.No.	Load (%)	Torque (Nm)	Brake power (kW)	Fuel rate (kg/Sec) × 10^{-4}	Air flow rate (m^3/hr)	Exh. gas temp. (°C)	HC (PPM)	CO (%Vol)	CO_2 (%Vol)	Lambda	Speed (RPM)
1.	0	0	0	1.806	4.2	490	593	6.29	4.43	1.282	2,750
2.	25	1.95	0.686	1.914	4.2	508	580	5.23	5.73	1.41	2,665
3.	50	3.95	1.29	2.76	5.2	509	374	3.8	6.82	1.53	2,644
4.	75	5.92	1.96	2.96	6.7	516	262	3.22	8.22	1.423	2,630
5.	100	7.9	2.56	3.297	8.5	546	142	1.32	8.3	1.39	2,575

Table 4 Concentration of formaldehyde emission

S.NO.	Load (%)	Concentration (ppm)	Absorbance
1.	0	5.5	0.687
2.	25	3.7	0.501
3.	50	2.0	0.259
4.	75	6	0.079
5.	100	3.5	0.041

5 Results and Discussions

5.1 *Methanol-Gasoline Blends*

5.1.1 Variation of CO Emission with Load

Figure 9 shows that the CO emissions for gasoline and methanol blend decreases with increase of varying loads. There is a significant reduction in CO emission by using methanol-gasoline blends and pure methanol Table 5.

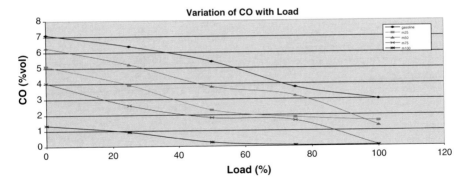

Fig. 9 Variation of CO with load

Table 5 Readings with M75. fuel: 75 % methanol + 25 % gasoline

S.No.	Load (%)	Torque (Nm)	Brake power (kW)	Fuel rate (kg/sec) $\times 10^{-4}$	Air flow rate (m³/hr)	Exh. gas temp. (°C)	HC (PPM)	CO (%Vol)	CO$_2$ (%Vol)	Lambda	Speed (RPM)
1.	0	0	0	2.38	3	458	739	4.04	3.56	1.2	2,750
2.	25	1.95	0.687	3.19	6.1	464	522	2.615	4.86	1.32	2,770
3.	50	3.95	1.392	3.57	5.2	522	273	1.824	5.983	1.227	2,800
4.	75	5.92	2.17	4.05	7.3	542	186	1.642	6.864	1.641	2,912
5.	100	7.9	2.88	4.33	9.2	574	39	0.056	7.036	1.122	2,900

6 Conclusion

It is concluded from the experimental results that the methanol fuel blends with gasoline is one of the best alternative fuel for S.I. Engine, because it has good octane rating and is very volatile which are the essential requirements for a S.I. Engine fuel. It reduces the emission of HC and CO and is less harmful to the environment than gasoline.

References

1. Chan WH, Shuang S and MMF Choi (2001) The analyst paper 126(5):720–723
2. Lodge JP (1989) Determination of formaldehyde content of the atmosphere (MBTH Method). Lewis Publishers
3. Heywood JB (1998) Internal combustion engine fundamentals. McGraw-Hill, New York
4. Degobert P (1995) Automobiles and pollution. SAE/Editions Technip, Paris
5. Roux and JH, Loubser NH (1986) Carburetor corrosion: the effect of alcohol–petrol blends. In: Proceedings of the VII International Symposium on Alcohol Fuels Technology, Paris
6. Cortney RL, Newhall HK (1984) The properties and performance of modern automobiles fuels. SAE 841210
7. Mathur ML, Sharma RP (2005) Internal combustion engines. Dhanpat Rai Publications, New Delhi pp 214–249
8. Davis GW, Heil ET (2000) The development and performance of a high blend ethanol fuelled vehicle. SAE 2000-01-1602
9. Bresenham D, Reisel J (1999) The effect of high ethanol blends on emissions from small utility engines. JSAE 9938100
10. Bardon MF, Rao VK Battista V (1995) Flammability tests of alcohol/gasoline vapours. SAE 950401

Performance Characteristics of Diesel Engine Operating with Preheated Palm Biodiesel

Sumedh Ingle, Vilas Nandedkar and Madhva Nagarhalli

Abstract Increase in price of diesel fuels, stringent emission regulation, and foreseeable future depletion of petroleum reserves forces us to research novel technologies to meet human demands for environment and energy. World's petroleum supplies are getting constrained, attention has been directed to find out alternative sources of fuels for engines. The nonrenewable nature and limited resources of petroleum fuels have become a matter of great concern. The present reservation of fuels used in internal combustion engines including diesel will deplete within 40 years if consumed at an increasing rate estimated to be of an order of 3 % per annum. This work aims to find out the prospects and opportunities of using methyl esters of palm oil as fuels in an automobile. Thus, suitability of such fuels in transportation vehicles will helps in saving foreign exchange. Tests were conducted on a four stroke, single cylinder, D.I diesel engine with Diesel and various blends of Biodiesel at various preheating temperature. The results of performance tests are compared with various blends of palm oil biodiesel with that of neat diesel. The result indicates that at blend B20 with preheat temperature 60 °C, the brake thermal efficiency is maximum while minimum brake specific fuel consumption is observed as compare to other blends of biodiesel.

Keywords Brake thermal efficiency • Brake specific fuel consumption • Eddy current dynamometer • Transesterification • Etc

---

S. Ingle (✉)
Department of Mechanical Engg, SRES's College of Engineering, Kopargaon 423603, India
e-mail: sumedhi@rediffmail.com

V. Nandedkar
Department of Production Engg, SGGS Institute of Engg. & Technology, Nanded 431606, India
e-mail: vilas.nandedkar@gmail.com

M. Nagarhalli
Navsahyadri Education Society's Group of Institutes, Faculty of Engg, Naigaon, Pune, India
e-mail: nagarhalli@rediffmail.com

S. Sathiyamoorthy et al. (eds.), *Emerging Trends in Science, Engineering and Technology*, Lecture Notes in Mechanical Engineering, DOI: 10.1007/978-81-322-1007-8_11, © Springer India 2012

Abbreviation (for Figures)

B0	Diesel
B20	Palm Biodiesel 20 % +Diesel 80 %
B40	Palm Biodiesel 40 % +Diesel 60 %
B60	Palm Biodiesel 60 % + Diesel 40 %
B80	Palm Biodiesel 80 % + Diesel 20 %
B100	Palm Biodiesel 100 %
BMEP	Brake mean effective pressure
BSFC	Brake specific fuel consumption
η_{bth}	Brake thermal efficiency

1 Introduction

Study reveal that due to increasing cost of petroleum in the international market and increased demand of fuel due to increasing number of vehicles, an alternative source for diesel is the need of the day. Vegetable oils, edible, and nonedible, hold special promise in this regard, as they are locally produced and can be grown on barren land also. Vegetable oils such as soybean, coconut, sunflower, groundnut, castor oil, etc., have been used and their performance was reported by many researchers. These oils pose some problems when they are used without any treatment. Due to their long chain hydrocarbon structure [1, 7] they have good ignition characteristics but have higher viscosity and have problem of carbon deposits, gum formation and poor thermal efficiency. Hence use of neat vegetable oils is not encouraging in the long run. Another drawback of edible oils is their high cost and shortage. Hence use of nonedible oils such as Jatropha curcas (Jatropha), Pongamia Pinnata (Karanja), Deccan hemp, castor oil, etc., can be made upon by improving their properties [2, 3]. To improve the properties of fuel such as viscosity and breaking down of higher hydrocarbons, the oil is esterified with low molecular weight alcohols. Generally, methyl alcohol is used for esterification and hence these fuels are known as methyl esters [4]. These oils (methyl esters) have properties similar to that of petroleum diesel and hence are also known as 'biodiesel'. These esters have much lower viscosity than parent oils. Esterification is a simple process in which the oil is mixed with methyl alcohol and heated in presence of a catalyst [5]. The products of the reaction are methyl ester (biodiesel) and glycerin. Glycerin being heavier, settles down and oil which is lighter is removed from the top and used after washing and removal of moisture. Biodiesel is a clean burning, renewable, nontoxic, biodegradable, environment friendly, transportation fuel. Biodiesel can be blended with petroleum diesel in different proportions and used or it can be used as a neat fuel (100 %). A blend of 20 % (by volume) biodiesel with petroleum diesel is designated as B20 fuel and when neat biodiesel is used it is denoted as B100. No engine modifications are required to be done when biodiesel is used.

Table 1 Properties of palm oil biodiesel [2]

Density @ 15 °C	0.8751 g/cm^3
Viscosity at 40 °C	4.1 mm^2/s
Flash point	175 °C
Pour point	−12 °C
Cloud point	Not applicable
Specific gravity @15 °C	0.8722
Calorific value	37253.62 kJ/kg
Visual appearance	Dark brown liquid
Ash content	0.001 %
Cetane number	52
Density @ 15 °C	0.8751 g/cm^3

Though palm biodiesel is costly, for research purpose it is used in diesel engine by preheating at 50, 55, and 60 °C before injected in to the engine. The various blends of palm biodiesel are preheated and tested for performance characteristics.

The properties of Palm oil biodiesel are given in Table 1 [6].

2 Experiments

The present research was carried out to investigate the performance characteristics of palm oil biodiesel. Diesel, Bio-diesel (B100) and its blends B20, B40, B60, B80 were used to test the engine of the specifications mentioned in Table 2. The experiments were conducted on a single cylinder, 4 stroke D.I diesel engine. No engine modifications were done. The engine was loaded using the Eddy current dynamometer. The engine speed in rpm was sensed using a sensor preinstalled in the dynamometer and was recorded from the display on the control panel of the dynamometer.

Table 2 Specifications of engine used

Make	Kirloskar
Type	Single-cylinder, four-stroke, compression ignition diesel engine
Stroke	110 mm
Bore	80 mm
Compression ratio range	16.5:1
BMEP at 1500 rpm	5.42 Bar
Rated output	3.7 kW
Rated speed	1500 rpm
Dynamometer	Eddy current, water-cooled with loading unit

The performance characteristics of the engine were studied at different preheating temperature and different engine loads (25, 50, 75, 100, and 115 % of the load corresponding to the load at maximum power at an average engine speed of 1500 rpm). At each load, the engine was stabilized for 60 min and then performance parameters were measured. The various graphs were plotted between Brake thermal efficiency and Brake mean effective pressure, Brake specific fuel consumption, and Brake mean effective pressure.

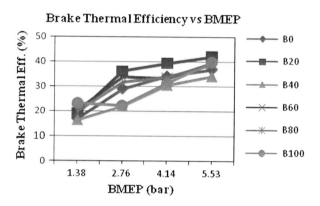

Fig. 1 η_{bth} versus BMEP for 50 °C preheating temperature

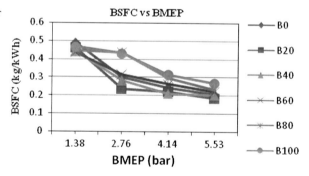

Fig. 2 BSFC versus BMEP for 50 °C preheating temperature

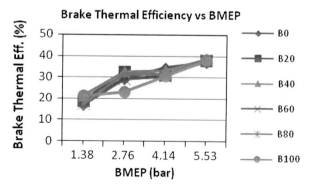

Fig. 3 η_{bth} versus BMEP for 55 °C preheating temperature

3 Results and Discussions

The variation of Brake thermal efficiency with brake mean effective pressure for different blends is shown in Figs. 1, 3, 5. The results are somewhat similar to research work carried by Chakrabarti [7]. In all cases it increases with increase

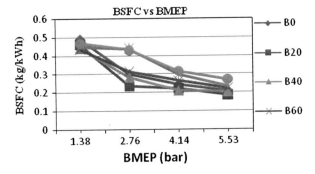

Fig. 4 BSFC versus BMEP for 55 °C preheating temperature

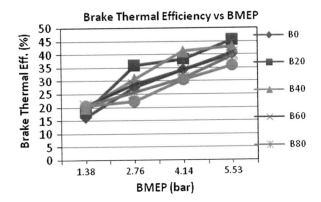

Fig. 5 η_{bth} versus BMEP for 60 °C preheating temperature

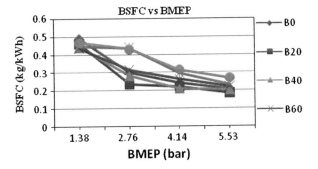

Fig. 6 BSFC versus BMEP for 60 °C preheating temperature

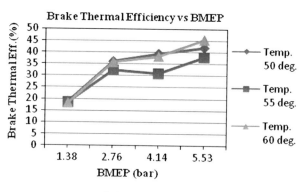

Fig. 7 η_bth versus BMEP for various preheating temperature

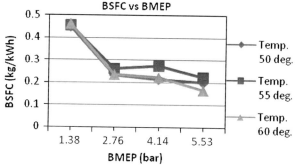

Fig. 8 BSFC versus BMEP for various preheating temperature

in brake mean effective pressure up to full load (BMEP 5.53 bar). This is due to a reduction in heat loss and increase in power with increase in load. The trend is same for all blends but is slightly less as compared to blend B20 at 60 °C. The lower brake thermal efficiency obtained for B40 at 50 °C and B100 at 60 °C is due to reduction in calorific value and an increase in fuel consumption. As shown in Fig. 7, the *maximum Brake Thermal Efficiency* is observed for B20 at 60 °C at full load as compare to other preheating temperature.

The variation of brake specific fuel consumption with BMEP is shown in Figs. 2, 4, 6. For all blends tested, brake-specific fuel consumption decreases with increase in brake mean effective pressure up to full load (BMEP 5.53 bar). The overall characteristics of palm oil biodiesel and diesel are similar. The lowest brake specific fuel consumption is predicted at blend B20 with 60 °C preheating temperature at full load as compare to other blends. Also, as shown in Fig. 8, the *minimum brake specific fuel consumption* is observed at B20 with 60 °C. This is due to the combined effect of low heating value and high density of palm oil biodiesel.

4 Conclusion

This chapter focuses on study and analysis of the performance characteristics such as brake thermal efficiency and brake specific fuel consumption of a diesel engine using palm biodiesel. The maximum brake thermal efficiency and lowest brake specific fuel consumption is observed for blend B20 at 60 °C at full load as compared to other blends and preheating temperatures.

In totality B20 at 60 °C blend of palm oil biodiesel shows an optimized trend of performance characteristics. Therefore, use of palm oil creates a need to increase palm oil sources. This in turn will increase the use of waste land productivity and generate rural employment and increase the GDP. Thus, biodiesel is a 'New Era Fuel' which will reduce our dependence on diesel or petrol.

References

1. Babu AK, Devarao G (2003) Vegetable oils and their derivatives as fuels for CI engine: an overview. SAE 2003-01-0767
2. Ali Y, Hanna MA (1994) Alternative diesel fuels from vegetable oils. Bioresour Technol 50(2):153–163 paper no-0960-8524(94)00072-7
3. Nagarhalli MV, Nandedkar VM, Mohite KC (2010) Emission and performance characteristics of Karanja biodiesel and its blends in CI engine and its economics. ARPN J Eng Appl Sci 5(2):52–56
4. Raheman H, Phadtare AG (2004) Diesel engine emissions and performance from blends of karanja methyl ester and diesel. Biomass Energ 27(4):393–397
5. Alamu OJ, Akintola TA, Enweremadu CC, Adeleke AE (2008) Characterization of palm-kernel oil biodiesel produced through NaOH-catalysed transesterification process. Sci Res Essay 3(7):308–311 ISSN 1992-2248
6. Kalam MA, Masjuki HH (2002) Biodiesel from palm oil–an analysis of its properties and potential. Biomass Bioenergy 23:471–479
7. Chakrabarti MH, Ali M (2009) Performance of compression ignition engine with indigenous castor oil biodiesel in Pakistan. NED Univ J Res 6(1):10–18

Recent Trends in Development of Hydraulic Press for Damper Flap Notching

K. Vinoth Raj

Abstract The current project is aimed at improving the quality of damper flaps and also to increase the productivity by designing a special purpose hydraulic press with notching tool for making the notches in the damper flap plates to replace by individual marking and oxy-acetylene cutting done manually. The present manufacturing has inherent quality and productivity problems, which need to be studied and resolved to meet the growing production targets of Industry. The project was executed based on the methodology that involves seven phases such as, information gathering and analysis, concept for project generation, concept evaluation and selection, and product architecture. For gathering information the industry was visited and studied the manufacturing process, various drawbacks and difficulties faced by workers and notes had been prepared after discussion. By identifying the problem, it was suggested to replace individual marking and oxy-acetylene cutting were replaced by means of hydraulic press. The required suitable tool was selected from the available option based on design calculation. The project was executed based on methodology that involves the following stages configuration design, parametric design, testing and refinement, and detail design. The press structure was analyzed by using ansys.

Keywords Hydraulic press • Damper flap • Ishikawa diagram • Hydraulic circuits • Press structure

1 Introduction

1.1 Design Process

Design is the profession in which knowledge of mathematical and natural sciences gained by study, experience and practice are applied with judgment to develop a

---

K. Vinoth Raj (✉)
Department of Mechanical Engineering, Kingston Engineering College, Vellore, India
e-mail: nivi619@gmail.com

S. Sathiyamoorthy et al. (eds.), *Emerging Trends in Science,*
Engineering and Technology, Lecture Notes in Mechanical Engineering,
DOI: 10.1007/978-81-322-1007-8_12, © Springer India 2012

way to utilize economically the material and forces of nature for the benefit of mankind.

"Design establishes and defines solutions to and pertinent structure for problems not solved before, or new solutions to problems which have previously been solved in a different way". "The ability to design is both a science and an art". The science can be learned through techniques and methods, but the art is best learned by doing design. It is for this reason for our design experience must involve some realistic project experience. To become proficient in design is a perfectly attainable goal for an engineering student, but its attainment requires the guided experience.

Design should not be confused with discovery. E can discover what has already existed but has not been known before. But a design is the product of planning and work. Good design requires both analysis and synthesis. The challenges presented by the design environment are to think of the C's of design creativity, complexity, choice and compromise.

1.2 Fans for Boiler Applications

In engineering application fans are used mainly to produce a flow from one point to another, movement of air or gas itself may be the primary object, but often the gas is merely medium for carrying heat, cold, moisture etc., or solid substances like ash or saw dust. Fan can be defined as a volumetric machine which pumps like, moves quantities of gas or air from one place to another.

Physically a fan is a machine fitted with a rotating wheel (called as impeller), which does work upon gas or air flowing through it and a housing to collect the incoming air or gas and direct it flows. Fans are used in boilers for different applications such as supplying air for combustion, removable of combustion products, air for cooling of equipment working at hot zones and so on.

Fans are designed according to the function they do in boiler induced draft fan, forced draft fan, primary air fan, gas re-circulation fan, sealing air fan, and so on. Depending upon the direction of flow of the medium at outlet with respect to rotational axis, fans are broadly classified into axial fans and radial fans. Axial flow may be described as a fan in which the flow of air is substantially parallel to the axis of the impeller. In radial flow fans, the air is drawn into the center of the revolving wheel or impeller turns through 90° i.e. in perpendicular to the axis of rotation.

1.3 Damper Assembly

Damper assembly (Fig. 1) consists of damper frame, flap assembly, flap axes, levers, and linkages. Damper frame is a rectangular steel casing with houses flap assembly and actuating mechanisms. Damper frame end flanges are connected to

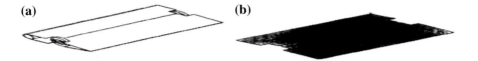

Fig. 1 a Flap assembly 1. b Flap plate

Fig. 2 Shows the problem occur in damper flap notching

the ducts. These dampers are used at specific locations in duct system to have necessary flow control.

These dampers are commonly single flap, parallel flap, or opposed type. Pneumatic, electric, and manual actuators are available to control the dampers.

1.4 Damper Flap Assembly

Flap assembly (Fig. 2) is a fabricated component with top and bottom flap plates with suitable stiffeners and end flanges. Special notches are made on the individual flap plates on both sides to the specific dimensions by manual marking and gas cutting. Flap flanges are located and welded with the flap plate at these special notches made at the ends of the flap assembly.

A flap axle is a shaft that extends from the flap through the frame. Flap axles are mechanically fastened to the both ends of the flap assembly. Flap axles having end flanges which are mechanically fastened with welded on both the sides of the flap.

Damper flaps are slats that generally rotate inside the damper frame anywhere from the fully open to fully closed position, or any other desired percentage of flap opening. Flap rotations are available in parallel or opposed mode. Flaps are generally categorized as flat, triple –V or aerofoil.

1.5 Problem Definition

At present, boiler manufacturing company is manufacturing 800 blade assemblies annually which require 3,200 notches to be made in 1,600 blade plates of various (Fig. 3).

Fig. 3 Existing manufacturing practices

Dimensions. To meet the targets for the ensuing years, the production rates have to be increased by 100 % (i.e. 6,400 notches). With the existing manufacturing practices and the associated quality problems, it may not be possible to achieve the production targets. The various problems associated with quality and productivity are analyzed thoroughly in the brain storming sessions and the problems are categorized below by the ISHIKAWA DIAGRAM.

2 Objectives

To identify the alternate process to make the notches.
To increase production rate by improved method of manufacturing.
To reduce rework and rejection present in the existing manufacturing process.
To ensure product reliability by eliminating the causes for failure.
To maintain a uniform quality level.
To ensure customer satisfaction.

3 Limitations of New Process

Different notching tools are needed for varying notch profiles.

Based on the merits and demerits of the various manufacturing processes, the concept of notching tool and power press may be ideally suited to our problem, which requires very less investment and very high productivity among other processes. Hence, we decided to select this concept for further study and implementation.

4 Advantages of Hydraulic Press

The hydraulic press continues to be the press of choice for today's modern manufacturers. The advantages of the hydraulic press versus mechanical press are realized and utilized by more and more manufacturers. Today's modern hydraulic presses provide; performance, reliability, and unlimited capability in almost any application such as stamping, punching, blanking, drawing, bending and so on. All standard press designs and circuits can be configured to match even the most non-standard press requirements.

5 Hydraulic-Circuits

Hydraulic-circuit diagrams are complete drawings of a hydraulic circuit. Included in the diagrams are a description, a sequence of operations, notes, and a components list. A graphical diagram, the short-hand system of the industry, is usually preferred for design and troubleshooting. Simple geometric symbols represent the components and their controls and connections.

5.1 Regenerative Hydraulic Circuits

A regenerative circuit combines pump delivery and rod end discharge of a differential cylinder to obtain rapid speed when expending. Pressure is equal at both the head and rod ends during regenerative movement. Pump delivery and rod end discharge both are directed to the cylinder head end through the directional valve, equal pressure on the difference in areas creates a larger force at head end to extend the cylinder.

The circuit shown below is a regenerative circuit. The circuit contains two directional control valves (V1, V2), three relief valves (R1, R2, R3) and three non retune needle check valves (C1, C2, C3). In the neutral position 4/3 DCV V1 remains in the center position and the 2/2 DCV V2 is in the position 'A' causing the cylinder to stay in a stand still position. This also avoids creeping action (lowering of ram on load). At this position, the piston will be at limit switch-F. DCV V1 solenoid 'A' is and 2/2 DCV V2 solenoid 'b' this is done with the intention for fast lowering the ram up to the limit switch-P, when ram reaches LS-p solenoid 'B' of 2/2 DCV V2 is switched off to facilitate the ram with necessary power for notching.

Thus the ram will reach the limit switch-R, this cause the solenoid 'B' of 4/3 DCV V1 will be enabled at the same time 2/2 DCV V2 solenoid 'B' is deactivated this causes the ram to return to the limit switch –F where the 4/3 DCV V1 solenoid 'B' is deactivated and spool is positioned in the center position (Figs. 4, 5).

Fig. 4 Shows the design of hydraulic circuit

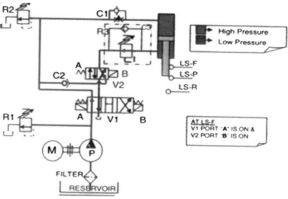

Fig. 5 Shows the design of regenerative circuit

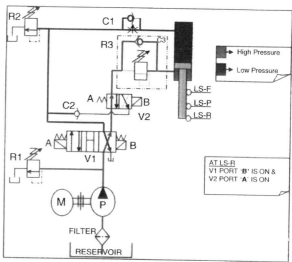

5.2 Design Calculations

5.2.1. Capacity of Hydraulic Press

$$\text{Load required for notching} = L \times t \times Fs \times FS \quad (1)$$

where

L	Periphery of maximum size Notch (282.6 mm)
t	Maximum thickness of flap plate (4 mm)
Fs	Ultimate shear strength of flap plate material (30 kg/mm^2)
FS	Factor of safety (3)

$$\begin{aligned}\text{Load required for notching} &= 282.6 \text{ mm} \times 4 \text{ mm} \times 30 \text{ kg/mm}^2 \times 3 \\ &= 10{,}176 \text{ kg} \\ &= 101 \approx 100 \text{ Tonnes}\end{aligned}$$

Recent Trends in Development of Hydraulic Press

$$\text{Effective area } A = F/\Delta p \times \eta \text{ hm} \qquad (2)$$

where,

A effective area (cm^2)
F cylinder force (kg)
Δp fluid pressure (kg/cm^2)
η hm   cylinder hydraulic efficiency (0.85–0.95)

$$\begin{aligned}
\text{Effective area} \quad A &= F/\Delta p \times \eta \text{ hm} \\
&= 100{,}000/140 \times 0.95 \\
&= 751.87 \text{ cm}^2
\end{aligned}$$

$$\text{Cylinder diameter } D_p = \sqrt{(A/0.00785)} \qquad (3)$$

where,

D_p piston diameter (mm)
A effective area (cm^2)

$$\begin{aligned}
\text{Cylinder diameter } D_p &= \sqrt{(A/0.00785)} \\
&= \sqrt{(751.87/0.00785)} \\
&= 301.478 \text{ mm} \\
&= 300 \text{ mm}
\end{aligned}$$

$$\text{Cylinder speed } V = L/t \qquad (4)$$

where,

V   Stroke speed (mm/sec)
L   Cylinder stroke length (mm)
T   Time allowed to stroke cylinder (sec)

$$\begin{aligned}
\text{Cylinder speed } V &= L/t \\
&= 200/0.4 \\
&= 500 \text{ mm/sec} \\
&= 0.5 \text{ M/s}
\end{aligned}$$

$$\text{Flow rate } Q = A \times v \times 0.2597/\eta \text{ vol} \qquad (5)$$

where

Q   Flow rate required (cm^3/sec)
A   Effective area (cm^2)
V   Stroke speed (cm/sec).

$$\begin{aligned}
\text{Flow rate } Q &= A \times v \times 0.2597/\eta \text{ vol} \\
&= 751.87 \times 50 \times 0.2597/0.95 \\
&= 10276.87 \text{ cm}^3/\text{sec} \\
&= 0.01 \text{ M}^3/\text{sec}
\end{aligned}$$

5.2.1 Design and Tool Specification

In designing tool specification, it is important to recognize several factors. These factors are quantity or product life cycle, rates of production, tooling support, maintenance, safety and the particular design itself, it is good practice to keep the tooling specification user friendly and require only the attributes needed to meet yours company expectations.

5.3 Press Structure

The various types of structures for press available and particular frame have been chosen depends upon on many factors such as for our application we need to select a suitable structure which meets our requirements for notching of damper flaps plate. For this we need the provision for mounting of hydraulic cylinder, mounting for notching tool, and space for other hydraulic accessories. Also the manufacturing cost and time should be less (Fig. 6).

Based on the above criteria, the C-FRAME structure is most suited for our application, hence we decided to develop C-FRAME structure for this press.

Pressure load is the ratio of upward force and product of plate thickness and cutting clearance.

The plate thickness is taken from the calculated value of stripped plate thickness.

The cutting clearance value is taken from ideal cutting clearance formula.

Pressure load = upward force/Ts × cutting clearance
$$= 100 \times 100 \times 9.81/36 \times 0.001$$
$$= 2.725 \text{ MN/m}^2$$

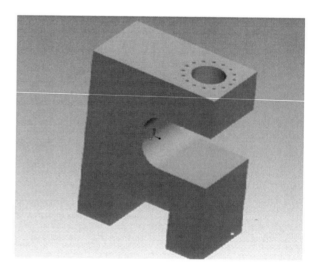

Fig. 6 Shows the arrangement C frame press structure

The upward force, the thickness, and cutting clearance values are taken from calculated values approach model = upward force 100 tonnes.

$$\text{Force on each bolt} = 100 \times 100 \times 9.81/16 \text{ (bolts)}$$
$$= 6131.25 \text{ N}$$

The Young's modulus is the property of the material which is referred as

Young's modulus	2.065 MN/mm²
Density	7.85 × 10³ kg/M³
Yield for IS-2062	250 N/mm²

2 Results and Discussions

The analysis was done on the initial press structure with all plates are of 25 mm thickness, to validate the design for structural stability, the structure is subjected to loading as specified earlier in Chap. 5. The stresses and deformation developed in the structure were found at the critical areas. (Figs. 7, 8, 9, Table 1)

Press structure internal press structure A.

In order to strengthen the structure two additional stiffener plates were introduced as shown in the Figs. 10, 11.

Structure with stiffener plate to strengthen the stress in the top plate. Initial press structure C.

Structure with additional stiffener plate to reduce stresses in the bed area.

After addition of stiffener plates the structure was analyzed, stresses and displacement were found very minimum. With the intention of optimization, the plate thickness was reduced to 20 mm from the initial thickness of 25 mm which we reduced. It was found that the stresses and displacement were well in the limits. Hence, we adopted final structure with 20 mm plate thickness. Results of the analysis of the final structure were shown.

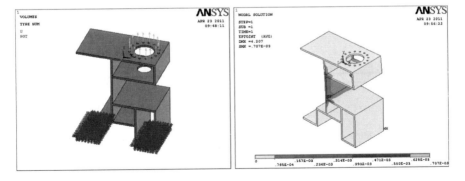

Fig. 7 Shows the final analysis on press structure

Fig. 8 Shows the design of design tool

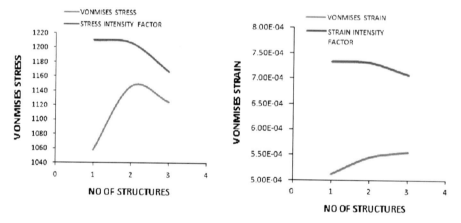

Fig. 9 Comparison graph between press structures A, B, and C

Table 1 Comparison between press structure A, B, and C

	A-structure	B-structure	C-structure
Initial press structure (25 mm)			
Pressure load			
Vonmises stress	0.240×10^9	0.112×10^9	0.112×10^9
Vonmises strain	116.444	54.466	54.443
Stress intensity	0.246×10^9	0.114×10^9	0.114×10^9
Strain intensity	148.715	68.780	54.443

(continued)

Table 1 (continued)

	A-structure	B-structure	C-structure
Force load			
Vonmises stress	629.9	670.065	697.026
Vonmises strain	0.305×10^{-3}	0.324×10^{-3}	0.338×10^{-3}
Stress intensity	722.265	716.868	700.46
Strain intensity	0.437×10^{-3}	0.434×10^{-3}	0.424×10^{-3}
Initial press structure (20 mm)			
Pressure load			
Vonmises stress	0.355×10^9	0.164×10^9	0.164×10^9
Vonmises strain	171.818	79.298	79.229
Stress intensity	0.364×10^9	0.169×10^9	0.169×10^9
Strain intensity	220.444	102.460	102.393
Force load			
Vonmises stress	1,058	1,147	1,125
Vonmises strain	0.512×10^{-3}	0.545×10^{-3}	0.555×10^{-3}
Stress intensity	1,211	1,207	1,167
Strain intensity	0.733×10^{-3}	0.731×10^{-3}	0.707×10^{-3}

Units of stress is N/mm^2

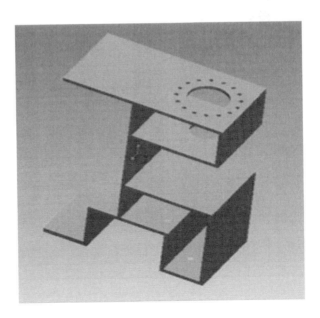

Fig. 10 Press structure

Fig. 11 Press structure C

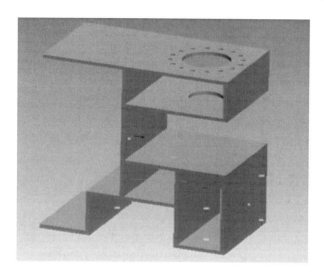

3 Conclusion

We have designed the press structure and notching tool for the damper flap plate. The press structure we have design and named as initial press structure A, initial press structure B, and an initial press structure C. Comparing the following three structures the values of vonmises stress, strain and stress intensity factor, and strain intensity factor is comparatively high. Comparing the structure A with press structure B the values of following criteria have been reduced by means of adding stiffener plate to the top plate. Comparing the above two structures with press structure C the values of vonmises stress, vonmises strain, stress intensity, and strain intensity are comparatively found less. So initial press structure C is very much suitable for doing this process. Implementation of this project will increase the production rate and reduces the overall production cycle time and improve quality of the damper assembly by eliminating the problems inherent in the present method of manual notching in the damper flap. We also selected suitable hydraulic cylinder for the press.

References

1. Lee DC, Lee JI (2003) Structural optimization concept for the design of aluminum control arm. J Automobile Eng 217(8):647–656
2. Wong EK, Leung TE, Chuen CW (1996) Object-oriented cad for practical hydraulic system design. Eng Appl Artif Intelligences 9(5):499–514
3. Altintat Y, Lane AJ (1997) Design of an electro-hydraulic CNC press brake. Mach Tools Manuf 37(1):45–49

4. Hambli R, Potiron A, Kobi A (2003) Application of design of experiment technique for metal blanking processes optimization. Mec Ind 4(3):175–180
5. Hamilton A (2002) Considering value during early project development: a project case study. Int J Proj Manage 20(2):131–136
6. Burr AH, Cheatham JB (1995) Mechanical analysis and design. Prentice Hall, NJ
7. Dieter GE (1999) Engineering design
8. Shigley JE, Mischike CR (2002) Standard handbook of the die maker handbook

Theoretical Analysis of Overall Heat Loss Coefficient in a Flat Plate Solar Collector with an In-Built Energy Storage Using a Phase Change Material

R. Sivakumar and V. Sivaramakrishnan

Abstract Flat Plate Solar Heater is one of the most widely used devices to harness solar energy available in abundance. The collector efficiency can be improved by reducing the overall losses. Efficiency of the collector depends on overall loss coefficient which is the sum of Top loss, Bottom loss, and Edge loss. The present theoretical analysis is on Overall Heat Loss Coefficient of a solar collector with and without in-built Phase Change Material (PCM). Theoretical results show a reduction in overall loss coefficient with decrease in distance between the absorber plate and PCM surface, and decrease in mean absorber plate temperature.

Keywords PCM • Top loss coefficient • Overall loss coefficient and efficiency

Nomenclature

A_c	Collector area (m^2)
c_p	Specific heat capacity (Jkg^{-1}k^{-1})
I	Intensity of solar radiation (Wm^{-2})
S	Solar radiation reaching the absorber plate (Wm^{-2})
L	Number of glass covers
M	Mass flow rate (Kgs^{-1})
T	Temperature (K)

R. Sivakumar (✉)
Department of Mechanical Engineering, MAM College of Engineering and Technology,
621 105 Tiruchirapalli, TamilNadu, India
e-mail: ramalingam.sivakumar@gmail.com

V. Sivaramakrishnan
Department of Mechanical Engineering, Roever Engineering College,
621 212 Perambalur, TamilNadu, India
e-mail: vsmp1967@yahoo.com

S. Sathiyamoorthy et al. (eds.), *Emerging Trends in Science,*
Engineering and Technology, Lecture Notes in Mechanical Engineering,
DOI: 10.1007/978-81-322-1007-8_13, © Springer India 2012

h_w	Wind loss coefficient (Wm^{-2}K^{-1})
U_L	Overall heat loss coefficient (Wm^{-2}K^{-1})
U_T	Top loss coefficient (Wm^{-2}K^{-1})
U_B	Bottom loss coefficient (Wm^{-2}K^{-1})
U_E	Edge loss coefficient (Wm^{-2}K^{-1})
Q_T	Top heat loss (W)
Q_B	Bottom heat loss (W)
Q_E	Edge heat loss (W)
Q_o	Is the heat loss (W)
Qu	Useful heat energy collected (W)
FR	Collector heat removal factor
F'	Collector efficiency factor

Greek

α	Absorptive
τ	Transitivity
η	Efficiency
σ	Stefan Boltzmann's constant $= (5.67 \times 10^{-8}\,\text{Wm}^{-2}\text{K}^{-4})$
β	Collector tilt angle

Subscripts

a	Ambient
c	Collector
i	Inlet
o	Outlet
g	Glass cover
p	Absorber plate

1 Introduction

Time delay in availability of solar energy and utility of solar energy reiterates the need for use of Phase Change Materials (PCM) in Thermal Energy Storage Systems (TES). A conventional liquid flat plate collector consists of an absorber plate, tube, transparent cover, insulation, and collector box. The advantages are: it utilizes both diffused and direct component of solar radiation, does not require orientation towards sun, and requires minimum maintenance. The disadvantages are: it has no optical concentration, hence heat is lost from large area and low efficiency as a result, and can be used only during day time. A natural circulation solar water heater works on the principle of conduction, convection and radiation, and flow is because of density difference developed in the heater. A forced circulation solar water heater is provided with a pump to circulate water at a constant flow rate.

The above mentioned Flat plate solar water heaters have limitation of usage only during sun shine. An inbuilt storage of heat in the Solar water heater by

Table 1 Properties of Phase change materials considered [6]

PCM	Melting point °C	Heat of fusion kJ/kg	Thermal conductivity Wm^{-1}K^{-1}	Density kg/m^3 solid, liquid	Specific heat capacity—solid, liquid in kJkg^{-1}K^{-1}
Paraffin wax Sunoco P116	50	190	0.21	930, 830	2.1, 2.1
Glauber's salt Na$_2$SO$_4$.10H$_2$O	32	251	2.25	1,460, 1,330	1.73, 3.3

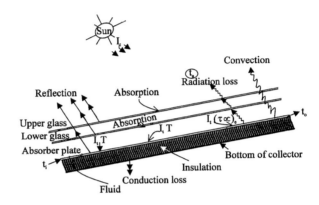

Fig. 1 Energy balance of a flat plate collector adapted from Garg 1987 [2]

providing PCM under the absorber surface enables charging of heat during sunshine and discharge of heat to water even during night time.

The properties of PCMs studies are given in Table 1 [1]. The volume changes of the PCMs on melting would also necessitate special volume design of the containers. This necessitates the provision of the expansion volume gap between the absorber plate and the top surface of the solid PCM in the storage tank. More the distance between the absorber plate surface and PCM surface, more will be the resistance to heat transfer by radiation and convection mode. This resistance tends to increase the average absorber plate temperature and hence the losses.

Figure 1 explains the energy balance of a flat plate collector. Solar radiation falling on the glazing cover is partly absorbed, partly reflected and the balance is transmitted through on to the absorber plate. The absorber plate absorbs the radiation and a fraction is reradiated particularly in the long wave length range.

The solar radiation that reaches the absorber plate is given by:

$$S = \tau_g \cdot \alpha_p \cdot I \quad (1)$$

The energy losses that take place in a flat plate solar collector are Top loss, Bottom loss, and Edge loss. Top loss is the heat energy from the top surface which is mainly by radiation and convection. Bottom loss is the heat energy lost from the bottom surface of the housing of the collector which is mainly by conduction.

Edge loss is the heat energy lost from the sides of the collector which is also mainly by conduction and subsequent convection to the surrounding.

$$Q_L = Q_T + Q_B + Q_S \tag{2}$$

$$Q_L A_p U_L \left(T_p - T_a\right) \tag{3}$$

$$U_L = U_T + U_B + U_E \tag{4}$$

Top loss coefficient is calculated using the empirical equation developed by Klein [2], following the basic procedure of Hottel and Woertz:

$$U_T = \left[\frac{N}{\frac{C_a}{T_p}\left[\frac{T_p - T_a}{N+f}\right]^e} + \frac{1}{h_w}\right]^{-1} + \frac{\sigma\left(T_p + T_a\right)\left(\left(T_p^2 + T_a^2\right)\right)}{\left[(\varepsilon_p + 0.00591 N h_w)^{-1} + \frac{[2N+f-1+0.133\varepsilon_p]}{\varepsilon_g} - N\right]} \tag{5}$$

where C_a and f and e and h_w are given by:

$$C_a = 520(1 - 0.00005\beta^2) \tag{6}$$

$$f = \left(1 + 0.089 h_w - 0.1166 h_w \varepsilon_p\right)(1 + 0.07866 N) \tag{7}$$

$$e = 0.43\left(1 + \frac{100}{T_p}\right) \tag{8}$$

The useful heat gained by the absorber plate is given by:

$$Q_u = A_p S - Q_L \tag{9}$$

$$h_w = 5.7 + 3.8v \tag{10}$$

The bottom loss is assumed to take place one dimensionally through the insulation at the bottom. The bottom loss coefficient is given by

$$U_B = \frac{K_s}{L_s} \tag{11}$$

The edge loss is negligible for a very small collector perimeter to area ratio [3].

The loss coefficient decreases with increase in the gap between the absorber plate and the glass cover. A gap width $>= 5$ cm is recommended for optimum loss coefficient [1].

Double layers of glass cover reduces the loss coefficient by 44 % [1] since glazing cover is transparent to thermal radiation but opaque to long wave radiations from the absorber plate surface.

Top loss coefficient increases with increase in emissivity of the absorber plate. Top loss coefficient decreases with increase in ambient temperature and hence efficiency increases. There is no significant effect of tilt angle β of the collector on top loss coefficient.

The thermal losses increase as temperature difference between collector and ambient air rises. In order to decrease the thermal losses, the idea proposed is to place the absorber plate on the surface of the PCM thereby heat transfer takes place by direct conduction and not by radiation and convection mode resulting in reducing the temperature of the absorber plate. The absorber plate should be designed movable upward and downward to always be in contact with the PCM surface as the volume of PCM changes with phase change. The sides along the edges of the absorber plate should designed such that the plate is movable and also leak proof to enable tilting the plane of the Solar water heater.

With the increase in absorber plate temperature, simultaneously the value of U_T and U_L increases at different tilt angle. Thus the value of U_T and U_L are found increasing due to increased convection and radiation losses [4].

2 Estimation of Overall Heat Loss Coefficient

The present work deals with an analytical method for predicting the heat loss coefficient of Flat plate solar water heaters without PCM and with PCM for inbuilt thermal energy storage with an objective of suggesting ways to reduce the heat loss coefficient.

A solar water heater of dimension 0.4×0.8 m with PCM to a depth of 0.1 m is assumed. The insulation thickness assumed is 0.05 m of Polyurethane. The glass cover is one in number and of 0.003 mm thickness. The glazing cover is at a distance of 0.05 m from the absorber plate. The absorber plate material is chosen as copper of thickness 0.00112 m and thermal conductivity 401 $Wm^{-1}K^{-1}$. The collector housing is designed with insulation at the bottom and sides by Polyurethane material of thickness 0.05 m to minimize convection heat losses to the surrounding. The emissivity and absorptivity of the glazing cover are taken as 0.1 and transmissibility of the glazing cover is taken as 0.9. The emissivity of the absorber plate is taken as 0.8 and emissivity for reradiation is 0.18. The flow rate is assumed to be constant as 10 $kghr^{-1}$. Wind velocity is assumed as 3.5 m/s for wind loss calculation. The ambient temperature is assumed.

The energy equations that govern heat transfer in the physical system are subjected to the following assumptions:

- The PCM is homogeneous and isotropic;
- The thermo physical properties of the PCM and the Heat Transfer Fluid are independent of temperature. However, the thermal conductivities of the liquid and solid phases of the PCM are different;
- The flow is Newtonian, incompressible, and fully developed dynamically but the local convective heat transfer coefficient for the Heat Transfer Fluid varies along the length of tube for laminar flow;
- The axial conduction is negligible comparatively to the heat convection in the flow;
- The effect of natural convection during the melting of PCM is taken into account by using the effective thermal conductivity.

2.1 Estimation of Top Loss Coefficient

The fundamental experimental data of Yasin Varol et al 2010 [5] is taken for theoretical analysis of overall loss coefficient. The absorber plate is assumed to be in thermal equilibrium considering the fact that its thermal capacity is small and the calculations of overall heat loss coefficient is a function of average absorber plate temperature. The solar radiation received by the absorber plate is calculated. Maximum possible absorber plate temperature neglecting losses is calculated. Based on the T_p maximum value, Loss coefficients are calculated. From the calculated losses, Useful energy transferred through the absorber plate is calculated. From the useful energy calculated, the absorber plate temperature is recalculated and the iteration is continued until repeated values of Loss coefficients have negligible difference. Top loss coefficients is calculated using Eqs.1–10 for (a) Conventional FPSWH, (b) FPSWH using Glauber's salt as PCM with expansion gap of 0.004 m between the PCM surface and absorber plate and (c) FPSWH using Paraffin without expansion gap by providing a movable absorber plate to ensure direct contact with the PCM always.

2.2 Estimation of Bottom Loss Coefficient

The bottom loss coefficient varies with temperature of the PCM and the ambient temperature. U_B is calculated using Eq. (11) at different time instants for the three cases mentioned above.

3 Results and Discussion

The overall loss coefficient has been calculated for a conventional flat plate solar collector and that with a flat plate solar collector with inbuilt PCM storage material which is in direct contact with the absorber plate and the results were compared.

3.1 Variation of Bottom Loss

The Bottom loss coefficient is a constant. The change in its value is small and decreases with decrease in solar radiation in the evening.

3.2 Variation of Overall Loss Coefficient

The variation of overall loss coefficient illustrated in Fig. 4 with respect to time and variation of solar radiation which is illustrated in Fig. 3.

Fig. 2 Experimental setup adapted from Yasin Varol [5]

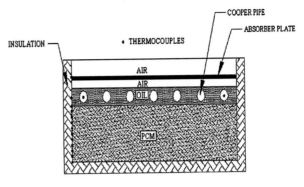

Fig. 3 Variation of total solar radiation with time in a day

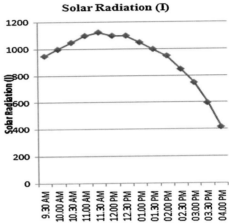

It is observed that the overall coefficient is found to vary a little with respect to time till noon for a conventional FPSWH. Its value decreases with decrease in solar radiation in the evening. The loss coefficient is found to be optimum since the absorber plate temperature is about 10–15 °C only above the fluid temperature. Since the radiation received by the absorber plate is transferred to the fluid by conduction mode of heat transfer, and the material being Copper of small thickness, the temperature difference between the fluid heated and the absorber plate is small (Fig. 4).

It is observed that for FPSWH using Glauber's salt as PCM with expansion gap between the absorber plate and the PCM top surface, the overall loss coefficient is found to vary in a similar trend in comparison with the conventional FPSWH, but the losses are comparatively higher. The reason for this is found to be the expansion gap where radiation takes place by convection and radiation mode. Higher the distance between the Absorber plate surface and the heat gaining fluid, greater is the resistance to heat transfer which results in increasing the average absorber plate temperature to transfer the useful energy available. Increased absorber plate temperature results in increased overall losses.

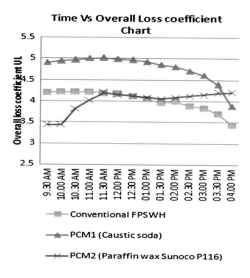

Fig. 4 Variation of overall loss coefficient U_L with time for three different cases

It is observed that for FPSWH using Paraffin wax (Sunoco P116), the heat loss coefficient is very low when compared to the other two cases, since the entire surface of the absorber plate is in direct contact with the PCM which results in pure conduction mode of heat transfer in the beginning. U_L increases till the PCM reaches its melting point. Once melting starts, heat transfer takes place by conduction and convection mode. High thermal conductivity and small thickness of the plate in the direction of heat transfer results in absorber plate temperature to be marginally above the melting point of the PCM and the temperature is nearly maintained from there onwards, even though the radiation decreases in the evening because of isothermal property of PCM which is in contact with it. This is the reason for the value of U_L not decreasing with decrease in solar radiation.

The average value of overall loss coefficient for FPSWH using PCM with expansion gap is greater than that of conventional FPSWH which is marginally greater than that of FPSWH. Higher the area of contact of the fluid with the absorber plate surface, lower is the overall loss coefficient and higher the distance between the absorber plate and the PCM, higher is the overall loss coefficient.

4 Conclusions

A theoretical analysis has been done to find the overall losses in three different cases of FPSWH mentioned in the Table 2. The theoretical analysis shows that, the overall loss coefficient increases with increase in average absorber plate temperature. Average absorber plate temperature depends on area of contact of the fluid or the PCM which receives heat from it.

Theoretical Analysis of Overall Heat Loss Coefficient

Table 2 Estimated overall loss coefficient at different time period in a day for flat plate solar collector of type

Time	I Wm^{-2}	Conventional FPSWH	PCM1 Nacl$_2$	PCM2 paraffin sunoco P116
		U_L	U_L	U_L
9:30 AM	950	4.20	4.91	3.433
10:00 AM	1,000	4.21	4.95	3.433
10:30 AM	1,050	4.21	4.97	3.823
11:00 AM	1,100	4.22	5.01	4.023
11:30 AM	1,125	4.19	5.02	4.193
12:00 PM	1,100	4.16	4.99	4.153
12:30 PM	1,100	4.12	4.97	4.123
01:00 PM	1,050	4.05	4.93	4.083
01:30 PM	1,000	3.96	4.86	4.053
02:00 PM	950	4.00	4.82	4.083
02:30 PM	850	3.89	4.72	4.123
03:00 PM	750	3.84	4.61	4.153
03:30 PM	600	3.71	4.39	4.193
04:00 PM	420	3.44	3.88	4.213

(a) conventional (b) having PCM with gap between PCM surface and absorber plate (c) having PCM without gap between PCM surface and absorber plate

FPSWH using PCM for inbuilt storage of heat without expansion gap has the entire surface of the absorber plate in direct contact with the heat transfer fluid and hence has least overall loss coefficient.

Conventional FPSWH has a portion of the surface of the absorber plate in direct contact with the heat transfer fluid and hence has medium overall loss coefficient.

FPSWH using PCM for inbuilt storage of heat with expansion gap has the entire surface of the absorber plate detached from the PCM/the heat transfer fluid and hence has maximum overall loss coefficient.

It is concluded that energy efficient FPSWH using PCM for inbuilt energy storage can be provided with a movable absorber plate such that it is always in direct full contact with the PCM top surface. The plate shall move up during expansion of the PCM during melting process and down during contraction of PCM during solidification process.

References

1. Agbo SN, Unachukwu GO (2006) Performance evaluation and optimization of the NCERD thermo-siphon solar water heater. Proceedings World Renewable Energy Congress, Florence, pp 19–25
2. Agbo SN, Okoroigwe EC (2007) Analysis of thermal losses in a flat plate collector of a thermo siphon solar water heater. Res J Phys 1:37–45

3. Garg HP (1987) Advances in Solar Energy Technology, vol 1. D. Reidel Publishing Company, Holland
4. Bhatt MK, Gaderia SN, Channiwala SA (2009) Experimental investigations on top loss coefficients of solar flat plate collector at different tilt angle. World Acad Sci Eng Technol 79:432–436
5. Varol Y, Koca A, Oztop HF, Avci E (2010) Forecasting of thermal energy storage performance of phase change material in a solar collector using soft computing techniques. Expert Syst Appl 37:2724–2732
6. Farid MM, Khudhair AM, Razack SAK, Al-Hallaj S (2004) A review on phase change energy storage: materials and applications. Energ Convers Manage 45:1597–1615

Part II
Design and Manufacturing Techniques

Design and Development of Nonautomatic Tabletop Mini Lathe

T. T. M. Kannan and P. Marimuthu

Abstract Miniaturization of machine tools to size compatible to target products without compromising machining tolerance leads to enormous saving in energy, space, and resources. Recently model techniques are findings an increasingly wider application in the design of small structures and models of actual machines that are made from appropriate materials. tabletop lathes are preferred over lathes by some professionals like lock smith, jewelers, and designers, engineers for prototyping and fabrication work. This chapter describes the design and development of tabletop mini lathe through 3D modeling and actual fabrication made by design calculations based on machine tool Design procedure that utilize a variable speed electric motor for getting various cutting velocities. It is specially designed to turn metals like steel, copper, aluminum, silver, and gold for carrying operations of external turning, facing, and step tuning. This model is analyzed by study behavior of actual behavior predicted from the knowledge of models. It is the first step to fabricate the mini lathe (150 mm length) for turning of small components with degree of accuracy.

Keywords Tabletop lathe • Modeling • Fabrication • Prototyping • Analyzing

1 Introduction

Micro machine is the foundation of the technology to produce miniature components with high relative accuracy requirements in 3 dimension features and made in wide range of engineering [2, 4] material. Some of the field such as machine tool fabrication have in corporate with application of micro product requirements [8, 10]. A single point micro diamond cutting tool has been applied to the cutting of

---

T. T. M. Kannan (✉)
Department of Mechanical Engineering, J.J. College of Engineering and Technology,
Tiruchirappalli, Tamilnadu, India
e-mail: ttmk_8@yahoo.com

P. Marimuthu
Syed Ammal Engineering College, Ramanathapuram, Tamilnadu, India
e-mail: pmarimuthu69@gmail.com

S. Sathiyamoorthy et al. (eds.), *Emerging Trends in Science,*
Engineering and Technology, Lecture Notes in Mechanical Engineering,
DOI: 10.1007/978-81-322-1007-8_14, © Springer India 2012

various shapes and cutting forces has been investigated using Mini lathes are related to lower rigidity (full-sized) lathes but are distinguished by their small size and differing capabilities, application, use, and locations [1, 3]. Sometimes referred to as desktop lathes or tabletop lathes, mini lathes can be comfortably used in areas where a full-sized lathe would be impractical. The idea of desktop machines was initiated for the concept of 'Micro factories'. May researchers in Japan have already developed micro machines that can be mounted on tabletop. Three directional forces to improving the working accuracy are applied to micro parts have been developed. In order to develop a machine tool for micro machine elements and kinetic accuracy have been evaluated quantitatively [5, 14]. Despite a more available price point and than their larger counterparts, mini lathes still operate within the required tolerances of many tasks. This makes them particularly useful in applications where small precision pieces are needed for projects like miniature engines that require fine detail and mechanical accuracy [6, 10]. Mini lathes are designed at the request of turners having limited space but high quality. Research and Development of micro machines for different applications is widely available mainly at academic and experimental level [7, 11]. The present study deals with detailed design and development of precision tabletop mini lathe for the purpose of machining small components.

2 Engineering Design Process Applied in Machine Tool

The design process for designing a new machine tool is presented in the form of block diagram and design process is carried out by three important stages like Design proposal, Preliminary design, and Detailed Design. At the end of each stage the Design must be subjected to a critical feasibility analysis and technical report is prepared and submitted (Fig. 1).

3 Material Selection for Machine Tool Parts

Material selection is a matter of quality and cost, the properties of machine tool material must be adequate to meet design requirements and service conditions. The procedure for material selection of machine tool components must consider both mechanical and technological properties. Low carbon steel is selected to fabricate the machine should have specific chemical composition and mechanical properties (Tables 1 and 2).

4 Design Calculations of Mini Lathe Components

Design calculations cover the design of major components of machine tool, such as Electric motor, Base, bed, spindle, Cross slide, saddle, and tool post etc., these calculations are done in accordance with design procedures only found most,

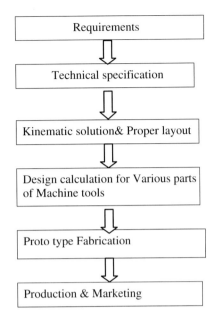

Fig. 1 Block diagram of design process in machine tool

Table 1 Chemical composition of low carbon steel (1018)

Grade	Fe	C	Mn	P	S
1,018	98.81	0.18	0.9	0.04	0.05

Table 2 Mechanical properties of low carbon steel (1018)

Grade	UTS (PSI)	Yield (PSI)	Elongation (%)	Rockwell hardness (RHN)
1,018	63,800	53,700	15	B71

suitable on the basis of proceeding analysis [12]. After design process the machine components are drawn to suitable scale with dimension, tolerance, and manufacturing specification in detailed drawings and then assembled as far as possible [13].

4.1 Selection of Electrical Motor

Induction motors are generally used as the source of power in machine tools and power rating is calculated as $Nm = Nc$; Nm = Power rating of electrical motor (Kw), Nc = Power required for removing the metal (Kw) = co-efficient of efficiency.

4.2 Design of Base

Machine tool bases are analyzed as plates on elastic foundation and dimensions are determined from consideration that maximum deflection due to load acting on the plate. It is calculated as $\Theta = \Theta q1 + \Theta q2 + \Theta m$, $\Theta1$ and $\Theta2$ = distribution of load on bed force, Θm = angle of slope due to bending moment $\Theta q1 = q1$ m.b/kq2, $\Theta q2 = q2$.m.b/k $\times$ kq2, Θm = m.m^3/k $\times$ km, K = stiffness of elastic foundation = stiffness co-efficient, Kq1, Kq2, Kq3 = co-efficient of geometrical parameters.

4.3 Design of Bed

The machine tool beds are generally fabricated by closed rectangular section with required stiffness is calculated as **N = SE Iw**, S = co-efficient of stiffness, E = Modulus of Elasticity, Iw = Moment of Inertia of Bed (mm 4).

4.4 Design of Saddle and Carriage Cross Rail

Cross rails, saddle, and carriage are supporting the cutting tools are imparting to it movement in 2 direction and calculated as **S = EB2/1−μ × t1−t2/ t1 + t2**, μ = poison ratio, E = Modulus of elasticity, b = width of saddle, t1 and t2 = thickness of saddle carriage, n = $\tilde{O}$ f/p (1–s), F = Frequency, P = No of poles.

5 Specification of Miniature of Lathe

1. Net Weight = 5 kg
2. Bed Length = 125 mm
3. Centre Height = 50 mm
4. Spindle Speed = 1,500 RPM
5. Longitudinal Feed = 75 mm
6. Cross Feed = 20 mm
7. Motor Power = 68 Watts.

See Figs. 2 and 3.

6 Performance Test on Desktop Mini Lathe

Most of machine tools carry out a machining by means of rotational force of spindle. The accuracy of spindle affects the roundness and surface roughness of work piece. The final product accuracy depends on the main spindle accuracy. The

Design and Development of Nonautomatic Tabletop Mini Lathe

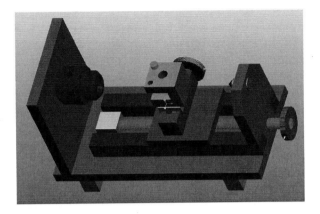

Fig. 2 Pro-E modeling of mini lathe

Fig. 3 Assembled view of mini lathe

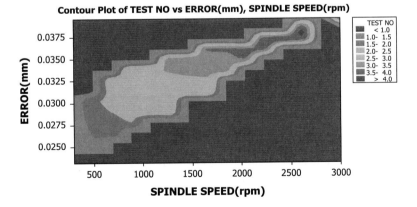

Fig. 4 Contour plot for spindle error

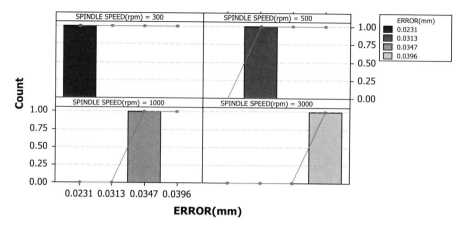

Fig. 5 Pareto chart of spindle error [15]

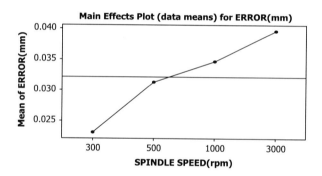

Fig. 6 Main effects plot for spindle error

Table 3 Spindle error measurement using gap sensor

Test number	Spindle speed (RPM)	Error (mm)
1	300	0.0231
2	500	0.0313
3	1,000	0.0347
4	3,000	0.0396
Average error		0.03217

accuracy of spindle can monitored the spindle using oscilloscope. In this method Eddy current type gap sensors which can measure the main spindle accuracy by non contacting method and produce average rotational error was 0.03217 mm (Figs. 4, 5, 6 and Table 3).

7 Main Features of Ultra Precision Tabletop Mini Lathe

- High precision axis of motion and control with degree of accuracy.
- Small space, resources, and energy efficient.
- Machine tool structure with high loop stiffness and good damping capacity.
- High thermal stability and mechanical stability with low Vibrations.
- Variable spindle speed and feed can be achieved.
- Portable and low power consumption.

8 Conclusion

This paper describes the concept of micro factory and its design basics. The micro factory is a desktop type miniature of machine tools which is actually fabricated as per design calculations and 3D modeling using Pro-E diagrams and performance is measured by average rotational spindle error measuring method. It is first step to fabricate the miniature of lathe (150 mm length) for turning of small components in concept of "small components for small machine tools". The objective of this effort is Miniaturization of Machine tool without compromising machining tolerance lead to saving energy, space, and basic resources.

References

1. Ehman KF, DeVor RE, Kapoor SG, Cao J (2008) Design and analysis of a micro/meso-scale machine tools. In: Smart devices and machines for advanced manufacturing, Springer, London, p 28
2. Kussul E, Baidyk T, Ruiz-Huerta L, Caballero-Ruiz A, Velasco G, Kasatkina L (2002) Development of micro machine tool prototypes for micro factories. J Micromech Micro Eng 12:795–812
3. The micro factory system in national R&D project of MITI. In: Proceedings of IWMF 1998, p 193–196
4. Kawahara N, Suto T, HiranoT, Ishikawa Y, Kitahara T, Ooyama N, Ataka T (1997) Microfactories: new applications of micromachine technology to the manufacture of small products. Microsys Technol 37–41
5. Moriwaki T, Shamoto E (1991) Ultra precision diamond turning of stainless steel by applying ultrasonic vibration. Ann CIRP 40(1):559–562
6. Mishima N, Ishii K (1999) Robustness evaluation of a miniaturized machine tool. In: Proceedings for ASME design engineering technology conference, Las Vegas, DETC/DAC-8579
7. Mishima N, Asida K, Tanikawa T, Maekawa H, Kaneko K, Tanaka M (2000) Micro-factory and a design evaluation method for miniature machine tools. In: Proceedings for ASPE 15th Annual Meeting, Phoenix, p 567–570
8. Kitahara T, Ishikawa Y, Terada K, Nakajima N, Furuta K (1996) Development of micro-lathe. J Mech Eng Lab 50(5):117–123

9. Mishima N, Ashida K, Tanikawa T, Maekawa H (2002) Design of a micro factory. In: Proceedings of ASME/DETC 2002, Montreal
10. Mishima N, Ishii K (1999) Robustness evaluation of a miniaturized machine tool. In: Proceedings of ASME/DETC99, Las Vegas
11. Kitahara T, Ishikawa Y, Terada K, Nakajima N, Furuta K (1996) Development of micro-lathe. J Mech Eng Lab 50:117–123
12. Okazaki Y, Mishima N, Ashida K (2004) Micro factory concept, history and developments. J Manufact Sci Eng 126:837–844
13. Kawahara N, Suto T, Hirano T, Ishikawa Y, Kitahara T, Ooyama N, Ataka T (1997) Micro factories: new applications of micro machine technology to the manufacture of small products. Microsys Technol 37–41
14. Lucca DA, Seo YW, Rhorer RL (1994) Micro cutting in the micro lathe turning system. Int J Mach Tools Manuf 39:1171–1183
15. Lin CS, Chang FM (1998) A study on the effects of the surface finish using a surface topography simulation model for turning. Int J Mach Tools Manuf 38(7):763–780 Taiwan

Experimental Analysis of the Pressure Distribution on Different Models of Carbody in the Windtunnel

K. Selvakumar and K. M. Parammasivam

Abstract Vehicle body aerodynamics has become increasingly important over the past three decades because of its vital contribution to improving a vehicle's overall performance. This was the motivation for the research, concerning the determination of the overall pressure distribution on the vehicle body of different models. Static pressure measurements have taken under various test conditions. The Wind Tunnel experiments were carried out simulating various roads like cruciform. The analyzed vehicle configurations include wheel rim-tire and body modifications. The results deal with static pressure measurements, drag and lift coefficients, ride heights, pitch, and roll angles. The acquired data were used to examine the underbody flow topology and determine to rectify the flow represents the real on-road conditions affect its pressure distribution. This research focus is to a correlate between Computational Fluid Dynamics (CFD) and physical scales behavior of a conventional car.

Keywords Aerodynamics · Drag · Reduction · Resistance

1 Introduction

Constant need for better fuel economy, greater vehicle performance, reduction in wind noise level, and improved road handling and stability for a vehicle on the move, has prompted vehicle manufactures to investigate the nature of air resistance or drag for different body shapes under various operating conditions.

However, as the passenger car must have enough capacity to accommodate passengers and baggage in addition to minimum necessary space for its engine

---

K. Selvakumar (✉)
Dept of Automobile Engineering, MIT, Anna University, Chennai, India
e-mail: kselvakumar@mitindia.edu

K. M. Parammasivam
Dept of Aerospace Engineering, MIT, Anna University, Chennai, India
e-mail: mparams@mitindia.edu

S. Sathiyamoorthy et al. (eds.), *Emerging Trends in Science,
Engineering and Technology*, Lecture Notes in Mechanical Engineering,
DOI: 10.1007/978-81-322-1007-8_15, © Springer India 2012

and other components, it is extremely difficult to realize an aerodynamically ideal body shape. The car is, therefore, obliged to have a body shape that is rather aerodynamically bluff, not an ideal streamline shape as seen on fish and birds.

Aerodynamic drag is usually insignificant at low speed but the magnitude of air resistance becomes considerable with raising speed. Fig which compares the aerodynamic drag force of a poorly streamlined, and a very highly streamlined medium sized car against its constant rolling resistance over a typical speed range.

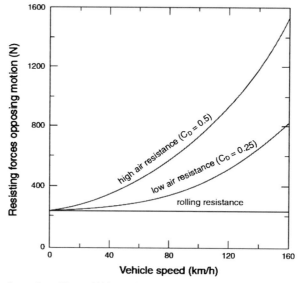

Comparison of low and high aerodynamic drag forces with rolling resistance

In the present era, optimization of car aero dynamics, more precisely reduction of associated drag coefficient (C_D), which is mainly influenced by exterior profile of a car [1].

2 Geometry Creation

2.1 Modeling

The tool which is used to create the geometry is CATIA V5. The image of the car with the overall dimension is taken for reference to create the geometry.

The image is taken in the drawing mode of CATIA, and created two points having distance of 2,400 mm (Wheel base of the vehicle) the image is expanded to match these points with that wheel base.

The analysis of car aerodynamics can present a significant challenge, the deployment of CFD FLUENT 6.1, GAMBIT, and CATIA within the design process, enables such studies to be carried out with relative case.

Experimental Analysis of the Pressure Distribution

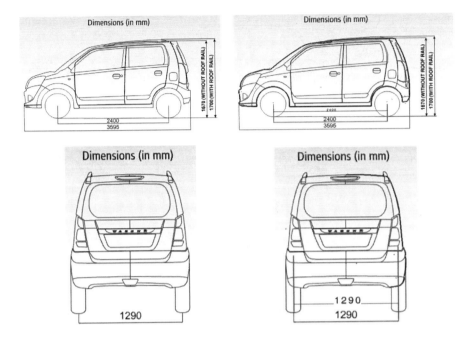

The image is with the actual dimension size and using the curves the profile of the car has been made which matches the image. The same operation followed for the other views also have the other curves for the front and top views [3].

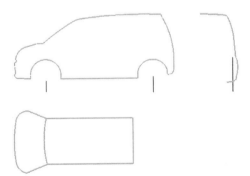

After creating the curves the Drawing file is saved in .dxf format. The same file opened and the Curves form the .dxf file copied and pasted in the sketch mode of CATIA for the all views. 3D view of actual dimensions of model geometry and 3D multiple view of actual dimensions of model.

The surfaces are created using various options in CATIA for the Windshield, bonnet, roof, side, etc.

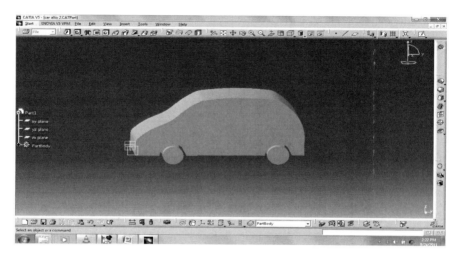

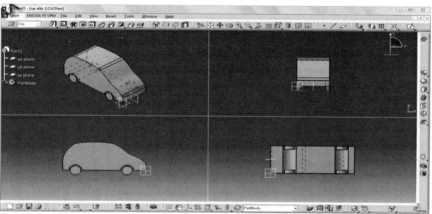

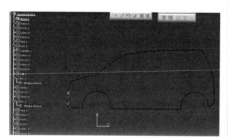
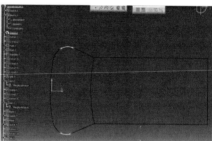

Experimental Analysis of the Pressure Distribution

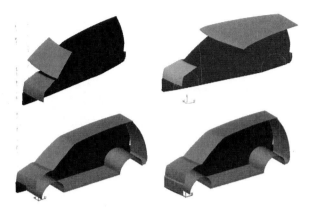

Created the block more than the size of the car, and the solid is trimmed using the various surfaces created earlier. After trimming the necessary blinds are made to match the profile in the edges. The door, wind shield rear glass, and window outer profile are created.

The above operations are made to create the half model; the solid model is mirrored to have the full car. Various stages of the trimmed bodies are shown below.

The Geometry which is created is detailed and checked the overall dimension to confirm the actual size.

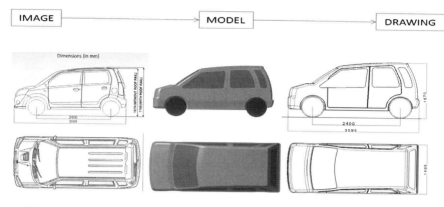

Created car model is saved in .iges format. This .iges file containing the geometry is the input for the CFD [2].

The tool which is used for meshing and CFD analysis are ICEM CFD and CFX 12.

When the geometry was defined in the creation of the computational mesh, all faces of the domain were assigned names mentioned below.

Inlet—Velocity inlet
Wall—Top, bottom, side
Body—Car
Outlet—Pressure outlet

2.2 Dimensions of the Domain

Height = 20 m
Length = 35 m
Breadth = 20 m

When the geometry was defined in the creation of the computational mesh, all faces of the domain were assigned names. The names of the inlet and outlet planes (at x = 0

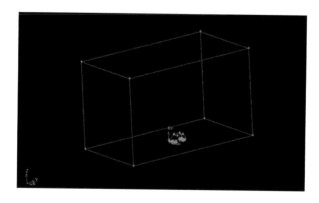

and x = L) are front face and back face of domain as velocity inlet and pressure outlet respectively. The names of the planes at y = L, z = 0, and z = L are outer wall as wall. The names of the model are car as a wall. And bottom face is defined as road.

3 Computational Fluid Dynamic Analysis Creation

3.1 Geometry Cleanup

The imported .iges file into ICEM CFD, and deleting the unwanted surfaces and closing the void created without affecting the flow over the vehicle.

3.2 Meshing

Meshing the half car model individually in ICEM CFD with prism cells, creating the box required for the external domain [4].

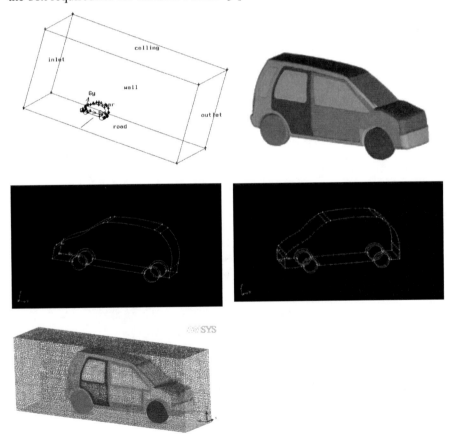

Dividing the main box into small boxes and meshing individually joining all the boxes into single external domain in ICEM CFD [4].

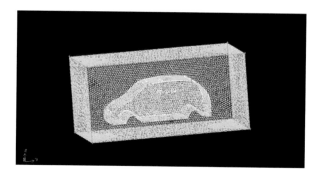

3D view of final mesh with volume source visible around the scaled model shown below.

Define the materials as air and it has properties of

Density = 1.225 kg m^{-3}
Viscosity = 1.464e^{-5} kg m^{-1} s^{-1}

Those values correspond to the ICAO norm. Fluent means dynamic viscosity as we consider air as incompressible and are not looking for problematic heat transfer, we do not need to specify properties.

3.3 Specify the Boundary Conditions

Boundary—in	
Type	Inlet
Location	In
Flow Regime	Subsonic
Mass and momentum	Normal speed
Normal speed	8.4 m/s
Boundary—top	
Type	Top
Location	Wall
Flow direction	Normal boundary condition
Flow regime	Subsonic
Turbulence	Medium intensity and Eddy viscosity ratio
Boundary—out	
Type	Outlet
Flow regime	Subsonic
Mass and momentum	Average static pressure

Boundary—bottom	
Type	Wall
Location	Base
Mass and momentum	No slip wall
Wall roughness	Smooth wall
Boundary—car body	
Type	Body
Location	Base
Wall roughness	Smooth wall
Boundary—side	
Type	Wall
Location	Side (Left and Right)
Mass and momentum	No slip wall
Wall roughness	Smooth wall

3.4 External Domain Details

The dimensions of the Domain are 12 L, 6 W, and 4 H where the L, W, H are the Length, Width, and Height of the car model respectively [5].

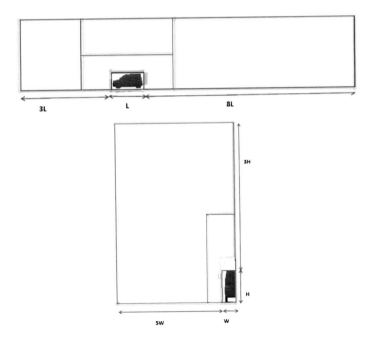

3.5 Mesh Details

Total Element Size:

Car = 2.7 million
External = 0.4 million

Type of mesh:

Tetrahedral = Car
Hexahedral = External domain
Prism cells = Near the wall region
Mesh quality = 0.3 and above

3.6 Details

Analysis Type: Steady-state analysis
Inlet: Flow speed 50 km/h
Outlet: One atmospheric pressure
Top and right side of the external domain: Opening (1 atmos)
Left side of the external domain: Symmetry boundary
Bottom side of the external domain: Wall with free slip
Car surface: Wall with no slip
Advection scheme: High resolution

3.7 Physic Setting and Solving

Physic setting of the external domain, car, and interfaces are created to ensure connectivity between the individual meshes and solver run was given.

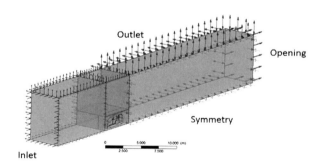

Experimental Analysis of the Pressure Distribution

3.8 Post Processing

The estimation of the C_D of the baseline model.

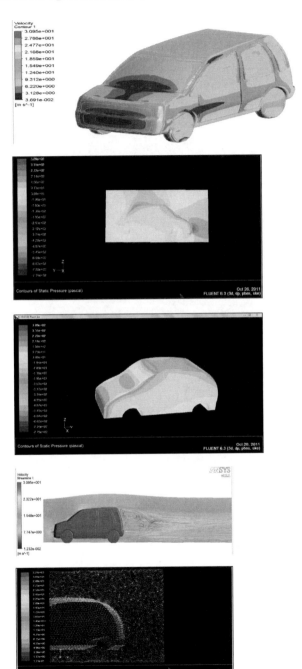

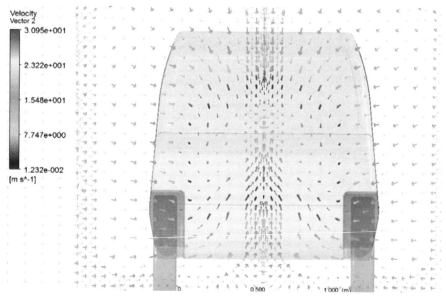

Streamline around the vehicle [11].
Observation: The pressure contours shows the stagnation regions in the bumper.
Section behinds the vehicle.

Experimental Analysis of the Pressure Distribution

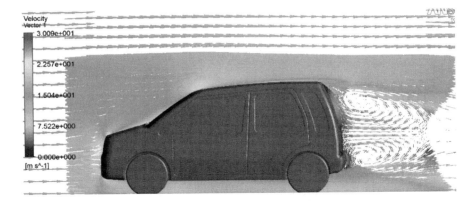

At mid-section of the vehicle.

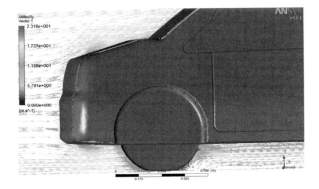

Front of the vehicle.

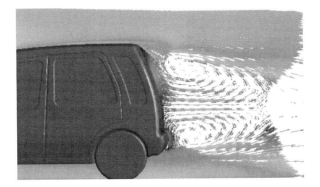

Rear of the vehicle [11].
Windshield and roof of the vehicle

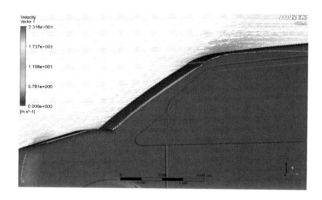

4 Drag Coefficient Calculation

The drag force obtained for the half model = 54.262 N
Drag force for full model = 2 × 54.262 N = 108.524 N
Coefficient of drag = (2 × Drag force)/($\rho V^2 A$)

Here,

ρ—Density of the air (1.105 kg/m^3)
V^2—Free stream velocity (16.67 m/s)
A—Projected area of the vehicle (2.073 m^2)
Coefficient of drag: 0.340975

4.1 Air Velocity Versus Drag Coefficient (V vs CD) [10]

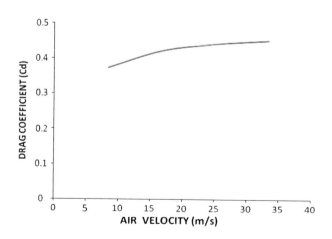

4.2 Air Velocity Versus Drag Force (V vs. CD)

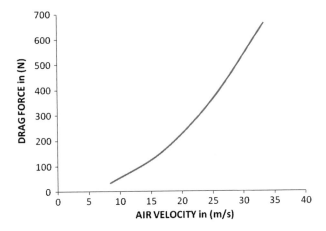

5 Experimental Method

In this approach the scaled model is instrumented with pressure tapings along centerline, over its profile, will be tested with different velocities static pressure distribution obtained will be represented in terms of a nondimensional parameter—pressure coefficient (Cp). Variation of Cp along center line at a particular air velocity and at a particular position along center line at different velocities will be plotted.

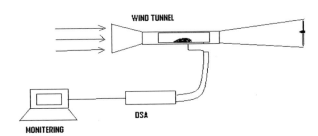

5.1 Making of Scaled Model

There are several materials have been used for making the scaled model as follows, here taken aluminum and following process will be carried out to making scaled model. CNC Machining, Holes for pressure tapings Painting and Stickering:

Experimental Analysis of the Pressure Distribution 181

5.2 Experimental Testing of Scaled Model

The scaled model with tubes is placed in the test section, the other ends of the tubes are connected with the ports available in the pressure scanner. The pressure scanner is connected with the computer, to find the pressure in the scanner and it is measured by using the valve scan software. The measurement is carried out with the different velocities of air and the pressure values are tabulated [6, 7].

Pressure distribution matches with prediction that pressures would be low in the regions with streamlined profiles such as nose, base of the windshield, etc.

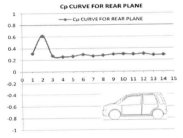

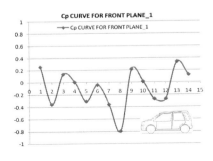

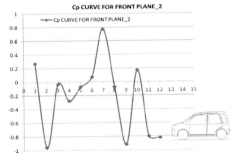

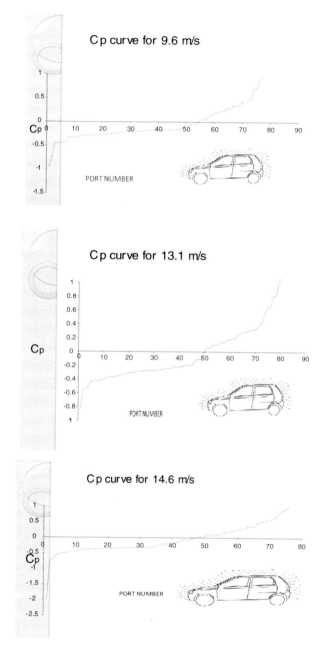

Almost identical nature of graphs of variation of Cp along car profile at different air velocities also verified that Cp is independent of speed [8, 9].

Graph shows C_D variation with air velocity of without base bleed. Pressure distribution matches with prediction that pressure would be low in the regions with streamlined profiles, such as nose, base of the windshield, etc., and almost

Experimental Analysis of the Pressure Distribution

identical nature of figs of variation of Cp along car profile at different air velocities also verified that Cp is independent of speed. Computational predictions of external aerodynamics of car showed quite a good agreement with and hence validated experimental once.

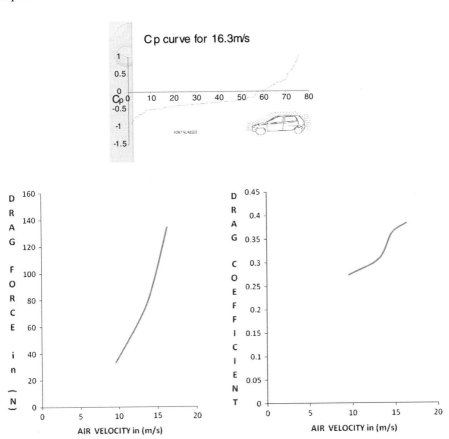

6 Conclusion

The C_D of the vehicle is around 0.34. It is observed that the Bonnet area is having the high pressure zone compared to others. The result shows the high velocity vector zone in the bonnet and the rear side. From the results we infer that the high pressure is the region were flow get staggered. A special aerodynamic attention should be given to avoid the aerodynamic grieving. The optimization, the rear body having some aerodynamic deserts like vortex generators helps to restrict the floundering of conventional vehicle, high vehicle. Literally the overall paper suggests that the aerodynamic drag reduction helps us to reduce the air load acting on the surface of the vehicle which in turn reduce the fuel consumption. The work can be elaborated when we have some stable vortex stabilizer to trick the flow not getting separated from the body.

References

1. Hucho WH Aerodynamics of road vehicles from fluid mechanics to vehicle engineering
2. Gillespie T (1997) Fundamentals of vehicle dynamics, ASME
3. Heinz H, Butterworth H (2002) Advanced vehicle technology, 2nd edn. London, p 584–634
4. Jenkins LN An experimental investigation of flow over the rear end of a notchback automobile configuration, NASA, Langley research center
5. White AS Twenty years of projects on vehicle aerodynamics, Middlesex University, Bounds Green Road, London N11 2NQ
6. Koike M, Nagayoshi T, Hamamoto N (2004) Research on aerodynamic drag reduction by vortex generators, 16:11–16
7. Ashgriz N, Mostaghimi J An introduction to computational fluid dynamics
8. Desai M, Channiwala SA, Nagarsheth HJ (2008) A comparative assessment of two experimental methods for aerodynamic performance evaluation of car, J Sci Ind Res 67:518–522
9. Hoerner SF (1958) Fluid-dynamic drag, Hoerner fluid dynamics, Brick Town, USA
10. Shibata H (1983) MMC's vehicle wind tunnel, automobile research review (JARI) 5(9):1983
11. Hucho WH (1998) Aerodynamics of road vehicles, 4th edn. SAE International

Neuro-Genetic Hybrid System for 2D Fixture Layout

M. Vasundara, K. P. Padamanaban, M. Sabareeswaran and K. Kalaivanan

Abstract Workpiece deformation may cause dimensional errors in machining. To reduce these errors, machining fixtures are used to position the workpiece accurately. Fixture layout design plays an important role to ensure sustained quality of manufacture. This paper is concerned with minimizing the workpiece deformation during machining by designing the fixture layout. The finite element method (FEM) has been used to determine the workpiece deformation. The fixture layout model has been developed using Neuro-genetic hybrid system.

Keywords Fixture layout design • Finite element method • Neuro-genetic hybrid system

1 Introduction

Quality of manufacturing is high only when the workpiece is accurately positioned. Fixtures are used to locate and constrain a workpiece during a machining operation to reduce the workpiece deformation thereby reducing the dimensional errors. The position of the locators and clamps, the number of locators and clamps, and the value of clamping force are critical in minimizing the workpiece deformation. Several approaches have been made in the design and optimization of fixture layout.

Necmittin Kaya [1] applied genetic algorithm (GA) to optimize the fixture layout. He used 3-2-1 locating scheme where the number of locators and clamps are fixed. The locator and clamp positions were considered as the design parameters for optimization. Nicholas Amaral et al. [2] optimized the support locations using

---

M. Vasundara (✉) · K. Kalaivanan
Department of Mechanical Engineering,
PSNA College of Engg. & Tech, Dindigul, Tamilnadu, India
e-mail: jvasundara@rediffmail.com

K. P. Padamanaban · M. Sabareeswaran
SBM College of Engineering & Technology, Dindigul, Tamilnadu, India

---

S. Sathiyamoorthy et al. (eds.), *Emerging Trends in Science,*
Engineering and Technology, Lecture Notes in Mechanical Engineering,
DOI: 10.1007/978-81-322-1007-8_16, © Springer India 2012

finite element analysis (FEA) after analyzing modular fixture–workpiece deformation. Shawki and Abdel-Aal [3], Hockenberger and De Meter [4], Hockenberger and De Meter [5], Hurtado and Melkote [6, 7] have examined the role of elastic fixture–workpiece contact point deformation, micro-slip and slip during both clamping and machining. Prabhaharan et al. [8] optimized the fixture layout to minimize the dimensional and form errors. FEM was used to predict the workpiece deformation. GA and ant colony algorithm (ACA) were adopted separately for optimization. It was concluded that ACA solutions are better than that of Kashyap and Devries [9] developed an optimization algorithm to minimize workpiece deformation due to machining loads about fixture locations. They used FEA to simulate the deformation at selected locations. Padmanaban et al. [10] optimized the machining fixture layout to minimize the workpiece elastic deformation by applying (ACA) based discrete and continuous optimization methods. The dynamic response of the workpiece with respect to machining and clamping forces was determined using FEM. Weifang Chen et al. [11] established a multi-objective optimization model to minimize the deformation and improve the uniform distribution of deformation by considering the friction and chip removal effects. The optimization process was performed through the integration of GA and FEM using clamping forces as design variables. Li and Melkote [12] developed an algorithm to optimize fixture layout and clamping force considering the workpiece dynamics while machining. Mohsen Hamedi [13] proposed a hybrid learning system of artificial neural network (ANN) and GA in the design of machining fixture layout. The clamping forces were optimized to minimize the deformation during machining. FEA was used to find workpiece deformation under cutting and clamping forces.

In this paper, 2D workpiece geometry [1] is considered. The workpiece deformation has been determined using FEM. The design variables for the system are the position of the locators and clamps. Each and every possible fixture layout has the position of three locators and two clamps. The values of all the five design variables are supposed to lie within a specified range. To develop a fixture layout model, a neuro-genetic hybrid system has been employed. The proposed system involves both ANN and GA where GA is used to calculate the weights for ANN. The workpiece deformation is predicted using the proposed system and also using ANN alone. By comparing the predicted results of ANN and neuro-genetic hybrid system, it has been found that the neuro-genetic hybrid system gives better results.

2 An Overview of Ann, GA and Neuro-Genetic Hybrid System

2.1 Artificial Neural Networks

Neural network is an interconnected assembly of simple processing elements, units or nodes, whose functionality is loosely based on the animal neuron. The processing ability of neuron is stored in the inter unit connection strength, or weights. The neurons are grouped into different layers:

- Input layer-Receives input from the external environment;
- Output layer-Communicates the output of the system to the user or external environment;
- Hidden layers-Layers between the input and output layers.

ANN solves problems by learning the relationship between inputs and outputs. The widely used learning algorithm is backpropagation network (BPN) which is a gradient descent technique with backward error propagation.

2.2 Genetic Algorithm

Genetic algorithms are computerized search and optimization algorithms based on the mechanics of natural genetics and natural selection. GA consists of the following steps:

a. A way of coding the solutions to the problem.
b. A fitness function which gives a rating for each chromosome.
c. Initializing the population of chromosomes.
d. Operators (crossover and mutation) to be applied to the parents when they reproduce to alter their genetic composition.

The reproduction process will continually improve the population, converging finally on solutions close to a global optimum.

2.3 Neuro-Genetic Hybrid System

In the neuro-genetic hybrid system, the weights of a multilayer feed forward network are determined using GA. ANN needs sufficient training to learn the input—output relationship whereas overtraining the network may lead to undesired effects. The weights are determined based on a gradient search technique and hence there is a risk of encountering the local minimum problem. GA has been found to be good at finding acceptably good solutions to problems acceptably quickly. By hybridizing BPN and GA, the necessary weights can be found in order to enhance the speed of training.

3 Illustration

The case study [1] considered in this paper is to demonstrate the neuro-genetic hybrid system based fixture layout method. The two dimensional workpiece geometry, in which the end milling operation is performed, is shown in Fig. 1. The

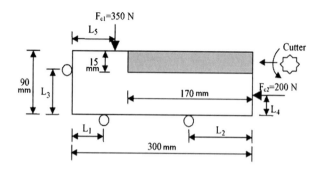

Fig. 1 2D workpiece geometry with force system

workpiece-fixture system consists of three locators L_1, L_2, L_3 and two clamps L_4 and L_5. The machining forces for performing end milling operation are 100 N ($\leftarrow$) and 286 N ($\downarrow$). The clamping forces acting at clamp L_4 and L_5 are 200 N and 350 N respectively.

The workpiece elastic deformation is obtained using FEA for different positions of all the fixturing elements and they are used as input for ANN training. The range of values for the design variables are as follows:

5 mm < L_1 < 148 mm
5 mm < L_2 < 148 mm
5 mm < $L3$ < 85 mm
5 mm < L_4 < 65 mm
5 mm < L_5 < 125 mm

4 Neuro-Genetic Hybrid System for Fixture Layout Design

A BPN is considered with one input layer consisting of five nodes, one hidden layer of five nodes and one output layer with one node. The BPN is trained with 50 sets of input and output data. Then the network is tested with 10 sets of input data. The output and the error were noted down. Next, GA is used to find the optimum weights of BPN. These optimum weights are applied to BPN to infer the output i e. In the testing phase, the inputs are fed to the BPN with the optimized weights to determine the output. The output and error of neuro-genetic hybrid system are compared with that of BPN. The parameters used to train and infer the neuro-genetic hybrid system are given in Table 1.

4.1 Coding

The parameters representing the solution of a problem are called as genes. These genes are joined together in the form of string of values referred as a chromosome. For a BPN

Neuro-Genetic Hybrid System for 2D Fixture Layout

Table 1 Parameters for neuro-genetic hybrid system

Parameter	Value
Transfer function of neurons	Tan sigmoidal
Learning co-efficient	0.45
Momentum factor	0.55
Encoding	Real
Chromosome length	150
Population size	100
Weight initialization routine	Random
Stopping criterion	Max. iteration = 3,000
Fitness normalization	Rank
Selection operation	Rank
Crossover probability (two-point)	0.9
Mutation probability	0.01

with l-m-n network configuration i.e., l input neurons, m hidden neurons and n output neurons, the number of weights to be determined is $(l + n)$ m. If the gene length (the number of digits in the weight) is assumed to be d, then the chromosome (string S) length is $L = (l + n)$ md. In the problem considered, there are five input nodes, five hidden nodes and one output node. Therefore, the total number of weights to be calculated is 30. Each weight is represented by five digits so that the length of the chromosome is 150. An initial population of 100 chromosomes is generated randomly.

4.2 Weight Extraction

The weights are extracted from each of chromosomes to determine the fitness values for each of the chromosomes. The weight extraction is performed using the following equations:

$$
W_k = \begin{cases} +\dfrac{X_{kd}10^{d-2} + X_{kd+3}10^{d-3} + \ldots\ldots X_{(kd+1)}}{10^{d-2}}, if\, 5 \le X_{kd+1} \le 9 \\[2ex] -\dfrac{X_{kd}10^{d-2} + X_{kd+3}10^{d-3} + \ldots\ldots X_{(kd+1)}}{10^{d-2}}, if\, 0 \le X_{kd+1} \le 5 \end{cases} \tag{1}
$$

where W_k is the actual weight, $X_1, X_2, \ldots\ldots, X_d$ represent a chromosome and $X_{kd+1}, X_{kd+2}, \ldots\ldots, X_{(k+1)d}$ represent the kth gene in the chromosome.

4.3 Fitness Function

A fitness function must be calculated for each problem to be solved. It is determined using the root mean square of the errors. The root mean square of the error is

$$
E = \sqrt{\sum E_i / N} \tag{2}
$$

where E_i is the error value for the sets, $i = 1, 2, 3, \ldots\ldots, N$ is the number of sets of input–output pairs for the problem.

The fitness value for each of the individual chromosome is

$$F_i = 1/E \qquad (3)$$

4.4 Reproduction

Reproduction is the selection operator applied on the population to select the chromosomes. The selected chromosomes act as parents to crossover and produce offspring. Before reproduction, a mating pool is formed with good strings from the population. In the mating pool, the chromosome with least fitness is excluded and replaced with a duplicate copy of the chromosome with highest fitness value. After forming the mating pool, the parents are selected in pairs at random to apply other operators.

4.5 Crossover

Crossover operator is applied to the mating pool to create a new better string from the good strings produced by reproduction. Two parent chromosomes are selected randomly to create a new better chromosome. Here a two-point crossover operator has been used where two random positions are chosen and the strings are exchanged between these two positions to create an offspring.

4.6 Mutation

The strings are subjected to mutation after performing crossover. Mutation of a bit involves flipping it, changing 0–1 and vice versa with a mutation probability P_m. A number between 0 and 1 is selected at random. If the random number is smaller than P_m, then the outcome is true or otherwise false. The bit is altered if the outcome is true; otherwise the bit is kept unaltered. The best individuals (with high fitness value) are selected from the existing population to form a new generation of possible solutions to the problem. The new generation formed contains better characteristics than their parents. Proceeding in this way, after many generations, the entire population inherits the best characteristics due to exchange of good characteristics. The final population gives the best fit solutions to the problem.

5 Flow-Chart of Neuro-Genetic Hybrid System

See Figure 2.

Fig. 2 Flow-chart of neuro-genetic hybrid system

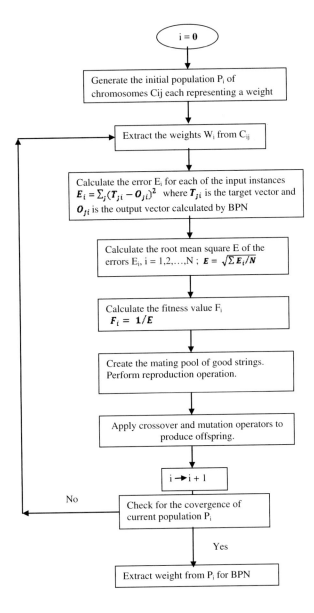

6 Discussion of the Results

First, the actual output which is the workpiece deformation for workpiece-fixture layout system is calculated using FEA for 50 random sets of inputs that lie within a specified range. These 50 layouts with inputs and outputs are used to train the BPN. Then the trained BPN is tested with 10 sets of inputs alone. The output and error are noted and plotted in a graph as shown in Fig. 3.

Next, GA is used to find the optimum weights of BPN for the fixture layout. These optimum weights are applied to BPN to infer the output. The optimized weights are fed to the BPN for the same 10 testing layouts to determine the output.

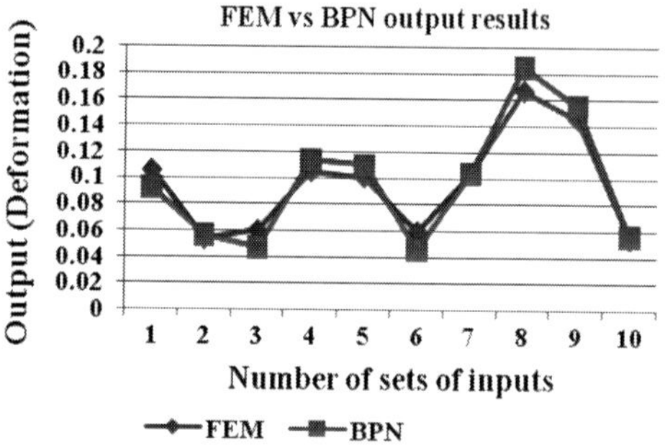

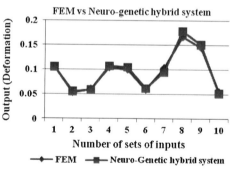

Fig. 3 Comparison of FEM and BPN output results

Fig. 4 Comparison of FEM and neuro-genetic hybrid system output results

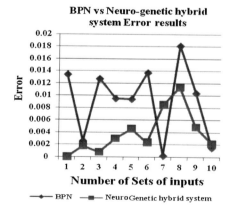

Fig. 5 Comparison of BPN and neuro-genetic hybrid system error results

The output and error of neuro-genetic hybrid system are noted and plotted in a graph as shown in Fig. 4.

From the above graphs, it is shown that the neuro-genetic hybrid system gives outputs that are much closer to the simulated outputs when compared to that of BPN. The comparison of error results for both BPN and neuro-genetic hybrid system is shown in Fig. 5.

7 Conclusion

All fields of human activities involve a number of real time problems that demand decision-making features. Soft computing techniques like GA provide good results. It can be used with other soft computing system to solve the non-linearity. In this work, GA is used as a complementary tool for ANN to find the optimum weights for the network. Using neuro-genetic hybrid system provides better results in lesser time. It has been observed from the graphs that the neuro-genetic hybrid system provides more accurate results.

References

1. Kaya N (2005) Machining fixture locating and clamping position optimization using genetic algorithms. Int J Comput Ind 57:112–120
2. Amaral N, Rencis JJ, Rong Y (2005) Development of a finite element analysis tool for fixture design integrity verification and optimization. Int J Adv Manuf Technol 25:409–419
3. Shawki GS, Abdel-Aal MM (1966) Rigidity considerations in fixture design- contact rigidity at locating elements. Int J Mach Tool Design Res 6:31–43
4. Hockenberger MJ, De Meter EC (1996) The application of meta functions to the quasi-static analysis of workpiece
5. Hockenberger MJ, De Meter EC (1997) The impact of rigid body workpiece displacement on the geometric errors of milled surfaces. Int J Comput Appl Technol 10(3–4):170–182
6. Hurtado J, Melkote SN (1998) A model for the prediction of reaction forces in a 3–2–1 machining fixture. Trans NAMRI SME 26:335–340
7. Hurtado J, Melkote SN (1997) An experimental study of the effects of fixture and work-piece variables on workpiece–fixture contact forces in end milling. ASME J Manuf Sci Eng 2:237–243
8. Prabhakaran G, Padmanaban KP, Krishnakumar R (2007) Machining fixture layout optimization using FEM and evolutionary techniques. Int J Adv Manuf Technol 32:1090–1103
9. Kashyap S, DeVries WR (1999) Finite element analysis and optimization in fixture design. Struct Optimi 18:193–201
10. Padmanaban KP, Arulshri KP, Prabhakaran G (2009) Machining fixture layout design using ant colony algorithm based continuous optimization method. Int J Adv Manuf Technol 45:922–934
11. Chen W, Ni L, Xue J (2007) Deformation control through fixture layout design and clamping force optimization. Int J Adv Manuf Technol
12. Li B, Melkote SN (2001) Optimal fixture design accounting for the effect of workpiece dynamics. Int J Adv Manuf Technol 18:701–707
13. Hamedi M (2005) Intelligent fixture design through a hybrid system of artificial neural network and genetic algorithm. Artif Intell Rev 23:295–311

Part III
Materials and Machining Processes

Basic Properties and Performance of Vegetable Oil-Based Boric Acid Nanofluids in Machining

P. Vamsi Krishna, R. R. Srikant, R. Padmini and Bharat Parakh

Abstract Machining is one of the most basic and essential processes in manufacturing industry. Quality of the workpiece essentially depends on the heat generated in the cutting zone during machining. The usage of cutting fluids poses a question of risk to ecology and health of workers, in spite of their extensive use in dissipating the heat. This paves search for user-friendly and eco-friendly alternatives to conventional cutting fluids. Keeping this aspect in perspective, the present work deals with the application of nano solid lubricant suspensions in lubricating oil in turning of AISI 1,040 steel using carbide tool. Boric acid solid lubricant suspensions of 100 nm particle size are added to coconut oil. Particle size of nano-boric aicd is confirmed through particle analyzer. XRD analysis is done to assess the purity of the sample. On weight-percentage basis boric acid nanoparticles are added to the oil. The variation of its basic properties like flash point, fire point, viscosity, thermal conductivity, and heat transfer coefficient is evaluated. Cutting tool temperatures, average tool flank wear, and the surface roughness of the machined surface is examined in machining.

Keywords Nanolubricant · Solid lubricant · Boric acid · Machining

---

P. V. Krishna (✉) · R. Padmini · B. Parakh
Department of IPE, GITAM University, Visakhapatnam, India
e-mail: vamsikrishna@gitam.in

R. Padmini
e-mail: Padhmini.r@gmail.com

B. Parakh
e-mail: bharatparakh@gmail.com

R. R. Srikant
Department of Mechanical Engg, GITAM University, Visakhapatnam, India
e-mail: srikant_revuru@yahoo.com

S. Sathiyamoorthy et al. (eds.), *Emerging Trends in Science,*
Engineering and Technology, Lecture Notes in Mechanical Engineering,
DOI: 10.1007/978-81-322-1007-8_17, © Springer India 2012

1 Introduction

During machining, friction between the tool and workpiece give rise to high temperatures, which influences dimensional accuracy and surface quality of the workpiece. Cutting fluids play the dual role of lubricants and coolants by reducing the friction between tool-workpiece, tool-chip interface. Environmental pollution, dermatitis to operators, soil contamination during disposal, and water pollution [1, 2] are the adverse effects due to application of conventional cutting fluids. Besides this, cutting fluids also incur a major portion of the total manufacturing cost. These factors have prompted investigations to probe into the use of eco-friendly coolants or coolant free machining. Hence, researchers have started to experiment with dry machining, coated tools, cryogenic cooling, vegetable oils, minimum quantity lubrication (MQL), and solid lubricants as alternatives to conventional cutting fluids.

MQL is a promising and reliable technique adopted by the researchers which results in reduction in tool wear, improvement in dimensional accuracy and surface finish [3]. Comparing MQL with dry and wet machining, performance improved in terms of cutting forces, tool life, cutting temperature, and surface finish [3–6]. Effective implementation of MQL requires fluids with high thermal conductivity. Due to this reason nanofluids have gained ample importance. The utilization of vegetable oil in metalworking applications may reduce problems faced by workers, such as inhalation of toxic mist in the work environment and skin cancer, owing to their higher biodegradability and lower environmental impact.

2 Literature Review

Many researchers have been working to achieve eco-friendly sustainable manufacturing. As an alternative to cutting fluids, solid lubricants like MoS_2, graphite, etc., were used. Shaji and Radhakrishnan [7] investigated the effect of graphite in surface grinding and improvement in surface finish is observed. In another study [8, 9], they reported improvement in process using graphite, CaF_2, BaF_2, and MoS_2 in grinding. Then investigations on solid lubricant molded grinding wheels were tried [10]. The normal force and tangential force components due to the frictional effects are reduced due to the effective lubrication by lubricant molded wheels. Venugopal and Rao [11]outlined the surface finish improvement with the application of graphite in grinding Jianhua et al. [12] studied the friction coefficient at the tool-chip interface in dry cutting of hardened steel and cast iron with an $Al_2O_3/TiC/CaF_2$ ceramic tool and reported reduction in friction coefficient by adding CaF_2 solid lubricant. An attempt was made to use graphite as a solid lubricant to reduce the heat generation at the milling zone [13], resulting in remarkable reduction of cutting force, specific energy, and surface roughness.

Reddy and Rao [14] reported that graphite and MoS_2 assisted end milling process showed considerable improvement compared to machining with a cutting fluid in terms of cutting forces, surface quality and specific energy. Rao and Singh [15] studied the use of solid lubricants during hard turning while machining bearing steel with mixed ceramic inserts at different cutting conditions and tool geometry. Results showed 8–15 % improvement in the surface finish with the use of solid lubricants compared to dry hard turning. Krishna and Rao [16, 17] investigated the performance of boric acid in machining experimentally. Machining performance is improved with reduced particle size while using dry solid lubricants. Then, a 50 µm particle size boric acid suspension in SAE-40 oil is tested and performance improvement is observed compared to conventional cutting fluids and dry machining. Cutting forces, cutting temperatures, and tool flank wear were reduced and surface finish improved.

Ioan et al. [18] presented the first experimental results on lubricating capacity of rapeseed oil compared to mineral oil. Belluco and De Chiffre [19] made an investigation on the effect of new formulations of vegetable oils on surface integrity and part accuracy in reaming and tapping operations with AISI 316 L stainless steel. All vegetable-based oils produced better results than the commercially available mineral oil in terms of tool life improvement and reduction in thrust force. Skerlos and Hayes [20] studied canola, soybean, and rapeseed vegetable oil as cutting fluids and reported that performance of vegetable based cutting fluids is comparable or better than that of traditional petroleum based metal working fluids. Jayadas and Prabhakaran [21] compared the cooling behavior, thermal and oxidative stabilities of coconut oil with sesame oil, sunflower oil, and a mineral oil (Grade 2T oil). Coconut oil showed better oxidative stability in comparison to other vegetable oils, comparatively lesser weight gain under oxidative environment because it has very high pour point (23–25 °C) due to the predominantly saturated nature of its fatty acid constituents precluding its use as base oil for lubricant in temperate and cold climatic conditions.

Kulkarni et al. [22] investigated the properties of nanocoolants in another application than machining. Different properties of nanofluids like specific heat and thermal conductivity are calculated as given below:

$$k_{\text{nf}} = k_f \left[\frac{k_p + (n-1)k_f - (n-1)\phi(k_f - k_p)}{k_p + (n-1)k_f + \phi(k_f - k_p)} \right] \tag{1}$$

where k_{nf} is the nanofluid thermal conductivity, k_f is the base fluid thermal conductivity, k_p is the bulk solid particle thermal conductivity, φ is the particle volume fraction, and n is an empirical scaling factor that takes into account how different particle shapes affect thermal conductivity. The effective density of nanofluids is given by:

$$\rho_{\text{nf}} = (1 - \phi)\rho_f + \phi\rho_s \tag{2}$$

where ρ_{nf} is the nanofluid density and ρ_s and ρ_f are the densities of the solid particles and base fluid, respectively.

The specific heat of nanofluids, C_{pnf}, can be calculated using the standard equation based on the volume fraction.

$$C_{pnf} = \phi C_{ps} + (1 - \phi)C_{pf} \tag{3}$$

where C_{ps} is specific heat of solid particles and C_{pf} is specific heat of base fluid.

With the calculated properties of nanofluids, the heat transfer coefficients were obtained. It was concluded that nanofluids have high thermal conductivities and heat transfer rates compared to the conventional fluids.

Based on the available literature, it can be concluded that suspensions of boric acid particles in vegetable or other lubricating oils, provide better lubrication compared to the conventional fluids. Further, reduced particle size provides better lubricating action. By combining the advantages of solid lubricants, nanoparticles, and MQL, an attempt is made in the present work to investigate the affect of nano solid lubricants in turning under MQL. Boric acid particles of 50 nm particle size are used as suspensions in coconut oil and machining was carried out with varying proportions of solid lubricant suspensions i.e., 0.25, 0.5, 0.75, and 1 % by weight. Influence of solid lubricant to oil proportion on cutting temperatures, tool flank wear, and surface roughness was studied with respect to cutting conditions.

3 Experimentation

Experiments are conducted to evaluate the performance of nano-boric aicd suspension in coconut oil during turning. All the experiments are conducted three times and average value is taken as response value. Boric acid particles of 100 nm particle size are obtained through mechanical milling with high energy ball mill. XRD analysis is done to assess the purity of the sample and particle size is confirmed through particle size analyzer. Manual mixing of solid lubricant particles of 100 nm size in coconut oil in different weight proportions at room temperature is done, followed by mixing with a sonicator for 1 h.

These suspensions along with pure coconut oil are used as lubricants in machining. Flash and fire points were measured by standard method using a Cleveland open cup flash and fire point apparatus. The viscosity of the nano-lubricant is measured using a Redwood Viscometer. Thermal conductivity, specific heat and heat transfer coefficient of nanolubricants are calculated from Eqs. 1–3 and presented in Table 1 to understand their performance. 'n' in Eq. 1, is taken as 3, a value typical for spherical nanoparticles [19]. For calculation of heat transfer coefficient, Nusselt number, Nu, is obtained from the Hilpert equation for flow over cylinders, due to its analogy with turning process.

$$Nu = h \times D/k_{nf} = C. Re^{m} Pr^{1/3} \tag{4}$$

where, Re and Pr are Reynold's number and Prandtl number respectively, h is the heat transfer coefficient, D is the diameter of the workpiece and C and m are constants that depend on the value of Re [23].

Basic Properties and Performance of Vegetable Oil-Based Boric Acid

Table 1 Properties of nanofluids

(a) Fire point and flash point, ^{0}C

Percentage of nano boric acid suspensions	0%	0.25%	0.5%	0.75%	1%
Flash point	220	245	266	270	310
Fire point	250	280	282	290	327

(b) Kinematic viscosity, st (cm^2/sec)

Temperature (°C)	Percentage of nano boric acid suspensions				
	0%	0.25%	0.50%	0.75%	1%
30	0.310	0.350	0.522	0.574	0.608
35	0.207	0.227	0.326	0.374	0.431
40	0.169	0.190	0.245	0.278	0.314
45	0.085	0.126	0.152	0.173	0.237
50	0.039	0.092	0.119	0.143	0.161

(c) Thermal Conductivity, kW/m-K

Percentage of nano boric acid suspensions	0%	0.25%	0.5%	0.75%	1%
Thermal conductivity	0.5	0.5004	0.5009	0.5014	0.5018

(d) Specific heat, J/kg-K

Percentage of nano boric acid suspensions	0%	0.25%	0.5%	0.75%	1%
Specific heat	3500	3494.5	3489.09	3483.6	3478.2

(e) Heat transfer coefficient, W/m^2-K

Percentage of nano boric acid suspensions		0%	0.25%	0.5%	0.75%	1%
Heat transfer coefficient	60 m/min	662.38	662.69	663.0	663.32	663.61
	80 m/min	727.75	728.09	728.42	728.78	729.09
	100 m/min	782.97	783.34	783.7	784.04	784.42

All the cutting tests are performed on PSG-124 lathe with cemented carbide tool (SNMG 120408) and heat treated AISI 1,040 steel of 30 ± 2 HRC workpiece. The temperature is sensed by the embedded thermocouple by placing at the bottom of the tool insert in the tool holder. The temperature measured by the thermocouple is only a representative figure for comparison purpose as this does not measure the cutting zone temperature. Calibration of the thermocouple is carried out in a water bath with a thermometer and a maximum of 2 °C difference is noted over a range from 40 °C to 95 °C. Cutting tool is analyzed under an optical projector to measure tool flank wear. The obtained tool profiles are compared with the virgin tool profile and flank wear was determined. Talysurf with stylus radius 0.0025 mm and cut-off length 0.8 mm is employed for measuring average surface roughness (R_a). An average of three measurements is taken as a response value. The experimental setup is developed for liquid lubricant supply at the machining zone (Fig. 1). Lubricant oil with solid lubricant suspensions is stored in tank and placed above the axis of machining. Lubricant storage tank is open to atmosphere; hence flow of lubricant is due to its self weight and atmospheric pressure. Flow rate of lubricant mixture is controlled by a regulating valve. Initially lubricant mixture is collected in a vessel at different positions of the valve and flow rate is calibrated by measuring the volume of the lubricant collected in certain amount of time. In the trial tests it is observed that 10 ml/min flow rate is sufficient for selected cutting conditions and tool-work piece combination. Hence this flow rate is kept constant. After ensuring the flow rate of 10 ml/min, experiments are conducted.

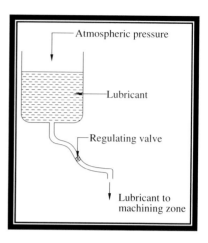

Fig. 1 Lubricant supply system for nanofluid

4 Results and Discussion

4.1 XRD and Particle Size

In the present work nano-boric aicd particles are investigated and characterized by X-Ray Diffraction technique which reveals their material property. XRD analysis confirms the purity of the nano-boric acid. Presence of single peak in XRD pattern symbolizes the purity of the sample as shown in Fig. 2. Multiple peaks in XRD graph indicate the presence of other materials in the sample considered. The purity of the sample is thus confirmed through XRD analysis. Particle size analyzer is used to determine the domain size of nano engineered boric acid particles. The particle size distribution is shown in Fig. 3. The average size of nano-boric aicd particles was found to be 100.

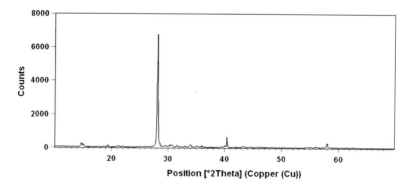

Fig. 2 XRD pattern of nano-boric acid particles

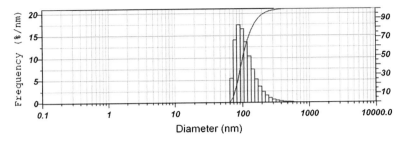

Fig. 3 Average particle size of ball milled boric acid particles

4.2 Basic Properties

Table 1a shows that both flash point and fire point are increased with addition of boric acid suspensions in coconut oil. As expected, the viscosities generally decrease with the increase in temperature. From the experiments it is observed that the kinematic viscosity increased to some extent by the addition of nanoparticles (Table 1b). This is because of the combined effect of coconut oil and solid lubricant powder.

Thermal conductivity of nanolubricants increased slightly with percentage increase of nanoparticles compared to base oil (Table 1c), whereas, from Table 1d it is observed that the specific heat decreased with increase in nanoparticles percentage in base oil. Heat transfer coefficient increased slightly with percentage increase of nanoparticles in base oil at specific cutting speed (Table 1e). However, significant improvement is observed in heat transfer coefficient with increase in cutting speed at particular quantity of nanoparticle suspensions.

4.3 Cutting Temperatures

Variation of cutting temperatures with cutting speed is presented in Fig. 4. Cutting temperatures increased with cutting speed irrespective of the lubricant. The high lubricating property of coconut oil is due to by the fundamental composition of the vegetable oil molecules as well as the chemical structure of oil itself. Lubricity of vegetable oils arises from the oiliness property of its constituents; and its properties are the result of long, heavy, and dipolar molecules. The polar heads of the molecules have great chemical affinity for metal surfaces and attach themselves to the metal like magnets. The result is a dense, homogenous alignment of vegetable oil molecules, perpendicular along the metal surface that creates a thick, strong, and durable film layer of lubricant. Also, at elevated temperatures solid lubricant softens and forms a film, more over in nano level these solid lubricant particles increases the heat transfer capacity of the lubricating oil. This combined effect

Fig. 4 Variation of cutting temperatures with cutting speed (feed = 0.2 mm/rev, d.o.c = 1 mm, time = 5 min)

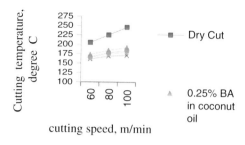

of coconut oil and nano solid lubricant particles is reason for reducing the cutting tool temperatures. It can be seen that though the heat transfer coefficients are not very high for suspensions, compared to the respective base oils, cutting temperatures reduced significantly. This may be due to reduced friction by the use of nano-boric acid suspensions compared to the base oils. Among the coconut oil, lubricating oil with 0.5 % nano-boric acid particle suspensions performed well. This may be because, 0.25 % boric acid cannot provide the adequate lubricating effect compared to 0.5, 0.75, and 1 % inclusions may reduces the flowability of the lubricant and prevents it from entering the cutting zone, thus decreasing its effectiveness.

4.4 Tool Flank Wear

Tool flank wear measured at different lubricating conditions with cutting speed is shown in Fig. 5. Flank wear increased gradually with increase in speed. During machining, heat is generated at the primary deformation zone and secondary deformation zone, and induces high cutting temperatures. Under such high cutting temperatures, the solid lubricant creates a thin lubricating film on the workpiece and tool. The particles of solid lubricant flow at the interface with the oil and decrease the plastic contacts, leading to reduction of flank wear. Low coefficient of friction, sliding action, and low shear resistance within the contact interface reduce flank wear. The combined effect of solid lubricant and vegetable oil, as explained above, leads to reduction in flank wear with 0.5 % nano-boric acid

Fig. 5 Variation of tool flank wear with cutting speed (feed = 0.2 mm/rev, d.o.c = 1 mm, time = 5 min)

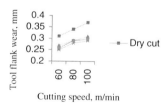

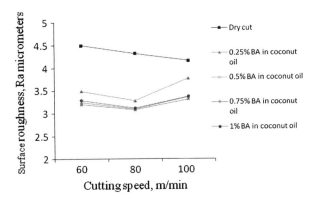

Fig. 6 Variation of surface roughness with cutting speed (feed = 0.2 mm/rev, d.o.c = 1 mm, time = 5 min)

particles suspensions in coconut oil compared to remaining conditions, due to similar reasoning as in case of cutting temperatures.

4.5 Surface Roughness

Surface roughness initially reduced and then increased with increase in cutting speed at all the lubricating conditions (Fig. 6). Among the selected lubricating conditions, surface roughness reduced with coconut oil with 0.5 % nano-boric acid suspensions. This is because of the lubricating action of the solid lubricant and coconut oil. The reduction in surface roughness in case of nano-boric acid suspensions in coconut oil may be attributed to its better lubricating action, which reduced the frictional forces between the tool and workpiece there by reducing the temperatures developed and ultimately preventing tool wear, thus prolonging tool life, resulting in surface quality improvement.

5 Conclusion

XRD pattern of nano engineered boric acid particles reveals the nanostructure of the sample taken, particle size analyzer reflects the presence of nano sized particles. Nano-boric acid suspensions in coconut oil provide adequate lubrication in machining. Flash and fire points increased with increase of nano-boric acid content. Kinematic viscosity increased with increase in nano-boric acid content. Thermal conductivity of nanolubricants increased slightly with percentage increase of nanoparticles. Heat transfer coefficient increased with addition of nano-boric acid. Specific heat decreased with addition of nano-boric acid. Compared to 0.25, 0.5, 0.75, and 1 % nano-boric acid inclusions, 0.5 % showed better performance in terms of machining parameters like cutting temperatures, tool wear, and surface roughness.

References

1. Howes TD, Ktoenshoff H, Heuer W (1991) Environmental aspects of grinding fluids. Ann CIRP 40(2):623–629
2. Byrne G, Scholta E (1993) Environmentally clean machining processes- a strategic approach. Ann CIRP 42(1):471–474
3. Khan MMA, Dhar NR (2006) Performance evaluation of minimum quantity lubrication by vegetable oil in terms of cutting force cutting zone temperature tool wear job dimension and surface finish in turning AISI-1060 steel. J Zhejiang Univ Sci 7(11):1790–1799
4. Bagchi H, Mukharjee NP, Basu SK (1972) Investigation of MEtal cutting using molybdenum disulphide as a cutting Fluid. Ind lubr Tribol 239–243
5. Lathkar GS, Basu SK (2000) Clean metal cutting process using solid lubricants. In: Proc. 19th AIMTDR Conf, IIT Madras, Narosa Pub, 15–31
6. Varadarajan AS, Philip PK, Ramamoorthy B (2002) Investigations on hard turning with minimal cutting fluid application and its comparison with dry and wet turning. Int J Mach Tools Manuf 42(2):193–200
7. Shaji S, Radhakrishnan V (2002) An investigation of surface grinding using graphite as lubricant. Int J Mach tools Manuf 42 (6):733–740
8. Shaji S, Radhakrishnan V (2002) Investigations on the application of solid lubricants in grinding. Proc Inst Mech Eng B J Eng Manuf 216 (10):1325–1343
9. Shaji S, Radhakrishnan V (2003) Analysis of process parameters in surface grinding with graphite as lubricant based on the taguchi method. J Mater Proc Technol 141(1):51–59
10. Shaji S, Radhakrishnan V (2003) An investigation on solid lubricant moulded grinding wheels. Int J Mach Tools Manuf 43:965–972
11. Venugopal A, Rao PV (2004) Performance improvement of grinding of SiC using graphite as a solid lubricant. Mater Manuf proc 19 (2):177–186
12. Deng J, Cao T, Yang X, Lin J (2005) Self-lubrication of sintered ceramic tools with CaF_2 additions in dry cutting. Int J Mach Tools Manuf 45:1–7
13. Suresh Kumar N, Reddy P, Rao V (2005) Performance improvement of end milling using graphite as a solid Lubricant. Mater Manuf Proc 20(4):673–686
14. Kumar Reddy NS, Rao PV (2006) Experimental investigation to study effect of solid lubricants on cutting forces and surface quality in end milling. Int J Mach Tools Manuf 46:189–198
15. Dilbag Singh, Rao PV (2007) Performance improvement of hard turning with solid lubricants. Int J Adv Manuf Technol 32:170–177
16. Vamsi Krishna P, Nageswara Rao D (2008) The influence of solid lubricant particle size on machining parameters in turning. Int J Mach Tools Manuf 48 (1):107–111
17. Vamsi Krishna P, Nageswara Rao D (2008) Performance evaluation of solid lubricants in terms of machining parameters in Turning. Int J Mach Tools Manuf 48(10):1131–1137
18. Ioan IS0, Camelia C, George C (2002) On the future of biodegradable vegetable lubricants used for industrial trybosystems. Ann Univ "Dunarea De Jos" of Galati Fascicle 8:1221–4590
19. Belluco W, De Chiffre L (2002) Surface integrity and part accuracy in reaming and tapping stainless steel with new vegetable based cutting oils. Tribol Int 35:865–870
20. Skerlos S, Hayes K Vegetable based cutting fluid—A new dimension in research. http://www.engn.umich.edu/alumni/engineer/03FW/research/holography/index.html
21. Jayadas NH, Prabhakaran Nair K (2006) Coconut oil as base oil for industrial lubricants-evaluation and modification of thermal, oxidative and low temp properties. Tribol Int 39:873–878
22. Kulkarni Devdatta P, Namburu Praveen K, Das Debendra K (2007) Comparison of heat transfer rates of different nanofluids on the basis of the mouromtseff number. Electr Cooling 13(3):28–32
23. William SJ (2000) Engineering Heat Transfer, 2nd edition. CRC Press LLC, Boca Raton

Cuckoo Search Algorithm for Optimization of Sequence in PCB Holes Drilling Process

Wei Chen Esmonde Lim, G. Kanagaraj and S. G. Ponnambalam

Abstract Optimization of drill path can lead to significant reduction in machining time which directly improves productivity of manufacturing systems. Most electronic manufacturing industries use computer numerical controlled machines for drilling holes on printed circuit board. To increase PCB manufacturing productivity, a good option is to minimize the drill path route using an optimization algorithm. In order to find the best sequence of operations that achieve the shortest drill path, Cuckoo search algorithm is proposed. The performance of the proposed algorithm is tested and verified with two case studies from the literature. The computational experience conducted in this research indicates that the proposed algorithm is capable to efficiently find the optimal route for PCB holes drilling process.

Keywords Drill path optimization • PCB holes drilling process • Cuckoo search algorithm

1 Introduction

Nature inspired meta-heuristic algorithms are pioneered in solving the problems of the modern global optimization, most notably the NP-hard optimization that includes the traveling sales man problem [1, 2]. The power and beauty of modern meta-heuristics comes from the capability of emulating the best feature in nature, specifically biological systems evolved from natural selection over millions of years via two important characteristics, which are selection of the fittest, and adaptation to the environment. Statistically, these features can be interpreted into

---

W. C. E. Lim (✉) · G. Kanagaraj · S. G. Ponnambalam
School of Engineering, Monash Univeristy Sunway Campus, Bandar Sunway, Malaysia
e-mail: wclim17@student.monash.edu

G. Kanagaraj
e-mail: kanagaraj.ganesan@monash.edu

S. G. Ponnambalam
e-mail: sgponnambalam@monash.edu

S. Sathiyamoorthy et al. (eds.), *Emerging Trends in Science,*
Engineering and Technology, Lecture Notes in Mechanical Engineering,
DOI: 10.1007/978-81-322-1007-8_18, © Springer India 2012

two parts of the modern meta-heuristics features: intensification and diversification [3–5]. Intensification intends to search around the best existing solutions and choose the best solutions, while diversification makes sure that the algorithm can explore the search space more efficiently, often via randomization [2].

The cuckoo search (CS) is one of the latest nature inspired meta-heuristic algorithms developed by Xin-She Yang and Suash Deb in [1]. It was inspired by the obligate brood parasitism of some cuckoo species by laying their eggs in the nests of other host birds (of other species). Some host birds can engage direct conflict with the intruding cuckoos. For example, if a host bird discovers the eggs are not their own, it will either throw these alien eggs away or simply abandon its nest and build a new nest elsewhere. Some cuckoo species such as the new world brood parasitic *Tapera* have evolved in such a way that female parasitic cuckoos are often very specialized in the mimicry in colors and pattern of the eggs of a few chosen host species. CS idealized such breeding behavior, and thus can be applied for various optimization problems. Recent studies show that CS is potentially far more efficient than particle swarm optimizers and genetic algorithms [2].

CS algorithm contains a population of nests or eggs (solution). Each cuckoo lays one egg at a time, and dumps it in a randomly chosen nest. The best nest with high quality of eggs (solutions) will carry over to the next generations. The quality of the solutions is improved by generating a new solution from an existing solution and modifying certain characteristics. The number of available host nests is fixed, and a host can discover an alien egg with a probability Pa. In this case, the host bird can either throw the egg away or abandon the nest so as to build a completely new nest in a new location. Based on these rules, the basic steps of the CS can be summarized and the corresponding pseudo code is shown in Fig. 1. In this paper CS is proposed to find the optimal drill path to minimize the distance of drill tool movement in order to drill all holes at given locations.

Fig. 1 Pseudo code of the cuckoo search algorithm

Begin
Objective function f(x)
Generate initial population of n host nest
Evaluate fitness
        while *(t<MaxGeneration)*
        t = t + 1
        Randomly generate new solutions from best cuckoo
        Evaluate fitness
        Choose a nest among n (say, j) randomly
if *(Fi <Fj)*
    Replace j by the new solution
end if
        A fraction (Pa) of worse nests are abandoned
        Randomly generate new solutions from all cuckoo
        Evaluate fitness
        Keep the better solutions
        Rank the solutions and find current best
        end while
Post process results and visualization
End

Most electronic manufacturing industries use computer numerical controlled (CNC) machines for drilling holes on printed circuit board (PCB). PCB is a laminated flat panel of insulated material designed to provide electrical interconnections between electronic components attached to it. High accuracy in hole location is required in PCB hole drilling, because the holes determine where the electronic components must be accurately placed to make the PCB circuitry perform well. Due to a significant amount of time required for moving the drill bit from one point to another point, holes drill routing optimization problem attracts a great interest among the academicians, researchers and engineers to solve for. This concern for calculation of the minimum tool path length between holes, the drilling device has to be steered to the location of each hole exactly once, is similar to a very well-known problem from the operational research field called the (symmetric, single objective, and Euclidean) traveling salesman problem (TSP).

Numerous researches have been oriented toward the development of algorithms for calculating the minimum drill path between holes [6–8]. Specifically in CNC machine, one of the earliest routing problems in holes drilling is a paper written by Sigl and Mayer [9]. They introduced the 2-Opt Heuristic Evolutionary algorithm in solving drill routing for CNC machine. Using CNC as the subject, Quedri et al. [10] employed genetic algorithm (GA) in searching the optimized route for holes cutting process in CNC machine tool. Also, Ghaiebi and Solimanpur [11] have introduced an ant algorithm (AA) for holes drilling of multiple holes sizes.

Despite the importance of drilling path optimization, few researchers worked on this problem in the literature. First, Kolahan and Liang [12] have formulated the problem as TSP and worked on applying Tabu search algorithm to solve the problem, and then they extended their research to a more complex case [13]. El-Midany et al. [14] have considered the application to EDM drilling. Ghaiebi and Solimanpur [15] have used ant colony algorithm. Krishnaiyer and Cheraghi [16] used same algorithm and they proposed a Web-based system. Zhu [17] has used particle swarm algorithm, and then Zhu and Zhang [18] have extended the research and applied to new sample problems. Onwubolu and Clerc [19] have proposed a usage of particle swarm algorithm.

The rest of the paper is organized as follows: mathematical formulation of the route optimization in PCB holes drilling problem is presented in Sect. 2. The proposed cuckoo search algorithm is illustrated in Sect. 3. Section 4 describes the implementation and verification results for two case study problems. Section 5 outlines the conclusion. A comprehensive list of references ends the paper.

2 Routing Problem in PCB Drilling Process

When drilling a group of holes in a PCB using a CNC milling machine, the machine table is driven back and forth in the X–Y directions, so that each hole is to be drilled in its designed position. The optimum drilling sequence can minimize the total table movements, thus shorten the no-cutting time and lengthen working life of the table's driving system. Given a drilling pattern, the drilling head will start from its initial position, visit and drill every specified location on the PCB, and finally return to its initial position. The drill path can be visualized as a TSP, where a salesman (drill-bit) must visit each city (drilling locations) in his/her designated area and then returns home. The worktable of the numerical control machine tool moves in the x and y axes and when the worktable reaches an exact drilling position, the drilling bit moves in the negative z direction to carry out the drilling operation on the workpiece. Suppose there are two hole locations i and j of the PCB, their coordinates are $(x_i, y_i), (x_j, y_j)$, then the rectilinear distance between the two holes is define as follows.

$$d_{ij} = |x_i - x_j| + |y_i - y_j| \tag{1}$$

During the machining operation, the total distance moved by the worktable is the sum of the absolute values of the distances that the worktable moved in the x and y directions and also it vary for different drilling paths selected. The goal is to find the optimal sequence to reduce the time/distance taken by the CNC machine to complete its task. Hence, the objective function of PCB drilling process can be expressed as

$$\min \sum_{i=1}^{n} \sum_{j=1}^{n} d_{ij} x_{ij} \tag{2}$$

$$\text{subject to} \sum_{j=1}^{n} x_{ij} = 1; \quad \forall i \, \& \, \forall j \tag{3}$$

where n is the number of holes to be drilled. Let x_{ij} is the decision variable related to the movement of the robotic arm from hole i to hole j. If there is a movement of the robotic arm from hole i to hole j, $x_{ij} = 1$, otherwise, $x_{ij} = 0$.

Assumptions:

1. All drilling holes are in same size.
2. The distance from initial position to the first hole and last hole to the initial position are ignored.

For solving this kind of problems, one of the simple approach is to list all the possible paths, then compare all the path lengths to find out which one is the shortest. Unfortunately, there are too many paths. The number of possible paths increases if the number of holes in the PCB increases. For example, 10 holes to be drilled in a PCB. The number of possible paths are $(10-1)!/2 = 181,440$, and for 20 holes $(20-1)!/2 = 60,822,550,204,416,000$, and so on. When the number of holes increases, it can easily be seen that exhaustive exploration of all possible paths becomes impracticable, despite the performance of actual computers.

3 Cuckoo Search Algorithm

The following step illustrates the procedural steps of CS algorithm for drilling path optimization problem. Zhu's 9 holes drilling problem shown in Fig. 2 has been chosen to illustrate the CS algorithm. Figure 2 shows the workpiece-1 in which 9 holes (the holes are labeled in the figure) to be drilled using a CNC drilling machine such that the rectangular coordinates (x_i, y_i) are (12.75, 69.75), (0, 45), (12.75, 20.25), (62.25, 20.25), (76.88, 39.64), (62.25, 69.75), (99.50, 82), (90.04, 58.53), (99.50, 8), respectively. The corresponding drilling path evaluation matrix is shown in Fig. 3.

Step 1: Initialize the cuckoo search algorithm Parameters
The number of nests (n), the step size parameter (α), discovering probability (pa), and maximum number of generations (GEN) are initialized as: 50, 0.1, 0.125 and 10,000.

Step 2: Generate nest or eggs of host birds
A nest/egg represents the solution vector of size [1 × n], where n is the number of holes to be drilled. In our case, there are 9 holes to be drilled. So a vector of 9 normally distributed random numbers is generated:

Nest/Egg: [−1.3376 −0.7472 1.2529 0.7576 1.0708 −0.7724 −1.8793 −0.4416 −1.5338]

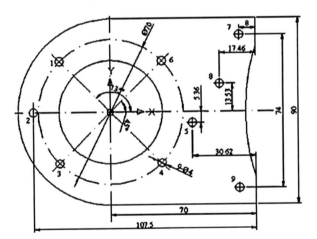

Fig. 2 Workpiece-1

Fig. 3 Drilling path evaluation matrix (9 × 9)

$$\begin{pmatrix} 0 & 37.50 & 49.50 & 98.99 & 94.24 & 49.50 & 99 & 88.91 & 148.5 \\ 37.50 & 0 & 37.5 & 87 & 82.24 & 87 & 136.5 & 103.6 & 136.5 \\ 49.50 & 37.5 & 0 & 49.50 & 83.52 & 98.99 & 148.5 & 115.6 & 99 \\ 98.99 & 87 & 49.5 & 0 & 34.02 & 49.50 & 99 & 66.07 & 49.50 \\ 94.24 & 82.24 & 83.52 & 34.02 & 0 & 44.74 & 64.98 & 32.05 & 54.26 \\ 49.50 & 87 & 98.99 & 49.50 & 44.74 & 0 & 49.50 & 39.01 & 99 \\ 99 & 136.5 & 148.5 & 99 & 64.98 & 49.5 & 0 & 32.93 & 74 \\ 88.91 & 103.6 & 115.6 & 66.07 & 32.05 & 39.01 & 32.93 & 0 & 59.99 \\ 148.5 & 136.5 & 99 & 49.5 & 54.26 & 99 & 74 & 59.99 & 0 \end{pmatrix}$$

Only the sign decimal part of the number will be considered:

−0.3376 −0.7472 0.2529 0.7576 0.0708 −0.7724 −0.8793 −0.4416 −0.5338

The fitness value of all eggs is calculated as follows:

- The solution (egg) vector will be sorted in ascending order
- The drill path length is calculated by referring the evaluation matrix
- From the drill path length the longest hole-to-hole distance will be subtracted and assign that hole becomes the first hole to be drilled.
- The calculated drill path length is fitness value of the egg.

[−0.3376 −0.7472 0.2529 0.7576 0.0708 −0.7724 −0.8793 −0.4416 −0.5338]

For example, the egg/solution vector will be sorted in an ascending order and the first number is repeated again at the end of the vector in order to complete the closed path.

Sequence: [7 6 **2 9** 8 1 5 3 4 7]

Drill path length = 49.50 + 86.99 + **136.5** + 59.99 + 88.50 + 94.24 + 83.52 + 49.49 + 99 = 747.75

The longest hole-to-hole distance of 136.5 is subtracted from the drill path length and assign the corresponding hole number is the starting hole position, then modify the sequence. The modified sequence and corresponding fitness are given below.

Modified sequence: [**9** 8 1 5 3 4 7 6 **2**] and fitness: 611.2499

Step 3: Generate new cuckoos by simple random walks

All the nests except for the best one are replaced based on fitness value by new cuckoo eggs are generated by random walks.

$$\text{nest}^{(t+1)} = \text{bestnest}^{(t)} + \alpha.\,\text{rand} \tag{4}$$

where $\text{rand} \in [−1, 1]$.

For example, the best egg is:

−0.4101 −0.6590 0.2762 0.8280 0.0639 −0.8499 −0.9266 −0.5194 −0.5317

The cuckoo moved by adding a random number with a range of $[−1, 1]$ with a step size of 0.1:

−0.4092 −0.6655 0.2750 0.8120 −0.0027 −0.8791 −1.0015 −0.4287 −0.4730

Only the sign decimal part of the number will be considered:

−0.4092 −0.6655 0.2750 0.8120 −0.0027 −0.8791 −0.0015 −0.4287 −0.4730

Step 4: Alien egg discovery

The alieneggs are discovered with a probability of Pa. The value of Pa can be calculated by Pa = 1/(no of holes-1) = 0.125

$$\text{nest}^{(t+1)} = \text{nest}^{(t)} + P.\,\alpha.\,\text{rand} \tag{5}$$

$$P = \begin{cases} 1 & \text{if } \text{rand} < \text{Pa} \\ 0 & \text{else} \end{cases} \tag{6}$$

where $\text{rand} \in [−1, 1]$.

Current solution:

0.4092 −0.6655 0.2750 0.8120 −0.0027 −0.8791 −0.0015 −0.4287 −0.4730

The new solution is generated by moving the cuckoo by adding a random number with a range of $[-1, 1]$ with a step size of 2 and a probability of Pa:

0.8032 −0.6655 0.2750 **1.4232** −0.0027 −0.8791 −0.0015 −0.4287 −0.4730
0.8032 −0.6655 0.2750 0.4232 −0.0027 −0.8791 −0.0015 −0.4287 −0.4730

Step 5: Termination

Steps 3 and 4 are alternatively performed until a termination criterion is satisfied. The maximum number of generations as 10,000, but the algorithm converges very quickly within 100 generations. The optimal drill path sequence obtained is: [3 2 1 6 7 8 5 4 9] with the corresponding total distance of 322.502525.

4 Implementation and Verification

The proposed CS algorithm is verified by solving two sample problems by Zhu and Zhang [20]. In Zhu and Zhang's paper, they solved the workpieces with the basic PSO [19] and their Global Convergence PSO. The ACS was also applied on the second workpiece. In solving the sample problems, the generation number was set to 10,000, and the number of nests was set to 100. In finding the average number of iterations for global convergence, the CS is made to run 1,000 times before the average is taken.

4.1 Verification of Workpiece 1

The global optimum drill path sequence for the workpiece 1 shown in Fig. 2 is 3 2 1 6 7 8 5 4 9. The results and comparisons are shown in Table 1.

Table 1 Data of verification of workpiece 1

	Basic PSO			GC PSO			CS
	$\omega = 0.0$	$\omega = 0.5$	$\omega = 1.0$	$\omega = 0.0$	$\omega = 0.5$	$\omega = 1.0$	–
The least iteration number while global convergence	7	3	9	1	5	4	1
The average iteration number while global convergence	7	7	20	1251	646	1620	27
Length of optimal path (mm)	322.5	322.5	322.5	322.5	322.5	322.5	322.5
Average fitness after computing 50 computations (mm)	348.07	344.89	338.05	332.25	331.62	327.57	366.88

When compared to the basic PSO and the global convergence PSO, the CS algorithm proved to be clearly superior. It consistently converges to the global optimum after only a small number of iterations. The higher value of average fitness after 50 computations is due to the global search of CS.

4.2 Verification of Workpiece 2

The verification of workpiece 2 is shown in Fig. 4. The results and comparisons are shown in Table 2 and Fig. 5 shows a sample curve of the fitness value. There

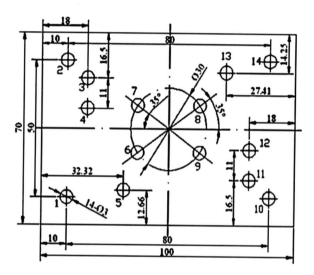

Fig. 4 Workpiece 2

Table 2 Data of verification of workpiece 2

	Basic PSO			GC PSO			ACS	CS
	ω = 0.0	ω = 0.5	ω = 1.0	ω = 0.0	ω = 0.5	ω = 1.0	–	–
The least iteration number while global convergence	–	–	93	815	10	110	193	3
The average iteration number while global convergence	–	–	847	3806	1620	1764	1037	706
Length of optimal path (mm)	–	–	280.0	280.0	280.0	280.0	280.0	280.0
Average fitness after computing 50 computations (mm)	362.3	344	299.2	304.4	306.4	291	283.6	426.03

Cuckoo Search Algorithm for Optimization

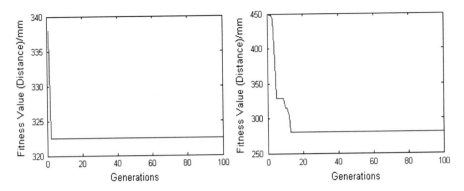

Fig. 5 The sample curve of the fitness value for workpiece 1 and 2

are four global optimums for this problem: 10 11 12 9 6 5 1 2 3 4 7 8 13 14, 14 13 8 7 4 3 2 1 5 6 9 12 11 10, 1 5 6 9 12 11 10 14 13 8 7 4 3 2, and 2 3 4 7 8 13 14 10 11 12 9 6 5. The convergence speed of CS is also shown to be faster in this workpiece when compared to the basic PSO, GSPSO, and ACS. Although the average fitness after 50 computations is relatively large, it simply shows that the CS is constantly exploring for global minima while the exploitation of the current solutions is not compensated.

5 Conclusion

In this paper, CS algorithm was proposed and implemented for PCB holes drilling route optimization problems with objective to minimize the distance of the route chosen by the CNC machine. The proposed CS is quite straightforward and easy to implement. The verification results indicate that CS able to find the optimal drill path at a faster rate compared to PSO and ACS.

References

1. Yang XS, Deb S (2009) "Cuckoo search via L´evy flights," in world congress on nature and biologically inspired computing (NaBIC) India pp 210–214
2. Yang XS (2010) Firefly algorithm L´evy flights and global optimization. In: Bramer M, Ellis M, Petridis M (eds.) Research and development in intelligent systems XXVI 209–218
3. Blum C, Roli A(2003) "Metaheuristics in combinatorial optimization" ACM Comput vol 35
4. Gazi K, Passino KM (2004) "Stability analysis of social foraging swarms" IEEE Trans Sys vol 34
5. Yang XS (2009) "Harmony search as a meta-heuristic algorithm," Music-inspired harmony search: Theory and Appli 191:1–14
6. Aoyama E, Hirogaki T, Katayama T, Hashimoto N (2004) Optimizing drilling conditions in printed circuit board by considering hole quality optimization from viewpoint of drill-movement time. J Mater Process Technol 155:50–1544

7. Changa PC, Hsiehb JC, Wanga CY (2007) Adaptive multi-objective genetic algorithms for scheduling of drilling operation in printed circuit board industry. Appl Soft Comput 7(3):6–800
8. Liu J, Linn R, Kowe PSH (1999) Study on heuristic methods for PCB drilling route optimization. Int J Ind Eng Theory Appl Pract 6(4):96–289
9. Sigl S, Mayer HA (2005) Hybrid evolutionary approaches to CNC drill route optimization. Proc Comput Intell Modeling, Control Autom 3(1):905–910
10. Qudeiri JA, Yamamoto H, Ramli R (2007) Optimization of operation sequence in CNC machine tools using genetic algorithm. J Adv Mech Design Syst Manuf 1(2):272–282
11. Ghaiebi H, Solimanpur M (2007) An ant algorithm for optimization of hole-making operations. Comput Ind Eng 2(52):308–319
12. Kolahan F, Liang M (1996) Tabu search approach to optimization of drilling operations. Comput Ind Eng 31(1–2):371–374
13. Parker S, Kolahan F, Liang M (2000) Optimization of hole-making operations: a tabu-search approach. Int J Mach Tools Manuf 40(12):1735–1753
14. EL-Midany TT, Kohail AM, Tawfik H (2007) A proposed algorithm for optimizing the tool-point path of the small-hole EDM-drilling. Proc Geom Model Imaging 11(1):25–32
15. Ghaiebi H, Solimanpur M (2007) An ant algorithm for optimization of hole-making operations. Comput Ind Eng 52(2):308–319
16. Krishnaiyer K, Cheraghi SH (2006) Ant algorithms: web-based implementation and applications to manufacturing system problems. Int J Comput Integr Manuf 19(3):264–277
17. Zhu G -Y (2006). Drilling path optimization based on swarm intelligent algorithm. Proc IEEE Int Conf Robotics Biomimetics 3:193–196
18. Zhu G-Y, Zhang W-B (2007) Drilling path optimization by the particle swarm optimization algorithm with global convergence characteristics. Int J Prod Res 46(8):2299–2311
19. Onwubolu GC, Clerc M (2004) Optimal path for automated drilling operations by a new heuristic approach using particle swarm optimization. Int J Prod Res 3(42):473–491
20. Zhu GY, Zhang WB (2008) Drilling path optimization by the particle swarm optimization algorithm with global convergence characteristics. Int J Prod Res 46(8):2299–2311

Efficient Cooling of Building Using Phase Change Materials Along with Coolant

S. Mathana Krishnan, M. Joseph Stalin and P. Barath

Abstract Due to upsurge in the global warming day by day, ambient temperature escalates imperceptibly and hence the residence in the tropical region experiences severe hot climate which induces the need of Air Conditioning and Refrigerants. This paper focuses on theoretical analysis on cooling of building using Phase Change Material. In existing systems, the backup time for Phase Change Material gets over by the day, so in night time it can't be used. This is the main shortcomings of this system. In order to overcome this, coolant is used to promulgate in aluminium bent shaped tube kept inside the slab of Phase Change Material in a well designed manner. In this paper, water is used as a coolant and is promulgated from the overhead water tank. In overall process, the room heat is carried away by the coolant from the overhead water tank. This leads to cool the Phase Change Material and in night time also it is possible to operate the system. The results of this paper show the time of cooling for a pondered volume of room and also its cost analysis. Factors like design, volume and selection of Phase Change Material etc., are carefully acclimatized to achieve effective forced convective heat transfer and efficacious cooling takes place. Finally this could become a great surrogate for Air conditioning.

Keywords Coolant • Phase change material • Reduce carbon emissions • Energy efficient

---

S. Mathana Krishnan (✉) · M. Joseph Stalin · P. Barath
Thiagarajar College of Engineering, Madurai, India
e-mail: mathanakryshnan@gmail.com

M. Joseph Stalin
e-mail: stalin661@gmail.com

P. Barath
e-mail: barath93ajith@gmail.com

S. Sathiyamoorthy et al. (eds.), *Emerging Trends in Science,*
Engineering and Technology, Lecture Notes in Mechanical Engineering,
DOI: 10.1007/978-81-322-1007-8_19, © Springer India 2012

1 Introduction

As demand for Air conditioning and Refrigerants are emphatically enhanced during last decennium, it should be economically abated in order to obtain comfort cooling in buildings. Zain [1] studied about the thermal comfort of a building in 2007. During scorching summer, peak demands of electric power is escalating year by year and limited reserves of fossil fuels have led to a surge of interest with efficient energy application. Usually, day time load has become double that of the night time load. This is due to usage of Air Conditioner both in day as well as in the night. Therefore our system would be the best way to eradicate the peak demand of using electricity during night as well as day time, thereby reducing air conditioner load. Salyer [2] indulged in the work of refrigeration and air conditioning in the residential buildings using unconventional energies and latent heat storage in 1990.

This paper focuses on cooling of building by using slabs of Phase Change Material which is very attractive because of its high thermal storage capacity. In the past years, Abhat [3] confabulate the concept of latent heat storage materials in low temperature in 1983 itself. Choice of Phase Change Material plays a substantial role in addition to heat transfer mechanism in Phase Change Material. Chen [4] discussed and experimentally proved the possibility of cold storage using the thermal storage tank in 2000. A Phase Change Material of desired operating range is picked out and when the ambient temperature is beyond the melting point of Phase Change Material, it absorbs the heat by melting since it has amble of latent heat storage. Khudair [5] had done a case study of how phase change material is installed and effectively used for the residential building using its latent heat storage in 2004. The molecules of Phase Change materials are encapsulated in a glassy substance and hence it stores the lavish amount of heat energy than in other previously existing thermal storage devices. Latent heat storage is a recent area of study and it received hefty attention during early 1970s and 1980s. Dincer [6] studied and confabulated about thermal energy storage and also he shared the applications of cold storage in 2002. Zabla [7] undergone a experimental study on cooling of the building using phase change material in the early 2004. In this paper, Phase Change materials are arranged in such a way that there should be air gap between each slab and mass flow rate of air was made forcefully by means of external agency and the hot air in the leeway will promulgate through the air gaps. Hence the heat is transferred to Phase

Change Material which in turn heat is carried away by the water coolant from the overhead water tank. This can be made possible by passing the Aluminium tubes of calculated diameter with number of bents which are made to enlarge the surface area of contact. Air is made to pass through each slabs of Phase Change Material. By this way this system extends the usage of air coolness in the night time with minimum load as possible. This system can be extended to cool the leeway within a short span of time by using a better coolant and with a proper design. In general, there will be carbon emissions due to the usage of air conditioning which could be completely eradicated by fixing this method and this could also completely abates the carbon emissions due to decrease of peak power generation supplied by oil thermal power plants. Although the initial cost for mounting of Phase Change Material and its arrangements and construction are high, it will cool the leeway to a comfort level by free of cost (by neglecting the cost of electric current consumed by the fan which induces the forced convection process). This system not only reduces the cost of electricity due to air conditioner load but also it abates the huge investments in new electricity generating plants. Voelker [8] and Tyagi [9] confabulated the effective temperature reduction due to the application of the thermal heat storage for residential buildings in 2007.

2 Description

In India, there is scorching summer and all of us are longing of comfort cooling. This can be accomplished only by means of investing conventional energies. Alternative solutions for cooling of leeway have appeared in practice as a counterweight to expensive energy wasting conventional systems. Arkar [10] and [11] investigated about cooling of building using the latent heat storage in the ventilation system in 2007. Researchers are working for utilizing the night coolness to impart cooling effect during day, without using enormous energy. Recently Hiroshi [12] has done an experimental set up for increasing effective heat transfer using the combination of PCM and copper foam for effective thermal storage in 2010. The system is designed for absorbing night coolness and gives it back in day time. Phase change materials are used for the purpose of imparting cooling effect. These are one of the efficient thermal energy storage materials. At the beginning, Phase change materials are in idle mode when they are purchased from the chemical industry. Soon after it was incinerated to the temperature of 50 °C using water bath so as the passive crystals of PCM melts. At this moment, the PCM is made active. Then the aluminium packets are used to seal the PCM and its better conductivity is the ultimate reason for its preference.

Figure 1 for example, the room size 20 × 20 × 20 feet is pondered for calculating its cooling effect. It consists of five PCM slabs with six air gaps in it. The design of the system is based on the heat load acting in the room. Various types of PCM are commercially available in the shops. Selection of PCM depends upon its latent heat and melting point since they are the major parameters for absorbing room heat. By considering several parameters, HS 29 type phase change material is selected. The property of PCM is absorption of latent heat during change of phase. The shape of the PCM slab is rectangular in cross section. The selection of this shape is for efficacious heat transfer from room to PCM slab. The dimension of one slab is 1,000 × 303 × 60 mm. These are calculated by using heat load of room, mass and volume of PCM. The air gap is taken as 15 mm. The hot air from the leeway passes through the air gap where it transfers heat to the PCM. The maintenance of constant mass flow rate is mandatory and so we are in need of mounting fan. We are affixing one fan for every two air flow gaps. These fans regulate the mass flow rate of inlet air. The RPM of fan is calculated

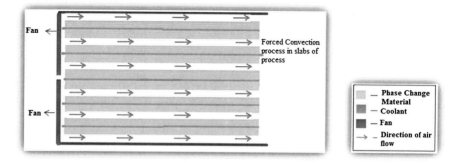

Fig. 1 Cooling of air using phase change materials

by considering the diameter of fan. Then the RPM is set constant for maintaining constant mass flow rate. There are five PCM slabs with four air gaps and so total height of PCM is 363 mm. The required floor space for mounting the PCM is 1,000 × 303 × 363 mm. We are using water as a coolant for transferring heat from the PCM to outside of the room. The small aluminium tubes are fitted which has the inlet from water tank of the residence and it has the outlet to the environment. The aluminium material is selected for its good conductivity and the heat absorbed from the room by the PCM is taken away by the water from the tank. The insulated pipe is fitted from water tank to the inlet of the PCM and it is coupled with aluminium tubes which are fitted inside the PCM. Self regulating valves are fitted in the inlet of the insulated piping for controlling the mass flow rate of water. According to the prevalence of temperature in the PCM, the heat is efficaciously transferred by the water within certain time. The backup time is made infinity by installing heat exchanger. The diameter of tube is 10 mm with 10 bents in one slab. The reason for bending of tube is to make contact of larger surface area with PCM for effective heat transfer. A pump is needed for recirculation of water from the tank. The tank must be insulated to maintain water at low temperature for increased efficiency. The power for pump could also be taken by installing solar panel.

3 Working

Our ultimate aim of the system is to impart coolness to the leeway by absorbing night coolness. Overall working of the system is dependent on the effective heat transfer in the system. Five PCM slabs and six air gaps constitute the system. The PCM slabs are designed in such a way that it absorbs maximum amount of heat from the room. Imparting cooling effect to the room indicates extracting and transferring the heat from leeway to alfresco environment. The selected PCM has the melting point of 29 °C. Temperature of room is chosen as 34 °C Naturally heat flows from higher temperature to lower temperature. So air from the leeway passes through the air gap and the heat is efficaciously transferred to the PCM Fig. 2.

Efficient Cooling of Building Using Phase Change Materials

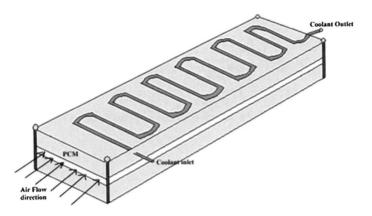

Fig. 2 Working of PCM along with coolant

The mass flow rate of air is kept constant for efficient heat transfer. For this purpose, a fan is mounted in the inlet of the air gap with constant RPM. Constant mass flow rate is given by the fan. The mounting of one fan is for two air gaps and the fan diameter is 90 mm. Heat transfer takes place until it attains equilibrium and when the backup time is over, it emits heat in the night. This is the hefty problem in all of the existing systems. In order to avoid this problem, we are using small aluminium tube of 10 mm which has 10 bents is sealed inside the PCM. The tank and the aluminium tube from the tank to the inlet of the PCM slab are well insulated. Water flows inside the tube and it carries away the heat in the PCM. The inlet to the tube is from the water tank. PCM is in high temperature and the water is in low temperature and so heat is transferred to the water. The mass flow rate of water is set constant by means of regulating Valve. A pump is required for recirculation of water in the tank. If our system should not depend on electricity, solar panels could be installed for the required power. In total, we absorb the night coolness and give it back in day time. This is one of the efficient methods of bringing cooling effect to room without spending energy. Hed [13] derived the equations of the PCM air heat exchanger in 2006.

4 Model Calculation

Assumptions

- The size of the room is 20 × 20 × 20 feet (Fig. 3).
- The wall of the room is made up of Brick (plastered on one side) which has a thickness of 40 cm.
- The floor of the room is made up of concrete (22 cm) with insulation (2.6 cm).
- The Ceiling of the room made of plaster (concrete).

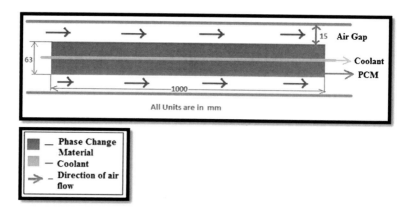

Fig. 3 PCM in single slab with coolant in side view

- Room maximum temperature in day time is chosen to be 307 K (In Bangalore)
- At night time the minimum temperature is assumed to be 295 K (In Bangalore)
- Bent tube circulating water is assumed to have 10 bends per slab of PCM
- Backup of 2 h is chosen since coolant is used, there will not be any backup time and the heat will be continuously carried out by the coolant water.

Load Calculations

- Heat transfer Coefficient Values for various materials (Table 1).

Wall

$Q1 = U \times A \times dT$
Heat received by the wall = 2.051 kW

Floor

$Q2 = U \times A \times dT$
Heat received by the Floor = 1.962 kW

Ceiling

$Q_3 = U \times A \times dT$
Heat comes from the ceiling = 0.601 kW
Human who rejects heat while doing light work $Q_4 = 0.183$ kJ/s
Heat rejected by the Equipment in the room:
$Q_5 = $ kW $\times 3{,}600 \times$ Use factor

Table 1 Heat transfer coefficient for different materials

Materials	U in W/m2 K
Wall brick (plastered on one side)	1.0
Plaster ceiling	1.15
Floor concrete with insulation	1.0

Efficient Cooling of Building Using Phase Change Materials

Table 2 Properties of PCM

Property	Values
Name	HS 29
Melting point	29 °c
Density	1,550 kg/m^3
Latent heat	190 kJ/kg

Power = 225 kW
Use factor = 0.68
Q_5 = 0.153 kW
Total heat Load = $Q_1 + Q_2 + Q_3 + Q_4 + Q_5$
Total heat Load = 3.078 kW = 0.86 tons

PCM Properties
For 2 h backup

Load = 1.72 ton Table 2
Load = 21,776.92 kJ/day
Mass of the PCM required
m = Load/Latent heat of PCM
m = 128 kg
Density of PCM ó PCM = 1,400 kg/m^3
Calculate Volume of PCM V = 0.0914 m^3
PCM slab dimensions is assumed as
Length l = 1 m
Breadth b = 0.303 m
Height h = 0.065 m
Thus we should have five slabs to attain the calculated volume

Mass flow rate of air

M = Load/sensitive heat (Latent heat)
M = 0.01779 kg/s
Since forced convection has to be made a fan is fitted.
For a fan diameter,
d = 93 mm
$M = \dot{p} \times N \times d^3$
Rpm of the fan is calculated as
N = 1,084 rpm

Heat Exchanger Calculation

The maximum temperature in Bangalore during day time is 307 K

$$q = h \times A \times dT \tag{1}$$

The minimum night time temperature in Bangalore is 295 K
The difference in temperature dT = 12 K
Convective heat transfer coefficient of water h = 500 W/m^{2}K.

From Eq. (1),
Surface area of contact of bent tube circulating water
$A = 0.5 \text{ m}^2$

$$D \times \pi \times L = A \tag{2}$$

The length of the bent tube contact with PCM is calculated as $L = 15$ m
The diameter of the tube is calculated as

$$D = 10^{-2} \text{ m}$$

[by Eq. (2)]

Heat transfer by the PCM = Heat absorbed by the PCM

$$h \times A \times dT = M_w \times C_p \times dT$$

The mass flow rate of water is calculated as

$$M_w = 0.015 \text{ kg/s}$$

The graph plotted between Temperatures of the room versus Time is performed by using the psychometric chart and the following equation.

$$m_1 w_1 + m_2 w_2 = m_3 w_3$$
$$m_1 h_1 + m_2 h_2 = m_3 h_3$$

5 Results and Discussion

Results obtained by calculation are shown below in terms of graph for better and quick understanding. The ultimate goal of this work is to attain cooling effect using coolant without spending huge amount of conventional energy. In India, we are exposed to hot summer and lavish solar radiation and all of us are in need of comfort. We are presenting about the maximum and minimum temperatures prevailing in India. Average temperatures are also indicated in the graph corresponds to respective month's average temperature. The Graph 1 is drawn with respect to the month and corresponding temperature.

In order to balance the scorching summer people begins to use air conditioning system. The production and sales of AC are increased in our day to life. But AC requires lot of power consumption and it also affects the environment by releasing hazards gases. Global warming and ozone depletion are major problems caused by these kinds of equipments. The following Graph 2 shows that sales of AC with respect to year.

Efficient Cooling of Building Using Phase Change Materials

Graph 1 Temperature versus months

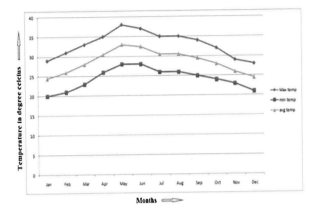

Graph 2 No. of air conditioner versus year

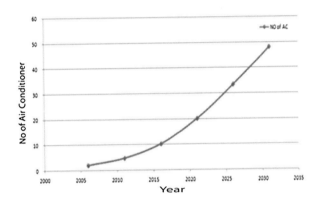

Among all of the home appliances AC requires huge amount of power. We are collecting some information about average power consumption in home by various appliances. The Graphs 3 and 4 given below clearly indicates that average power consumption by various equipments per year.

But today there is a huge demand in electricity and majority of problems arising because of energy demand. Ozone depletion has to be eradicated. These are the familiar topics which are flashing in many TV channels and news papers. In order to avoid certain amount of power consumption, we have designed eco friendly and non hazardous system. The system consists of sealed PCM slabs and air gaps. The main criterion in selection of PCM is melting point and latent heat of PCM. Air from room at room temperature comes in contact with PCM at certain mass flow rate. The effective heat transfer and efficiency of the system depends on mass flow rate of air. So we have pondered the various types of PCM and calculated the corresponding mass flow rate. The Graph 5 are drawn with mass flow rate with respect to various types of PCM.

Graph 3 Load versus equipments in 2006

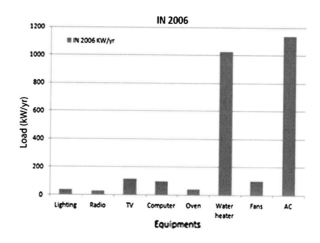

Graph 4 Load versus equipments in 2011

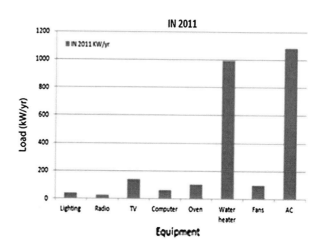

Graph 5 Mass flow rate of air versus PCM

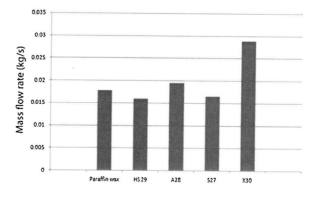

Efficient Cooling of Building Using Phase Change Materials 227

Graph 6 Temperature of the room versus time

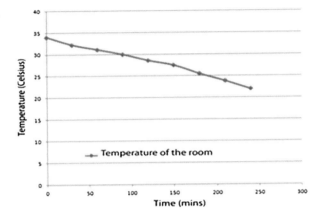

Graph 7 Cost of air conditioner versus months

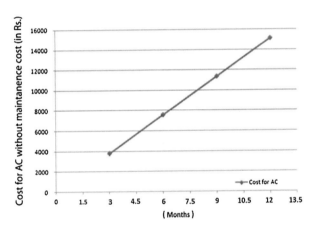

Graph 8 Cost of air conditioner with maintenance versus month

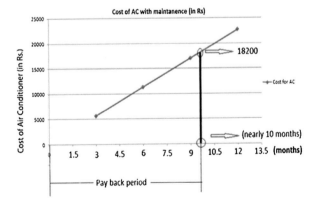

Our goal is to maintain the room with comfort temperature. The time for various comfort temperatures are calculated. The Graph 6 is plotted for comfort temperature with respect to time. Finally concentration is made on cost of the system. The overall cost for our system is Rs. 18,200 for room size 20×20 feet. The AC cost for same room of size 20×20 feet is also analyzed.

The Graphs 7 and 8 plotted below shows the AC cost with and without maintenance and payback period is also indicated in the graph. By considering the cost of AC with maintenance the payback period is nearly 10 months. So in total it must be a better alternative for AC.

6 Conclusion

As our demand for Air Conditioning and Refrigeration are escalating day by day, the amount of fossil fuel will become scarce on one day. By implementing this method, it eradicates both power consumption of Air Conditioner and the carbon emissions due to Air Conditioning. The key behind this efficacious solution is the Phase Change Material. In previous papers, the overall system will work only during the daytime, but in most of the houses there will be need of Air Conditioning during night time. The usage of coolant to carry the heat energy in the PCM extends the system to work during night time. Thus this could bring extravagant changes to this hectic world. This method could greatly reduce the usage of Air Conditioning and thus saves fuel energy. Initial investment may be high in case of using Phase Change Material but maintenance cost and installation of Air Conditioner will be higher and it is also shown that within a short span of time it is able to recover the loss of initial investment of Phase Change Material. The effects of thermo physical properties of PCM, installation methodology, location of PCM are the scope for future work.

References

1. Zain ZM, Taib MN, Mohd SBS (2007) Hot and humid climate: prospect for thermal comfort in residential building. Desalin 209:261–268
2. Salyer IO, Sircar AK (1990) Phase change materials for heating and cooling of residential buildings and other applications. In: Proceedings of 25th intersociety energy conversion engineering conference, pp 236–243
3. Abhat A (1983) Low temperature latent heat thermal energy storage, heat storage materials. Sol Energ 30:313–332
4. Chen SL, Chen CL, Tin CC et al (2000) An experimental investigation of cold storage in an encapsulated thermal storage tank. Exp Therm Fluid Sci 23:133–144
5. Khudhair A, Farid M (2004) A review on energy conservation in building applications with thermal storage by latent heat using phase change materials. Energ Conv Manage 45:263–275
6. Dincer I, Rosen MA (2002) Thermal energy storage, systems and applications. Wiley, Chichester

7. Zalba B, Marin JM, Cabeza LF, Mehling H (2004) Free cooling of building with phase change materials. Int J Refrig 27:839–849
8. Voelker C, Kornadt O, Ostry M (2008) Temperature reduction due to the application of phase change materials. Energ Bldg 40:937–944
9. Tyagi VV, Buddhi D (2007) PCM thermal storage in buildings: a state of art. Renew Sustain Energ Rev 11:1146–1166
10. Arkar C, Medved S (2007) Free cooling of a building of PCM heat storage integrated into the ventilation system. Sol Energ 81(9):1078–1087
11. Arkar C, Vidrih B, Medved S (2007) Efficiency of free cooling using latent heat storage integrated into the ventilation system of a low energy building. Int J Refrig 30(1):134–143
12. Isa MHM, Zhao X, Yoshino H (2010) Preliminary study of passive cooling strategy using a combination of PCM and copper foam to increase thermal heat storage in building facade. Sustainability 2:2365–2381
13. Hed G, Bellander R (2006) Mathematical modeling of PCM air heat exchanger. Energ build 38:82–89

Fatigue Damage Mechanisms in Fiber Reinforced with Al₂O₃ Composites Under Cyclic Reversed Loading

K. Mohamed bak and K. Kalaichelvan

Abstract The present chapter deals with investigation of damage mechanisms in fiber reinforced with alumina oxide specimen (Hybrid composite) under cyclic reversed loading. Two different percentages (5 and 10 %) of alumina oxide particle are added with composites by weight proportions and are formed to dog-bone-shaped specimen with total thickness 3.2 mm according to ASTM standard D3479-76, to determine the fatigue strength. The magnification view of scanning electron microscope (SEM) was used to identify the failure mechanism of composite specimen such as matrix failure, fiber-matrix debonding and fiber failure in the direction of fiber. It is observed that fiber failure is a additional fatigue damage and growth of all these damages leads to final fatigue failure of the hybrid composite specimen, but the GFRP with 10 % Al_2O_3 particle exhibits an improved fatigue life over that of GFRP with 5 % Al_2O_3 particle.

Keywords Composite (glass-epoxy) • Hybrid composite specimen • Al_2O_3 particle • Fatigue test (tension-compression cyclic loads) • Scanning electron microscope

1 Introduction

High-performance composite structures are increasingly used to meet the demand for lightweight, high strength, stiffness, and corrosion-resistant materials in all engineering structures. Applications of composite materials have been extended to

---

K. M. bak (✉)
Department of Production Technology, Anna University,
MIT Campus, Chennai 44, India
e-mail: mohamedaero@yahoo.com

K. Kalaichelvan
Department of Production Technology, Anna University,
MIT Campus, Chennai 44, India
e-mail: kalaiselvan@mitindia.edu

S. Sathiyamoorthy et al. (eds.), *Emerging Trends in Science,*
Engineering and Technology, Lecture Notes in Mechanical Engineering,
DOI: 10.1007/978-81-322-1007-8_20, © Springer India 2012

various fields, including aerospace structures, marine structures, automobiles, and robot systems. As composite materials are replacing metal materials in many areas, the need for a reliable database of structural design has become more important [1]. Continuous fiber-reinforced composites, such as those consisting of glass fibers in an epoxy resin, offer an attractive potential for reducing the weight of high-performance aerospace structures [2, 3]. However, one of the major factors limiting the design of structures made from glass/epoxy system is the susceptibility of the material to impact damage in the form of multiple delamination through the thickness. However, the fatigue damage of composite is a complex process that involves many different damage mechanisms, such as matrix cracking, fiber breakage, fiber/matrix debonding, and delamination [4, 5], Khelifa [6] investigated fatigue study of Glass E glass fiber reinforced composite under reversed loading and spectrum loading. Freire [7] investigated damage mechanism and failure prevention in E glass polyester resin composites, his work was limited to bidirectional and stacked bidirectional woven fabric textile. Extensive research has been conducted to investigate the degradation of residual strength and stiffness at different maximum stress loads.

Al_2O_3 is the most cost effective and widely used material in the family of engineering ceramics. Today, the material has been developed into a high quality technical grade ceramic with very good mechanical properties. It is used in the Gas laser tubes, structural members, high temperature electrical insulators, jet engine parts, and high-performance applications. Al_2O_3 is the strongest and stiffest of all the oxide ceramics. Its high hardness, resistant to high temperature creep, excellent dielectric properties, refractoriness, and good thermal properties make it as the material of choice for a wide range of applications [8].

The fatigue life of composites depends on a large number of variables, stress state, mode of cycling, and environmental conditions. Fatigue cracking is one of the primary damage mechanisms of composite laminate [9]. Van and Degrieck [10] investigated fatigue test on the specimen which have eight layer woven glass epoxy composite, they concluded the bending performance of the composite from experimental tests on straight specimens. The stress-time relationship for the models of fluctuating or alternating stresses, repeated stresses, and reversed stresses shown in Fig. 1.

The experimental work on fatigue damage mechanisms in fiber reinforced with alumina oxide composites under cyclic reversed (fully reversible tension-compression loads) loading is investigated and the failure mechanisms of composite specimen is identified using scanning electron microscope.

2 Fabrication Work

2.1 Hand Lay-up Process

The FRP composite laminates were prepared by hand layup process. The unidirectional glass fiber is weighed and the resin is taken as 1:1 ratio by the weight

Fatigue Damage Mechanisms in Fiber Reinforced with Al$_2$O$_3$

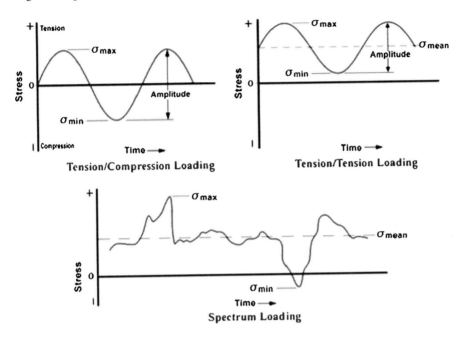

Fig. 1 Forms of stress cycles in fatigue loading

of the fiber [11]. The resin with Al$_2$O$_3$ particle and hardener is completely mixed which forms the matrix. Glass fiber reinforced in matrices by weight proportions to prepare the composite specimens A and B respectively. It is identified that the distribution of Al$_2$O$_3$ particle inside the matrix of epoxy resin is uniform over the matrix, which is maintained by stirring it for a time period of 10 min and the uniformity is verified in the microstructure and average size of the alumina oxide particle visualized is 500 μ. The detailed compositions of composite specimen are given in the Table 1.

2.2 Specimen Preparation

The unidirectional glass fiber is placed over the matrix (LY 556 with hardener HY951 and Al$_2$O$_3$ particle), by a stippling action using resin wetted brush; the

Table 1 Detailed composition of Composite specimen

Specimen	Material	Resin	Particle	Thickness (mm)
A	60 % E-glass fiber-0° orientation	35 % Epoxy LY 556	5 % Al$_2$O$_3$	3.2
B	50 % E-glass fiber-0° orientation	40 % Epoxy LY 556	10 % Al$_2$O$_3$	3.2

resin is squeezed to the top surface. After the first layer is laid up; subsequent layers are laid up in a similar manner. This procedure was repeated till the required thickness has been built up. The symmetric ply [0°/0°]$_4$ laminates were then cured by a compression molding machine at 100 kPa pressure and room temperature. The dog-bone specimen of suitable dimensions (90 × 30 × 3.2 mm) have to be cut using a diamond cutter to avoid machining defects and maintain good surface finish for fatigue test from the fabricated laminates shown in Fig. 2 and 3. The ASTM D 3,479 standard determines the fatigue testing of GFRP with Al$_2$O$_3$ particle specimen subjected to cyclic reversed loading [12].

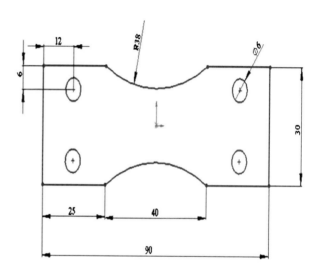

Fig. 2 Fatigue specimen (All dimensions are in mm)

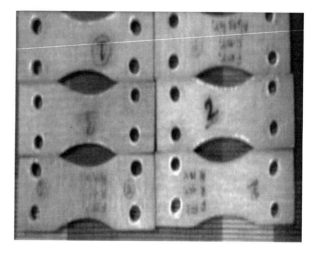

Fig. 3 ASTM D3479-76 test specimen

3 Testing

3.1 Experimental Setup and Methodology

An Avery fatigue testing machine type 6,301 as shown in Fig. 4 used to test the composite specimen and calculate the number of cycles to failure. The power of motor (0.5 HP) is connected along with the testing unit. The grips are provided for bend test where the load is applied at one end of the specimen by an oscillating spindle motor. The eccentric attachment is adjustable to give the necessary range of bending angle. Then the dial gauge is adjusted to zero. The tip is adjusted for the dial gauge to measure the angle of twist as shown in the Fig. 5. Then the motor is started on to the fatigue, the specimen attains fracture after long time as shown in Fig. 6. The revolution counter is fixed to the motor to record the number of cycles; the cycling rate was 1,420 rpm.

The number of cycles required for attaining fatigue fracture and fatigue life is calculated using formula

$$\text{No of Cycles: Time of Fracture (in Min)} \times \text{Motor RPM}$$

Bending moment is calculated using formula

$$M = (\sigma bh^3/6)$$

where

σ Stress applied
b Lowest Breadth of the specimen (15 mm)
h Thickness of the specimen (3.2 mm)

Fig. 4 AVERY fatigue machine 6,301 with specimen

Fig. 5 Dial gauge mounted in the machine

Fig. 6 Three regions of the fatigue failure mechanism observed in the specimen during loading process

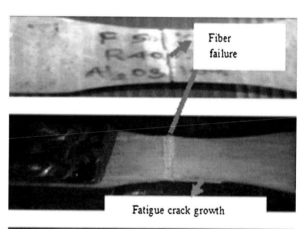

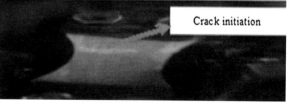

From the formula, the corresponding dial gauge reading is 90° and bending moment is 5 Nm, hence the eccentric should be 45°and it causes the deflection of 25 mm.

4 Results and Discussion

4.1 Fatigue–life Diagram

The AVERY Fatigue machine performs the fatigue tests under cyclic reversed loading with stress ratio R and also S–N data is used to predict the fatigue behavior of the composite specimen as shown in the Fig. 7 and 8. The three regions of the damage behavior were found such as crack initiation, fiber-matrix debonding, fiber failure or fatigue growth during loading process [6] as shown in the Fig. 6. In the region I, the damage does not develop due to initial cyclic loading process and the composite materials obeys perfectly elastic manner in initial cyclic loading. Below the 2E + 05 cycles, initial damage of matrix cracking occurs due to reversed cyclic loading as shown in Fig. 7.

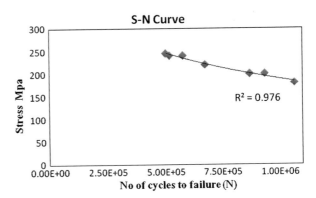

Fig. 7 S–N fatigue data for GFRP with 5 % Al_2O_3

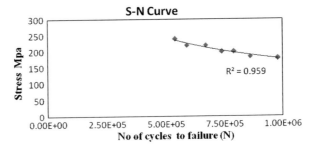

Fig. 8 S–N fatigue data for GFRP with 10 % Al_2O_3

Matrix cracking and fiber crack growth region was continuously monitored during the fatigue test. In the region II, the fiber failure occurs due to increasing number of cycles and combined matrix and fiber damages predominately giving rise to the rapid weakening of the composite specimen. The failure damage mechanism and fatigue crack growth depends on the applied cyclic load level [13, 14]. In final region three, the failure load above 10^5 cycles is expected to produce the damage to the fiber (fracture) failure within an infinitely long time and, hence underneath the composite specimen for an expected endurance limit using S–N fatigue data as shown in Fig. 8. It is observed that, as the Al_2O_3 particle percentage decreases in the epoxy matrix, the total deformation increases gradually [15]. From the Fig. 8 observed that, as the proportion of Al_2O_3 particle increases inside the epoxy resin matrix, the fatigue life of the specimen increases gradually and also the number of cycles slightly increases.

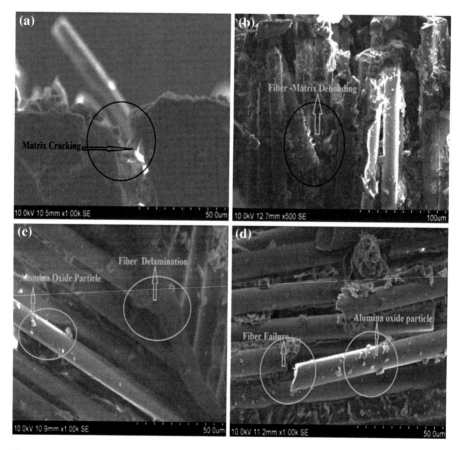

Fig. 9 Micrographs of fatigue failure observed in GFRP with Al_2O_3 composite specimen **a** Matrix cracking **b** Fiber-Matrix debonding **c** Fiber delamination **d** Fiber failure and Al_2O_3 particle

4.2 Images of Scanning Electron Microscope Analysis

The studies of failure/fractured surfaces are done using Scanning electron microscope. The fractured surface is cut to a suitable size, cleaned from any organic residues, and mounted on a specimen holder for viewing in the SEM. Figure 9 shows, the microstructure of the specimen which has been taken at two range of feed rate of 100 μm/min and 50 μm/min. The fractured surface of composite specimen was examined; then microstructure indicated that damage modes are observed in composite specimen. The corresponding images taken using SEM is shown in Fig. 9. Scanning electron microscope examinations of sample specimens subjected to fully reversed tension- compression test indicated that the failure mechanisms associated with cyclic loading along the fiber direction, are matrix cracking, fiber/matrix debonding and fiber failure as shown in Fig. 9a–d. It can be seen in Fig. 9a the matrix cracking was observed below 2E + 05 cycles. It may be clearly noted Fig. 9b, c that fiber debonding and visible fiber delamination was observed for the same number of applied load cycles. The further fatigue crack growth of such delamination to critical size, leads to final fiber failure above 10^5 cycles. The Fig. 9d has been shown that fiber failure is additional fatigue damage and growth of all these damages leads to final fatigue failure of the hybrid composite specimen but the GFRP with 10 % Al_2O_3 particle exhibits an improved fatigue life over that of GFRP with 5 % Al_2O_3 particle.

5 Conclusion

The fiber reinforced with alumina oxide composite specimen (Hybrid) were tested under cyclic reversed loading as fatigue testing, to estimate the fatigue life and SEM an examination was carried out for fractured areas.

The following conclusions were obtained from this work.

i. The three stages of the damage performance are predicted at reversed (tension -compression) cyclic load.

ii. The fatigue life of GFRP with 10 % Al_2O_3 particle was slightly increased than that of 5 % Al_2O_3 particle. As the alumina oxide percentage increases in the epoxy matrix, the total number of cyclic time increases gradually, and also as the proportion of alumina oxide increases (5–10 %) inside the epoxy resin matrix, the fatigue life of the specimen slightly increases.

iii. Scanning electron microscope techniques show that during the application of the fatigue load, the damage is constituted by the fracture of the fibers and matrix and identify their growth mechanism. The SEM observations show that the Al_2O_3 particles are dispersed uniformly in the matrix which is essential for enhancement of the properties of the composite specimen.

References

1. Barbero EJ (1998) Introduction to composite materials design. materials science and engineering series, 1st edn. Taylor and Francis, London
2. McCullough RL (1971) Concepts of fiber–resin composites. Marcel Dekker, NY, p 12
3. Edwards KL (1998) An overview of fiber reinforced plastics for design purposes, J Mater Des, 19
4. Dyer KP, Issac DH (1998) Fatigue behavior of continuous glass fibre reinforced composites. Compos Part B 49B:725–733
5. Caprino G, Giorleo G (1999) Fatigue life time of glass fabric/epoxy composites. Composites Part A 30:299–304
6. Khelifa MZ, Al-shukri (2008) Fatigue study of E glass fiber reinforced polyester composite resin under fully reversed loading and spectrum loading. J Eng Technol, 26: 10
7. Freire R, Acquino C (2005) Fatigue damage mechanism and failure prevention in fiber glass reinforced plastic. Mater Res 8:1
8. Chung S, Im Y, Kim H, Park S, Jeong H (2005) Evaluation for micro scale structures fabricated using epoxy-aluminum particle composite and its application. J Mater Process Technol 16:168–173
9. Demers CE (1998) Tension-tension axial fatigue of E glass fiber reinforced polymeric composites: tensile fatigue modulus J constr Build Mater, 12 pp 51–58
10. Degrieck J, Van Paepegem W (2001) Fatigue damage modeling of fiber reinforced composite materials review. Appl Mech Rev 54:279–300
11. (2007) http://mihd.net/, ASTM D 5687/D 5687 M-95. standard guide for preparation of flat composite panels with processing, guidelines for specimen preparation, 15: 03
12. (2007) ASTM D 3459/D 3479 M. standard test method for tension–tension for fatigue of polymer matrix composite materials. Rev. 76, http://www.mihd.net
13. El-assal AM, Khashaba UA (2007) Fatigue analysis of unidirectional GFRP composites under combined bending and torsional loads. Compos Struct 79:599–605
14. Rita R, Bose NR (2001) Behaviour of E glass fibre reinforced vinyl ester resin composites under fatigue condition. Bull Mater Sci 24:137–142
15. Zunjarrao SC, Singh RP (2006) Characterization of the fracture behavior of epoxy reinforced with nanometer and micrometer sized aluminum particles. Compos Sci Technol 66:2296–2305

Investigation of Chip Morphology and Tool Wear in Precision Turning Process

R. Vinayagamoorthy and M. Anthony Xavior

Abstract The objective of the work is to perform precision turning using conventional lathe on Ti6Al4V under dry working conditions. Various parameters that affect the machining process were identified and a consensus was reached regarding its values. The proposed project is to perform machining under these conditions and parameters and to compare the chip morphology and tool. This thesis work aims to optimize the machining performance in precision turning operations. In finishing operations, Tool wear and Chip Morphology are major concerns. Hence, to quantify the machining performance in precision turning operations, two criteria are used in this thesis; Chip breakability and Tool Wear. Chip breakability takes care of chip shape and size, and chip side flow. By finding optimal depth of cut and feed in each segment through the profile, the machining performance in precision turning can be improved.

Keywords Titanium alloy • Precision machining • Tool wear and chip morphology • Cutting forces dry conditions

1 Introduction

Manufacturing industries all over the world are focusing more on achieving higher quality and reliability for existing products and introducing new products into the market. One of the major initiatives associated with this trend is the development of precision manufacturing systems. The issues associated with the scaling down of a macro scale manufacturing system to precision levels are assessed. A brief overview of the developments in precision machining, in terms of the manufacturing processes, equipment, and the techniques used, is given in this report. The operational requirements and issues faced in the scaling down of conventional manufacturing systems into micro, meso, and nano levels are discussed [1]. Some

---

R. Vinayagamoorthy (✉) · M. Anthony Xavior
School of Mechanical and Building Sciences, VIT University, Vellore 14, Tamil Nadu, India
e-mail: vinayagamoorthy.r@vit.ac.in

S. Sathiyamoorthy et al. (eds.), *Emerging Trends in Science,*
Engineering and Technology, Lecture Notes in Mechanical Engineering,
DOI: 10.1007/978-81-322-1007-8_21, © Springer India 2012

specific issues associated with non-orthogonal single point diamond facing of Ti-6AL-4V are addressed in detail. The machining parameters ranged from micro levels to meso and nano levels. The cutting energy associated with the process is analyzed to compare the cutting mechanism at different scales. At small depths of cut, the edge preparation of the tool significantly affected the cutting process. Non cutting plastic work on the material including plowing and flank face rubbing had more impact on the unit energy than the cutting process at lower depths of cut.

Manufacturing systems have advanced to higher levels of precision to cater to the growing needs of the 'soft manufacturing' sectors dominated by the electronics, computer, and biomedical industries. One of the main features of this development is the scaling down of the operational characteristics of manufacturing from macro levels to micro, meso, and nano levels. The chapter looks at some of the developments that have occurred in the precision manufacturing sector. A deeper look into the characteristics of manufacturing at a scaled down level is attempted. Manufacturing dominates world trade. It is the main wealth creating activity of all industrialized nations and many developing nations. A manufacturing industry based on advanced technologies with the capability of competing in world markets can ensure a higher standard of living for an industrial nation [2]. The soft manufacturing sectors aim to achieve greater miniaturization and packing densities for the components. For the electronics and computer industries smaller sizes implies less time for information transfer and higher input/output rates [3]. These urge manufacturers to focus on the scaling down of manufacturing processes to achieve the required miniaturization. The manufacturing processes thus scaled down fall under the general category of precision manufacturing.

This takes a look at some of the issues in the manufacturing system from a macro level to micro, meso, and nano levels. The relevance of precision machining in the current manufacturing scenario is explained. Some of the important aspects associated with precision machining are described in detail. The difference in characteristics and capabilities between the existing 2D fabrication processes and the 3D precision machining systems is explained. The history of precision machining is reviewed by detailing the development of machining processes and its equipment. They also explains the characteristics of the 3D precision machining systems by looking at the scaling down issues associated with the different operational characteristics of a machining system like machine tool, tooling/production engineering, material removal process, product design, and assembly/material handling. The need for precision machining exists in areas other than electronics. The attempts to define precision machining can be traced back to the 1970s [4]. Some of the definitions that closely capture the nature of the machining processes covered by the precision regime are given below. Precision machining is "the process by which the highest possible dimensional accuracy is achieved at a given point in time" according to the definition given by Taniguchi [5]. The machine accuracy capabilities Taniguchi predicted along with the processes or tools used to achieve it.

The target material used for the experimentation is Ti-6Al-4V. The high toughness to mass ratio of titanium alloys and excellent resistance to corrosion has made this titanium alloy a very suitable component in the industry [5–7]. Gedee Weiler MLZ

250 V variable speed adjusting capstan lathe is used for the experiment. PVD coated carbide tool with 98 HRC hardness, nose radius of 0.1, 0.2, and 0.4 are used for the turning operation. Surface roughness was measured using mitutotyo surf test SJ-301 portable surface roughness tester with a sampling length of 4 mm. The cutting temperature is measured using a thermocouple. The cutting parameters were so selected after comparison with different literature surveyed. The design of experiments and analysis of variance was done using Minitab 15 software. This is similar to a situation where the tool is rigidly supported and cuts the workpiece under a stress such that no median vents are generated, but the material below the tool is plastically deformed due to large hydrostatic pressure. Due to this, a large nose radius is theoretically desirable, but the waviness control of the large nose radius is very expensive Fang [8]. A negative degree rake angle tool with a large nose radius will have an effective negative rake angle that could be much higher, creating excessive pressure that could damage the surface.

2 Design of Experiments and Observations

Design of Experiments is a highly efficient and effective method of optimizing process parameters, where multiple parameters are involved. The design of experiments using Taguchi approach was adopted to reduce the number of trials. The time and cost for doing an experiment is very high, therefore it is necessary to select an orthogonal array with minimum number of trials. In this research work L27 orthogonal array is chosen which is a multilevel experiment Feed, depth of cut, cutting speed, nose radius are the feed four factors considered for the experiment.

2.1 Chemical Composition of Titanium Alloy (Grade 5)

These treatments Ti-6al-4v is an example of β alloys. Alpha + Beta alloys contain compositions which support a mixture of α and β phases. These alloys may contain from 10 to 50 % of β phase at room temperature (Table 1).

2.2 Mechanical Properties of Ti-6Al-4 V

These treatments Ti-6AL-4V is an example of β alloys. Alpha + Beta alloys contain compositions which support a mixture of α and β phases. These alloys may contain from 10 to 50 % of β phase at room temperature (Table 2).

Table 1 Chemical composition of titanium alloy

Alloy	Al (%)	V (%)	Fe (%)	C (%)	Ti (%)
Ti-6Al-4V	6.40	3.89	0.16	0.002	Balance

Table 2 Mechanical properties of Ti-6Al-4V

Hardness (HRA)	Hardness knoop	Tensile strength, ultimate	Elongation (%)	Poisson's ratio	Modulus of elasticity	Density
70	363	950 Mpa	14	0.342	113 Gpa	4.43 g/cm3

2.3 Tool Specifications of CCGT09T301F Coated Carbide Insert

Table 3 Tool specifications of CCGT09T301F Coated carbide insert

Composition	80 % Al_2O_3 and 20 % TiC
Grain size	3.0 μm
Transverse rupture strength	551–786 MPa
Average density	3.90–3.99 g/cm^3
Youngs modulus	641 GPa
Hardness	91–94 HRA
Coefficient of thermal expansion	Good

The cost of machining a Ti6Al4V sample is very high and a highly time consuming process. For a four factor three level experiment more than 80 experiment have to be carried out leading to a very huge expenditure and waste of time. Taguchi [9] designed certain standard orthogonal arrays by which the simultaneous and independent evaluation of two or more parameters for their ability to affect the variability of a particular product or process characteristics can be done in a minimum number of tests Table 3. Taguchi's method of experimental design provides a simple, efficient, and systematic approach for the optimization of experimental designs for performance quality and cost. Table 4 shows the machining parameter and their levels. Table 5 shows the machining parameters and observation for each trail of experiment.

2.4 Machining Parameters and Trail Level

Table 4 Machining parameters and trail level

Cutting parameter	Level 1	Level 2	Level 3
Feed (mm/rev)	0.02	0.04	0.06
Depth of cut (mm)	0.05	0.10	0.15
Cutting speed (m/min)	30	60	90
Nose radius (mm)	0.1	0.2	0.4

3 Experimental Observations

Fig. 1 Experimentation setup

4 Results and Discussion

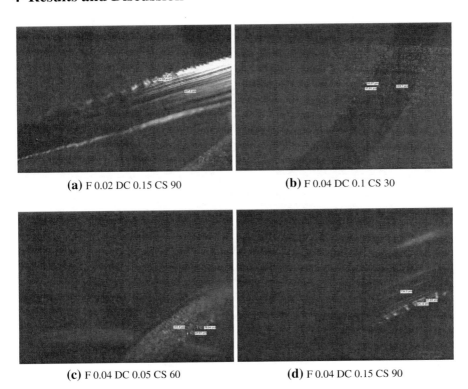

Fig. 2 Saw tooth height and cutting speed. Saw tooth height and Depth of cut

(a) f 0.02, d 0.1, v_c 90 (b) f 0.02, d 0.05, v_c 30

Fig. 3 Tool Wear

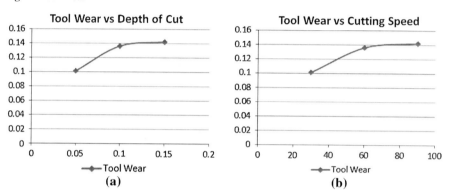

Fig. 4 Graphs of Tool Wear versus Depth of Cut and Cutting Speed

In addition, the highly non linear thermo mechanical behaviors of a work material are heavily coupled during precision machining. Even if these two issues had been resolved, the exact wear mechanisms would not have been identified and described quantitatively. During machining, the cutting tool directly interacts with a work material.

In this experimental work effect of cutting conditions on tool wear and chip morphology was examined in the machining of Titanium alloy. The work is of interest because of its relevance to increasing hard machining implementation as a quicker, cleaner, and practical alternative to finish grinding. Cutting condition was quantified to measure the effect of Chip Morphology and Tool wear. The following conclusions can be drawn based on this work: Saw Tooth Height increases with an increase in Depth of Cut and Cutting Speed at constant Feed and Nose Radius. Tool Wear increases with an increase in Depth of Cut and Cutting Speed at constant Feed and Nose Radius. No monotonic relation can be established between the Machining Parameter and Chip Width (Fig. 1).

Chip is generated by shearing the work material while the generated heat from plastic deformation of the work material and the interfacial friction between work

Investigation of Chip Morphology and Tool Wear in Precision Turning Process

Table 5 Experimental observations

S no	Feed	Depth of cut	Cutting speed	Nose radius	Saw tooth		Chip width	Max tool wear
					Width	Height		
	mm/rev	mm	m/min	mm	mm	mm	mm	
1	0.02	0.05	30	0.1	0.012	0.013	0.149	0.038
2	0.02	0.05	60	0.2	0.031	0.023	0.170	0.046
3	0.02	0.05	90	0.4	0.021	0.035	0.201	0.239
4	0.02	0.1	30	0.2	0.032	0.025	0.203	0.101
5	0.02	0.1	60	0.4	0.015	0.021	0.272	0.117
6	0.02	0.1	90	0.1	0.023	0.020	0.186	0.129
7	0.02	0.15	30	0.4	0.013	0.014	0.145	0.222
8	0.02	0.15	60	0.1	0.027	0.017	0.267	0.142
9	0.02	0.15	90	0.2	0.024	0.025	0.207	0.142
10	0.04	0.05	30	0.1	0.007	0.011	0.087	0.134
11	0.04	0.05	60	0.2	0.016	0.023	0.161	0.142
12	0.04	0.05	90	0.4	0.021	0.024	0.220	0.173
13	0.04	0.1	30	0.2	0.026	0.017	0.223	0.150
14	0.04	0.1	60	0.4	0.026	0.037	0.269	0.141
15	0.04	0.1	90	0.1	0.024	0.009	0.068	0.160
16	0.04	0.15	30	0.4	0.047	0.060	0.276	0.202
17	0.04	0.15	60	0.1	0.023	0.025	0.236	0.229
18	0.04	0.15	90	0.2	0.033	0.026	0.134	0.163
19	0.06	0.05	30	0.1	0.027	0.021	0.089	0.288
20	0.06	0.05	60	0.2	0.026	0.025	0.190	0.266
21	0.06	0.05	90	0.4	0.015	0.023	0.177	0.270
22	0.06	0.1	30	0.2	0.021	0.017	0.240	0.221
23	0.06	0.1	60	0.4	0.020	0.024	0.238	0.231
24	0.06	0.1	90	0.1	0.013	0.018	0.147	0.150
25	0.06	0.15	30	0.4	0.076	0.039	0.282	0.195
26	0.06	0.15	60	0.1	0.017	0.012	0.165	0.022
27	0.06	0.15	90	0.2	0.025	0.021	0.049	0.283

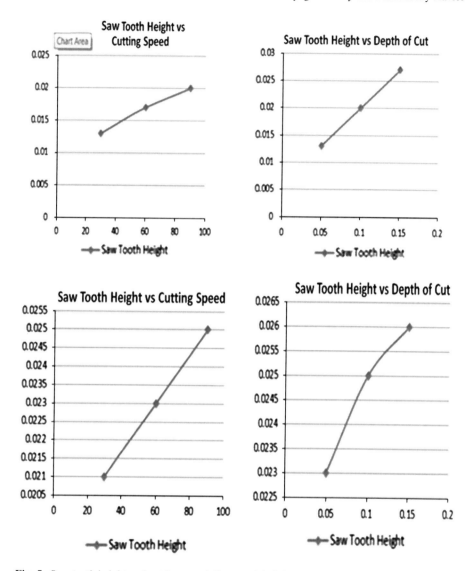

Fig. 5 Saw tooth height and cutting speed. Saw tooth height and Depth of cut

material and cutting tool transfers into a cutting tool. The temperature in both work material and cutting tool increases substantially as the cutting condition becomes more severe. This is the motivation for this paper so that the fundamental mechanisms of tool wear can be revealed to the researchers in this field. The readers should be cautioned that this paper will claim neither to be a complete review paper on this topic nor to represent a complete understanding of the tool wear Fig. 2. The purpose of this paper is to cite the literatures that have delineated the physics behind the tool wear to provide the fundamental tool wear mechanisms in machining. Tool wear is of foremost importance in metal cutting. Owing to its direct impact on the surface quality and precision machining economics, tool wear is commonly used to evaluate the performance of a cutting tool. Many research studies to understand and predict tool wear have been carried out. However, most of these studies are considered to be an empirical approach to tool wear. Consequently, many fundamental issues have not been resolved mainly due to the complex physics behind tool wear (Fig. 3 and 4).

The complexity surrounding tool wear stems from many factors including work material, machine tool, cutting tool, coolants, and cutting conditions. Because of the coupled effects of these factors, the tool chip and tool work interfaces have almost unidentifiable contact conditions with highly localized interfacial temperatures and tractions (Fig. 5).

5 Conclusions

In this study effect of tool wear and chip morphology on the cutting conditions was examined in the machining of Titanium alloy. The work is of interest because of its relevance to increasing hard machining. Implementation as a quicker, cleaner, and practical alternative to finish grinding. Cutting condition was quantified to measure the effect of Chip Morphology and Tool wear. The following conclusions can be drawn based on this study.

- Saw Tooth Height increases with an increase in Depth of Cut and Cutting Speed at constant Feed and Nose Radius.
- Tool Wear increases with an increase in Depth of Cut and Cutting Speed at constant Feed and Nose Radius.
- No monotonic relation can be established between the Machining Parameter and Chip Width.

References

1. Corbett J, McKeown PA, Peggs GN, Whatmore R (2000) Nanotechnology: international developments and emerging products. Ann CIRP 49(2):523–544
2. Ei Baradie MA (1993) Surface roughness model for turning grey cast iron, J. Eng Manuf 207:43–54
3. Machado AR, Wallbank J, (2005) Machining of titanium and its alloys: a review Proceedings of the institution of mechanical engineers part B, J Eng Manuf. 204:11: 53–60
4. McKeown PA (1986) High precision manufacturing and the british economy, James clayton lecture, Inst Mech E Roc, Vol 200, No 76
5. Taniguchi N (1983) Current status in and future trends of precision machining and ultrafine material processing. Ann CIRP 32(2):1–8
6. Thomas TR (1998) Trends in surface roughness. Int J Mach Tools Manuf 38(5–6):405–4114. Thomas TR (1999) Rough Surfaces, 2nd edn. Imperial College Press, London
7. Trent EM, Wright PK (2000) Metal Cutting, 3rd edn. Butterworth Heinemann, Boston
8. Vajpayee S (1981) Analytical study of surface roughness in turning. Wear 70:165–175

Investigation of Micromachining on CNC

D. Rajkumar, P. Ranjithkumar and C. Sathiyanarayanan

Abstract This paper presents a method for material removal in micron level with high accuracy and surface finish using a CNC machine. In this work, commercially available pure aluminum 19,000 rods are micro-machined with TNMG 1,60,404 (TiN) Inserts, as cutting tool. The Al19000 rod is electroplated with Nickel for plating thickness of 21.5 μm. While machining the actual material removed is the difference between the depth of cut given and the electroplating thickness. Since the electroplating thickness is less than 100 μm and minimum possible depth of cut given is 100 μm, the actual work-piece metal removal is in the order of microns. Thus micromachining is achieved using CNC turning machine. The experiments were conducted with various combinations of cutting speed, feed, and depth of cut. Performance responses such as Surface Roughness (SR) and Metal Removal (MR) for different conditions were measured and reported. Optimum machining conditions were identified. The machined surface was viewed using Scanning Electron Microscope (SEM) and SEM image were correlated with micro scratches, worn surface, dirty layer formed for different cutting condition.

Keywords CNC turning · Al19000 · Micromachining · Surface roughness

---

D. Rajkumar (✉)
Department of Production Engineering, JJ College of Engineering and Technology, Tiruchirappalli, Tamilnadu, India
e-mail: profdrajkumar@gmail.com

P. Ranjithkumar
Department of Mechanical Engineering, MAM School of Engineering, Sriganur, Tamilnadu, India
e-mail: ranjjith@mamce.org

C. Sathiyanarayanan
Department of Production Engineering, National Institute of Technology, Tiruchirappalli, Tamilnadu, India
e-mail: csathiyanarayanan@yahoo.co.in

S. Sathiyamoorthy et al. (eds.), *Emerging Trends in Science, Engineering and Technology*, Lecture Notes in Mechanical Engineering, DOI: 10.1007/978-81-322-1007-8_22, © Springer India 2012

1 Introduction

Machining is an important process in any manufacturing Industry. Machining removes unwanted material from the workpiece to give the required size and shape of the component. Many kinds of machining processes have been developed in the manufacturing field to suit the requirement of industries and to satisfy the quality requirement of the components produced. The quality requirement includes the dimensional accuracy and surface finish of the components. The machining processes are classified as conventional machining and unconventional machining processes. As a recent development in the machining, micromachining paves its way to produce micro features in the components or miniature of components. Micromachining is the basic technology for production of miniaturized parts and components, fabrication of miniature components in terms of micron range dimensions [1]. Applications of micro component to enhance health care, quality of life, and growth grabbed huge attention of the commercial industries. Some examples are micro channels for micro fuel cell, lab-on chips, shape memory alloy, fuel cell applications, miniature actuators and sensors, medical devices, etc. [2, 3]. The micromachining is generally used to define the material removal process for parts having dimensions that lie between (1–999) μm [4]. In Micromachining, a micro turning process is a conventional material removal process that has been miniaturized. The tool wears mostly affected by the Form accuracy and dimensional stability of the production of micro machined parts [5]. In micro turning of copper, response of copper to the tool wear and surface Roughness are good. Machining can be obtained with medium cutting velocity (24 m/min), feed (8 μm/rev) and depth of cut (10 μm) [6]. A multipurpose machine tool is mostly used for fabrication of micro pin and micromachining. Micro turning is the one of the process for tool based micromachining. The dimension of fabricated micro pin are the minimum Ø 276 μm, maximum Ø 475 μm and length is 2 mm while machining used both the straight micro turning and taper micro turning on brass as workpiece [7]. There are immeasurable uses of aluminum for the fabrication of micro devices and micro features for generic micromachining technology. Since aluminum is the main material used in IC technology, high-aspect-ratio and also processing tool to develop the fabrication technologies, automobile body components. The aluminum micro gear was fabricated in thickness of 45 μm, with an outer diameter of 300 μm, an inner diameter of 50 μm, and a tooth width of 40 μm using electroplated aluminum material and micromachining process was achieved on micro molding [8]. The response of the paper is surface roughness while machining with different machining conditions such as cutting speed, feed, and depth of cut, etc. Most of the researchers studied the relation between the input parameters and surface roughness. Surface roughness (SR) are one of the main results of process parameters and cutting conditions [9, 10]. The quality of the surface is important in the evaluation of machine tool productivity. Hence, it is important to achieve a dependable tolerance and surface finish. Recent examination has shown that increasing the cutting speed can help to maximize productivity and at the same time, improves surface quality [11]. Bharathi et al. [12] developed the higher cutting speed, lower feed, and depth of cut is advisable to obtain better surface roughness for the given range. The Feed has mostly affected the machining time and surface roughness when compared to cutting speed and depth of cut. The author proved that the surface roughness is directly proportional to feed and inversely proportional to cutting speed. Cemal et al. [13] used a set of procedure to determine the machining conditions for turning operation with lower production cost as the aimed function. Using Similar machining variable, calculate the production time and cost for different work piece and tool material. Meng et al. [14] described a machining theory to calculate optimal machining condition when turning are the minimizing cost or maximizing production rate. Dilbag et al. [15] investigated that the feed is the foremost factor affecting the surface roughness, followed by the nose radius, cutting velocity, and effective rake angle while the machining of steel as workpiece and the aluminum oxide and titanium carbo nitride, having different nose radius and different effective rake angles, were used as the insert. Most of the researchers had studied the surface roughness using Taguchi's techniques [16–19] wide application in the engineering field. In order to obtain the behavior of a given process, first design the experiments with limited data and processing the experiments, analyzing the data. After the completion of the experiment, the data from all the experiments in the set are analyzed to determine the effect of the various design parameters. Conducting Taguchi experiments in terms of orthogonal arrays allows the effects of several parameters to be determined efficiently and it is an important technique for a robust design. The treatment of the experimental results is based on the analysis average and the analysis of variance.

The micromachining removes material from the components in a very little amount that is in micron level. The micromachining is accomplished with very good surface finish and dimensional accuracy. An important issue in going for micromachining in the cost of the machine tool comparing to the cost of the conventional machine tool. The cost of machine tool for micromachining is very high whereas the cost of the conventional machine tool is about at least 50 times less. This cost is not affordable for many of the small scale industries. A method which can perform micromachining in CNC machines will be a great gift to such small scale industries.

2 Experimental Details

The main objective of the work is to study the feasibility of performing micromachining in Fanuc CNC lathe and the co-relation between cutting speed, feed and depth of cut on SR and MR. The electroplated aluminum is shown as Fig. 1. The electroplated AL19000 material is taken as the work piece materials for all trials diameter of 8.4777 mm and machined length of 135 mm with incremental 15 mm for each experiment. Totally nine experiment conducted on workpiece. The nickel

Fig. 1 Aluminum before plating and after plating

plating thickness is 25.1 μm. The chemical composition of the work piece is given in Table 1. The work piece material's qualities are confirmed to grade AL19000. The experiments are conducted in Fanuc CNC lathe. The work piece positioning accuracy is ±0. 010 μm and run out is tested by a dial gauge heavy duty magnetic base. The range of cutting parameters is selected based on tool manufactures handbook. In this investigation cutting insert TNMG 160404 (TiN) are taken for turning process. SR is measured by the SURF TEST 211 and its denoted by Ra. The machined diameter is measured using 3D Co-ordinate measuring machine (CMM) by the METRONIC 3DPlus. Measurements are taken by the proper setting of work pieces and the instrument. In the present work, three parameters each set at three levels are chosen for experimentation. The turning parameters and their levels chosen in all cases are presented in Table 2. In order to have a complete study of turning process, the range of parameters selected, and an appropriate plan of experimentation is essential to reduce the cost and time consuming. Hence, an experimental plan based on Taguchi's L9 orthogonal array has been selected. The factors are assigned to columns 1–3, respectively, and nine experiments are carried out under dry condition with different combinations of parameter levels. The machined diameter details are shown in Table 4.

Table 1 Chemical composition of Al19000

IS	Cu	Mg	Si	Fe	Mn	Zn	Ti	Cr	AL
19,000	0.1	–	0.5	0.6	0.1	–	–	–	99

Table 2 Machining parameters and levels

Parameter	Designation	Level 1	Level 2	Level 3
Cutting speed (m/min)	V	800	1200	1600
Feed (mm/rev)	F	0.07	0.14	0.21
Depth of cut (μm)	D	20	25	30

Investigation of Micromachining on CNC

Table 3 Taguchi analysis sr versus V, F, and D

Level	V	F	D
1	−1.70297	1.89420	−0.05761
2	−2.27557	−2.15433	−7.77449
3	0.35539	−3.36303	4.20894
Delta	2.63096	5.25724	11.98343
Rank	3	2	1

Table 4 Machined diameter details

Experiment number	Doc (mm)	Machined diameter (mm)	Metal removal (MR) (mm)
1	0.03	8.4175	0.0602
2	0.025	8.4278	0.0499
3	0.02	8.4373	0.0404
4	0.025	8.4264	0.0513
5	0.02	8.4363	0.0414
6	0.03	8.4186	0.0591
7	0.02	8.4382	0.0395
8	0.03	8.4165	0.0612
9	0.025	8.4285	0.0492

There are three categories of quality characteristic in the analysis of the S/N ratio:

1. The-lower-the-better
2. The-higher-the-better
3. The-nominal-the-better

Since the quality characteristic is to be minimized, the-lower-the-better category is used to calculate the S/N ratio for SR Eq. (1) shows the smaller the better characteristic.

$$\eta = -10 \log 10 \, 1/n \sum_{i}^{n} y^2 \qquad (1)$$

where η is the signal to noise ratio, n is the number of repetitions of experiment and y is the measured value of quality characteristic. Minitab 14 statistical software has been used for the analysis of the experimental work. The software analyses the experimental data and then provide the calculated results of signal-to-noise ratio. The effect of different process parameters on SR is given in Table 3. The main effect plot for S/N ratio for SR is shown in Fig. 2. From these figures, process parameter changes from one level to another. The average value of S/N ratio has been calculated to find out the effects of different parameters and their levels. In addition, a statistical ANOVA is performed to see those process parameters that significantly affect the responses. The experimental results are analyzed with ANOVA which is used for identifying the factors which significantly affecting the performance measures. The result of the ANOVA for SR is shown in Table 5.

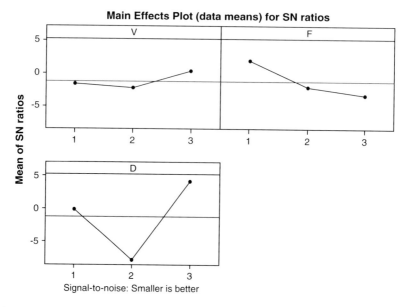

Fig. 2 The main effect of plot for S/N ratio for SR

Table 5 Anova for Sr

Source	DF	SS	MS	F	P
V	2	0.0964	0.0964	0.13	0.766
F	2	0.7288	0.7288	2.31	0.302
D	2	5.6204	5.6204	17.81	0.053
Error	2	0.3155	0.3155		
Total	8	6.7611			

S = 0.397171 R−Sq = 95.33 % R−Sq (adj) = 81.34 %

This analysis is carried out for a significance level of $\alpha = 0.05$, i.e., for a confidence level of 95 % (Table 4).

The plating thickness varies from the edge of the rod to the center of the rod in the range of ± 10 μm. Initially the lower value of the electroplated rod diameter is machined and secondly machining the actual material removed is the difference between the depth of cut given and the electroplating thickness. The average of metal removal is calculated in terms of micron. The required metal removals are in 40 μm, 50 μm, and 60 μm that can be achi CNC machines CNC machine.

3 Results and Discussion

3.1 Optimal Setting of Machining Parameters

The response table of S/N ratio for SR by Taguchi analysis is shown in Table 3. It shows the rank value of machining parameters and it is stated by the depth of cut has the strongest influence on SR followed by Feed and cutting speed. From Table 3, shows the S/N ratio at each level of machining parameters, and the rank level changes when the machining parameters change from one level to another. The main effect plot for SR is shown in Fig. 2. The surface Roughness can be determined using machining conditions which in turn determine the quality of the Surface.

The main effects plots are used to determine the optimal design conditions to obtain the optimum SR. The graph shows the reduced level of SR, with depth of cut should be set to its lowest level (30 µm), Cutting speed to its high level (1,600 m/min), and feed to its low level (0.07 mm/rev). Table 5 shows the results of ANOVA for SR. It is observed that the depth of cut (p value $= 0.053$) is the most significant parameter followed by feed (p value $= 0.302$), while the effect of cutting speed has not been found statistically significant (p value $= 0.766$). A larger F value shows the greater impact on the machining performance characteristics. Larger F-values are observed for depth of cut and feed.

4 Surface Integrity

During machining, the associated surface structures are subjected to changes which are the result of plastic deformation. The surface integrity is dependent on the machining conditions, and the machining environments prevailed during machining. The machined surface topography was examined using scanning electron microscopy (SEM). It is observed from the SEM micrographs that the machined surface consists of various irregularities even though they appear smooth and shiny. The alterations such as, micro scratches, pit holes, worn surfaces, and dirty layer are seen on the surfaces. The quality of the machined surfaces above can be proved from the quantitative roughness assessment. Though the Surfaces show a development of a variety of modifications, these are even seen at a few locations on the surfaces. This can also be confirmed from the lower values of surface roughness as evident from the experimental examination.

The surface roughness value is shown for three different machining conditions; see Fig. 3 (Ra $= 1. 54$ µm), Fig. 4 (Ra $= 0.92$ µm), and Fig. 5 (Ra $= 3.05$ µm). The lower surface roughness value described when machining condition is lower feed, higher cutting speed. The formation of intensely messy excess metal layers and stitch line along the feed marks is an evidence of severe plastic deformation induced during machining. It is evident from the machined surfaces shown in Fig. 4. These parameter conditions introduce better surface integrity during when turning of AL19000. The higher surface value is high, when machining is done with lower cutting speed, higher feed conditions. The result is in poor surface integrity.

Fig. 3 SEM micrograph of × 300 (D = 0.02 mm, F = 0.14 mm/rev, V = 1,200 m/min)

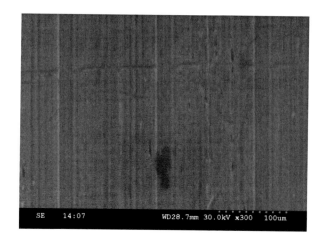

Fig. 4 SEM micrograph of × 300 (D = 0.02 mm, F = 0.07 mm/rev, V = 1,600 m/min)

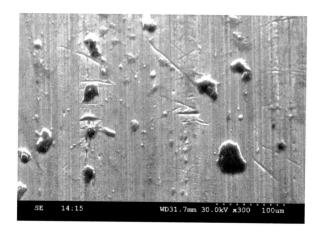

Fig. 5 SEM micrograph of × 800 (D = 0.025 mm, F = 0.14 mm/rev, V = 800 m/min)

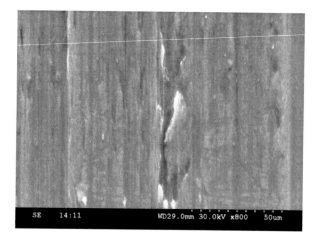

5 Conclusions

This paper proved that micromachining is possible with CNC machine and optimal machining condition analyzed. From the study of micromachining using Taguchi's techniques and ANOVA, the following can be concluded,

1. Micromachining can be performed on the Fanuc CNC machine.
2. From the results of ANOVA, the Depth of cut and Feed are the significant cutting parameters for affecting the Surface Roughness.
3. Minimum Surface Roughness are obtained at a cutting speed of 1,600 m/min, feed rate of 0.07 mm/Rev and depth of cut 0.03 mm.
4. Micro scratches, worn surfaces, and dirty layer are observed from the on SEM images of machined surface.

References

1. Taguchi N (1993) Current status in, and future trends of ultraprecision machining and ultrafine materials processing. Ann CIRP 32(2):573–582
2. Madou M (1997) Fundamentals of micro fabrication. CRC Press, Ohio
3. Corbett J, McKeon PA, Peggs GN, Whatmore R (2000) Nanotechnology, international developments and emerging products. Ann CIRP 49(3):473–478
4. McGeough JA (2002) Micromachining of engineering materials. Dekker, NY
5. Gowri S, Ranjith Kumar P, Vijayaraj R, Balan ASS (2007) Micromachining: technology for the future. Int J Mater Struct Integrity 1(2/3):161–179
6. Ranjith Kumar P, Balan ASS, Gowri S (2010) Monitoring and prediction of tool wear in micro turning of copper. Int J Precision Technol, 1: 3/4
7. Rahman MA, Rahman M, Senthil Kumar A, Lim HS, Asad ABMA (2006) Development of micro pin fabrication process using tool based micromachining. Int J Adv Manuf Technol 27:939–944
8. Frazier AB, Allen MG (1997) Uses of electroplated aluminum for the development of microstructures and micromachining processes, J Micro Electromechanical Syst, 6: 2
9. Choudhury IA, EI-Baradie MA (1997) Surface roughness prediction in the turning of high-strength steel by factorial design of experiments. J Mater Process Technol 67(1–3):55–61
10. Pawade RS, Suhas S, Joshi (2011) Multi-objective optimization of surface roughness and cutting forces in high-speed turning of inconel 718 using Taguchi grey relational analysis (TGRA), Int J Adv Manuf Technol 56:47–62
11. Oezel T, Karpat Y (2005) Predictive modeling of surface roughness and tool wear in hard turning using regression and neural networks int. J Mach Tools Manuf 45(4–5):467–479
12. Bharathi Raja S, Baskar S (2011) Particle swarm optimization technique for determining optimal machining parameters of different work piece materials in turning operation'. Int J Adv Manuf Technol 54:445–463
13. Cakir MC, Gurarda A (1998) Optimization and graphical representation of machining conditions in multi-pass turning operations. Comput Integr Manuf Syst 11:157–170
14. Meng Q, Arsecularratne JA, Mathew P (2000) Calculation of optimum cutting conditions for turning operations using a machining theory. Int J Mach Tools Manuf 40:1709–1733
15. Singh D, Rao PV (2007) A surface roughness prediction model for hard turning process. Int J Adv Manuf Technol 32:1115–1124

16. Ross P (1988) Taguchi techniques for quality engineering-loss function, orthogonal experiments parameter and tolerance design. McGraw Hill, NY, pp 10–50
17. Taguchi G, Konishi S (1987) Taguchi methods, orthogonal arrays and linear graphs, tools for quality engineering. Am Supplier Inst, , pp 35–38
18. Taguchi G (1993) Taguchi on robust technology development methods. ASME, New York, pp 1–40
19. Yang W, Tarng Y (1998) Design optimization of cutting parameters for turning operations based on the Taguchi method. J Mater Process Technol 84:122–129

Machining Performance of TiCN/Al$_2$O$_3$ Multilayer and B-TiC Nano Multilayer Coated Inserts on Martensitic Stainless Steel in CNC Turning

Kamaraj Chandrasekaran, P. Marimuthu and K. Raja

Abstract Martensitic stainless steels (AISI410) are widely used in aerospace industries, turbines, compressor components, shafting, nuclear applications, etc., because of the excellent corrosion resistance and high strength. Machinability and surface roughness are poor in turning AISI410, because of low thermal conductivity, high ductility, high strength, and rate of work hardening. Coated carbides are basically a cemented carbide insert coated with one or more thinly layers of wear resistant materials. It is well known that coating can improve the SR. This chapter presents the optimum cutting parameters for dry CNC turning on AISI410 by different types of coated cutting tools using Taguchi technique. The cutting tools were used multilayered TiCN + Al$_2$O$_3$, and nano multilayered B-TiC. Experiments were carried out by using Taguchi's L$_{27}$ orthogonal array. The effect of cutting parameters on SR was evaluated and optimum cutting conditions for minimizing the SR was determined. Analysis of variance (ANOVA) was used for identifying the significant parameters affecting the responses.

Keywords CNC turning · AISI410 · Surface roughness · Taguchi's method · ANOVA

K. Chandrasekaran (✉)
Department of Mechanical Engineering, Anna University of Technology Madurai, Madurai, Tamilnadu, India
e-mail: kchandrusekaran1984@gmail.com

P. Marimuthu
Department of Mechanical Engineering, Syed Ammal Engineering College, Ramanathapuram, Tamilnadu, India
e-mail: pmarimuthu69@gmail.com

K. Raja
Department of Mechanical Engineering, Anna University of Technology Madurai, Ramanathapuram Campus, Ramanathapuram, Tamilnadu, India
e-mail: rajagce@gmail.com

S. Sathiyamoorthy et al. (eds.), *Emerging Trends in Science, Engineering and Technology*, Lecture Notes in Mechanical Engineering, DOI: 10.1007/978-81-322-1007-8_23, © Springer India 2012

1 Introduction

Modern industry goal is to manufacture high quality products in a short time. CNC machines are capable of achieving high accuracy and very low processing time. Turning is common method for cutting, especially for the finished machined parts [1, 2]. In machining surface quality is one of the most specified customer requirements. surface roughness is one of the main results of process parameters such as tool geometry and cutting conditions (cutting speed, feed rate, depth of cut, etc.) [3]. AISI410, 420, and 440 are all considered as martensitic stainless steel and can be hardened like other alloy steels. AISI410 are widely used in aerospace industries for bearings, water valves, pumps, turbines, compressor components, shafting, surgical tools, plastic moulds nuclear applications, etc., which demand high strength and high resistance to wear and corrosion [4]. Machinability is poor in turning AISI410 because of low thermal conductivity, high ductility, high strength, high fracture toughness, and rate of work hardening [5]. Coated carbides are basically a cemented carbide insert coated with one or more thinly layers of wear resistant materials such as titanium nitride, titanium carbide, and aluminum oxide [6, 7]. It is well known that coating can reduce tool wear and improve the surface roughness [8, 9]. Therefore, most of the carbide tools used in the metal cutting industries is coated while coating brings about an extra cost [10, 11]. Thamizhmani [12] was investigated the AISI410 by PCBN cutting tool. They found that SR value was low at high cutting speed with low feed rate. Gutakorskis [13] was performed turning test on AISI410 by using nono coated cutting tool. Noortin [14] conducted a dry turning test on AISI410 using coated cermets and coated carbide tool. Tamizhmani [15] analyzed surface roughness on AISI410 with CBN cutting tool by turning. Ihsan [16] carried out turning tests on stainless steel to determine the optimum machining parameters. They found a decrease in tool wear and an increase in the surface roughness with decreasing the cutting speed. Cebeli [17] studied the machining characteristics of on stainless steel in turning processes. They found that the SR was increased when the depth of cut and feed rate were increased, while the cutting speed had an inverse influence. Lin [18] performed a research to evaluate the behavior of the stainless steel in high speed turning. He found that the SR values were decreased along with the feed rate. Only limited numbers of research papers are available for turning of AISI410. Various composition of cutting tools was used by past researchers for turning. But comparisons of single and multi layered cutting tools for turning AISI410 were not carried out by them.

To produce a quality product, the manufacturing engineers are employing off-line techniques also apart from on-line quality control (QC) methods. The QC activities at the manufacturing stage are on-line QC methods. The QC methods conducting at the design stage are off-line QC methods. Considerable advantages can be obtained by achieving product quality at the initial stage instead of controlling quality at the manufacturing stage [19].The Taguchi method of off-line QC includes all stages of product development. However, the key element for achieving high quality and low cost are parameter design. Through parameter design optimal levels of machining parameters can be determined [20]. Taguchi concept is product must be produced at optimal levels and with minimal variation in its functional characteristics [21]. Two factors affect the product functional characteristics are control factors and noise factors. Control factors are easily controlled. Noise factors are nuisance variables these are difficult, impossible or expensive to control [22]. Noise factors are responsible for causing a product functional characteristic. Controlling of noise factors is very costly or difficult; if not impossible [23] Taguchi method is recommended for solutions in metal cutting problems to optimize the machining parameters [24]. So, in this work Taguchi technique is used to determine the optimum machining parameters for different cases. ANOVA has been performed to analysis the effect of these machining parameters on SR.

2 Description of Experiment

The main objective of the work is to establish a relation between cutting speed, feed, and depth of cut on surface roughness. AISI410 material is taken as the work piece materials for all trials diameter of 32 mm and machined length of 60 mm. The chemical composition of the work piece is given in Table 1. The work piece material's qualities are confirmed to grade AISI410. The experiments are conducted in Fanuc CNC lathe. A work piece positioning accuracy is ± 0.010 and run out is tested by dial gauge heavy duty magnetic base. The range of cutting parameters is selected based on tool manufactures handbook. In this investigation four types of cutting inserts are taken for turning process. Coated cutting tool geometry, coating composition, layered thickness, coating methods and tool holders are shown in Table 2. Electron gun vacuum physical vapor deposition machine was used for B-TiC coating on uncoated carbide tool. Ultimate Vacuum of 1×10^{-6} m·bar can be achieved in clean, cold, empty degassed chamber after back filling the chamber with pure and dry nitrogen. The Electron Beam Gun is 4 Source, 8 KW 270 DEG.BENT Beam Gun with 10 kW Power supply. Operating pressure is below 5×10^{-4} m·bar. Surface roughness is measured for both cases by the SURF TEST 211 and it is denoted by Ra. Measurements are taken by the proper setting of work pieces and instrument. In the

Table 1 Chemical composition of AISI 410

C	Si	Mn	P	S	Cr
0.095	0.341	0.680	0.040	0.0063	12.170

Table 2 CNS inserts details

Case	Composition	Thickness	Coating method	Tool geometry	Tool holder
Type-I	TiCN/Al$_2$O$_3$	14 µm	CVD	CNMG 120408	PCLNR 25 × 25 M12.1
Type-II	B-TiC	500 nm	PVD	CNMG 120408	PCLNR 25 × 25 M12.1

Table 3 Machining parameters and levels

Parameter	Designation	Level 1	Level 2	Level 3
Cutting speed (m/min)	V	110	160	210
Feed (mm/rev)	F	0.1	0.2	0.3
Depth of cut (mm)	D	0.7	1.4	2.1

present work, three parameters each set at three levels are chosen for experimentation. The turning parameters and their levels chosen for all cases are presented in Table 3. In order to have a complete study of turning process, the ranges of parameters are selected, and an appropriate planning of experimentation is essential to reduce the cost and time consuming. Hence, an experimental plan based on Taguchi's L_{27} orthogonal array has been selected. The factors are assigned to columns 1–3, respectively, and 27 experiments are carried out under dry condition with different combinations of parameters levels.

There are three categories of quality characteristic in the analysis of the S/N ratio:

1. The-lower-the-better
2. The-higher-the-better
3. The-nominal-the-better.

Since the quality characteristic is to be minimized, the-lower-the-better category is used to calculate the S/N ratio for surface roughness. Equation (1) shows the smaller the-better characteristic.

$$\eta = -10 \log 10 \, 1/n \sum_i^n y^2 \tag{1}$$

where η is the signal-to-noise ratio, n is the number of repetitions of experiment and y is the measured value of quality characteristic.

Minitab 14 statistical software has been used for the analysis of the experimental work. The software studies the experimental data and then provides the calculated results of signal-to-noise ratio. The effect of different process parameters on surface roughness for type-I is given in Table 4, type-II in Table 5. The main effect plot for S/N ratio for surface roughness for type-I is shown in Fig. 1, type-II in Fig. 2. From

Table 4 Taguchi analysis surface roughness for type-I versus V, F, and D

Level	V	F	D
1	−5.099	−2.743	−3.960
2	−4.514	−4.035	−5.973
3	−6.016	−8.851	−5.696
Delta	1.503	6.108	2.013
Rank	3	1	2

Machining Performance of TiCN/Al$_2$O$_3$ Multilayer

Table 5 Taguchi analysis surface roughness for type-II versus V, F, D

Level	V	F	D
1	−2.80738	−0.09912	−2.79359
2	−4.35384	−3.51382	−4.64561
3	−4.35227	−7.90055	−4.07429
Delta	1.54646	7.80143	1.85202
Rank	3	1	2

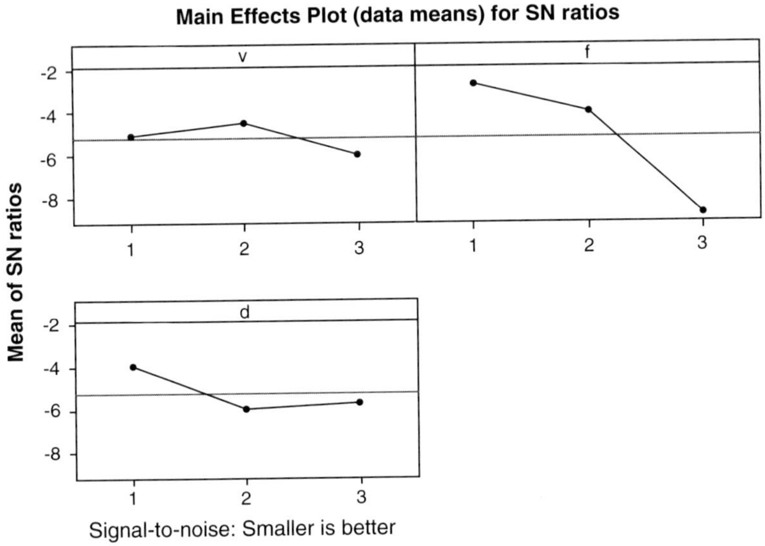

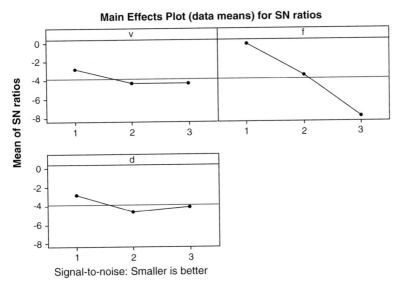

Fig. 1 The main effect of plot for S/N ratio for surface roughness of type-I

Fig. 2 The main effect of plot for S/N ratio for surface roughness of type-II

these figures, process parameters changes from one level to another. The average value of S/N ratio has been calculated to find out the effects of different parameters and their levels. In addition, a statistical ANOVA is performed to see those process parameters that significantly affect the responses. The experimental results are analyzed with ANOVA which is used for identifying the factors which significantly affecting the performance measures. The results of the ANOVA for surface roughness of all types are shown in Tables 6 and 7 respectively. This analysis is carried out for significance level of $\alpha = 0.05$, i.e., for a confidence level of 95 %.

3 Results and Discussion

3.1 Optimal Setting of Machining Parameters for Type-I

The response table of S/N ratio for surface roughness of type-I by Taguchi analysis has been shown in Table 4. It shows the rank value for machining parameters and it is stated by feed has the strongest influence on surface roughness followed by depth of cut and cutting speed for type-I. From Table 4, it is clear that the S/N ratio at each level of machining parameters and how it is changed when settings of each machining parameters are changed from one level to another. The main effect plot for surface roughness of type-I is shown in Fig. 1. In the plots, the x-axis indicates the value of each machining parameters at three level and y-axis the response value. Horizontal line indicates the mean value of the response. The main effect plots are used to determine the optimal design conditions to obtain the optimum surface roughness. The graph showed that reducing the level of surface roughness, feed should be set to its lowest level (0.1 mm/rev), depth of cut to its low level (0.7 mm) and cutting speed to its middle level (160 m/min). Table 7 shows the results of ANOVA for surface roughness of case-I. It is observed that feed (p value $= 0.00$) is the most significant parameter followed by depth of cut (p value $= 0.00$), while the effect of cutting speed has not been found statistically significant (p value $= 0.357$). A larger F value shows the greater impact on the machining performance characteristics. Larger F-values are observed for feed and depth of cut.

Table 6 ANOVA for surface roughness of type-I

Source	DF	SS	MS	F	P
V	2	0.2800	0.1400	1.09	0.357
F	2	9.3044	4.6522	36.06	0.000
D	2	0.4441	0.2220	1.72	0.204
Error	20	2.5806	0.1290		
Total	26	12.6090			

$S = 0.359205$ R$-$Sq $= 79.53$ % R$-$Sq (adj) $= 73.39$ %

Table 7 ANOVA for surface roughness of type-II

Source	DF	SS	MS	F	P
V	2	0.2665	0.1332	1.99	0.163
F	2	9.7220	4.8610	72.63	0.000
D	2	0.2045	0.1022	1.53	0.241
Error	20	1.3385	0.0669		
Total	26	11.5314			

S = 0.258700 R−Sq = 88.39 % R−Sq (adj) = 84.91 %

3.2 Optimal Setting of Machining Parameters for Type-II

Similarly, the response table for surface roughness of type-II is given in Table 5. According to the response table, the rank value represents that feed has the strongest influence of SR followed by depth of cut and cutting speed. The main effect plot for surface roughness of type-II has been shown in Fig. 2. The graph showed that reducing the level of surface roughness, feed should be set to its lowest level (0.1 mm/rev), cutting speed to its low level (110 m/min) and depth of cut to its low level (2.1 mm). Table 7 shows the results of ANOVA for surface roughness of type-II. It is observed that feed (p value = 0.00) is the most significant parameter followed by depth of cut.

3.3 Performances for all Types

Comparisons of TiCN + Al_2O_3 and B-TiC in CNC turning AISI410 are presented in Figs. 3, 4, 5, 6, 7, 8, 9, 10, 11 at various cutting range. Figures 3, 4, 5 shows that the surface roughness for variation with cutting speed at a feed rate of 0.1 mm/rev and depth of cut 0.7, 1.4 and 2.1 mm respectively. It clearly shows that the best result obtained from the B-TiC than TiCN/Al_2O_3. From Fig. 3 surface roughness is decreased for TiCN/Al_2O_3 at low and medium cutting speed.

Figures 6, 7, 8 shows that the surface roughness for variation with cutting speed at a feed rate of 0.2 mm/rev and depth of cut 0.7, 1.4, and 2.1 mm, respectively. It clearly shows that the best result obtained from the B-TiC than TiCN/Al_2O_3. From Fig. 7, surface roughness is decreased for TiCN/Al_2O_3 at high cutting speed.

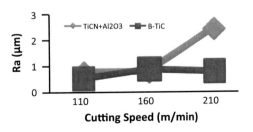

Fig. 3 Cutting speed versus surface roughness—feed rate of 0.1 and DOC 0.7 mm

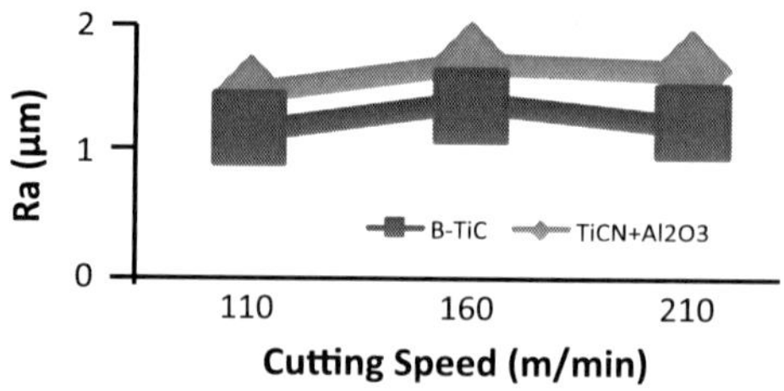

Fig. 4 Cutting speed versus surface roughness—feed rate of 0.1 and DOC 1.4 mm

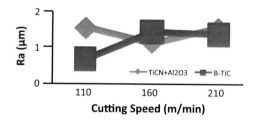

Fig. 5 Cutting speed versus surface roughness—feed rate of 0.1 and DOC 2.1 mm

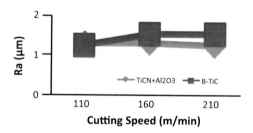

Fig. 6 Cutting speed versus surface roughness—feed rate of 0.2 and DOC 0.7 mm

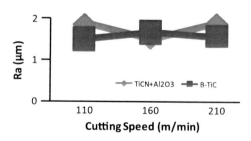

Fig. 7 Cutting speed versus surface roughness—feed rate of 0.2 and DOC 1.4 mm

Fig. 8 Cutting speed versus surface roughness—feed rate of 0.2 and DOC 2.1 mm

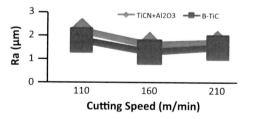

Fig. 9 Cutting speed versus surface roughness—feed rate of 0.3 and DOC 0.7 mm

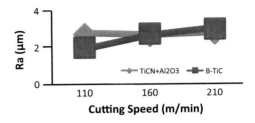

Fig. 10 Cutting speed versus surface roughness—feed rate of 0.3 and DOC 1.4 mm

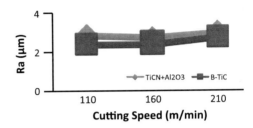

Fig. 11 Cutting speed versus surface roughness—feed rate of 0.3 and DOC 2.1 mm

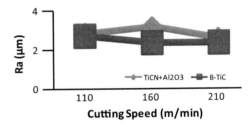

Figures 9, 10, 11 shows that the surface roughness for variation with cutting speed at a feed rate of 0.3 mm/rev and depth of cut 0.7, 1.4, and 2.1 mm, respectively. It clearly shows that the best result obtained from the B-TiC than TiCN/Al$_2$O$_3$. From Fig. 7, surface roughness is decreased for TiCN/Al$_2$O$_3$ at high cutting speed.

The analysis of these figures B-TiC is best performance then the TiCN/Al$_2$O$_3$. The low feed and cutting speed minimize surface roughness. The high feed and cutting speed surface roughness increased. The feed is significantly affecting surface roughness.

4 Conclusions

The current investigation is focused on optimization and analysis of CNC turning AISI410 during change of cutting parameters for different cases. From the study of result in turning is using Taguchi's techniques and ANOVA. These following can be concluded by present studies.

1. Optimum parameter setting for minimization of surface roughness is obtained at a cutting speed of 160 m/min, feed rate 0.1 mm/rev, and depth of cut 0.7 mm, i.e., v2f1d1, for $TiCN/Al_2O_3$ is used.
2. Optimum parameter setting for minimization of surface roughness is obtained at a cutting speed of 110 m/min, feed rate 0.1 mm/rev, and depth of cut 0.7 mm, i.e., v1f1d1, for B-TiC is used.
3. From the results of ANOVA, the feed rate and cutting speed are the significant cutting parameters for affecting the surface roughness with B-TiC.
4. From the results of ANOVA, the feed rate and depth of cut are the significant cutting parameters for affecting the surface roughness with $TiCN/Al_2O_3$.
5. From this analysis of figures, B-TiC is best performance than the $TiCN/Al_2O_3$.

References

1. Abburi NR, Dixit US (2006) Knowledge-based system for the prediction of surface roughness in turning process. Rob Comput Int Manuf 22(4):363–372
2. Nalbant M, Goekkaya G, Sur H (2007) Application of Taguchi method in the optimization of cutting parameters for surface roughness in turning. Mater Des 28(4):1379–1385
3. Oezel T, Karpat Y (2005) Predictive modeling of surface roughness and tool wear in hard turning using regression and neural networks. Int J Mach Tool Manuf 45(2):467–479
4. Kwok CT, Lo KH, Cheng FT, Man HC (2003) Effects of processing condition on the corrosion performance of laser surface melted AISI 440 C martensitic stainless steel. Surf Coat Technol 166(2):84–90
5. Attanasion A, Gelfi M, Giardini C, Remino C (2006) Minimum quantity lubrication in turning: effect on tool wear. Wear 260(1–3):333–338
6. Groover MP (1996) Fundamentals of modern manufacturing—materials, processes and systems. Prentice-Hall Inc., New Jersey
7. Sarwar M, Zhang X, Gillibrand D (1997) Performance of titanium nitride coated carbide tipped circular saws when cutting stainless steel and mild steel. Surf Coat Technol 94:617–621
8. DeGarmo EP, Black JT, Kohser RA (1997) Materials and processes in manufacturing. Prentice-Hall Inc., New Jersey
9. Lim CYH, Lim SC, Lee KS (1999) The performance of TiN-coated high speed steel tool inserts in turning. Tribo Int 32(7):393–398
10. Ezugwu EO, Okeke CI (2001) Tool life and wear mechanisms of TiN coated tools in an intermittent cutting operation. J Mater Process Technol 116(1):10–15
11. Nouari M, List G, Girot F, Coupard D (2003) Experimental analysis and optimization of tool wear in dry machining of aluminum alloy. Wear 255(7–12):1359–1368
12. Thamizhmanii S, Hasan S (2011) Machinability of hard martensitic stainless steel and hard alloy steel by CBN and PCBN tools by turning process. In: Proceedings of the world congress on engineering, vol I. London, pp 554–559
13. Gutakovskis V, Bunga G, Torims T(2010) Stainless steel machining with nano coated duratomictm cutting tool. In: Proceedings of 7th international DAAAM baltic conference, Estonia, pp 171–176
14. Noordin MY, Venkatesh VC, Sharif S (2007) Dry turning of tempered martensitic stainless tool steel using coated cermet and coated carbide tools. J Mater Process Technol 185(1–3):83–90
15. Thamizhmanii S, Bin Omar B, Saparudin S, Hasan S (2008) Surface roughness analyses of hard martensitic stainless steel by turning. J Ach Mater Manuf Eng 26(2):139–142

16. Ihsan K, Mustafa K, Ibrahim C, Ulvi S (2004) Determination of optimum cutting parameters during machining of AISI 304 austenitic stainless steel. Mater Des 25(4):303–305
17. Cebelli O (2006) Turning of AISI 304 austenitic stainless steel. J Eng Nat Sci 2:12–117
18. Lin WS (2008) The study of high speed fine turning of austenitic stainless steel. J Ach Mater Manuf Eng 27:2
19. Ross Philip J (1996) Taguchi techniques for quality engineering. McGraw-Hill Book Company, New York
20. Peace GS (1993) Taguchi methods: a hand book- on approach. Addison-Wesley, New York
21. Yang WH, Tarng YS (1998) Design optimization of cutting parameters for turning operations based on the Taguchi method. J Mater Process Technol 84(2):22
22. Diniz AE, Oliveira AJ (2004) Optimizing the use of dry cutting in rough turning steel operations. Int J Mach Tool Manuf 44(12–13):1061
23. Lee BY, Tarng YS (2000) Cutting-parameter selection for maximizing production rate or minimizing production cost in multistage turning operations. J Mater Process Technol 105(1–2):61
24. Ghani JA, Choudhury IA, Hasan HH (2004) Application of Taguchi method in the optimizations of end milling operations. J Mater Process Technol 145(1):84–92

Metallurgical Test on 20MnCr5 Steel: To Suggest as a Suitable Crankshaft Material

M. Umashankaran, J. Hari Vignesh, A. Saravana Kumar, R. Shyamaprasad and R. Saravana Prabhu

Abstract The crankshaft is the part of an engine which translates reciprocating linear piston motion into rotation. In converting the linear motion of the piston into rotational motion; crankshafts operate under high loads and require high strength and stiffness to withstand the high loads in modern engines, and to offer opportunities for downsizing and weight reduction. We have come with an idea of suggesting a new crank shaft material 20MnCr5 steel, which has high strength and stiffness and satisfies all the requirements of the crank shaft materials. We did the various important testing like hardness test (Rockwell Hardness), impact test, torsion test, fatigue test, metallography test on this 20MnCr5. Our chapter deals with our experimental works on this 20MnCr5. The objective of this testing is to suggest that 20MnCr5 alloy steel is one of the suitable material for making crankshafts for the modern engines.

Keywords Crankshaft · Strength · Stiffness · Weight reduction · 20MnCr5 alloy steel

1 Introduction

Crankshaft is among the largest component in internal combustion engines. The crankshaft converts reciprocates motion to rotational motion. It contains counter weights to smoothen the engine revolutions. Crankshafts are one of the most critically

---

M. Umashankaran (✉) · J. Hari Vignesh · A. Saravana Kumar · R. Shyamaprasad ·
R. Saravana Prabhu
V.S.B Engineering college, Karur, Tamilnadu, India
e-mail: umashan83@gmail.com

J. Hari Vignesh
e-mail: harimetly@gmail.com

A. Saravana Kumar
e-mail: 2mailsaravana@gmail.com

R. Shyamaprasad
e-mail: shyamprasad10@gmail.com

R. Saravana Prabhu
e-mail: saravanasp14@gmail.com

S. Sathiyamoorthy et al. (eds.), *Emerging Trends in Science,*
Engineering and Technology, Lecture Notes in Mechanical Engineering,
DOI: 10.1007/978-81-322-1007-8_24, © Springer India 2012

Fig. 1 Crank shaft

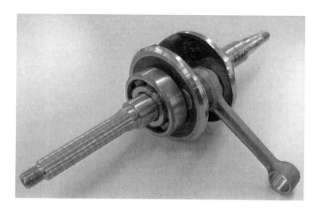

loaded components and experience a high bending moment and torsion. Typical automotive crankshaft consisting of main journals (located on shaft centreline), connecting rod journals (located off-centreline), a trust bearing journal, cranks that connect the journals and hold the component together, and a number of auxiliary parts [1]. Cyclic loads in the form of bending and torsion during its service life (Fig. 1).

2 Crankshaft Materials

Crankshafts materials should be readily shaped, machined and heat-treated, and have adequate strength, toughness, hardness, and high fatigue strength. The crankshafts are manufactured from steel either by forging or casting. The main bearing and connecting rod bearing liners are made of Babbitt, a tin and lead alloy. Forged crankshafts are stronger than the cast crankshafts, but are more expensive. Forged crankshafts are made from SAE 1045 or similar type steel [2]. Forging makes a very dense, tough shaft with a grain running parallel to the principal stress direction [3]. Crankshafts are cast in steel, modular iron or malleable iron. The major advantage of the casting process is that crankshaft material and machining costs are reduced because the crankshaft may be made close to the required shape and size including counterweights. Generally automobile crankshafts were forged in past to have all the desirable properties. However, with the evolution of the nodular cast irons and improvements in foundry techniques, cast crankshafts are now preferred for moderate loads. Only for heavy duty applications forged shafts are favored. The selection of crankshaft materials and heat treatments for various applications are as follows.

3 Methodology

See Chart 1

Metallurgical Test on 20MnCr5 Steel

Chart 1 Methodology chart

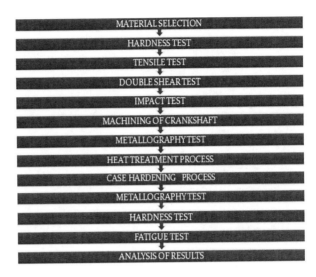

3.1 Material Selection

The selected material for experimental work is SAE 5120 (DIN EN 20MnCr5).

3.1.1 Chemical Composition of SAE 5120 (DIN EN 20MnCr5)

See Table 1

3.2 Hardness Test

The Hardness test was done according to the ASTM Standards in Rockwell hardness machine [4].

Table 1 Chemical composition of SAE 5120 (20MnCr5)

Chemical composition	% Percentage
C	0.205
Si	0.281
Mn	1.21
P	0.0095
S	0.031
Cr	1.21
Mo	0.043
Ni	0.216
Al	0.0210
Cu	0.157
N	0.007

1. Applied load 60 kg
2. Diamond Indenter is used
   (a) Surface = 69 RHN
   (b) Core = 84 RHN

3.3 Tensile Test

The Tensile test was done according to the ASTM Standards in Universal Testing machine (Fig. 2a, b) [4, 5].

1. Length of the rod = 300 mm
2. Diameter of the rod = 12 mm
3. Ultimate load of the material = 70 kN
4. Breaking point of the material = 56 kN
5. Yield point of the material = 48 kN
6. Ultimate tensile strength of the material = 619.24 N/mm^2

3.4 Double Shear Test

The Double Shear test was done according to the ASTM Standards in Universal Testing machine [4].

1. Diameter of specimen = 12 mm
2. Ultimate load = 130 kN
3. Ultimate shear strength = 575.01 kN/mm^2

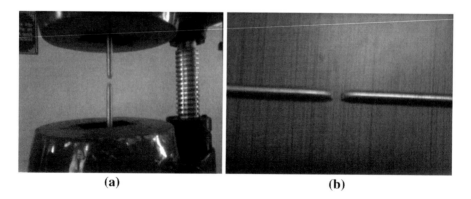

(a)   (b)

Fig. 2 Photography of the (**a**) tensile test and (**b**) tensile facture photography

Metallurgical Test on 20MnCr5 Steel

3.5 Impact Test

The Impact test was done according to the ASTM Standards in Impact Testing machine [4].
Length of the specimen = 50 mm
Diameter of the specimen = 10 mm

1. Charpy's test:
   (a) Impact strength = 0.22 J/mm^2
2. Izod method
   (a) Impact strength = 0.18 J/mm^2

3.6 Design of Crank Shaft

Crank shaft was designed according to withstand a sufficient Bending moment and torsion moment (Fig. 3).

1. Length of the crankshaft = 2.42 M
2. Crank throw = 0.68 M
3. Crank webs diameter = 0.31 M
4. Diameter or thickness = 0.25 M
5. Length of the main journal = 0.62 M

3.7 Heat Treatment

The heat treatment was done according to the ASTM Heat treatment Standards [6].

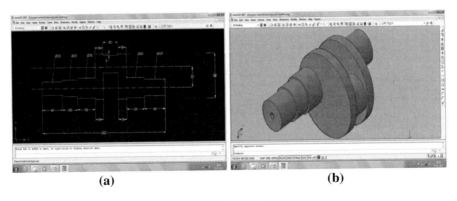

Fig. 3 Design of the crankshaft (a) 2D view and (b) 3D view

3.7.1 Normalising

The Normalising was done in an electrical bogie Furnace at 920 °C (Table 2) [6].

3.7.2 Gas Carburising

The Gas Carburising was done in a Seal Quenched Furnace at 440 °C (Table 3) [7].

3.7.3 Tempering

The Tempering was done in an Electrical Furnace at 440 °C (Table 4).

3.8 Metallography Test

The Metallography test was done according to the ASME Standards [4]. The metallography structure for different heat treatment is given in below Figs. 4, 5, 6, 7 (as received).

Table 2 Normalizing process

Furnace	Electrical bogie furnace
Temperature	920 °C
Time	3 h
Quenched	Air cooled

Table 3 Gas carburizing process

Furnace	Seal quenched furnace
Temperature	440 °C
Time	(3 + 4) h
	3-boosting
	4-diffusing
Quenched	Oil quenched (hi quench sa)
Depth	1.5 mm

Table 4 Tempering process

Temperature	440 °C
Time	2 h
Quenched	Air cooled

Metallurgical Test on 20MnCr5 Steel 279

Fig. 4 Wide man Staten structure

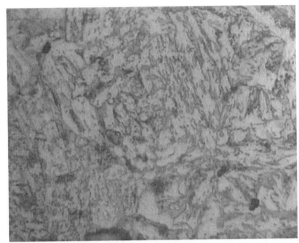

Fig. 5 Coarse ferrite and pearlitic

Fig. 6 Fine tempered martensite (surface)

Fig. 7 Low carbon martensite (core)

3.8.1 Before Treatment

The received micrographic structure was Long plates of ferrite cutting across pearlite

Grain Structure: 100 μ
Magnification: 100 x

3.8.2 After Normalizing

The received micrographic structure was Coarse Ferrite and Pearlitic

Magnification: 100 x
Grain Structure: 100 μ

3.8.3 Case Hardening and Tempering

1. At Surface of the specimen

   The received micrographic structure was Fine Tempered Martensite

   Magnification: 100 x
   Grain Structure: 100 μ

2. At the Core of the specimen

   The received micrographic structure was Low Carbon Martensite

   Magnification: 100 x
   Grain Structure: 100 μ

3.9 Hardness Test

The Hardness test was done according to the ASTM Standards in Rockwell hardness machine. And the conversions are made according to the ASTM Standard conversions (Standard Hardness Conversion Tables for Metals Relationship Among Brinell Hardness, Vickers Hardness, Rockwell Hardness, Superficial Hardness, Knoop Hardness, and Scleroscope Hardnessin 2005 as E140–05) (Table 5).

Microhardness of case hardened material:

1. Surface $= 659$ HV
2. Core $= 438$ HV

3.10 Fatigue Test

3.10.1 Life Estimation Using the S–N Approach

An circular rod with a diameter of 30 mm is subjected to constant amplitude bending at room temperature, with $S_m = 200$ MPa. The material SAE5120 with $S_u = 1{,}240$ MPa, $S_y = 1{,}170$ MPa, and $S_{y'} = 1{,}000$ MPa. If the rod is commercially polished, estimate the values of S_a, S_{max}, S_{min} and R for a median fatigue life of 50,000 cycle and no yielding [8].

Since a mean stress other than zero is involved, a constant life diagram similar to figure is needed. This diagram is also represented by equation [1]. We could use either the modified Goodman or the morrow mean stress equation, but since σ_f is not given, we choose the finite life modified Goodman equation.

$$\frac{S_a}{S_{Nf}} + \frac{S_m}{S_u} = 1$$

We also assume Basquin's equation and a noncorrosive environment. Therefore

$$S_{Nf} = A(N_f)^B$$

Where S_{Nf} is the fully reversed, $R = -1$, fatigue strength at N_f cycles to failure. For yielding we use equation

$$\frac{S_a}{S_{y'}} + \frac{S_m}{S_y} = 1$$

Table 5 Hardness number after treatment

As received	277 BHN
After normalising	262 BHN
After gas carburising	Surface $= 58$ HRC
	Core $= 42$ HRC

Since $S_u, S_y, S_{y'}, S_m$ and $N_f = 50,000$ cycles are known. we need to determine S_{Nf} at 50,000 cycle, along with A and B, before we can solve for S_u, S_{max}, S_{min} and R. We assume that the line represented by Basquin's equation passes through $S_u = 1,240$ MPa at one cycle and S_f at 10^6 cycle (this could also be 10^7 cycles) for steels. From the first condition, $A = S_u = 1,240$ MPa. From the second condition [1].

$$S_{Nf} = A\left(N_f\right)^B = S_u\left(N_f\right)^B \text{ (or) } S_f = S_u\left(10^6\right)^B$$

Then

$$\text{Log}\left(S_f/S_u\right) = B \log 10^6 = 6b \text{ (or) } B = (1/6)\log\left(S_f/S_u\right)$$

For highly polished, small rotating beam steel specimens

$$S_f = 0.5 S_u \quad \text{if } S_u =< 1,400 \text{ MPa}$$

and therefore

$$S_f = 0.5 \times 1,240 = 620 \text{ MPas}$$

For commercial polishing, a correction factor for S_f from the figure is needed. With $Su = 1,240$ MPa(180ksi), this correction factor is 0.87. For the diameter equal to 32 mm, no size effects adjustment is needed and therefore

$$S_f = 0.87 \times 620 = 540 \text{ MPa}$$

And

$$B = (1/6)\log(540/1,240) = -0.06$$

At 50,000 cycles and for $R = -1$

$$S_{Nf} = 1,240 \times (50,000)^{-0.06} = 648 \text{ MPa}$$

And

$$\frac{S_a}{648} + \frac{200}{1,240} = 1$$

Resulting in

$$S_a = 543 \text{ MPa}$$

$$\frac{S_a}{S_{yr}} + \frac{S_m}{S_y} = \frac{543}{1,000} + \frac{200}{1,170} = 0.714$$

Which is <1, and therefore yielding does not occur

$$S_{min} = S_m - S_a = 200 - 543 = -343 \text{ MPa}$$

$$S_{max} = S_m + S_a = 200 + 543 = 743 \text{ MPa}$$

$$R = (S_{min}/S_{max}) = -343/743 = -0.46$$

If a few experimental tests were performed using the calculator value of $S_a = 543$ MPa and the given value of $S_m = 200$ MPa, failure would most likely not be 50,000 cycle due to the several approximations Assume and scatter inherent in fatigue tests. However, this is our best median estimate [1].

4 Conclusion

Tempered martensite is a microstructure which we observed after the surface treatment the hardness which maintains 58 HRC in case and 42 HRC in core the metastable martensite improve the endurance strength of the material. The higher endurance strength will cause the metal to fail after a large number N of applications of the stress. Results of the fatigue tests are shown that there is no discontinuity in the relation between lifetime and stress amplitude beyond 10^7 cycles, meaning that a Very High Cycle Fatigue effect is observed from these materials SAE 5120—20MnCr5 so we conclude and suggest that 20MnCr5 can be used as a crankshaft material for the transportations.

References

1. Williams J, Fatemi A, (2007) Fatigue performance of forged steel and ductile cast iron crankshafts, SAE technical paper No. 2007-01-1001, society of automotive engineers
2. Pollack HW (1988) Material science and metallurgy. Prentice-Hall, Englewood Cliffs
3. Dieter GE, Bacon D (1988) Mechanical Metallurgy. McGraw-Hill, London
4. Annual Book of ASTM Standards (1989) Section 3, metals test methods and analytical procedures: metals—mechanical testing; elevated and low-temperature tests. Metallography
5. ASTM Standard E8-04 (2004) Standard test methods for tension testing of metallic materials. Annual book of ASTM standards, vol 03.01. West Conshohocken, PA 2004
6. Totten GE, Howes MAH (1988) Steel heat treatment hand book. Marcel Dekker Inc., CRC Press, Prentice-Hall, NY, FL, Englewood Cliffs
7. Hassell PA (2009) Induction heat treating of steel, ASM handbook, vol 4. Heat treatment, pp 164–168, 171–173, 180–181, 184–185
8. Stephens RI, Fatemi A, Stephens RR, Hendry OF (2000) Metal fatigue in engineering. Wiley, New York

Multi-Objective Optimization and Empirical Modeling of Centerless Grinding Parameters

N. Senthil Kumar, C. K. Dhinakarraj, B. Deepanraj and G. Sankaranarayanan

Abstract This paper deals with the optimization and analysis of grinding parameters in external centerless grinding process. Taguchi's technique is used to analyze effects of grinding parameters on surface roughness of the workpiece, specific energy consumption, and the roundness error produced during grinding AISI 1040 steel using aluminum oxide-grinding wheel. Taguchi's design of experiment is used to design an experimental array, based on which experiments were conducted. For 3 parameters and 3 levels of each parameter, L_9 Orthogonal array (OA) is selected. Signal-to-Noise ratio is used to analyze the output quality characteristics. Based on the analysis, an optimum condition is chosen and a validation experiment is conducted and the output values were compared and it is found that the specific energy consumption, roundness error were reduced by more than 25 % and the surface roughness obtained is the least value. In order to predict the quality characteristics, an empirical modeling is also developed using general linear regression analysis.

Keywords Centerless grinding • Surface roughness • Taguchi's technique • Empirical modeling

1 Introduction

Grinding process is dissimilar to other machining processes such as turning and milling, as the multipoint cutting edges of the grinding wheel don't have uniformity, which act differently on the work piece at each grinding. These complexities and

---

N. Senthil Kumar (✉) · C. K. Dhinakarraj
Adhiparasakthi Engineering College, Melmaruvathur 603319, India
e-mail: nskumar_1998@yahoo.co.in

C. K. Dhinakarraj
e-mail: ckdinakar@yahoo.com

B. Deepanraj
National Institute of Technology, Calicut 673601, India
e-mail: babudeepan@gmail.com

G. Sankaranarayanan
Sri Muthukumaran Institute of Technology, Chennai 600069, India
e-mail: gs2000narayanan@gmail.com

S. Sathiyamoorthy et al. (eds.), *Emerging Trends in Science,*
Engineering and Technology, Lecture Notes in Mechanical Engineering,
DOI: 10.1007/978-81-322-1007-8_25, © Springer India 2012

difficulties of illustrating the grinding process also raise obstacles to the optimization of the grinding process and to the verification of the interrelationship between grinding parameters and outcomes of the process. High quality output on the centerless grinding machine is achieved through proper selection of grinding parameters. Improper selection of input parameters gives rise to out of roundness, poor surface finish etc. Quantification of surface roughness value, specific energy consumption and roundness error of the workpiece is necessary to determine the quality of the component undergoing grinding. The centerless grinding parameters have to be optimized in order to obtain best quality of machined component and to achieve less production cost. Figure 1 shows the schematic diagram of centerless grinding process [1].

Dhavlikar et al. [2] have minimized the roundness error of workpiece by applying both Taguchi and dual response methodology and has carried out regression analysis to model an equation to average out roundness error. Garitaonandia et al. [3] have predicted the setup conditions to analyze the dynamic and geometrical instabilities using root locus perspective, making it possible to study the influence of different machine variables in stability of the process. Asilturk and Cunkas [4] has developed a surface roughness model using regression analysis for turning and predicted it using ANN and has concluded that the feed rate is the dominant factor affecting the surface roughness, followed by depth of cut and cutting speed. Kwak [5] has evaluated the effect of grinding parameters on geometric error and has shown that depth of cut is the dominant parameter followed by grain size and has developed a second-order response model to predict it. Kwak et al. [6] have analyzed the grinding power and surface roughness of the workpiece by RSM and have developed a model to predict them and increasing the depth of cut affects the grinding power more than increasing the transverse speed.

Krajnik et al. [7] have minimized the surface roughness by optimizing the grinding process by RSM and by developing an empirical model for it. Nalbant et al. [8] have identified an optimal surface roughness performance for a particular combination of cutting parameters and has found that insert radius and feed rate are the main parameters that affect the surface roughness. Zhou et al. [9] have minimized the lobing effect by developing a stability diagram for workpiece and thereby selecting the grinding parameters and have found that characteristic root

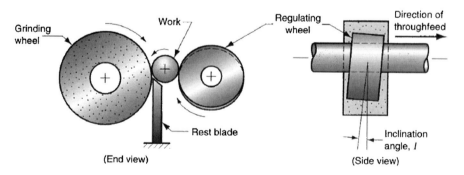

Fig. 1 Schematic of centerless grinding process

distribution of the lobing loop is periodic. Tamizhmani et al. [10] have showed that depth of cut has a significant role in producing lower surface roughness by 14.467 %, followed by feed rate contribution of 9.764 %. Xu et al. [11] have investigated the workpiece roundness based on process parameters by both simulation and experimental analysis and was found that a slower worktable feed rate and a faster workpiece rotational speed result in better roundness.

2 Materials and Methods

2.1 Workpiece and Grinding Wheel Material

The workpiece material chosen for our work is AISI 1040 steel, which was used in the applications of stressed pins, shafts studs, keys etc. The chemical composition of the workpiece material is shown in Table 1.

The SEM micrograph of the workpiece material is shown in Fig. 2. The microstructure of the matrix shows large grains of pearlite in a matrix of ferrite. The microstructure is typical of medium carbon steel that has been hot rolled and re-crystallized. The flow line of the pearlite grains clearly shows the hot working operation was carried out. The grain size of the matrix is ASTM grain size No: 6 as per ASTM grain size measurement. The Grinding wheel used for the experiment is an Aluminium oxide abrasive wheel which is designated as A60-L5-V20.

2.2 Taguchi's Technique

The Taguchi method is a powerful tool in quality optimization. Taguchi method makes use of a special design of OA to examine the quality characteristics through

Table 1 Chemical composition of AISI 1040 steel

Sl. no	Element	% Composition
1	Carbon	0.436
2	Silicon	0.198
3	Manganese	0.761
4	Molybdenum	0.012
5	Titanium	0.006
6	Vanadium	0.004
7	Tungsten	0.045
8	Phosphorous	0.023
9	Sulphur	0.025
10	Copper	0.056
11	Aluminium	0.045
12	Iron	98.389

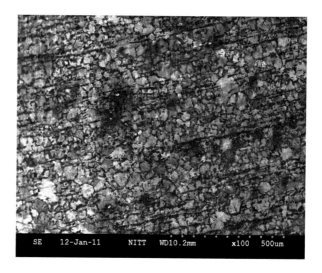

Fig. 2 SEM image of AISI 1040 steel

a minimal number of experiments. The experimental results based on the OA are then transformed into Signal-to-Noise ratios to evaluate the performance characteristics. Depending upon the minimization and maximization of the output parameters the lower the better or larger the better condition is chosen.

S/N ratio for lower-is-better is given by,

$$(S/N) = -10 \log \left[\frac{1}{n} \sum_{i=1}^{n} y_i^2 \right] \quad (1)$$

S/N ratio for bigger-is-better is given by,

$$(S/N) = -10 \log \left[\frac{1}{n} \sum_{i=1}^{n} y_i^{-2} \right] \quad (2)$$

where y_i represents the experimentally observed value of ith experiment and n is the no of replications of each experiment.

Taguchi's design of experiments is used to design the OA for 3 parameters like regulating wheel speed, feed rate, and depth of cut and for each parameter 3 different values that were taken, which is shown in Table 2. The three different values of regulating wheel speed chosen is 20, 40, and 80 m/min, feed rate is chosen as 2, 4, and 6°, whereas the depth of cut is selected as 0.02, 0.04, and 0.06 mm, respectively. The minimum number of experiments to be conducted for the parametric optimization was calculated as,

Minimum experiments = [(L−1) x P] + 1 = [(3−1) × 3] + 1 = 7 ≈ L$_9$.

For various combinations of the input control parameters and its 3 levels chosen, a L$_9$ OA is formulated using Taguchi's design of experiment which is shown in Table 3. The interactions between the factors were not considered.

Table 2 Control parameters and its levels

Sl. no	Parameter/level	Level 1	Level 2	Level 3
1	Regulating wheel speed (m/min)	20	40	80
2	Feed rate (°)	2	4	6
3	Depth of cut (mm)	0.02	0.04	0.06

Table 3 Experimental L_9 OA

Trial no.	Regulating wheel speed (m/min)	Feed rate (°)	Depth of cut (mm)
1	20	2	0.02
2	20	4	0.04
3	20	6	0.06
4	40	2	0.04
5	40	4	0.06
6	40	6	0.02
7	80	2	0.06
8	80	4	0.02
9	80	6	0.04

3 Results and Discussion

Based on the L_9 OA designed using Taguchi's design of experiment, experiments were conducted in a centerless grinding machine in which energy meter is connected. In this study, 9 different workpieces were taken and for each level a separate workpiece was used. While grinding was in progress, the specific energy consumed was calculated using the readings taken from the energy meter. The roundness error was calculated using a Bench center attached with a digital dial gauge and the surface roughness was measured with a SURFCODER SE500. The determined output quality characteristics were given in the Table 4.

Figure 3 shows the area graph for the output quality characteristics, roundness error, specific energy consumption, and surface roughness. It is observed that the

Table 4 Inner and outer array of designed experiment

Sl no	Inner array			Outer array		
	Wheel speed (m/min)	Feed rate (°)	Depth of cut (mm)	Specific energy (kW)	Surface roughness (µm)	Roundness error (mm)
1	20	2	0.02	0.211	2.825	0.020
2	20	4	0.04	0.248	0.982	0.016
3	20	6	0.06	0.351	1.646	0.023
4	40	2	0.04	0.256	2.791	0.010
5	40	4	0.06	0.334	2.885	0.020
6	40	6	0.02	0.296	2.024	0.010
7	80	2	0.06	0.365	2.166	0.013
8	80	4	0.02	0.292	2.937	0.016
9	80	6	0.04	0.344	2.273	0.013

surface roughness values are higher, while the roundness error and specific energy consumption are lower.

The specific energy consumption, roundness error, and surface roughness of the work piece has to be minimum for a given set of input parameters. Hence the Smaller-the-better condition is chosen as given in Eq. (2) for this analysis.

Based on the importance of the output quality characteristics, the specific energy consumption was given a weightage of 20, 50 % weightage to surface roughness, and 30 % weightage to roundness error. Surface roughness is given more weightage because; the quality of the grinding process lies in its finish. Roundness error is given next importance due to the accuracy of the component produced. Energy consumed is given the least importance since when compared to the finish and accuracy, energy consumed will have little impact. Table 5 shows the individual S/N ratio values of quality characteristics and then the combined S/N ratio, based on the weightage given to each individual output parameters.

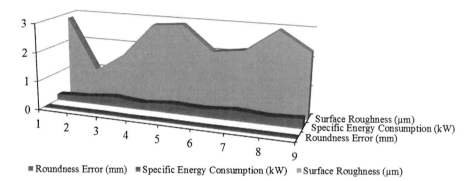

Fig. 3 Area graph for output quality characteristics

Table 5 Inner and outer array of designed experiment

Sl. no	S/N ratio of outer array			Combined S/N ratio (A*0.2 + B*0.5 + C*0.3)
	Specific energy	Surface roughness	Roundness error	
1	13.5144	−9.0204	33.9794	8.387
2	12.111	0.15777	35.9176	13.276
3	9.09386	−4.3286	32.7654	9.484
4	11.8352	−8.9152	40	9.909
5	9.52507	−9.2029	33.9794	7.497
6	10.5742	−6.1242	40	11.053
7	8.75414	−6.7132	37.7211	9.711
8	10.6923	−9.3581	35.9176	8.235
9	9.26883	−7.132	37.7211	9.604

Multi-Objective Optimization and Empirical Modeling

Table 6 Response table for combined S/N ratio

Level/parameter	Regulating wheel speed (m/min)	Feed rate (°)	Depth of cut (mm)
Level 1	**10.382**	9.336	9.225
Level 2	9.486	9.669	**10.93**
Level 3	9.183	**10.047**	8.897

The response table for the combined S/N ratio is obtained by averaging the values for corresponding grinding parameters, which are given in Table 6.

From the response table of combined objective, main effects plot is drawn, as shown in Fig. 4. From the response table of the combined S/N ratio and the main effects plot, the optimum grinding conditions obtained are regulating wheel speed of 20 m/min, feed rate of 6°, and depth of cut of 0.04 mm.

Figure 5 shows the effect of wheel speed and feed rate over combined S/N ratio. Only for a wheel speed of 20 m/min, a higher level of interaction is observed and for the other two wheel speeds, no interaction effect was observed.

Figure 6 shows the effects of wheel speed and depth of cut over combined S/N ratio. For all the values of wheel speed, a higher level of interaction is seen and the combined S/N ratio is lower only for a depth of cut of 0.06 mm.

Figure 7 shows the effects of feed rate and depth of cut over combined S/N ratio. A higher level of interaction is seen for a feed rate of 4° and a low level of interaction for other two feed rates. For 0.06 mm depth of cut, lower combined S/N ratio is observed.

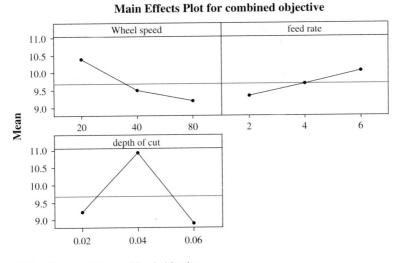

Fig. 4 Main effects plot for combined objective

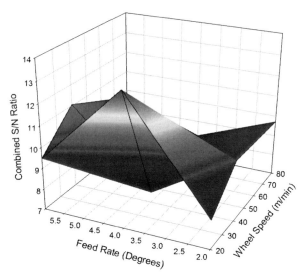

Fig. 5 Effect of wheel speed and feed rate

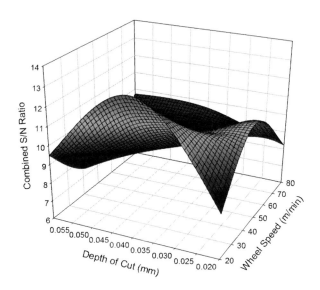

Fig. 6 Effect of wheel speed and depth of cut

Table 7 Output of optimum grinding parameters

Surface roughness (μm)	Specific energy consumption (kW)	Roundness error (mm)
0.692	0.298	0.01

Fig. 7 Effect of feed rate and depth of cut

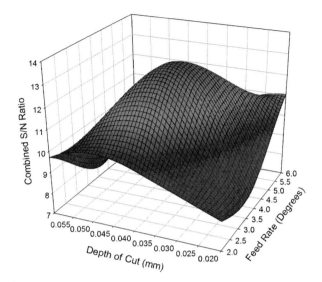

A validation experiment is conducted based on the optimum condition obtained and the measured quality characteristics are shown in Table 7.

4 Empirical Modeling

Multiple linear regression models [12] were developed for combined objective using Minitab-16, statistical Software. The control factors were the regulating wheel speed, feed rate and depth of cut, whereas the quality characteristics are combined objective of surface roughness, roundness error and specific energy consumption. The equations of the fitted model for the quality characteristics are given in Eq. (3).

$$\text{Combined Objective} = 10.1502 - 0.0182107 \text{ Regulating wheel speed} + 0.177833 \text{ Feed Rate} - 8.19167 \text{ Depth of Cut} \quad (3)$$

Table 8 Calculated and predicted combined S/N ratio

Trial no.	Calculated combined S/N ratio	Predicted combined S/N ratio
1	8.387	9.977819
2	13.276	10.16965
3	9.484	10.36148
4	9.909	9.449771
5	7.497	9.641604
6	11.053	10.32494
7	9.711	8.55751
8	8.235	9.240843
9	9.604	9.432675

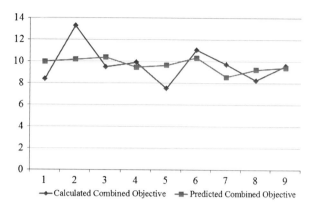

Fig. 8 Comparison of calculated and predicted combined objective

Table 8 shows the calculated combined S/N ratio and predicted S/N ratio using the fitted model using linear regression analysis as given in Eq. (3).

Figure 8 shows the graph drawn between the calculated combined S/N ratio and predicted combined S/N ratio using the fitted regression equation.

5 Conclusion

In this work, Taguchi's technique was used to optimize the grinding parameters like regulating wheel speed, feed rate, and depth of cut. The outcomes are,

1. The optimum condition obtained was regulating wheel speed of 20 m/min, feed rate of 6° and depth of cut of 0.04 mm.
2. When the regulating wheel speed increases, the surface roughness value increases by 35 %, with increase in specific energy consumption by 23.7 %, however the roundness error decreases by 30 %.
3. When the feed rate increases, the specific energy consumption increases by 19 % and the roundness error was increased marginally by 7 %. Surface roughness value was high when the feed rate was low and when the feed rate increases the surface roughness value decreases by 23.63 %.
4. When the depth of cut is high, the specific energy consumption is high, which is increased up to 31.58 %. For lower depth of cut, the surface roughness value is high and when the depth of cut increases, the surface roughness value decreases by 13.99 % and the roundness error value is increased by 20 % for increase in depth of cut.

Hence it is concluded that the cutting speed has to be low with higher feed rate and a moderate depth of cut.

References

1. Groover MP (2010) Fundamentals of modern manufacturing, materials, processes and systems, 4th edn. Wiley, New York
2. Dhavlikar MN, Kulkarni MS, Mariappan V (2003) Combined Taguchi and dual response method for optimization of a centerless grinding operation. Int J Mater Process Technol 132:90–94
3. Garitaonandia I, Fernandes MH, Albizuri J, Hernandez JM, Barrenetxe D (2010) A new perspective on the stability study of centerless grinding process. Int J Mach Tools Manuf 50:165–173
4. Asilturk I, Cunkas M (2011) Modeling and prediction of surface roughness in turning operations using artificial neural and multiple regression method. Int J Expert Syst Appl 38:5826–5832
5. Kwak JS (2005) Application of Taguchi and response surface methodologies for geometric error in surface grinding process. Int J Mach Tools Manuf 45:327–334
6. Kwak JS, Sim SB, Jeong YD (2006) An analysis of grinding power and surface roughness in external cylindrical grinding of hardened SCM440 steel using the response surface method. Int J Mach Tools Manuf 46:304–312
7. Krajnik P, Kopac J, Sluga A (2005) Design of grinding factors based on response surface methodology. Int J Mater Process Technol 162–163:629–636
8. Nalbant M, Gokkaya H, Sur G (2007) Application of Taguchi method in the optimization of cutting parameters for surface roughness in turning. Int J Mater Des 28(4):1379–1385
9. Zhou SS, Gartner JR, Howes TD (1996) On the relationship between setup parameters and lobing behavior in centerless grinding. Ann CIRP 45(1):341–346
10. Thamizhmani S, Saparudin S, Hassan S (2007) Analysis of surface roughness by turning process using Taguchi method. J Achiev Mater Manuf Eng 20(1–2):503–506
11. Xu W, Wu Y, Sato T, Lin W (2010) Effects of process parameters on workpiece roundness in tangential-feed centerless grinding using a surface grinder. Int J Mater Process Technol 210:759–766
12. Montgomery DC (2001) Design and analysis of experiments, 5th edn. Wiley, New York

Nanomaterials in Gas Sensors: A Big Leap in Technology!

G. Paulraj, A. Evangeline and G. Arun Prakash

Abstract Nanotechnology is the study of things very small, with the generally accepted definition that items smaller than 100 nm are nanomaterials, which can be inorganic, organic or biological. The important feature of nanomaterials is that as particle size decreases, the percentage of bulk material decreases and the surface area increases. The paper discusses the application of the following nanomaterials carbon nanotubes, metal oxides (MOs), organic molecules, quantum dots (QDs) and MEMS.

Keywords Nanomaterials • Carbon nanotubes • Metal oxides • Organic molecules • Quantum dots and MEMS

1 Introduction

There was an earlier saying that history cannot be changed. But the technology especially nanotechnology makes the revolutions in the history. There is no present and future, the development was such like a dream come true when one gets up from the sleep. The change was so rapid that no one realised when the pigeon has become electrons. It is really thought provoking that how nanotechnology brings out evolutions in telecommunication as well as computing and networking industries. Forthcoming developments in nanotechnology through which the impossible can be made possible are nanomaterials with novel optical, electrical and magnetic properties, compact as well as fast nonsilicon based chipsets for processors, quantum computing and DNA computing, development of telecom

---

G. Paulraj
Department of Production Engineering, JJ College of Engineering and Technology, Trichirapalli, India

A. Evangeline (✉) · G. Arun Prakash
Production Engineering,
JJ Colllege of Engineering and Technology, Tirichirapalli, India

S. Sathiyamoorthy et al. (eds.), *Emerging Trends in Science, Engineering and Technology*, Lecture Notes in Mechanical Engineering, DOI: 10.1007/978-81-322-1007-8_26, © Springer India 2012

switches which are fast and reliable, micro-electro-mechanical systems and above all the development of imaging and microscopic systems with high resolution.

2 Nanomaterials: Shrinking the World

Nanotechnology has been around for many years. Nanotechnology is the study of things very small, with the generally accepted definition that items smaller than 100 nm are nanomaterials, which can be inorganic, organic or biological. The body is full of nanotechnology: DNA is an excellent example of a nano programme for creating proteins (nanomaterials) to very specific design rules. With over $1 billion/annum research effort by the USA alone, nanotechnology [more correctly titled nanoscale science and engineering (NSE)] is the buzz word in research. The important feature of nanomaterials is that as particle size decreases, the percentage of bulk material decreases and the surface area increases, with the extreme being materials such as single walled carbon nanotubes (SWCNTs) which, being only one atom thick, are only a surface. As a material becomes mostly surface area, then quantum mechanics dominates, with electronic and thermal conduction and optical and mechanical properties no longer simply predicted from the elements' position in the periodic table.

Nanotechnology can be approached from the 'top down' attitude of shrinking processes: integrated circuit track widths are now below 100 nm and new, quantum mechanical problems appear at these small sizes [5]. The alternative "bottom up" approach uses atoms deposited or synthesised into a nanomaterial; an example is carbon nanotubes (CNTs), small tubes of pure carbon, six or more atoms in diameter and thousands of atoms long. Inorganic nanodots are spheres of atoms, with diameters from ten to hundreds of atoms, normally synthesised using toxic and exotic organometallic precursors.

3 Nanomaterials in Gas Sensors

How can nanomaterials be used to improve gas sensors? Gas sensor nanotechnology is not new: the platinum catalyst used in carbon monoxide electrochemical sensors has a primary particle size of three to 20 nm, classifying it as an extremely small nanomaterial. Figure 2 shows a nanorod of precious metal catalyst, with each atom easily identified. A major difficulty with nanomaterials is not the manufacture of very small particles, but trying to avoid the particles coalescing into a larger particle. Figure 1 shows how nanoparticles with an average size of 5 nm cluster into a larger, albeit still very small particle [8].

Research with other nanomaterials over the last seven years has produced results in gas detection that has yet to be commercialised. Attempts at Pennsylvania State to make nanoparticulate titanium have resulted in laboratory hydrogen sensors operating at 300 C. Further work to control particle size and anodisation is needed.

Fig. 1 Nano particle cluster

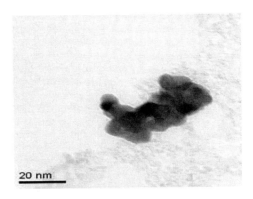

Fig. 2 Nanorod of precious metal catalyst

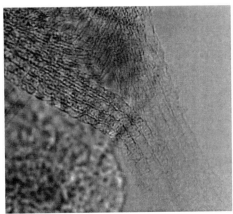

4 Carbon Nanotubes

CNTs are a focus of the popular press, but the underlying knowledge has been around for decades. Buckeyballs and fullerenes are spherical nanocarbon materials, which were identified decades ago, but rod-like CNTs are only ten years old. CNTs can be built using plasma, physical and chemical vapour deposition [3]. At £600/g, these materials are only a research material at this time, but new production methods will semantically reduce these prices to a price where they can be used as affordable sensing layers in gas sensors [2]. The University of Cambridge and Thomas Swan has teamed to manufacture a low cost source of CNT, and market costs should eventually approach the cost of platinum of £25/g.

CNTs are generally classed as two types.

- SWCNTs with single layered atoms of carbon, wrapped into a tube, thousands of atoms long [4].
- Multiwalled carbon nanotubes (MWCNTs) where several atomic layers of carbon are wrapped into tubes, with the tubes stacked together in various formations. The three-dimensional (3D) structure of MWCNTs is a research topic in itself. Figure 3 is a micrograph.

Fig. 3 MWCNT micrograph

CNTs are being studied as alternatives to MOs, where the response to gases such as nitrogen oxides, hydrogen, ammonia and oxygen is measured as a change of the electrical conductivity. The chemical sensitivity of CNT is much greater than its macro cousin, graphite and is potentially more selective than MOs to specific gases. The conduction and electrochemistry of CNTs can be modified by adding organic molecules or inorganic atoms to either inside or onto the surface of a CNT. Opportunities to improve selectivity and sensitivity by this "doping" of CNTs are being explored in laboratories worldwide.

A curious problem of CNTs is that two rods next to each other are not the same: one will be electrically conducting and the other will be semiconducting. If one can learn how to make 100 % semiconducting CNTs, then many more gas sensing opportunities will become available.

CNTs offer great opportunities for customised, repeatable gas sensors with better selectivity and repeatability, but the research has only begun, and, without a stepwise breakthrough, people may have to wait years before an improved sensor with CNT technology is commercially available.

5 Metal Oxides

MO sensors can be very sensitive, but their lack of selectivity has always let them down. MO sensors (also called mixed MOs: MMOs) are the most popular gas sensor. The ubiquitous tin oxide sensor is manufactured in millions per year, but lacks selectivity and stability. It was previously believed that shrinking MO particles to the nano scale would give greater selectivity, but apparently it only yields improved sensitivity, which allows ppb resolution for applications such as environmental monitoring and indoor air quality.

Researchers have tested ZnO, ITO, InZnO and CuO, using a combinatorial chemistry approach to see which nano MOs work best. If smaller particle size is combined with the chemistry of other MOs, then the combination of our elemental knowledge of materials such as gallium, vanadium and praseodymium cerium oxides, along with reduced particle size gives new possibilities for improved gas sensor selectivity. By producing new nano-MOs, researchers should have more selective and stable gas sensors in three to five years with the advantages of greater sensitivity; much work is required to optimise this improved selectivity and repeatability.

6 Organic Molecules and Polymers

Few tricks can be learned from biotechnology. Biotechnologists frequently create self-assembling structures using binding reactions such as thiols to gold layers. From this they get a 'field of wheat' type structure of organic chains that is highly ordered and hence give repeatable and well-characterised sensitivity and selectivity. By careful control of the active site, classes of materials such as pyrroles and porphyrins can selectively target gases such as NOx and ammonia [7]. But how to measure property changes which happen when gases interact with our highly ordered nano organic structure? Only the change of index can be measured of refraction with surface plasmon resonance, or the change in mass using surface acoustic waves and quartz crystal microbalances. A satisfying combination of MEMS technology and nanotechnology [6]. These methods can measure property changes as small as 10–6, 10–7.

7 Quantum Dots

Quantum dots (QDs) have now arrived. Being developed mainly as fluorescent markers to assist in biological and cancer identification, QDs are frequently made of toxic materials such as cadmium sulphide and cadmium selenide. To reduce toxicity, QDs are often enclosed in a larger, nontoxic particle [1]. These materials are now being offered by companies such as Nanoco for use in gas detection. The synthetic routes are now known, the QDs are there, and other QD materials including gallium arsenide and indium antimonide are also becoming available. These materials could be close to market for gas sensors, but it all depends on the QD performance, price and stability.

Using QDs as fluorescent markers have not yet been exploited for gas detection, but may offer a gas sensor with unique selectivity. Figure 4 shows how different particle sizes in Nanoco QDs emit light at different wavelengths.

8 MEMS: Learning from Silicon Technology

MEMS devices are more prosaic, but equally important, taking advantage of our knowledge of silicon processing gained over the last decades. Instead of producing simple electronic circuits, silicon foundries are now using silicon as a starting material for producing 3D structures such as suspended platforms, micro-hotplates, resonant cantilevers and micromirrors. Besides accelerometers in air-bags and pressure sensors in the latest car tyre technology, new MEMS products include infrared cameras, micromirrors for desktop projectors and self-correcting telescope optics. Turning to gas sensors, MEMS offers new opportunities including microhotplates for MO sensors, pellistors and microscopic resonant cantilevers to monitor mass changes of organic sensing layers as they absorb and adsorb various gases.

With electron beams ("e-beams") one can now write directly onto silicon. Figure 5 shows electrodes only 40 nm width for use as a gas sensor.

9 Combining the Two

Why not combine the two technologies and use nanomaterials as a sensing layer and MEMS as the transduction method? A Californian company, Nanomix has done just this. They use a platform where they create a field effect transistor (FET) and then grow CNTs directly onto the gate. The charge of the CNTs controls the gate voltage and the charge of the CNTs varies in response to different oxidising and reducing gases. A sensitive and potentially low cost method of making chemical sensors, this technology needs to develop the selectivity and sensitivity that the industry requires.

So the combination of silicon technology and advances in materials science is finally offering the gas sensor industry opportunities to improve sensor performance, which has been, until now, a very slow process.

10 Conclusions

Nanotechnology offers an extremely broad range of potential applications. Although it is not sure which pathway the nanotechnology can take and bring the fantasies real, it is certain that nanotechnology is penetrating into every aspects of our life and will make the world different from what we know now. As sure as the sun rises each day, the technology behind it continues to flow and transform at breakneck speed.

References

1. Colombelli R et al (2004) Fabrication technologies for quantum cascade photonic-crystal microlasers. Nanotechnology 15 675–681, Wang et al US patent no. 6,713,519 2004
2. Dresselhaus MS, Dresselhaus G, Eklund PC (1996) Science of fullerenes and carbon nanotubes. Academic Press, San Diego

3. Kong J et al (2000) Nanotube molecular wires as chemical sensors. Science 287:622–625
4. Sumanasekera GU et al (2000) Effects of gas adsorption and collisions on electrical transport in single-walled carbon nanotubes. Phys Rev Lett 85(5):1096–1099
5. Kong J et al (2001) Functionalized carbon nanotubes for molecular hydrogen sensors. Adv Mater 13(18):1384
6. Modi A (2003) Miniaturized gas ionization sensors using carbon nanotubes. Nature 424:171–174
7. Bradley CC et al (2003) Charge transfer from ammonia physisorbed on nanotubes. Phys Rev Lett 91(21):218301–218304
8. Cao G (2004) Nano structures and nanomaterials Synthesis properties and applications. Imperial College Press, London

Statistical Analysis of Tool Wear Using RSM and ANN

A. Arun Premnath, T. Alwarsamy and T. Abhinav

Abstract The research reported herein is to model the tool wear during face milling of Hybrid composites using response surface methodologies (RSM) and artificial neural network (ANN). Aiming to achieve this goal, several milling experiments were carried out with polycrystalline diamond (PCD) inserts at different machining parameters namely cutting speed, feed, depth of cut, and weight fraction of Al_2O_3. Materials used for the present investigation are Al 6061-aluminum alloy reinforced with alumina (Al_2O_3) of size 45 microns and graphite (Gr) of an average size 60 μ, which are produced by stir casting route. Central composite face centered second order RSM was employed to create a mathematical model and the adequacy of the model was verified using analysis of variance. Comparison has been made between prediction capabilities of model based on RSM and ANN. The comparison clearly indicates that the models provide accurate prediction of tool wear in which ANN perform better than RSM.

Keywords Hybrid composites · Milling · PCD insert · RSM · ANN

1 Introduction

Composites are synthetically made novel class of materials consisting of distinct insoluble phases and deriving the advantage of each phase to enhance the overall performance of the material. Hybrid Metal Matrix Composites are new class

---

A. Arun Premnath (✉)
Sri Chandrasekharendra Saraswathi Viswa Mahavidyalaya University,
Enathur, Kancheepuram, Tamil Nadu 631561, India
e-mail: arun_premnath@yahoo.co.in

T. Alwarsamy
Directorate of Technical Education, Chennai, India
e-mail: alwar_samy@yahoo.co.in

T. Abhinav
Indian Institute of Technology, Chennai, India
e-mail: thiyagarajan.abhinav414@gmail.com

S. Sathiyamoorthy et al. (eds.), *Emerging Trends in Science,*
Engineering and Technology, Lecture Notes in Mechanical Engineering,
DOI: 10.1007/978-81-322-1007-8_27, © Springer India 2012

of composites, which has better mechanical and machining properties than conventional composites. Metal matrix composites (MMCs) combine metallic properties with ceramic properties, which increase the strength of these materials. The increase in strength finds them difficult to machine due to their extreme abrasive properties [1]. Also, they lead to improper surface finish and wear of tools. In order to overcome the above said difficulties during machining, a small amount of softer reinforcement is added to composites, which will favors machining process. The composites materials thus obtained, with more than one reinforcement are termed as hybrid composites.

Many researchers have studied the effect of machining parameters in the various machining process on tool wear. Rajesh Kumar Bhushan et al. [2]. has investigated the effect of cutting speed, depth of cut, and feed rate on surface roughness and tool wear rate during the machining of 7075 Al alloy and 10 % wt. SiC particulate metal-matrix composites; tungsten carbide and polycrystalline diamond (PCD) inserts have been used as cutting tools. Munoz-Escalona and Maropoulos [3] studied the tool performances when face milling of 416 stainless steel and concluded that cutting speed has the most influence on tool life. Ibrahim ciftci et al. [4] has investigated effects of cutting speed and effect of coating on tool wear during turning of Al-2014 alloy matrix composite. Saeed chavosi [5] has investigated the effects of feed, depth of cut, and cutting speed on flank wear of Tungsten carbide and PCD inserts in CNC turning of 7075 Al Alloy with 10 % wt SiC Composites. Karthikeyan et al. [6] has optimized the milling characteristics (Flank wear, Specific energy, and surface roughness) of Al-SiC particulate composites using goal programming technique and examined that volume fraction of SiC particles present in the aluminum alloy matrix has a significant effect on the milling characteristics increasing tool wear and specific energy and decreasing surface roughness. El-Gallab and Sklad [7] studied the surface integrity of Al/SiC particulate MMCs and concluded that surface roughness and flank wear increases with increase in the feed rate and cutting speed. Metin Kok [8] studied the machinability of Al_2O_3 particle reinforced aluminum alloy composite using TiN (K10) coated and HX uncoated carbide tools under different conditions and concluded that the tool life of TiN coated K10 tool was significantly longer than that of the HX tool and the tool life decreased with an increase in the cutting speed for both tools in all cutting conditions. Übeyli et al. [9] studied the effect of feed rate on tool wear in milling of B_4C_p reinforced aluminum MMCs, produced by liquid phase sintering method. Milling experiments were conducted with three different types of cementide carbide tools (uncoated, double coated (TiN + TiAlN) and triple coated (TiCN + Al_2O_3 + TiN)) for three different feed rates. They concluded that higher feed rates led to lower tool wear for all type of tools and coated tools exhibited better performance than uncoated tool with respect to the flank wear. Choudhury et al. [10] predicted the response variables flank wear, surface finish, and cutting zone temperature in turning operations using Design of Experiments and the neural network technique and the values obtained from both methods were compared with the experimental values of the response variables to determine the accuracy of the predictions. Srinivas and Ramakotaiah [11] developed a neural network model to predict tool wear and cutting force in turning operations for cutting parameters cutting speed, feed, and depth of cut. Brezak et al. [12] considered the application of the radial basis function neural network (RBFNN) for tool wear determination in the milling process. Tool wear (i.e. flank wear zone width) has been estimated in two phases using two types of RBFNN algorithms. In the first phase, a RBFNN pattern recognition algorithm is used in order to classify tool wear features in three wear level classes (initial, normal, and rapid tool wear). On behalf of these results, in the second phase, the RBFNN regression algorithm is utilized to estimate the average amount of flank wear zone widths. Most of the current literatures present experimental results when machining ceramic-reinforced MMCs or Gr reinforced MMCs. However, limited information is available on the milling of graphitic ceramic-reinforced composites. The main objective of the paper is to study the influence of feed rate, cutting speed, depth of cut, and alumina weight fraction on tool wear in face milling of hybrid composites using RSM and ANN.

2 Experimental

2.1 Materials and Methods

In this present work Al 6061 aluminum alloy is reinforced with alumina (Al_2O_3) of size 40 μm with 5, 10, and 15 % of weight fraction and graphite (Gr) of size 60 μm with 5 % weight fraction is produced through stir casting method. Thus three different weight proportion of Al_2O_3 based hybrid composites are obtained in size of 120 $\times$ 100 $\times$ 50 mm, which are used for experimentation. The chemical composition of Al 6061 is shown in Table 1. The face milling test is carried out on ARIX VMC 100 CNC Vertical machining center. The machining of the fabricated composites was performed at the different machining parameters namely feed, cutting speed, depth of cut, and weight fraction of alumina using PCD inserts. The tool wear is measured using tool maker microscope (model 1395 A, Metzer make).

2.2 Response Surface Methodology

RSM adopts both mathematical and statistical techniques which are useful for the modeling and analysis of problems in which a response of interest is influenced by several variables and the objective is to optimize the response [13].

Table 1 Chemical composition of 6061 aluminum alloy

Element	Si	Cu	Mg	Mn	Fe	Zn	Sn	Ti	Pb	Al
Wt%	0.80	0.35	0.8	0.02	0.01	0.008	0.01	0.01	0.02	97.9

The objectives of quality improvement, including reduction of variability and improved process and product performance, can often be accomplished directly using RSM. In the RSM, the quantitative form of relationship between the desired response and independent input variables is represented as follows:

$$Y = F\ (f, v, d, w) \tag{1}$$

Where Y is the desired response, F is the response function, and f, v, d, and w represent feed rate, speed, depth of cut, and weight fraction of Al_2O_3 respectively. In order to study the effect of the process parameters a second order polynomial response surface can be fitted into the following equation.

$$Y = \beta_0 + \sum_{i=1}^{k} \beta_i X_i + \sum_{i=1}^{k} \beta_i X_i^2 + \sum_i \sum_j \beta_{ij} X_i Y_j + \xi \tag{2}$$

where, 'Y' is the corresponding response and X_i is the value of the ith machining process parameter. The terms β are the regression coefficients, and ξ is the residual measure, resulting from the experimental error in the observations. This quadratic model works quite well over the entire factor space.

The necessary data required for developing the response models have been collected by designing the experiments based on central composite design (CCD). CCD is the most popular second-order design which was introduced by Box and Wilson. It is a factorial or fractional factorial design with center points and star points. The test was designed based on a four-factor-three levels CCD with full replication. A CCD consisting of 30 experiments was used in the experiments. Table 2 shows the process variables used in the experiments. Table 3 shows comparison of tool wear of experimental values with RSM and ANN. Using analysis of variance (ANOVA), the significance of input parameters is evaluated. Design-Expert 8.0 software was used to establish the design matrix, to analyze the experimental data and to fit the experimental data to a second-order polynomial. Sequential F test, lack-of-fit test, and other adequacy measures were used to check the model's performance.

Table 2 Process variables and experimental design levels used

Process variables	Notation	Unit	Limits		
			-1	0	$+1$
Feed rate	f	mm/tooth	0.04	0.08	0.15
Cutting speed	v	rpm	1,500	3,000	4,500
Depth of cut	d	mm	0.1	0.3	0.5
Weight fraction of Al_2O_3	w	%	5	10	15

Statistical Analysis of Tool Wear Using RSM and ANN

Table 3 Comparison of experimental values of tool wear with RSM and ANN

Exp No.	Feed rate f (mm/tooth)	Cutting speed v (m/min)	Depth of cut d (mm)	Weight fraction of Al_2O_3 w (%)	Flank wear (mm)		
					Exp.	RSM	ANN
1	0.04	1,500	0.1	5	0.075	0.076	0.079
2	0.15	1,500	0.1	5	0.126	0.129	0.126
3	0.04	4,500	0.1	5	0.107	0.107	0.106
4	0.15	4,500	0.1	5	0.157	0.153	0.156
5	0.04	1,500	0.5	5	0.082	0.080	0.083
6	0.15	1,500	0.5	5	0.131	0.134	0.137
7	0.04	4,500	0.5	5	0.119	0.120	0.119
8	0.15	4,500	0.5	5	0.169	0.169	0.164
9	0.04	1,500	0.1	15	0.088	0.089	0.087
10	0.15	1,500	0.1	15	0.137	0.135	0.137
11	0.04	4,500	0.1	15	0.122	0.118	0.122
12	0.15	4,500	0.1	15	0.152	0.156	0.152
13	0.04	1,500	0.5	15	0.094	0.097	0.093
14	0.15	1,500	0.5	15	0.143	0.144	0.150
15	0.04	4,500	0.5	15	0.136	0.134	0.136
16	0.15	4,500	0.5	15	0.177	0.175	0.163
17	0.04	3,000	0.3	10	0.094	0.095	0.091
18	0.15	3,000	0.3	10	0.148	0.142	0.148
19	0.08	1,500	0.3	10	0.12	0.109	0.120
20	0.08	4,500	0.3	10	0.133	0.139	0.133
21	0.08	3,000	0.1	10	0.103	0.102	0.103
22	0.08	3,000	0.5	10	0.118	0.114	0.118
23	0.08	3,000	0.3	5	0.115	0.112	0.115
24	0.08	3,000	0.3	15	0.124	0.122	0.126
25	0.08	3,000	0.3	10	0.105	0.114	0.111
26	0.08	3,000	0.3	10	0.112	0.114	0.111
27	0.08	3,000	0.3	10	0.11	0.114	0.111
28	0.08	3,000	0.3	10	0.115	0.114	0.111
29	0.08	3,000	0.3	10	0.113	0.114	0.111
30	0.08	3,000	0.3	10	0.113	0.114	0.111

2.3 ANN Modeling

ANN is considered to be a powerful computational technique that takes into account nonlinear relationship between the variables [14]. ANN eliminates the limitations of the classical approaches by extracting the desired information using the input data. They are successfully used to solve variety of complex engineering and scientific problems. An ANN may be seen as a black box which contains hierarchical sets of neurons (e.g., processing elements) producing outputs for certain inputs. Each processing element consists of data collection, processing the data and sending the results to the relevant consequent element. The whole process may be viewed in terms of the inputs, weights, the summation function, and the activation function. The neuro-intelligence software was used to create, train, validate, and predict the different ANN reported in this research.

3 Results and Discussions

3.1 RSM-Based Predictive Models

The RSM was performed to predict the tool wear in milling of Al hybrid composites. The ANOVA of the experimental data was done to statistically analyze the relative significance of the machining parameters such as feed rate (f), speed (v), depth of cut (d) and weight fraction of Alumina (w) on response variables tool

Table 4 ANOVA for tool wear

Source	Sum of squares	df	Mean square	F value	p value Prob $> F$	
Model	0.016779	14	0.001198	39.60	< 0.0001	Significant
f-feed	0.009248	1	0.009248	305.58	< 0.0001	
v-speed	0.003727	1	0.003727	123.14	< 0.0001	
d-depth of cut	0.000272	1	0.000272	8.99	0.0090	
w-weight fraction alumina	0.000722	1	0.000722	23.85	0.0002	
fv	4.56E−05	1	4.56E−05	1.50	0.2387	
fd	0.00015	1	0.00015	4.95	0.0417	
fw	0.000189	1	0.000189	6.24	0.0245	
vd	6.25E−08	1	6.25E−08	0.002	0.9644	
vw	1.41E−05	1	1.41E−05	0.46	0.5058	
dw	4.56E−05	1	4.56E−05	1.50	0.2387	
f^2	4.18E−05	1	4.18E−05	1.38	0.2581	
v^2	0.001429	1	0.001429	47.20	< 0.0001	
d^2	0.000315	1	0.000315	10.39	0.0057	
w^2	6.43E−05	1	6.43E−05	2.125	0.1655	
Residual	0.000454	15	3.03E−05			
Lack of fit	0.000385	10	3.85E−05	2.79	0.1339	Not significant
Pure error	6.88E−05	5	1.38E−05			
Cor total	0.017233	29				
R^2						0.95
Adj R^2						0.92

wear. From the Table 4, model F value of 39.94 indicates that the model is significant for tool wear. The values of "Prob > F" for model is less than 0.0500 which indicates that the model terms are significant. The "Lack of Fit F-value" of 3.13 implies the lack of fit is not significant for tool wear. There is only a 0.01 % chance that a "Model F Value" this large could occur due to noise. The R^2 (0.95) for tool wear value are high, close to 1, which is desirable. From the ANOVA results, it is concluded that the factors f, v, w, and their interactions fv and fw have significant effect on tool wear. The regression equations obtained for the response factors by using multiple regressions is given below.

$$\begin{aligned}
\text{Tool wear} = \ & +0.10810 + 0.76730 * f - 9.97932E - 004 * v + 0.17451\,d \\
& -6.50548E - 004 * w - 4.09091E - 004 * f * v \\
& +0.27841 * f * d - 0.012500 * f * w + 4.16667E - 006 * v * d \\
& -2.50000E - 006 * v * w - 1.68750E - 003 * d * w \\
& -1.32811 * f^2 + 4.17466E - 006 * v^2 \\
& -0.27544 * d^2 + 1.99298E - 004 * w^2
\end{aligned}$$
(3)

From the developed RSM-based mathematical model, the effect of milling parameter on tool wear is examined. The 3D surface plots are shown in the Figs. 1 and 2. From the results it is inferred that tool wear is influenced by speed and feed rate than other machining parameters [6, 15].It clear that as speed increases, the flank wear increases, it is because, the increase in cutting speed increases the impact load on the work material by the insert due to the increased kinetic energy which cause a steady state increase in wear with speed assisted by the increased temperature at the interface. Similarly, as the feed increases flank wear increases owing to increase in force and work done in turn causes increase in heat generation and thermal load per cutting edge

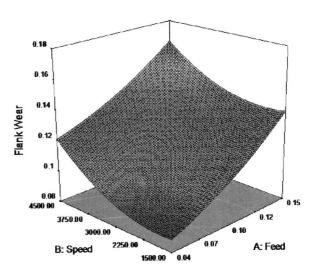

Fig. 1 3D graph shows the interaction effects of flank wear on feed rate versus Speed

unit length causing high wear. The increase in depth of cut increases tooth contact angle, in turn increases the uncut chip thickness and the idle time is reduced thereby resulting in increased thermal stress. This leads to increased wear with increase in depth of cut. Similarly, as the weight fraction of Al_2O_3, increases, the wear increases due to the increased hardness of the material and the abrasive Al_2O_3 particles sliding against the insert and also with the increased brittleness with low heat conduction increases abrasion and the concentration of chip pressure on a small area of the edge.

3.2 ANN for Prediction of Surface Roughness

In this work, back propagation neural network model have been developed to predict the tool wear on the face milling of composites by constructing non-linear functions between several inputs such as feed, speed, depth of cut, and weight fraction of alumina. The back propagation training algorithm is an iterative method to minimize the root mean square error between the actual output of the hidden layers and the desired output. The lower the roots mean square error, the better the ANN predictions. The experimental data set is divided into three data set namely train data set, testing data set, and validating data set.

The model with 12-6-1 architecture suits well for tool wear prediction which is shown in Fig. 3. It consists of 12 neurons in input layer, 6 in hidden layer and 1 neuron as output corresponding to the tool wear of the composites. The training data set is used to modify the weights between the interconnected neurons, until the desired error level is reached. The network training function updates the weight and bias values so as to minimize the error between the training data set and network prediction. There are 22 data sets used for training the network from the experimental results.

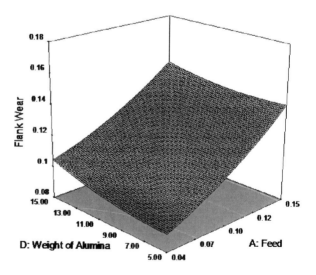

Fig. 2 3D graph shows the interaction effects of flank wear on weight of alumina versus feed

Fig. 3 ANN architecture for tool wear

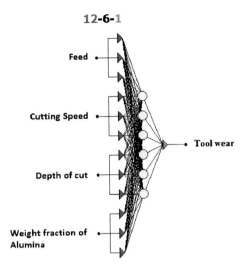

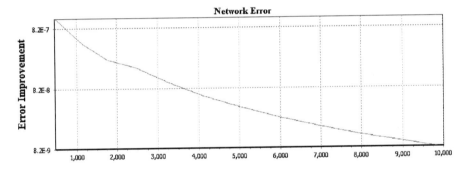

Fig. 4 Error improvement as iteration increases

The learning rate and momentum was set 0.1 by trial and error method to minimize the error. The predicted results of tool wear for RSM and ANN with the experimental values are presented in Table 3. From the Table, it is found that the results provided by the ANN are clearly accurate and predicted values have good agreement with the RSM and experimental values. Figure 4 shows the error improvement, which indicates that the error approaches to zero as the iteration increases.

3.3 Comparison of ANN and RSM

The RSM and ANN models were tested with 30 data sets of CCD. For each input combination, the predicted value of tool wear was compared with the respective experimental values and the absolute percentage error is computed as follows:

$$\%\text{Absolute error} = \frac{Y_{(\text{ExP})n} - Y_{(\text{Pred})n}}{Y_{(\text{ExP})n}} \times 100 \qquad (4)$$

where y_n, expt is the experimental value and y_n, pred is the predicted value of the response for the nth trial by the RSM and ANN models. The average absolute percentage error in the ANN prediction of tool wear was found to be around 0.97 % while for the RSM model, it is around 4.98 %. Thus, it can be concluded that the ANN predicts more accurately than the RSM model. Comparison of Target and Predicted values of ANN model is shown in Fig. 5. An R value of 1 indicates perfect correlation between targets and outputs. In the present work, the R values of ANN is more than that of RSM models, which clearly indicates that the prediction accuracy of the ANN model is higher compared to the RSM model. It is noted that ANNs perform better than RSM when highly non-linear behavior is the requirement. Also, this technique can build an effective model when using a small number of experiments however, the accuracy of ANN would be better when a larger number of experiments are used to develop a model. Generation of ANN model requires a large number of iterative calculations which is time consuming whereas RSM is only a single step calculation.

4 Conclusions

In this paper, the effect of machining parameters on tool wear was investigated while milling of hybrid MMCs using PCD insert. Four process parameters such as feed rate, cutting speed, depth of cut, and weight fraction of Al_2O_3 were considered for the model development. The experimentally determined tool wear values are compared with predicted values obtained from the RSM and ANN models. The following conclusions are drawn from the present investigation.

- Cutting speed and feed rate from the ANOVA table are found to be more significant when compared to other machining parameters affecting tool wear.
- Among the interaction, cutting speed-feed rate and Feed rate-Weight fraction of Al_2O_3 has greater influence compared with other interactions on tool wear on milling of Al Hybrid MMC composites.

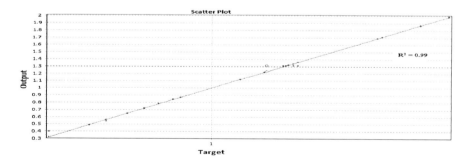

Fig. 5 Comparison of target and predicted values of ANN model

- The predicted values match the experimental values reasonably well, with R^2 of 0.95 for RSM and R^2 of 0.99 for ANN.
- The predictive ANN model is found to be capable of better predictions of tool wear within the range that they had been trained compared with RSM model.

Acknowledgments The author gratefully acknowledges the financial support of Sri Chandrasekharendra Saraswathi Viswa Mahavidyalaya, Kancheepuram, India for carrying out this research work.

References

1. Sahin Y, Kok M, Celik H (2002) Tool wear and surface roughness of Al_2O_3 particle—l reinforced aluminium alloy composites. J mater process Technol 128:280–281
2. Bhushan RK, Kumar S, Das S (2010) Effect of machining parameters on surface roughness and tool wear for AL alloy composite. Int J Adv Manuf Technol 50:459–469
3. Munoz-Escalona P, Maropoulos P (2010) Integrated optimisation of surface roughness and tool performance when face milling 416 SS. Int J Comput Integr Manuf 23(3):248–256
4. Ciftci I, Turker M, Seker U (2004) Evaluation of tool wear when machining SiCp-reinforced AL-2014 alloy matrix composites. Mat Des 25:251–255
5. Chavoshi SZ (2011) Tool flank wear prediction in CNC turning of 7075 Al alloy composite. Pro Eng Res Dev 5:37–48
6. Karthikeyan R, Raghukandan K, Naagarazan RS, Pai BC (2000) Optimizing the milling characteristics of Al-SiC particulate composites. Met Mater 6:539–547
7. El-Gallab M, Sklad M (1998) Machining of Al/SiC particulate MMCs part II: work piece surface integrity. J Mat Process Technol 83:277–285
8. Kok M (2008) A study on the machinability of Al2O3 particle reinforced aluminium alloy composite. 11th Int inorganic-bonded fiber composite conference, Spain
9. Übeyli M, Acir AM, Karakaş S, Bilgehan Ogel B (2008) Effect of feed rate on tool wear in milling of Al-4 %Cu/B_4 C $_p$ composite. Mat Manuf Process 23(8):865–870
10. Choudhury SK, Bartarya G (2003) Role of temperature and surface finish in predicting tool wear using neural network and design of experiments. Int Journal of Mach Tools Manuf 43:747–753
11. Srinivas J, Rama Kotaiah K (2005) Tool wear monitoring with indirect methods. Manuf Technol Today India 4:7–9
12. Brezak D, Udiljak T, Mihoci K, Majetic D, Novakovic B, Kasac J (2004) Tool wear monitoring using radial basis function neural network. IEEE T Neural Networ 1(04):1859–1862
13. Montgomery DC (2001) Design and analysis of experiments. Wiley, New York
14. Palau A, Velo E, Puigjaner L (1999) Use of neural networks and expert systems to control a gas/solid sorption chilling machine. Int J Refrig 22(1):59–66
15. Seeman M, Ganesh G, karthikeyan R, Velayudham A (2010) Study on tool wear and surface roughness in machining of particulate aluminum metal matrix composites—response surface methodology approach. Int J Adv Manuf Technol 48:613–624

Surface Integrity of Ti-6Al-4V Precision Machining Using Coated Carbide Tools Under Dry Cutting Condition

R. Vinayagamoorthy and M. Anthony Xavior

Abstract Aerospace components from titanium alloys require the greatest reliability and satisfied surface integrity requirement. However, during machining of titanium alloys, the machined surface can be easily damaged because of the difficult-to-machine material and poor machinability. The aims are to investigate the surface integrity of Ti-6 %Al-4 %V precision machining under dry cutting condition. The results showed that the surface roughness values recorded were more affected by feed rate and nose radius geometry. Surface roughness was high value at the first machining followed by decreasing. Work hardening beneath the machined surface caused higher hardness than based material; on the other hand, Changing orientation of microstructure and 2 μm of white layer on the machined surface was found when turning at cutting speed of 30 m/min, feed rate of 0.02 mm/rev, and depth of cut of 0.01 mm.

Keywords Precision turning • Ti-6Al-4V • Surface roughness • Dry machining • Cutting force

1 Introduction

Studies on surface finish of aerospace materials become a more critical issue, mainly, to produce a high quality of machined surface component, which requires high accuracy. The mechanical component, which designed from titanium alloys, has more difficult to produce a good machined surface because of these alloys are difficult to machine and high generated temperature when machining [1, 2]. The surface integrity of titanium alloys is also affected by selected condition of machining. Requirement on

---

R. Vinayagamoorthy (✉) · M. A. Xavior
School of Mechanical and Building Sciences, VIT University, Vellore, Tamil Nadu 632014, India
e-mail: vinayagamoorthy.r@vit.ac.in

S. Sathiyamoorthy et al. (eds.), *Emerging Trends in Science, Engineering and Technology*, Lecture Notes in Mechanical Engineering, DOI: 10.1007/978-81-322-1007-8_28, © Springer India 2012

the satisfied surface integrity is not only based on surface roughness but also focused on surface hardness, microstructure, plastic deformation of machined surface, residual stress, and surface defects such as porosity, micro crack, stress concentration etc. [3].

The surface finish determines the surface quality of machined component and the integrity obtained after machining. The surface integrity is defined as the inherent or enhances condition of a surface produced in machining [3]. Metal removal operations lead to the generation of surfaces that contain geometric deviation (deviation from ideal geometric) and metallurgical damage, which differs from the bulk material. The geometrical deviation refer to the various forms of deviations suck as roundness, straightness etc. Types of metallurgical surface damages that produced during machining include micro crack, micro pits, tearing, and plastic deformation of feed marks, redeposit materials etc. Therefore, control of its machining process to produce components of acceptable integrity is essential. Machined components for aerospace application are subjected to rigorous surface analysis to detect surface damage that will be detrimental to the highly expensive machined components [4]. Superior mechanical at elevated temperature and excellent corrosion resistance.

Cutting tool materials employed for machining titanium alloys usually have short tool life and most react with titanium materials. This disadvantage is due to the generation of heat temperature closer to the cutting edge of tool [2]. This phenomenon lead to rapid tool wears when machining titanium alloys. Hence, selection for suitable type of cutting tool and machined condition of machining titanium alloys is required to produce the good quality of machined surface. Cemented carbide tools selected, which are coated by TiN and TiCN layer(s) can reduce the wear on cutting tool, mainly on flank wear and crater wear [4]. Che Haron [5] found that the straight cemented carbide tools were suitable used in turning Ti-6Al-4V. The hard coating layer(s) on the surface of cutting tools can reduce the tool wear progression on the flank face. The thin layer(s) from TiN and TiCN material reduced the friction between the cutting edge and workpiece materials, so it will produce a smooth surface of titanium alloys and less surface damages. This paper investigates the integrity of machined surface by analyzing the surface roughness value recorded, surface damage, and surface texture after machining Ti-6Al-4V precision machining using coated carbide tools and dry machining condition.

2 Experimental Procedures

The workpiece material used in the machining trials was a titanium alloy alpha–beta Ti-6Al-4V Extra Low Interstitial (Ti-6Al-4V ELI), which is equated α phase and surrounded by β in the grain boundary. At least 3 mm of material at the top surface of workpiece was removed in order to eliminate any surface defects and residual stress that can adversely affect the machining result [6]. The machining trials under dry machining condition and high cutting The average flank wear (VB) was measured by using a Mitutoyo Tool Maker Microscope at 20x magnification, and the machining time was recorded using a stopwatch. The wear and the cutting time were

Surface Integrity of Ti-6Al-4V Precision Machining Using Coated Carbide Tools

Table 1 Chemical composition of titanium alloy

Alloy	Al (%)	V (%)	Fe (%)	C (%)	Ti
Ti-6Al-4V	6.40	3.89	0.16	0.002	Balance

recorded at regular interval of one pass turning operation. The wear mechanisms of inserts were observed under the optical microscope. The detailed investigation of the worn out tool was carried out using scanning electron microscope (SEM). Speed was carried out using the Gedee Weiler MLZ 250V variable speed adjusting capstan lathe is used for the experiment. Lathe machine. Tools and tool holders were selected based on the recommendation of the tool supplier (Kennametal 2012). Four layers of coating materials for each insert, which consist of TiN-Al$_2$O$_3$-TiCN-TiN, were selected. The parameters for turning operation are as shown in Table 1.

3 Design of Experiments and Observations

3.1 Chemical Composition of Titanium Alloy (Grade 5)

See Table 1

3.2 Tool Specifications of CCGT09T301F Coated Carbide Insert

Titanium and its alloys are generally considered as difficult to machine materials due to their poor thermal conductivity and high strength, which is maintained at elevated temperatures. This work examines the tool wear mechanisms involved in precision machining of titanium. In this study cc tools were used to machine commercial pure titanium (CP-Ti) and Ti–6Al–4V alloy. Industrial expectations for surface quality and tool life based on optical grade applications are presented. Results obtained from the characterization of the tool, chip, and workpiece led to the identification of graphitization as the mechanism that initiates tool wear. The parameters for turning operation are as shown in Table 3. Surface roughness for the specimen is found at three points on the specimen and the average is found out and tabulated. The process is continued for all the L27 specimens and the results

Table 2 Tool specifications of CCGT09T301F coated carbide insert

Composition	80 % Al$_2$O$_3$ and 20 % TiC
Grain size	3.0 µm
Transverse rupture strength	551–786 MPa
Average density	3.90–3.99 g/cm3
Youngs modulus	641 GPa
Hardness	91–94 HRA
Coefficient of thermal expansion	Good

Table 3 Machining parameters and trail level

Work piece	Ti–6Al–4V
Cutting speed m/min	30, 60, 90,
Feed μm/rev	2,4,6
Depth of cut mm	0.01, 0.1, 0.15
Lubricant	Dry

are tabulated. ANOVA is used to analyze the machining parameters and optimize the turning operations in order to reduce surface roughness (Table 2).

4 Effect of Cutting Parameter on Surface Roughness

In general surface roughness of machined component decreases with the increase in cutting speed. It may be suggested that adherence of the work piece material to the tool at cutting speeds are less pronounced, perhaps due to the high temperature generated. In machining, surface roughness decreases as cutting speed increases in dry machining, By Analysis on surface roughness proved that an increase in feed rate produced a general trend towards higher surface roughness and the machined surfaces consist of uniform feed marks in perpendicular to the tool feed direction. This is attributed by plastic flow of material during the cutting process. Plastic flow of material on machined surfaces results in higher surface roughness values. The analysis of the effect of feed rate on surface roughness (Fig. 4) shows that this parameter has a very significant influence, because its increase generates uniform ploughing effect on the surface of the workpiece which, results in narrow cavities, these cavities are deeper and broader as the feed rate increases (Figs 1, 2, 3).

Titanium has a relatively low modulus of elasticity, thereby having more "springiness" than steel. Work has a tendency to move away from the cutting tool unless heavy cuts are maintained or proper backup is employed. Slender parts tend to deflect under tool pressures, causing chatter, tool rubbing, and tolerance problems. Rigidity of the entire system is consequently very important,

Fig. 1 Cutting forces versus feed rate

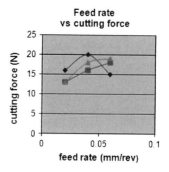

Fig. 2 Cutting forces versus cutting speed

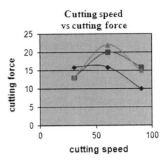

Fig. 3 Cutting forces versus depth of cut

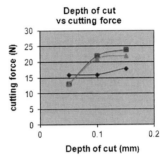

Fig. 4 Surface roughness versus feed rate

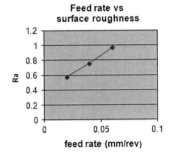

as is the use of sharp, properly shaped cutting tools. Characterization of surface topography is important in applications involving friction, lubrication, and wear [7]. In general, it has been found that friction increases with average roughness. Roughness parameters are, therefore, important in applications such as automobile brake linings, floor surfaces, and tires. The effect of roughness on lubrication has also been studied to determine its impact on issues regarding lubrication of sliding surfaces, compliant surfaces, and roller bearing fatigue. Finally, some researchers have found a correlation between initial roughness of sliding surfaces and their wear rate. Such correlations have been used to predict failure time of contact surfaces. Another area where surface roughness plays a critical role is contact resistance [8].

5 Results and Discussion

Thermal contact resistance is an important issue in space applications, such as satellites, where the heat generated by the electronic devices can only be driven away by conduction. Surface roughness is also a topic of interest in fluid dynamics [9–10]. The roughness of the interior surface of pipes affects flow parameters, such as the Reynolds number, which is used to evaluate the flow regime (i.e., laminar or turbulent). The performance of ships is also affected by roughness in the form of skin friction, which can account for 80–90 % of the total flow resistance. In addition, the power consumption can increase as much as 40 % during the service life of a ship as a result of increased surface roughness caused by paint cracking, hull corrosion, and fouling. The examples mentioned above are just a few of the applications in which surface roughness has to be carefully considered. However, the influence of roughness extends to various engineering concerns such as noise and vibration control, dimensional tolerance, abrasive

Table 4 Experimental observations

SL NO.	Feed mm/min	Depth of cut mm	Cutting speed m/min	Nose radius mm	Surface roughness	Cutting force(N)			Cutting tool temp
						Fx	Fy	Fz	48
1	0.02	0.05	30	0.1	0.45	16	13	13	47
2	0.02	0.05	60	0.2	0.42	16	20	22	49
3	0.02	0.05	90	0.4	0.47	10	16	15	54
4	0.02	0.1	30	0.2	0.47	16	22	21	59
5	0.02	0.1	60	0.4	0.42	18	22	25	64
6	0.02	0.1	90	0.1	0.65	16	14	13	59
7	0.02	0.15	30	0.4	0.58	13	17	22	63
8	0.02	0.15	60	0.1	0.64	21	17	16	64
9	0.02	0.15	90	0.2	0.43	18	24	22	49
10	0.04	0.05	30	0.1	0.76	20	16	18	51
11	0.04	0.05	60	0.2	0.67	20	22	24	53
12	0.04	0.05	90	0.4	0.6	11	16	19	52
13	0.04	0.1	30	0.2	0.69	10	17	15	62
14	0.04	0.1	60	0.4	0.61	10	14	11	59
15	0.04	0.1	90	0.1	0.79	18	17	20	69
16	0.04	0.15	30	0.4	0.57	10	16	12	76
17	0.04	0.15	60	0.1	0.81	22	18	25	72
18	0.04	0.15	90	0.2	0.71	14	15	17	52
19	0.06	0.05	30	0.1	0.97	15	18	19	57
20	0.06	0.05	60	0.2	0.82	12	18	11	63
21	0.06	0.05	90	0.4	0.68	12	16	18	68
22	0.06	0.1	30	0.2	0.87	16	19	17	69
23	0.06	0.1	60	0.4	0.57	10	14	11	77
24	0.06	0.1	90	0.1	1.12	17	20	22	76
25	0.06	0.15	30	0.4	0.69	12	16	18	83
26	0.06	0.15	60	0.1	1.19	19	20	22	82
27	0.06	0.15	90	0.2	0.89	15	20	21	48

processes, bioengineering, and geomorphometry (Thomas 1999). The parameters for turning operation are as shown in Table 4.

This shows that surface roughness will be minimum only for an optimum range of cutting speed between 50 and 65 m/min. The effect of depth of cut on cutting force in shown in Fig. 6 from this figure it is understood that for an increase in depth of cut there is a slight increase in surface roughness which indicates that depth of cut is not influencing as feed rate. Figure 7. In practice, the consequences of the influence of the feed rate on surface roughness are as follows: the increase in the feed rate from 0.02, 0.04 to 0.06 mm/rev correspondingly increases the criteria of roughness Ra, Fig. 5 shows the plot between cutting speed and surface roughness. From the graph it is understood that for a low cutting speed the surface

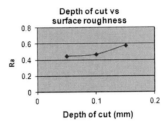

Fig. 5 Depth of cut versus surface roughness

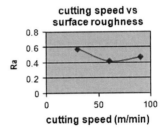

Fig. 6 Cutting speed versus surface roughness

Fig. 7 Experimental setup

roughness is more and as the cutting speed is increased the surface roughness value decreases. But beyond 60 m/min of cutting speed there is an increased trend in the surface roughness.

6 Conclusions

The following conclusions are based on the results of turning Ti-6Al-4V alloy with coated carbide tools under dry machining. Surface roughness values recorded when machining Ti-6Al-4V alloy under dry condition investigated were more affected by feed rate and also by nose radius. Three stages of trend line of surface roughness were high value at first machining followed by decreasing, Surface finish generated when machining Ti-6Al-4V alloy with coated carbide tools are generally acceptable and free of physical damage such as cracks and tears. Effects of machining on turned surface were micro pits, deformation of feed marks and red posited of titanium. The machined surface was found when cutting operation at cutting speed of 30, 60, and 90 m/min, feed rate of 0.02, 0.04, and 0.04 mm/rev and depth of cut of 0.01, 0.1, and 0.15 mm and at the end of tool life. The lower surface roughness value produces the smoother surface topography and it has a strength correlation to the surface roughness.

References

1. Boothroyd G, Knight WA (1989) Fundamental machining and machine Tools, 2nd edn. Marcel Dekker Inc, New York
2. Che Haron CH (2001) Tool life and surface integrity in turning titanium alloy. J Mat Process Technol 118:231–237
3. Field and Kahles (1971) Review of surface integrity of machined components. Ann CRIP 20 2:153–163
4. Ezugwu EO, Bonney J, Yamane Y (2003) An overview of the machinability of aeroengine alloys. J Mat Process Technol 134:233–253
5. Che Haron CH (2005) The effect of machining on surface integrity of titanium alloy Ti-6Al-4V. J Mat Process Technol 166:188–192
6. Kalpakjian S, Rchmid SR (2001) Manufacturing engineering and technology. Prentice Hall, New Jersey International Edition
7. Benardos PG, Vosniakos GC (2002) Prediction of surface roughness in CNC face milling using neural network and Taguchi's design of experiments. Rob Comput Integr Manuf 18:343–354
8. Ezugwu EO (2007) Surface integrity of finished turned Ti-6Al-4V alloy with PCD tools using conventional and high coolant supplies. Int J Mach Tools Manuf 47:884–891
9. Kennametal (2006) Kennametal lathe tooling catalogue, Kennametal Company
10. Thomas TR (1998) Trends in surface roughness. Int J Mach Tools Manuf 38(5–6):405–411

Surface Roughness Evaluation in Drilling Hybrid Metal Matrix Composites

T. Rajmohan, K. Palanikumar and G. Harish

Abstract The present study investigates the influence of machining parameters, such as cutting speed and feed rate, and wt % of reinforcement on surface roughness in drilling of hybrid metal matrix composites. Experiments are conducted on Al 356- aluminum alloy reinforced with silicon carbide of size 25 μ and Mica of size 45 μ. The experiments have been conducted on vertical computer numeric control (CNC) machining center. The Taguchi's orthogonal array and analysis of variance (ANOtVA) are employed to investigate the drilling parameters of hybrid composites. A second-order model has been established between the cutting parameters and surface roughness using response surface methodology. The experimental results reveal that the most significant machining parameter for surface roughness is feed rate followed by cutting speed. The influence of different parameters and their interactions are studied and presented in this study.

Keywords Taguchi method · Hybrid composites · Surface roughness · Response surface method

1 Introduction

Metal matrix composites (MMC) have become a large leading material in composite materials. Particle reinforced aluminum MMC have received considerable attention due to their excellent engineering properties. Several investigations into the machining aspects of MMCs have been carried out and reported. Hybrid MMCs are of recent kind and the literature available on their machining is inadequate. Noorul Haq et al. [1]

T. Rajmohan (✉)
Sri Chandrashekarendra Saraswathi Visva Mahavidyalaya University,
Kanchipuram 631561, India
e-mail: rajmohanscsvmv@yahoo.com

K. Palanikumar
Sri Sairam Institute of Technology, Chennai 600044, India
e-mail: palanikumar_k@yahoo.com

G. Harish
Indian Institute of Technology Madras, Chennai 600036, India

S. Sathiyamoorthy et al. (eds.), *Emerging Trends in Science,*
Engineering and Technology, Lecture Notes in Mechanical Engineering,
DOI: 10.1007/978-81-322-1007-8_29, © Springer India 2012

investigated the optimization of drilling parameters on drilling Al/SiC MMC with multiple responses based on orthogonal array with gray relational analysis. Drilling tests are conducted using TiN coated HSS twist drills of 10 mm diameter under dry condition. Further the drilling parameters namely cutting speed, feed, and point angle are optimized with the considerations of multi responses such as surface roughness, cutting force, and torque. Tosun [2] studied the statistical analysis of process parameters for surface roughness in drilling Al/SiC metal matrix composite using Taguchi method and suggested that Taguchi parameter design is an efficient and effective method for optimizing surface roughness performance. Bushan et al. [3] investigated the influence of cutting speed, depth of cut, and feed rate on surface roughness during turning of 7075 Al alloy and 10 wt % SiC particulate metal-matrix composites using tungsten carbide and polycrystalline diamond.

The Taguchi method and response surface methodology (RSM) are used in various areas including machining operations. Manna and Bhattacharyya [4] investigated the influence of cutting conditions on surface finish during turning of Al/SiC-MMC using Taguchi method. Routara et al. [5] have investigated the influence of machining parameters on the quality of surface produced in CNC and milling. Kok [6] investigated the effects of cutting speed, size, and volume fraction of particle on the surface roughness in turning of 2024 Al alloy composites reinforced with Al_2O_3 particles based on Taguchi method using coated carbide tools K10 and TP30. It has been reported that surface roughness increased with increasing the cutting speed and decreased with increasing the size and the volume fraction of particles for both cutting tools. Tsao [7] used RSM based on the Taguchi method to evaluate the influence of drilling parameters on delamination by compound core-special drills. Yahya Altunpak et al. [8] investigated the influence of cutting parameters on cutting force and surface roughness in drilling Al/20 % SiC/5 % Gr and Al/20 % SiC/10 % Gr hybrid composites fabricated by vortex method. The results indicate that inclusion of graphite as an additional reinforcement in Al/SiCp reinforced composite reduces the cutting force and surface roughness. Riaz Ahamed et al. [9] reported the application of EDM to machine cast aluminum–silicon carbide–boron carbide and cast aluminum–silicon carbide–glass hybrid MMC and how the metal removal rate and surface finish vary in response to the various EDM parameters using Taguchi method. Basavarajappa et al. [10] studied the influence of cutting parameters on thrust force, surface finish, and burr formation in drilling Al2219/15SiCp and Al2219/15SiCp-3Gr composites fabricated by liquid metallurgy method. Further it is reported that Graphitic composites exhibit lesser thrust force, burr height, and higher surface roughness when compared to the other material. Riaz Ahamed et al. [11] reported on the drilling of Al-5 % SiCp-5 % B4Cp hybrid composite with high-speed steel (HSS) and concluded that HSS drills only used with lower speed and feed combination. Rajmohan and Palanikumar have used RSM to predict thrust force in drilling hybrid composites using coated carbide drills. [12].

All the review of literature carried out done left the scope for the researcher to study the effect of the machining parameters on surface roughness in the drilling ceramic–Mica reinforced hybrid composites. Moreover adequate investigations have not been carried out to find out the effect of cutting parameters on surface roughness in drilling hybrid Al/SiC-Mica composites. This is significant because these factors play an important role in the performance of the machined component. Therefore, the main objective of this study is to discuss the use of RSM for predicting surface roughness in drilling hybrid MMC.

2 Experimental

2.1 Materials Used

Aluminum alloy Al 356 is used as a matrix material for fabricating composites and its chemical composition is illustrated in Table 1. The silicon carbide particles of size 25 μ and Mica, an average size of 45 μ were used as the reinforcement materials. Different combinations with 5–15 wt % of SiC particle in steps of the 5 wt % is used. The percentage of Mica used is 3 %. The composites are fabricated through stir casting method and it ensures uniform distribution of the reinforcements.

2.2 Taguchi Method

In this study Taguchi method—a powerful tool in parameter design is used. According to the Taguchi quality design concept L_{27} orthogonal array with 27 rows (corresponding to the number of experiments) are used for the experimentation. It provides a simple, efficient, and systematic approach to optimize the design for performance, quality, and cost. The methodology is valuable when design parameters are qualitative and discrete. Table 2 shows three drilling parameters used as control factors and their levels. Design expert 8.0.1 software is used for regression and graphical analysis of the experimental data.

Table 1 Composition of Al 356 alloy

Parameter	Copper	Silicon	Magnesium	Manganese	Iron	Titanium	Zinc	Al
%	<0.0005	7.27	0.45	<0.002	0.123	0.08	0.005	Remaining

Table 2 Machining parameter and levels

Control parameters	Units	Symbol	Level		
			1	2	3
Speed	m/min	V	18.85	37.74	56.65
Feed rate	mm/min	f	50	100	150
Wt percentage of sic	%	W	5	10	15

2.3 Response Surface Methodology

RSM is one of the important techniques for determining the relationship between various process parameters with the various drilling criteria and exploring the effect of these process parameters on the coupled responses [12]. Response-surface method is used to establish the mathematical relation between the output response 'y' and the various drilling parameters. In this study, the quantitative form of relationship between the desired response and independent input variables can be represented by

$$Y = f(X_1, X_2, X_3, X_4) \tag{1}$$

where, Y is the desired response and f is the response function (or response surface). The polynomial regression model called as quadratic model is more suitable for analyzing the interactive effects and square effects between the parameters and has been considered in this study. The quadratic model of Y can be expressed as follows:

$$Y = \beta_0 + \sum_{i=1}^{n} \beta i X i + \sum_{i=1}^{n} \beta i i X_i^2 + \sum_{i,j=1}^{n} \beta i j X i X j \tag{2}$$

where β_0 is constant β_i, β_{ii}, and β_{ij} represent the coefficients of linear, square, and interactive terms, respectively. In general the quadratic model of desired responses is written in the form including error term is as follows:

$$Y = Xa + E \tag{3}$$

where X is a matrix of model terms evaluated at the data points, E is an error vector

2.4 Experimental Procedure

Drilling tests are conducted on vertical CNC machining center. The machining samples were prepared in the form of $100 \times 100 \times 10$ mm sized blocks. The surface roughness of the work piece is measured by using Surf coder SE 1,200, manufactured by Kosaka Laboratory Limited Japan. The surface roughness is measured perpendicular to the circumference of the hole [10]. Solid carbide drills are used in these experiments having diameter of 6 mm, a helix angle of 30 ° and point angle of 118 ° manufactured by SANDVIK, INDIA. The surface roughness values given in this study is the mathematical average of the two measurements taken from the same hole surface. The surface roughness used in this study was the arithmetic mean average surface roughness (R_a), which is mostly used in the industry. The Schematic arrangement of experimental setup is presented in Fig. 1. The experimental results are presented in Table 3.

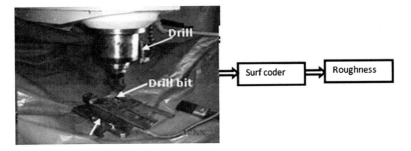

Fig. 1 Schematic arrangement of experimental setup

3 Results and Discussions

Table 3 Design of experiments based on L_{27} orthogonal array

Expt no.	Coded values			Actual setting			Roughness Responses
	V	f	R	V	f	R	Ra, μm
1	1	1	1	18.85	50	5	2.4
2	1	1	2	18.85	50	10	2.1
3	1	1	3	18.85	50	15	1.8
4	1	2	1	18.85	100	5	3.8
5	1	2	2	18.85	100	10	3.4
6	1	2	3	18.85	100	15	3.2
7	1	3	1	18.85	150	5	4.8
8	1	3	2	18.85	150	10	4.3
9	1	3	3	18.85	150	15	4
10	2	1	1	37.74	50	5	2
11	2	1	2	37.74	50	10	1.8
12	2	1	3	37.74	50	15	1.5
13	2	2	1	37.74	100	5	3.3
14	2	2	2	37.74	100	10	3
15	2	2	3	37.74	100	15	2.7
16	2	3	1	37.74	150	5	4
17	2	3	2	37.74	150	10	3.7
18	2	3	3	37.74	150	15	3.4
19	3	1	1	56.65	50	5	1.7
20	3	1	2	56.65	50	10	1.6
21	3	1	3	56.65	50	15	1.5
22	3	2	1	56.65	100	5	2.8
23	3	2	2	56.65	100	10	2.5
24	3	2	3	56.65	100	15	2.25
25	3	3	1	56.65	150	5	3.6
26	3	3	2	56.65	150	10	3.2
27	3	3	3	56.65	150	15	2.76

Surface roughness model is developed by RSM. The relative importance of the drilling parameters with respect to the surface roughness is investigated to determine more accurately the optimum combinations of the drilling parameters using analysis of variance (ANOVA). Table 4 gives the ANOVA results for the response surface quadratic model for surface roughness in drilling hybrid composites. The adequacy of the developed models is tested at 95 % confidence interval. The test for significance of the regression models, the test for significance on individual model coefficients and the lack-of-fit test are performed using statistical software package by selecting the step-wise regression method, which eliminates the insignificant model terms automatically. The adequacy measures such as R^2, adjusted R^2, and predicted R^2. The coefficient of determination R^2 indicates the goodness of fit for the models, which provides a measure of variability in the observed response values and can be explained by the controllable factors and their interactions. In this case, all the values of coefficient of determination R^2 are nearly equal to 1. The adjusted coefficient of determination R^2 is a variation of the ordinary R^2 statistic that reflects the number of factors in the model. The entire adequacy measures are closer to 1, which is in reasonable agreement and indicate adequate models.

The relative importance of the cutting parameters with respect to the surface roughness is investigated by ANOVA. Tables 4 give the ANOVA results for surface roughness. From the analysis of Table 4, it is observed that the feed rate factor (75.4 %) and speed (15.54 %) have statistical significance on the surface roughness. It can be observed from Table 4 that wt % of SiC also affects surface roughness by 6.81 %.

In analyzing surface roughness in drilling, statistical models play an important role. The models are used for prediction of results [12]. The mathematical relationship for correlating the surface roughness and the considered drilling parameters (cutting speed, feed rate, and wt % SiC) is obtained from coefficients resulting from using the Design Expert software. The following equations are the final empirical model in terms of uncoded factors for surface roughness (Ra):

$$\text{Roughness (um)} = + 0.83800 - 1.64000e - 004 * v + 0.045440 * f$$
$$- 0.036800 * w - 3.60000e - 006 * v * f + 9.00000e$$
$$- 006 * v * w - 4.20000e - 00 * f * w \ (\%) - 7.36000e - 005 * f^2$$
$$(4)$$

Table 4 ANOVA for surface roughness

Source	Sum of squares	df	Mean square	F value	p value prob > F	% of contribution
Model	22.5014	18	1.250078	207.2588	< 0.0001 Significant	
v-speed	3.473785	2	1.736893	287.9711	< 0.0001	15.39
f-feed	17.00179	2	8.500893	1409.42	< 0.0001	75.42
w-sic	1.556452	2	0.778226	129.0273	< 0.0001	6.87
vf	0.369348	4	0.092337	15.30918	0.0008	1.6
vw	0.022681	4	0.00567	0.940129	0.4880	
fw	0.077348	4	0.019337	3.206018	0.0754	
Residual	0.048252	8	0.006031			
Cor total	22.54965	26				

From the developed RSM-based mathematical model, the effect of drilling parameter on surface roughness is examined. The 3D surface plots for surface roughness are shown in Fig. 2. The response surface plots show the relationship between the two variables by keeping the third variable at constant middle level. From Fig. 2, it is observed that at higher values of the cutting speed, the surface roughness increases with the increase in feed rate at a constant wt % SiC of 10 %. Similarly, the 3D response surfaces for surface roughness in the case of two varying parameters by keeping the other parameters at constant middle level. By assessing these plots, it has been inferred that the increase of cutting speed and wt % of SiC decreases surface roughness.

4 Conclusions

This investigation has presented an application of Taguchi method and RSM for analysis the drilling parameters for surface roughness in drilling of hybrid MMCs. The conclusions drawn are as follows: (1) With proposed optimum conditions

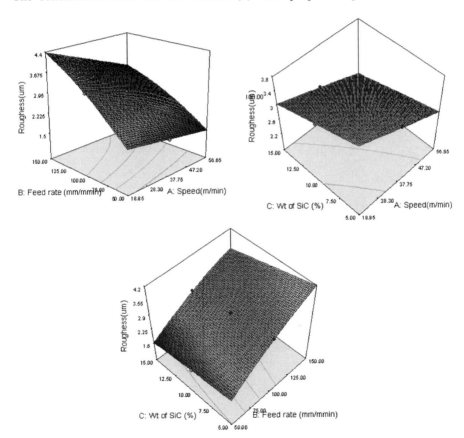

Fig. 2 3D surface graph of surface roughness as a function of machining parameters

using the Taguchi method and ANOVA, a better surface roughness is obtained. The level of the best of the cutting parameters on the surface roughness is determined using ANOVA. The results revealed that feed rate is main cutting parameter which influences surface roughness in drilling hybrid MMCs. (2) RSM has been used to determine the surface roughness attained by various drilling parameters. The quadratic modes developed using RSM is reasonably accurate and can be used for prediction within the limits of the factors investigated. (3) From RSM model and experiment results, the predicted and measured values are quite close, which indicates that the developed model can be effectively used to predict the surface roughness within the levels of factors studied and variable used.

References

1. Noorul Haq A, Marimuthu P, Jeyapaul R (2008) Multi response optimization of machining parameters of drilling Al/SiC metal matrix composite using grey relational analysis in the Taguchi method. Int J Adv Manuf Technol 37:250–255
2. Tosun G (2011) Statistical analysis of process parameters in drilling of AL/SICP metal matrix composite. Int J Adv Manuf Technol 55:477–485
3. Bhushan RK, Kumar S, Das S (2010) Effect of machining parameters on surface roughness and tool wear for 7075 Al alloy SiC composite. Int J Adv Manuf Technol 50:459–469
4. Manna A, Bhattacharyya B (2004) Investigation for optimal parametric combination for achieving better surface finish during turning of Al/SiC-MMC. Int J Adv Manuf Technol 23:658–665
5. Routara BC, Bandyopadhyay A, Sahoo P (2009) Roughness modeling and optimization in CNC end milling using response surface method: effect of workpiece material variation. Int J Adv Manuf Technol 40:1166–1180
6. Kok M (2011) Modelling the effect of surface roughness factors in the machining of 2024Al/ Al2O3 particle composites based on orthogonal arrays. Int J Adv Manuf Technol 55:911–920
7. Tsao CC (2011) Evaluation of the drilling-induced delamination of compound core-special drills using response surface methodology based on the Taguchi method. Int J Adv Manuf Technol. doi:10.1007/s00170-011-3785-5
8. Altunpak Y, Ay M, Aslan S (2011) Drilling of a hybrid Al/SiC/Gr metal matrix composites. Int J Adv Manuf Technol. doi:10.1007/s00170-011-3644-4
9. Riaz Ahamed A, Asokan P, Aravindan S (2009) EDM of hybrid Al–SiCp–B4Cp and Al–SiCp–Glassp MMCs. Int J Adv Manuf Technol 44:520–528
10. Basavarajappa S, Chandramohan G, Davim JP, Prabu M, Mukund K, Ashwin M, Prasanna Kumar M (2008) Drilling of hybrid aluminium matrix composites. Int J Adv Manuf Technol 35:1244–1250
11. Riaz Ahamed A, Asokan P, Aravindan S, Prakash MK (2010) Drilling of hybrid Al-5%SiCp-5%B4Cp metal matrix composites. Int J Adv Manuf Technol 49:871–877
12. Rajmohan T, Palanikumar K (2011) Experimental investigation and analysis of thrust force in drilling hybrid metal matrix composites by coated carbide drills. Mat Manuf Process 26:961–968

Thermomechanical Behavior of Commercially Pure Titanium (CP-Ti) During Isothermal Compressive Deformation

M. Vetrivel, T. Senthilvelan and G. Sriram

Abstract The thermomechanical deformation behavior of commercially pure Titanium (CP-Ti) was studied by isothermal compression up to 50 % height reduction of specimen at 773–1273 K with strain rate 0.01–1.0 s^{-1} with boron nitride as solid lubricant. The influence of Zener-Hollomon parameters, strain, strain rate imposing on the flow stress were analyzed in the high-temperature deformation of CP-Ti in the $\alpha + \beta$ phase region. The apparent activation energy of deformation calculated to be 343.69 kJ mol^{-1} and the processing map shows that the dynamic softening is accelerated with increase of deformation temperature and decrease of strain rate, and can be divided into three different regions viz., three-stage work hardening, two-stage work hardening, and flow softening. Geometric dynamic recrystallizations were found as a result of strain rate and temperature while describing the grain refinement process of CP-Ti during high-temperature compression.

Keywords Zener-Hollomon parameter · Activation energy · Processing map · Commercially pure titanium · Solid lubricant · Dynamic recrystallizations

1 Introduction

Commercially pure titanium (CP-Ti) is of great importance in many industrial applications due to its highly attractive properties, such as good deformability at high temperatures, low density, high biocompatibility, and excellent corrosion resistance [1]. It is chemically inert and biologically more compatible than

---

M. Vetrivel (✉) · G. Sriram
Sri Chandrasekharendra Saraswathi Vishwa MahaVidhalaya, Kanchipuram 631 561, India
e-mail: km_vetrivel@yahoo.com

T. Senthilvelan
Pondicherry Engineering College, Pudhucherry 605014, India

S. Sathiyamoorthy et al. (eds.), *Emerging Trends in Science,*
Engineering and Technology, Lecture Notes in Mechanical Engineering,
DOI: 10.1007/978-81-322-1007-8_30, © Springer India 2012

Ti-6Al-4V, which is currently the material of choice for forming bone-interfacing implants for orthopedic applications. However, it is hard for CP-Ti to strengthen to a level comparable to Ti-6Al-4V [2]; therefore, developing higher strength CP-Ti is an attractive work for medical applications. In order to increase the strength of CP-Ti, grain-size refining is an effective approach, and generally is being carried out from the dynamic recovery to dynamic recrystallizations.

Deformation behavior of CP-Ti during cold and hot working has been studied extensively by many researchers [3–9]. Zhipeng Zeng et al. [10, 11] investigated deformation temperatures below the transformation temperature in CP-Ti (Grade II). They recorded stress–strain curve between temperatures 673 and 973 K at strain 0.6, also developed the constitutive equation for CP-Ti.

One of the very common industrial processes in the $\alpha + \beta$ region is superplastic forming, which requires very fine grain sizes and slow speed of deformation [12]. Yong Niu et al. [13] studied the characteristics of superplastic deformation in Ti600 Titanium alloy by recording the stress–strain curves between temperatures 800 and 1,000° C. Rajagopalachary et al. [14, 15] investigated the superplastic deformation in IMI685 titanium alloy. The understanding of metals and alloy behavior at hot deformation condition has a great importance for designers of metal-forming processes; the success of the process depends on the application of correct magnitude of the flow stress. The basis and principles of processing map have been described in detail [16]. Its application to hot working of a wide range of materials is complied [17]. In brief, the dynamic materials model assumes the workpiece as a dissipater of the power supplied by a particular source. The total power dissipated P can be calculated as

$$P = \sigma\dot{\varepsilon} = \int_0^{\varepsilon} \sigma d\dot{\varepsilon} + \int_0^{\sigma} \dot{\varepsilon} d\sigma = G + J \tag{1}$$

where σ is the flow stress and $\dot{\varepsilon}$ is the strain rate. The first integral is defined as G content and represents the main power input dissipated in the form of a temperature rise. The second integral is defined as J co-content and is related to the power dissipated by metallurgical processes. The efficiency of power dissipation is obtained by comparing the dissipation of the workpiece with that of the ideal linear dissipater and is defined as

$$\eta = \frac{J}{J_{\max}} = \frac{2m}{1+m} \tag{2}$$

where m is the strain rate sensitivity of flow stress. The variation of η with deformation temperature and strain rate constitutes the power dissipation map. This map represents the pattern in which the power is dissipated by metallurgical processes. A continuum criterion for the occurrence of flow instability is obtained by utilizing the principle of maximum rate of entropy production and given by

$$\xi(\dot{\varepsilon}) = \frac{\partial \log[m/(1+m)]}{\partial \log \dot{\varepsilon}} + m < 0 \tag{3}$$

where $\xi(\dot{\varepsilon})$ is the instability parameter. Flow instabilities are predicted to occur when $\xi(\dot{\varepsilon})$ becomes negative. The processing map is obtained through superimposing an instability map on a power dissipation map, which not only exhibit the domains of different mechanisms but also the regions in microstructure instabilities.

In this research work, Isothermal compression tests [18] of CP-Ti have been conducted at different strain rates and the deformation temperatures in the $\alpha + \beta$ region. A solid lubricant Boron Nitride that is used between the forming tools and the workpiece yields a significant effect on material deformation. The characteristics of dynamic recovery to dynamic recrystallizations have been observed by developing a processing map mainly to study the influence of Zener-Hollomon parameter [19] which has been studied.

2 Materials and Experimental Procedures

The chemical composition of CP-Ti (Grade –II) used in this study, in weight percentages of 0.006 C, 0.069 Fe, 0.00023 H, 0.0042 N, 0.022 O, 0.0512 Al, and 99.84 Ti in balance. The specimens were heated with a heating rate of 5 °C/s and soaked for 2.0 min at deformation temperature prior to isothermal compression with the help of temperature controller used in experimental setup. The ring compression tests were carried out on a servo control universal testing machine equipped with vertical chamber furnace.

Servo control universal testing machine allows performing compression tests at constant strain rate. At the beginning of deformation, the temperatures of the tool set and specimen were equal to each other. The standard ratio (outer diameter: inner diameter: height) 6:3:2 was selected for each ring specimen. In the present investigation, the outer and inner diameter of the rings are 30 and 15 mm, respectively, and the height is 10 mm. The CP-Ti ring specimens are prepared with the specified dimensions. The isothermal ring compression tests were conducted on CP-Ti by deforming them in the temperatures ranging from 773 to 1273 K at an interval of 250 K and strain rates are 0.01, 0.1, and 1.0 s^{-1}. The height was reduced to 50 %. In order to reduce the frictional force between the die and the specimens, a solid lubricant boron nitride was used in the test.

3 Result and Discussion

3.1 Flow Stress During Hot Deformation

The true stress-true strain is calculated from force–displacement data. The curves of CP-Ti, compressed at 773, 1,023, and 1,273 K at different strain rates are shown in Fig. 1. It could be observed from Fig. 1 that the influences of temperature and

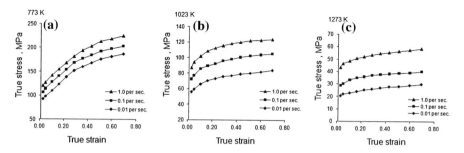

Fig. 1 True stress–true strain curves for CP-Ti during hot compression at various strain rate and temperatures of (a) 773 K, (b) 1,023 K, and (c) 1,273 K

strain rate on flow stress are significant. The flow stress increases with decrease in temperature and increases in strain rate. The flow curve typically shows work hardening region followed by dynamic softening due to dynamic recovery to dynamic recrystallizations.

The work hardening is predominant at lower temperatures and higher strain rates. On the contrary, the extent of dynamic softening is more at higher temperatures and lower strain rates. This is due to the fact that higher temperatures and lower strain rates offer higher mobility to the grain boundary and longer time for nucleation and growth of dynamically recrystallized grains.

3.2 Constitutive Equations for Flow Stress Prediction

The correlation between the peak flow stress (σ) at a strain of 0.5, temperature (T) and strain rate ($\dot{\varepsilon}$), particularly at high temperatures, could be expressed by an Arrhenius type equation. Further, the effects of temperatures and strain rate on deformation behavior could be represented by Zener-Hollomon parameter (Z) in an exponent-type equation. These are mathematically expressed as

$$Z = \dot{\varepsilon} \exp\left(\frac{Q}{RT}\right) = f(\sigma) = \begin{cases} A'\sigma^{n'} \\ A'' \exp(\beta\sigma) \\ A[\sinh(\alpha\sigma)]^n \end{cases} \quad (4)$$

Here, R is the universal gas constant (8.31 J mole^{-1} k^{-1}), T is the absolute temperature in K, Q is the activation energy (KJ mole^{-1}), $\dot{\varepsilon}$ represents the strain rate and A, $Á$, A'' n', n, β, and α ($\approx\beta/n'$) are material constants. In the above equations, the Zener-Hollomon parameter Z is temperature compensated strain rate, and Q is the deformation activation energy.

3.3 Determination of Hot Working Constants

Under the condition of constant temperature during the hot deformation process, the partial differentiation of Eq. (4) yields the following equations, respectively:

$$n' = [\partial \ln \dot{\varepsilon} / \partial \ln \sigma_p]_T \tag{5}$$

$$\beta = [\partial \ln \dot{\varepsilon} / \partial \sigma_p]_T \tag{6}$$

$$n = [\partial \ln \dot{\varepsilon} / \partial \ln [\sinh (\alpha \sigma)]]_T \tag{7}$$

It follows from these expressions that the slope of the plot of $\ln(\dot{\varepsilon})$ against $\ln(\sigma_p)$ and the slope of the plot of $\ln(\dot{\varepsilon})$ against σ_p can be used for obtaining the value of n' and β, respectively. These plots are shown in Fig. 2 a and b. The linear regression of these data results in the average value of 14.502 and 0.13073 for n' and β, respectively. This gives the values of $\alpha = 0.00901$. According to Eq. (7), the slope of the plot of $\ln(\dot{\varepsilon})$ against $\ln[\sinh(\alpha \sigma_p)]$ can be used for obtaining the value of n from Fig. 2c. The average value of n was determined as 9.49767.

3.4 Activation Energy for Deformation

The activation energy for deformation at a constant strain rate was derived by partial differentiation of Eq. (4) that yields the following equations, respectively:

$$Q = Rn' \left[\partial \ln \sigma_p / \partial (1/T)\right]_{\dot{\varepsilon}} \tag{8}$$

$$Q = R\beta \left[\partial \sigma_p / \partial (1/T)\right]_{\dot{\varepsilon}} \tag{9}$$

$$Q = Rn \left[\partial \ln [\sinh (\alpha \sigma)] / \partial (1/T)\right]_{\dot{\varepsilon}} \tag{10}$$

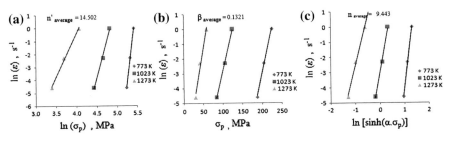

Fig. 2 a Plot of $\ln(\dot{\varepsilon})$ versus $\ln(\sigma_p)$, b Plot of $\ln(\dot{\varepsilon})$ versus σ_p and c Plot of $\ln(\dot{\varepsilon})$ versus $\ln[\sinh(\alpha \sigma_p)]$

These expressions the slope of the plots of $\ln(\sigma_p)$, σ_p, and $\ln[\sinh(\alpha\sigma_p)]$ versus the reciprocal of absolute temperature are shown in Fig. 3.

The linear regression of these data results in the average value of 369.03, 343.69, and 319.63 kJ/mol for activation energy from Eqs. (8), (9), and (10), respectively. There is a significant difference among these values. Analysis of the correlation coefficient (R^2) of these regression values reveals that Eq. (9) has better fit to experimental data. Therefore, the activation energy of hot working was considered to be 343.69 kJ/mol.

3.5 Peak Stress as a Function of the Zener-Hollomon Parameter

According to Eq. (4) through Eq. (7), the plotting of $\ln(Z)$ against $\ln(\sigma_p)$, σ_p and $\ln[\sinh(\alpha\,\sigma_p)]$ be used to find the relationship between Z and σ_p. The corresponding curves are shown in Fig. 4. Among these relations, the exponential equation has the highest correlation coefficient and has good fit. In summary, the peak stress of CP-Ti under deformation condition used in this study may be expressed as

$$Z = 0.1292 \times \exp 0.13073 \times \sigma_p \qquad (11)$$

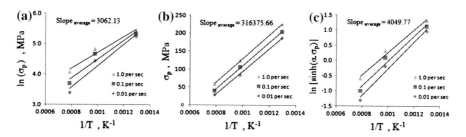

Fig. 3 Plot of (a) $\ln(\sigma_p)$, (b) σ_p, and (c) $\ln[\sinh(\alpha\sigma_p)]$ versus the reciprocal of absolute temperature

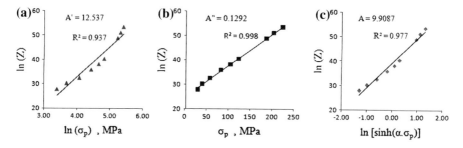

Fig. 4 Plots of $\ln(Z)$ against (a) $\ln(\sigma_p)$, (b) σ_p and (c) $\ln[\sinh(\alpha\,\sigma_p)]$

3.6 Deformation Condition Map

Plotting the data into a map with $\log(\dot{\varepsilon})$ and temperature as axes is shown in Fig. 5 was constructed by using Eq. (11) and it is possible to draw contour lines for different Z values. As seen, all data consistently fall into three domains viz., three-stage work hardening, two-stage work hardening, and flow softening. The SEM microstructure of tested CP-Ti has been shown in Fig. 6 and it can be seen from Fig. 6a that the dislocation reorganized by low-angle grain boundary elongated with refine grains at strain rate of $0.1~\text{s}^{-1}$ and deformed temperature of 773 K. Figure 6b shows the dislocation led to the merging of some grains and the low-angle grain boundaries transformed into high-angle grain boundary at strain rate of $0.1~\text{s}^{-1}$ and deformed temperature of 1,023 K and in Fig. 6c the grain boundary become straight it is clearly observed that the recrystallized grains grow through migration of dislocations and high-angle grain boundary at strain rate of $0.1~\text{s}^{-1}$ and deformed temperature of 1,273 K.

It is clearly observed work hardening is mainly caused by dislocations, multiplication, and interaction. The dynamic flow softening is mainly caused by dynamic recrystallizations.

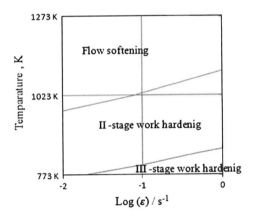

Fig. 5 Processing map of CP-Ti at a strain of 0.5

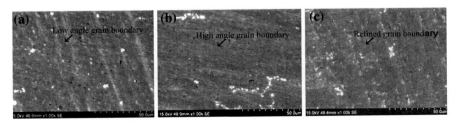

Fig. 6 Microstructure of CP-Ti at strain rate $0.1~\text{s}^{-1}$ under different deformation temperature (a) 773 K, (b) 1,023 K, and (c) 1,273 K

4 Conclusions

Isothermal compression tests are performed on CP-Ti by deforming them in the temperatures ranging from 773 to 1,273 K at an interval of 250 K and strain rates are 0.01, 0.1, and 1.0 s^{-1} at a strain of 0.5 a solid lubricant boron nitride was used and conclusions are summarized as follow:

1. The thermomechanical behavior affects the flow stress in the isothermal deformation of CP-Ti. The flow stress decreased with increase of deformation temperature and the decrease of strain rate.
2. The activation energy of hot working has been determined by using one of the three expressions of Z, namely, the power law, exponential law. For CP-Ti, the exponential law was found to be the appropriate relation, which resulted in the value of 343.69 kJ/mol^{-1}.
3. The processing map was developed at a strain of 0.5 and accordingly, the deformation behavior of CP-Ti can be divided into three regions, viz., three-stage work hardening, two-stage work hardening, and flow softening.
4. Microstructure of tested CP-Ti was analyzed, there was a geometric dynamic recrystallizations found at elevated temperatures.

Acknowledgments The authors thanks, the management, SCSVMV University, for the financial support to develop an experimental setup and the Librarian, IGCAR, Kalpakkam India.

References

1. Barbas A, Bonnet AS, Lipinski P, Pesci R, Dubois G (2012) J Mech Behav Biomed Mat 9:34–44
2. Stolyarov VV, Zhu YT, Alexandrov IV, Lowe TC, Valiev RZ (2003) Mater Sci Eng A 343:43
3. Nemet-Nasser S, Guo WG, Cheng JY (1999) Acta Mater 47:3705
4. Ray K, Poole WJ, Mitchjell A, Hawbolt EB (1997) In: Weiss I, Srinivasan R, Bania P, Eylon D, Semiatin SL (eds.) PA: TMS Warrendale 201
5. Tanaka H, Yamada T, Sato E, Jimbo I (2006) Scripta Mater 54:121
6. Ungar T, Glaicic MG, Balogh L, Nyilas K, Salem AA, Ribarik G (2008) Mater Sci Eng A 493:79
7. Nagasekhar AV, Chakkingal U, Venugopal P (2006) J Mater Process Technol 173:53
8. Stolyrov VV, Zhu YT, Lowe TC, Islamgaliev RK, Valivev RZ (1999) Nano Mater 11:947
9. Semiatina SL, DeLo DP (2000) Mater Des 21:311
10. Zeng Z, Zhang Y, Jonsson S (2009) Mater Des 30:3105
11. Zeng Z, Jonsson S, Zhang Y (2009) Mater Sci Eng A 505:116
12. Arieli A, Rosen A (1977) Metall Trans A8:1591
13. Niu Y, Hou H, Li M, Li Z (2008) Mater Sci Eng A 49:24
14. Rajagopalachary T, Kutumbarao VV (1996) Scripta Mater 35:305
15. Rajagopalachary T, Kutumbarao VV (1996) Scripta Mater 35:311
16. Prasad YVRK (2003) J Mater Eng Perform 12:368
17. Prasad YVRK, Sasidhara S (1997) Hot working guide. ASM Int Mater Park OH
18. Bernd-Arno B, Hans C, Pavel P (2009) Int J Mater Form 2:613
19. Mandal S, Rakesh V, Sivaprasad PV, Venugopal S, Kasiviswanathan KV (2009) Mater Sci Eng A500:114

Part IV
Intelligent Manufacturing

A Heuristic toward Minimizing Waiting Time of Critical Jobs in a Flow Shop

R. Pugazhenthi and M. Anthony Xavior

Abstract This paper deals with scheduling in permutation flow shop that is considered with the objective of minimizing the waiting time of priority based or critical jobs. Generally, priority based jobs are few in nature in all the flow shops. Hence, it is utmost important that we minimize the waiting time of critical or priority based jobs. All the existing heuristics are developed to minimize parameters like makespan time, flow time, etc. While processing jobs through machines, all machines are considered equally important. In this work, an attempt is made to improve the existing utilization of one or more jobs during processing. The above aim was achieved by reducing the waiting time of the critical or priority based jobs under consideration, and hence achieving zero waiting time for the same. The heuristic is proposed to improve the utilization of such critical jobs. The proposed heuristic was compared with the results of CDS method and validated.

Keywords Flow shop • Priority based jobs • Minimizing idle time of machines

1 Introduction

Production of finished product from raw materials needs several operations to be performed in a given sequence. Frequently, similar operations required for several products are performed at the same workstation, particularly when the products are not produced on a mass scale under such conditions; it is required to select a preferred order for products passing through a work center. The problem becomes complex when there are several workstations serving many products. In all such problems, the criterion of selection is the minimum processing time. It is generally

---

R. Pugazhenthi (✉) · M. Anthony Xavior
SMBS, VIT University, Vellore 632014, Tamil Nadu, India

S. Sathiyamoorthy et al. (eds.), *Emerging Trends in Science,*
Engineering and Technology, Lecture Notes in Mechanical Engineering,
DOI: 10.1007/978-81-322-1007-8_31, © Springer India 2012

not practicable to solve such problems by enumeration. Unfortunately, generalized sequencing models are also not available to deal with all cases. The solution method in such cases was first developed by Johnson [1] and the procedure is often referred as Johnson's rule.

Step 1. Find minimum (A_{i1}, A_{i2}), where $I = 1, 2..., n$.
Step 2. (a) If this minimum be A_{i1} for $I = 1$ process the Ith job first . (b) If this Minimum be A_{m2} for some $I = m$ process the mth job last of all.
Step 3. (a) If there is a tie, i.e., $A_{11} = A_{m2}$ process the Ith job first and m^{th} job in the last of last.

(b) If the tie for the minimum occurs among the A_{i1-s}, Choose the job corresponding to the minimum occurs among the A_{i2-s}, choose the job corresponding to the minimum of A_{i1-s} and process it in the last.
Step 4. Cancel the jobs already assigned and repeat steps 1–3 until all the jobs have been assigned.

All jobs are required to follow the same order viz., from the first machine to the second, the processing time in both the machines for each job is assumed to be known and is independent of jobs preceding it or following it. The flow shop scheduling problem has been a keen area of research for over 30 years ever since 1954, Johnson [1] has proposed the two stage scheduling problem with the makespan objective.

The optimizing algorithm has been proposed by Ignall et al. [2] to minimize makespan in 1965. The NP-completeness of the flow shop-scheduling problem has been discussed widely by Quan-Ke Pan and Ling Wang [3]. In the 1965, Palmer [4] has been the first to propose heuristic procedures. The first significant work in the development of an efficient heuristic is due to Campell et al. [5]. Their algorithm consists essentially in splitting the given m-machine problem into a series of an equivalent two-machine flow shop problem and solving by Johnson's rule.

Dannenbring [6] in 1977 has developed a procedure called 'rapid access', which attempts to combine the advantages of Palmer's slope index and CDS procedures. Although the procedure by Dannenbring is found to yield better quality solution than those by Palmer's and CDS methods; it requires much more computational effort. King and Spachis during the year 1980 [7] treat the makespan problem as equivalent to that of minimizing total delay and run-out delay. They have proposed heuristics that aim at matching the two consecutive job time-block profiles by considering these delays.

One of the heuristics turns out to be better than the CDS heuristic. In 1982, Stinson and Simith [8] have proposed a radically different approach. They treat the makespan problem as a traveling salesman problem and develop a procedure in two steps. The heuristic solution is found to yield better quality solution than those by Palmer [9] and CDS methods at the cost of increased computational effort. Nawaz et al. [10] have presented a heuristic based on the premise that a job with higher total processing time should be given higher priority than a job with less total processing time.

A schedule is developed by the job-insertion technique. The number of enumerations in the heuristic method is $\frac{1}{2} n (n + 1)-1$. The performance of this NEH heuristic was compared with CDS method, and the measure of evaluation, though deficient, has been the number of times a procedure yields a better solution. The NEH heuristic is found to fare better. A simple modification and extension of Palmer's [9] heuristic has been carried out by Hundal and Rajagopal [11]. Two sets of indices have been proposed and three sequences are hence obtained. It has been observed that while the proposed heuristic yields more number of times better quality solutions than those by CDS heuristic as the number of jobs increases, CDS heuristic fares better as the number of machine increases. Widner and Hertz [12] in 1989 have proposed a new heuristic called sequencing problem involving a resolution by integrated taboo search technique (SPIRIT).

Although the method is found to yield better quality solutions than those by the heuristic of CDS, Dannenbring and NEH, the quality of solution and the computational effort are heavily dependent on the setting of parameter values in the heuristic.

Osman and Potts [13] and Ogbu and Smith [14] have made use of simulated annealing; a heuristic technique adapted to a number of combinatorial optimization problems. Simulated annealing follows the same basic steps which are used in the conventional descent method with one exception: sometimes a new sequence is accepted as the current sequence, even though its objective function's value exceeds that of the old sequence.

2 Proposed Heuristic to Minimize Idle Time of Machine for a Critical Job

The proposed heuristic is explained stage by stage.

Stage I

Step 1: Select the jobs individually and assume it is processed at first

Step 2: Calculate the idle time of M/c A, M/c B and M/c C

Step 3: Select the job corresponding to the minimum idle time for the corresponding M/c this selected job is decided to be processed at first

Step 4: The selected job is eliminated from the list of jobs

Stage II

Step 5: All the other jobs are considered for it is second position in the sequence And their idle times on various machines are tabulated

Step 6: The job corresponding to the minimum idle time of the corresponding M/c is selected for processing as second job

Step 7: The process continues till all the jobs are over

The above problem will have a number of stages equal to that of the number of jobs. When all the stages are over, the final sequence will give the minimum idle for the particular machine. To validate our results, the same problem is selected and solved using the proposed heuristic and solved by the CDS method; and the idle time of the corresponding machine is compared and tabulated.

3 Numerical Example

The concept of the heuristic has been dealt in the methodology and algorithm. Now for the heuristic generated problem.

Problem

Job	J1	J2	J3	J4
M/c A	38	54	14	11
M/c B	22	26	25	56
M/c C	29	21	57	19

In this problem, four jobs are to be processed through three machines and their processing times are given. All the jobs must pass through these machines in the order A–B–C and job 'J2' required to process first. Give the optimal sequence for the jobs.

Solution

Stage I

According to the proposed heuristic, all jobs are assumed to be processed first. Their idle time and sum of the idle time is calculated. Here, the job J3 is having minimum idle of specified machines, but job J2 must be process earlier than other job. Hence, job 'J2' is placed first in the job sequence. From this stage onwards, J2 is not considered for further calculations (Table 1).

Table 1 Selection of less idle job in stage I

	M/c A		M/c B		M/c C		Idle time			Sum of idle time
Job	In	Out	In	Out	In	Out	A	B	C	
J1	0	38	38	60	60	89	0	38	60	98
J2	0	54	54	80	80	101	0	54	80	134
J3	0	14	14	39	39	96	0	14	39	53
J4	0	11	11	67	67	86	0	11	67	78

A Heuristic toward Minimizing Waiting Time

Stage II

In this stage, the starting times of the jobs J1, J3, and J4 are the completion time of the job J2 on machine A. For other machines, starting times of jobs are established according to the availability of machines and jobs. Now, the sum of idle time of M/c is computed as shown in the Table 2.

Job 'J3' is having minimum value, which is selected for minimum one. Now, 'J3' is selected to be placed 2nd in the job sequence and J3 is deleted from further calculations.

Stage III

The idle time and index value for the jobs 'J1' and 'J4' are calculated.

Here, J4 has the minimum sum of idle time and hence it is placed in the 3rd position in the sequence. As a result of which job 'J1' will be at the end of the sequence (Table 3).

Stage IV

The final optimal sequence obtained is *J2–J3–J4–J1* (Table 4).

Table 2 Selection of less idle job in stage II

Job	M/c A		M/c B		M/c C		Idle time			Sum of idle time
	In	Out	In	Out	In	Out	A	B	C	
J1	54	92	92	114	114	143	54	92	114	260
J3	54	68	80	105	105	162	54	80	105	239
J4	54	65	80	136	136	155	54	80	136	270

Table 3 Selection of less idle job in stage III

Job	M/c A		M/c B		M/C C		Idle time			Sum of Idle time
	In	Out	In	Out	In	Out	A	B	C	
J1	68	106	106	128	162	191	68	106	162	336
J4	68	79	105	161	162	181	68	105	162	335

Table 4 Schedule table

Job	M/c A		M/c B		M/c C		Idle time			Job waiting time
	In	Out	In	Out	In	Out	A	B	C	
J2	0	54	54	80	80	101	0	54	80	0
J3	54	68	80	105	105	162	0	0	4	66
J4	68	79	105	161	162	181	0	0	0	95
J1	79	117	161	183	183	212	0	0	2	123

Result

Total elapse time	212 h
Total idle time of machine B	54 h
Total idle time of machine C	86 h

4 Result Analysis and Conclusion

The proposed heuristic is applied to flow shop problems of various sizes and selecting any one or more machines for minimizing its idle time. The machines varying from 5, 10, and 15 and jobs varying from 5, 10, 15, 20, and 25 were solved and results are given. In these problems, one machine is selected for minimizing its idle time and the selected machine is also given in the Table 1. The processing times of jobs on machines are taken as random number varying from 0 to 100.

A particular problem is solved CDS heuristic and the idle time of the selected machine is noted down and the proposed heuristic reduced the idle time obtained from CDS method. The comparisons of the results for the selected machines idle time are shown as bar chart in Fig. 1. We can notice that the idle time of selected machines is minimized, compared to CDS method.

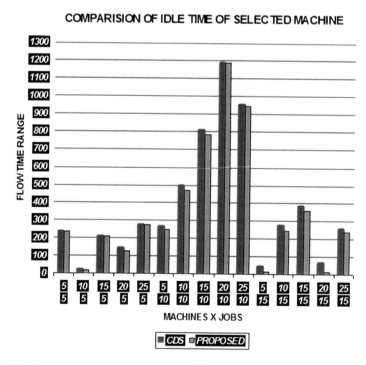

Fig. 1 Comparison of proposed heuristic versus CDS method

A Heuristic toward Minimizing Waiting Time

Table 5 Comparison of particular machine idle time

No. of M/c	No. of Job	No. of selected M/c [1]	Proposed heuristic's		CDS's heuristic's	
			Idle time	Flow time	Idle time	Flow time
5	5	4	237	1,627	237	1,627
5	10	2	16	6,190	19	6,025
5	15	3	206	9630	209	10,547
5	20	2	126	18,750	145	17,343
5	25	3	277	22,292	275	25,319
10	5	4	246	5,457	266	6,148
10	10	6	471	9,714	497	10,035
10	15	12	786	11,289	809	11,299
10	20	15	1,190	14,341	1,190	15,427
10	25	12	946	18,664	955	19,356
15	5	3	10	4,533	45	4,533
15	10	4	243	7,456	275	7,600
15	15	8	358	12,157	382	13,134
15	20	2	11	19,376	62	19,152
15	25	4	241	27,326	255	26,399

The proposed heuristic is applied to flow shop problems of various sizes and selecting any two machines for minimizing its idle time. The machines varying from 5 to 10 and jobs varying from 5, 10, 15, 20, and 25 were solved and results are given. In these problems, one machine is selected for minimizing its idle time and the selected machines are also given in the Table 5. The processing times of jobs on machines are taken as random number varying from 0 to 100. The programming is done in VB as back-end and MS Excel as front-end tool.

A particular problem is solved by CDS heuristic and the idle time of the selected machine is noted down and the proposed heuristic reduced the idle time obtained from CDS method. It is to be noted that from the table the sum of the idle

Table 6 Comparison of particular machine idle time

No. of jobs	No. of M/c	Proposed heuristic's				CDS's heuristic's		
		No. of selected P—job [6]	(Idle time of M/c)		Flow time	(Idle time of M/c)		Flow time
			Idle time P(1)	Idle time P(2)		Idle time P(1)	Idle time P(2)	
5	5	2, 4	80	156	1,533	83	189	1,982
5	10	4, 6	147	262	3,445	166	290	3,404
5	15	4, 10	170	539	4,874	200	545	4,900
5	20	7, 13	383	766	6,399	383	766	6,399
5	25	7, 13	526	759	8,248	532	837	8,134
10	5	2, 3	63	248	6,248	85	270	5,824
10	10	4, 6	371	494	8,361	454	595	9,514
10	15	6, 8	434	724	10,961	456	755	10,708
10	20	2, 15	116	1,006	14,166	128	1,021	14,468
10	25	10, 18	613	1,351	17,499	652	1,405	18,491

time of the selected machines are either equal or less than that obtained from CDS (Table 6).

References

1. Johnson SM, Balas E, Zawack D (1988) The shifting bottleneck procedure for the job shop scheduling. Manage Sci 34(3):391–401
2. Ignal S, Schrage, Lominick P (1965) Priority dispatching and job lateness in a job shop. J Ind Eng 6:228–237
3. Pan QK, Wang L (2012) Effective heuristics for the blocking flow shop scheduling problem with makespan Minimization. OMEGA 40(2):218–229
4. Palmer DS (1965) Sequencing jobs through a multi-stage process in the minimum total time—a quick method of obtaining a near optimum. Oper Res 16:101–107
5. Campell HG, Dudek RA, Smith ML (1970) A heuristic algorithm for the n job m machine sequencing problem. Manage Sci 16:630–637
6. Dannenbring DG (1977) An evolution of flow shop sequencing heuristics. Manage Sci 23:1174–1182
7. King JR, Spachis AS (1980) Heuristics for flow shop scheduling. Int J Prod Res 18(3):345–357
8. Stinson, Simith DT, Hogg GL (1982) A state of art survey of dispatching rules for manufacturing job shop operations. Int J Prod Res 20:27–45
9. Palmer K (1984) Sequencing rules and due date assignments in a job shop. Manage Sci 30(9):1093–1104
10. Nawaz M, Ensocore E Jr, Ham I (1983) A heuristic algorithm for the m- machine, n- job, flow-shop sequencing problem, OMEGA. Int J Manage Sci 11(1):91–95
11. Hundal TS, Rajgopal J (1988) An extension of Palmer heuristic for the flow shop scheduling problem. Int J Prod Res 26:1119–1124
12. Winder R, Hertz (1989) A survey of priority based scheduling. OR Spectr 11:3–16
13. Osmon IH, Potts CN (1989) A state of art survey of static scheduling research involving due dates. Omega 12:63–76
14. Ogbu FA, Smith DK (1990) Simulated annealing for the permutation flow shop problem. OMEGA 19(1):64–67

A Study on Development of Knee Simulator for Testing Artificial Knee Prosthesis

R. B. Durairaj, J. Shanker, P. Vinoth Kumar and M. Sivasankar

Abstract Nowadays the knee implant surgery is most often the solution for Bone sorption and Bone damage. This will cause huge demand on the requirement of knee simulator testing machine which is used to calculate the number of cycle times of the artificial knee prosthesis. Each and every prosthesis has to be tested at least 5–10 million cycles in the simulator which have to be implanted in the human body with expected lifetime of 10 years after the implant. This study deals with the basic anatomy of the human knee with previous simulators and their working methods.

Keywords Bone damage · Bone sorption · Knee simulator · Implant · Prosthesis

1 Introduction

The knee joint is nothing if not complicated. Long characterized a functionally as a hinge joint, the true motion in this joint complex is much more intricate. As with most biological hinge joints, the knee motion consists of combined gliding and rolling actions of the distal end of the femur with respect to the proximal articular surface of the tibia, as well as the sliding motion of the patella on the femur. Total knee replacement, or TKR, is the method of choice for the restoration of movement and the reduction of pain in those patients who show signs of destroyed articular cartilage, primarily due to

---

R. B. Durairaj · J. Shanker · P. Vinoth Kumar
SRM University, Kattankulathur, Chennai, India

M. Sivasankar (✉)
Department of Mechanical Engineering, SKP Engineering College, Tiruvannamalai, India
e-mail: mechanicprofile@gmail.com

S. Sathiyamoorthy et al. (eds.), *Emerging Trends in Science, Engineering and Technology*, Lecture Notes in Mechanical Engineering, DOI: 10.1007/978-81-322-1007-8_32, © Springer India 2012

the ravages of arthritis. The underlying concept behind TKR is the removal of the offending tissue and subsequent replacement with a man-made articular surface. According to research done by the CDC, osteoarthritis accounts for the greatest number of arthritis cases, affecting 21 million adults in 1998. Osteoarthritis, known also as degenerative arthritis or degenerative joint disease, is characterized by the degeneration of the articular cartilage in the joint, resulting in bone on bone wear. Osteoarthritis is usually thought of as occurring simply through natural wear and tear on the joints, but can also be caused by genetic disruptions in collagen formation, as well as injury or severe physical stress among the survey given by Arthritis foundation U.S. more than 67 % of Adults will have Arthritis by 2030 (Fig. 1). Reference Hootman JM [1] by this report it is much necessary to design the most feasible way of the simulator will be produced at low cost and with high strength.

2 Anatomy of Human Knee and Joint Types

The Basic Anatomy of human knee consists of Femur, Tibia, Fibula, Collateral Ligament, Articular Cartilage, Patella etc., shown in Fig. 2. The tibio-femoral joint, as the name implies, consists of the articulation between the distal surfaces of the femur and the proximal surfaces of the tibia. Though referred to as a single joint, technically this is a set of two different articulating surfaces: both the medial and lateral condyles articulate with the tibial plateau. The articular surfaces of the tibio-femoral joint are covered by articular hyaline cartilage, allowing for free movement under loading situations. This joint is the most responsible for the motions that are associated with the knee. Six degrees of freedom are associated with this joint; three of

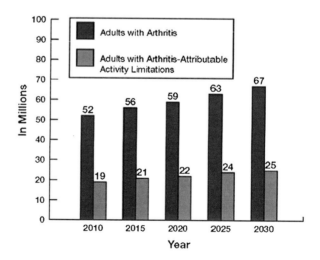

Fig. 1 Graph on adults with arthritis in U.S. expected by 2030

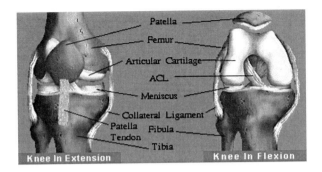

Fig. 2 Basic anatomy of human knee

translation, three of rotation. The tibio-femoral joint translates anterior-posterior, medial lateral, and through compression-distraction of the menisci. Rotation is accomplished through flexion–extension, internal-external, and varus-valgus movement. The primary rotational movement is the flexion–extension coupled rotation, from 0° at full extension to around 130–140° at full flexion, when the thigh-calf contact interferes with the movement. Internal-external rotation can be as much as 10° to each direction from neutral and increases when the joint is in flexion. The patello-femoral joint is the joint created by the interaction between the posterior surface of the patella and the anterior portion of the distal end of the femur. The path that this semi-solid bone traverses is controlled by the femoral groove found on the anterior portion of the femoral articular surface As the joint goes into extension, the patella moves proximally in relation to the femur. At full extension of the knee joint, the contact between the patella and the femur is located on the distal end of the patella and the most proximal portion of the femoral articular surface. When the joint is placed into flexion, the patella moves distally along the femur and fully places the sub-patellar ridge into the femoral groove. The patella bone acts as both a lever as well as a friction reduction device in the transition of force from the quadriceps to the tibial tuberosity. All of the articular surfaces in the patello-femoral joint are covered in articular hyaline cartilage, allowing for free movement of the patella relative to the femur [2].

3 Previous Knee Simulators Type

Several generations of knee simulators have been developed. Knee simulators are needed to determine the behavior of the human knee joint in vitro, and can be classified into two broad groups: quasi-static and dynamic simulators. These simulators are generally characterized by the use of muscle analogs to incite motion at the knee joint. There is great variation between simulators in choice of powering of muscle analog systems, what movements are fixed, controlled, or unconstrained, as well as whether or not the system is open or closed kinetic chain.

3.1 Purdue Mark II/Kansas Knee Simulator

The description of this particular simulator will focus primarily on the Purdue Mark II simulator, as the Purdue simulator was the original simulator, and the Kansas Knee Simulator was an almost exact copy created at the University of Kansas. As such, the paper that describes the creation and fundamental capabilities of this simulator deals with the Purdue Mark II. The Mark II, as seen in Fig. 3 was created as a natural outgrowth of other previous systems. Other simulators, including the MTS and EnduraTec were capable of physiologic load levels associated with walking or stairs, but did not seem capable of creating the quick movements and impact loads associated with jogging or other fast mobility movements. The Mark II was created expressly to study the effect of fast implant loading. The Mark II, as the name implies, is the second generation Purdue Knee Simulator. The original simulator utilized an unconstrained electrohydraulic loading system that was oriented such that the tested specimen was vertical, or in a standing position. This original system had an independently moveable femur and tibia that connected to the rig by means of moveable hip and ankle plates. Like the later generation Mark II, the independently moving bones and terminal plates allowed the knee to operate with little direct control. Indirect control was supplied through the soft tissues, ligamental structures, and applied loads to the quadriceps tendon and vertical loading on the hip slider. The Mark II was controlled through five load creation

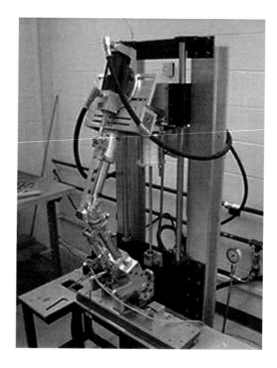

Fig. 3 Kansas knee simulator

devices [3]: vertical load, quadriceps load, torque about the global Y-axis (vertical line through the ankle joint), ankle flexion moment, and adduction/abduction of the tibia through X-axis (medial/lateral) translation of the ankle sled. It should be noted that the Purdue Knee Simulator: Mark II was considered a phenomenological simulator and therefore did not consider the use of knee flexor muscles, though the compressive load that should have been applied to the knee was compensated for in the vertical loading Starting at the most proximal point, the hip was constrained such that the only motions allowed were a vertical translation and a rotation about the local X-axis. The hip slider was limited to vertical translation by the use of vertical slide shafts, while load was created in the vertical direction by a hydraulic vertical load actuator. The rotation about the X-axis was an unconstrained motion, as this rotation would be induced as a result of the flexion of the knee. This X-axis rotation can best be described as a movement of the femur in the sagittal plane. The quadriceps load actuator was fixed to the proximal end of the femoral shaft, distal to the hip flexion joint. Connection from the quadriceps actuation device to the quadriceps tendon was created through the use of a three piece mechanical clamp when testing was conducted on cadaveric specimens, and a one-inch-wide Kevlar® strap was used to simulate the quadriceps tendon and patellar ligament when testing was conducted on prosthetic systems. The attachment of the hydraulic actuator was adjustable such that the line of action of the quadriceps tendon was variable. The ankle assembly provide for four degrees of freedom over all, three of which were controlled. The ankle itself was modeled by a universal joint, proximal to which the tibia, and assumedly the fibula, had been cut and potted in aluminum fixtures with bone cement. The universal nature of this joint allowed for flexion of the joint about the local X-axis, the adduction abduction roll of the foot with respect to the tibia, as well as a distal vertical rotation. The universal joint was connected distal to this vertical rotation to a plate which was able to move medial and lateral. This medial and lateral movement was controlled through the adduction-abduction actuator, named for the induced motion of the tibia with respect to the femur. Axial rotation of the tibia was primarily controlled through rotational motion of the vertical torque actuator, but as this torque was locked into rotating about the global Y-axis, it did not provide direct control. The true internal-external rotation of the tibia was coupled with the adduction-abduction actuator. The Mark II was upgraded with the addition of an ankle flexion moment actuator that could control the flexion of the ankle. This actuator was placed such that it would translate with the medial–lateral movement of the ankle plate and would rotate about the vertical axis to induce a true flexion moment.

3.2 Vermont Knee Simulator

The Vermont Knee Simulator (Fig. 4) was primarily designed and built by Dr. Kosmopoulos and Dr. Keller of the University of Vermont, Burlington. The primary goal of the Vermont Knee simulator was the design and validation of an

Fig. 4 Vermont knee simulator

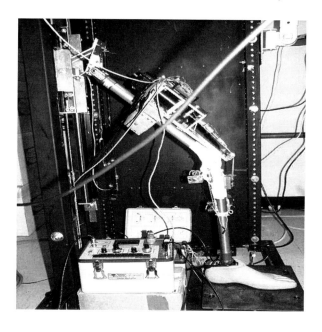

open- and closed-loop knee movement simulator with the described purpose of understanding the knee biomechanics to further the better design of artificial knee implants, fracture fixation, and rehabilitation. The aim of the study was to develop a robotic lower extremity knee simulator and to perform a validation on the actuation using a mechanically actuated prosthetic leg. Like almost all simulators, the Vermont system uses three main components: the frame, the muscle actuation method, and the motion analysis system. The primary frame consists of a set of two 38 inch long parallel rods used to constrain the hip into a vertical translational movement. The rods at 16 inches from the base, vertically. This limited the movement of the system into a range of 0 to approximately 90 ° of flexion.

Muscle actuation was enabled through the use of a micro stepper motor capable of 1.05 Nm holding torque, coupled with a planetary gearbox capable of 70 Nm static torque. The motor and gearing system is coupled with a miter gear set that connects to the winding drum. The drum, of unknown diameter, winds a polyethylene fiber that creates load in the line of action of the muscle. Control of the stepper motor system is achieved through the use of an ISA bus motion control card used in conjunction with a LabVIEW VI to simultaneously control motor movement and acquire data. Hamstring force in this system is a passive force, of a variable nature. The motion analysis was accomplished through the use of a Mac Reflex visual imaging system running at 50 Hz and utilizing 2D reflective markers. Using this system of motion analysis, only a 2D view of the system can be determined, and thus only 2D motion can be tracked, assumably in the plane of the primary flexion and extension of the knee. Validation of the system was conducted using a mechanical limb analog, comprised of a prosthetic leg with hip joint, knee, and ankle joint. As of this time, both open and closed kinetic chain testing has occurred with this system, and both styles of testing were conducted with the same prosthetic leg system. Preliminary results state that the joint functions similar to a normal human knee. Statements have been made in the online publication from the laboratory to the effect that further validations will be made utilizing the reflective markers to determine the joint angles and positions, as well as moving into cadaveric testing and simulations of total knee arthroplasty.

3.3 Hanover Knee Simulator

The primary purpose behind the work done by Ostermeier, is the in vitro study of the quadriceps loading required to extend the knee before and after a total knee arthroplasty in a isokinetic extension cycle. To accomplish this, a simulation was developed based on the concept of open kinematic chain movement. This particular simulator has been used in testing previously however, the data developed in this testing is not of particular use to the designer of a new knee simulator, so description of the simulator will be based off of the 2004 paper. The Hanover simulator is constructed such that only the distal end of the femur and the proximal end of the tibia are needed. The femur is oriented horizontal, parallel with the table, while the patella faced downward. The tibia was attached to one of the arms of the simulator at mid length by means of a rotational, sliding bearing allowing for axial sliding and turning, as well as transverse rotation to the tibial axis. The attachment to the arm of the simulator allowed for varus-valgus motion of the tibia with respect to the femur. The simulator was designed such that all degrees of freedom were unconstrained, other than the flexion and extension moment, which was controlled by the muscle simulation system and the position of the swing arm. Load cells were equipped onto the swing arm such that the externally applied resistive flexion moment could be directly determined. The actual movement of the system was enacted through the two hydraulic cylinders, one to provide the main loading of the quadriceps, and the other cylinder to power the resistive flexion moment in the swing arm. Motion in the swing arm were enacted through a rack and pinion system such that the swing arm was capable of loading the tibia to 31 Nm. Attachment to the tendon was made through the use of a tendon clamp, rated for 2,000 N, and a steel cable. Force measurements were made through a load cell attached to the quadriceps hydraulic cylinder. A feedback control was enabled through the use of this load cell on the quadriceps cylinder that allowed the quadriceps loading to be set to load control. A similar feedback circuit was enabled to allow the swing arm to maintain a constant torque state. The system would then be able to utilize the loading systems to develop a controlled motion of the extension or flexion of the knee in an open kinetic chain system. It should be noted that the data determined by this paper showed that for a constant flexion torque induced by a hamstring, that the minimum required quadriceps load level to induce movement occurred at 20° of flexion, while a maximum load level occurred at 109° of flexion. It should be noted that all of the data determined in the testing of the Hanover simulator only considered the flexion angles between 0 and 120° of flexion.

3.4 New York Knee Simulator

Interestingly, the New York Knee Simulator (NYKS), as seen in Fig. 5 is one of the first published simulators that seem to specifically focus on the determination of motion in the deep flexion of the knee. There has not been an overly large quantity of published data on the design and capacities of the NYKS, but some information can be called from the limited papers and pictures provided about the systems. Much like all of the other simulators, this simulator can be separated into its basic components: frame, muscle actuation, and motion measurement. The frame of the NYKS appeared to be close in design to that of any of the other simulators: vertical uprights constraining motion of the hip into a linear path, the ankle ridged, while what would be the toe of the foot is allowed to pivot. The foot system itself is attached to a translating plate that allows for medial/lateral translation of the foot. The researchers seemed to have used a intermedulary tibial implant as the method of fixating the tibia to the metallic lower tibia analog. Just distal to this implant is an unknown mechanism for applying an axial rotation to the tibia. A four bar linkage is used to develop a continuously vertical downward force that appears as a ground reaction in the foot. Counterweights are used to counteract the forces of the femoral and tibial assemblies, and appear to be connected to the hip slider mechanism. The muscle actuation is developed through what appears to be pneumatic cylinders, controlled by electronic air valves. Load cells are placed in line with the loading cables, and assumably create the feedback conditions for load control of the system. The cables pass through a set of retaining pulleys that

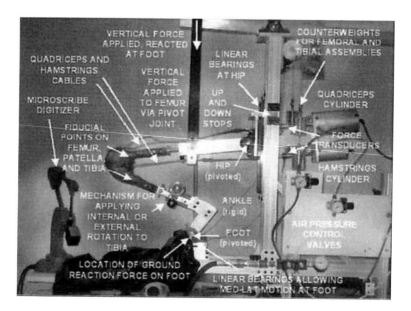

Fig. 5 New York knee simulator

[4] keep the straps in line with the anatomical line of muscle action. Linear movement of the hip, while constrained by the vertical bar system, only translates a small amount as this experiment was designed to examine the loading and contact areas in the knee joint at specific joint positions. In effect, this experiment is the ultimate in quasi-static joint testing. The flexion angulation that the system was tested at was 155°, with other testing taking place at 145 and 135° of flexion. When the joint had been loaded such that it moved to the upper stop in the allowed motion, a Microscribe G2X instrumented linkage was used to digitize the three sets of fiducial points (fixed reference points) on the femur, patella, and tibia. After the loading analysis was completed, points were digitized of all of the articular and bone surfaces. Information was then aligned to the reference points and assembled. A second generation of the NYKS was constructed at the end of 2006, and has begun in vitro testing. This next generation utilizes servomotors like the previous generation, but locates them in a different spot. The new construct places the servomotor assembly distal to the hip bearing, on the anterior aspect of the femoral construct. The hip is simulated through the use of a universal ball bearing, while the ankle was simulated as a hinge joint attached to polyurethane blocks. The tibial construct was allowed to rotate freely, while the hip was constrained to move in the vertical axis along guided rails. The prime difference, besides the lack of information on the freedom of movement of the ankle, was the substitution of two springs for the hamstring motor. The springs attached to either side of the posterior tibia and ran to the femoral fixture. The springs chosen simulated half of the quadriceps load and reduced force with the reduction of flexion angle. Motion data was taken through the use of a MicronTracker system, while a MicroScribe was used to digitize the ligamental connection points.

4 Conclusion

Thus the study of various types of knee simulators such as Kansas Knee Simulator, Vermont Knee Simulator, New York knee Simulator have been studied. This study will be more useful for further development of the knee simulator testing machine for the testing of Artificial Knee prosthesis. Also, the results have been taken for designing the simulator with reduced cycle time of the testing prosthesis with improved lifetime of the simulator.

References

1. Courtesy of the Arthritis Foundation and Hootman JM, Helmick CG (2006) Projections of U.S. Prevalence of Arthritis and associated activity limitations. Arthritis and Rheumatism 54(1), pp 226–9
2. Maletsky LP et al (2005) Simulating dynamic activities using a five-axis knee simulator. J Biomech Eng 127:123–133

3. Ostermeier S et al (2004) Quadriceps function after TKA: an in vitro study in a knee kinematic simulator. Clin Biomech 19:270–276
4. Yildirim G et al (2007) Next generation knee replacements: a new approach to replicate the function of the ACL, summer bioengineering conference, proceedings of BIO2007

An Integrated Methodology for Geometric Tolerance Analysis and Value Specification Based on Arithmetical, Graphical and Analytical Methods

R. Panneer and V. Sivaramakrishnan

Abstract With respect to Geometrical Tolerances Analysis and Value Specification, no standard approach is available. Generally, design engineers physically conduct the analysis manually or by using an automated method. Hence, design of Geometrical Tolerances largely depends on the designer's judgment and experience. In such a case, a true integrated functional tolerance specification is not guaranteed. Also, upon specification of geometrical tolerance values, they have to be checked for coherence, completeness, and feasibility of assembly in order to avoid rejections and reworks at the stage of assembly. In this paper, an integrated methodology is suggested for value specification and for checking the coherence and completeness of Position, Symmetry, and Perpendicularity tolerances based on distribution of minimum allowance. A selected mechanical assembly is used as a case study to verify the suggested methods.

Keywords Position tolerance: $\oplus$ Symmetry tolerance : $=$ Perpendicularity tolerance: $\perp$ • Maximum material condition: MMC • Minimum allowance: A_{mini} • Minimum clearance: C_{MINI} • Working clearance: C_e • Geometrical reference frames: GRF • Tolerance diagrams: TD • Virtual component: VC

R. Panneer (✉)
Anna University, Chennai, India
e-mail: panneer.ramalingam@gmail.com

R. Panneer
Department of Mechanical Engineering, Sarnathan College of Engineering,
Tiruchirapalli 620012, India

V. Sivaramakrishnan
Rover Engineering College, Perambalur 621212, India
e-mail: vsmp1967@yahoo.com

S. Sathiyamoorthy et al. (eds.), *Emerging Trends in Science,*
Engineering and Technology, Lecture Notes in Mechanical Engineering,
DOI: 10.1007/978-81-322-1007-8_33, © Springer India 2012

1 Introduction

The scope of Geometrical Tolerance Analysis includes specifying the types of geometrical tolerances and values for tolerances. In current practice, there are certain semi-automated and manual concepts/models available to establish the types of geometrical form or position features that have to be controlled in a particular linkage or mechanism.

These concepts known as technologically and topologically related surfaces (TTRS), minimum geometric datum element (MGDE) [1, 2] are based on theory of mechanisms, especially kinematics and more particularly on theory of group displacements.

As technology increases and performance requirements continually tighten, the cost and required precision of assemblies increase as well. There exists a strong need for increased attention to tolerance analysis to enable high-precision assemblies to be manufactured at lower cost. Methods for tolerance analysis of 2 and 3D assemblies have been developed and are in use in both research and commercial applications [3]. These methods are usually very complex, and are usually best implemented in some type of automated software. However, existing variation analysis software does not yet have an industry wide user base [4].

Also, no automated or semi-automated modules are available for tolerance value specification for geometrical tolerances. Reference [2] has made an extensive investigation and effort to relate various types of functional surfaces to tolerance values. But he found that an overwhelming amount of work is required to develop a tolerance value specification module, which can automatically propose appropriate tolerance values for many different types of linkages with many different types of forms and application domains.

Hence, with respect to specification of values, the designers have to manually suggest them either on a drawing or in a CAD system. Therefore, different designers will possibly arrive at different tolerance values for the same nominal geometry. Therefore, there must be a method or procedure to perform geometrical tolerance value specification functionally. Also upon specification of values, true geometric positions of the features and their tolerance zones must be established and the feasibility of the assembly of the features must be ensured before the drawings are issued for manufacturing in order to avoid rejection and reworks at the time of assembly [5].

The aim of this paper is therefore to review, discuss, and suggest methods and procedures for assigning Geometrical Tolerance values. The methodology will also suggest numerical, graphical, and analytical procedures for checking the values for their coherence and completeness and for checking the feasibility of assembly of the features for which the values are assigned.

2 Geometrical Tolerance Value Specifications

Values for geometric tolerances such as position, symmetry, and perpendicularity can be assigned based on the minimum allowance value (A_{mini}) that exists in the concerned linkage. A_{mini} value can be distributed between the shaft and hole type features as value for a relevant geometrical tolerance [6].

In order to match or satisfy the manufacturing capabilities of machines and processes, the ratio of distribution can be based on the process capability of machines that produce the whole feature and shaft feature.

Since A_{mini} is derived from the allowance which is basically decided by the function of the linkage, this method ensures a functional selection of geometrical tolerance value.

2.1 Study of Distribution of Minimum Allowance to Geometrical Tolerances

"The sum of tolerances of perpendicularity/concentricity/symmetry on a shaft feature and on a hole feature is equal to the A_{mini} between the hole and the shaft feature and not the maximum allowance" [7].

Hence, if we analyze the theory, we find that generally the minimum allowance is fully distributed as geometric tolerance to the hole and shaft features. The ratio of distribution of minimum allowance has greater influence on the virtual sizes (worst case) of the two features to which this A_{mini} is distributed as geometric tolerance.

2.2 Minimum Clearance to Ensure Assembly

The minimum clearance (C_{MINI}) between a set of mating features like a hole and shaft to ensure assembly is given by [7];

C_{MINI} Sum of geometry tolerances assigned on each of the mating features $+ C_e$ (Working clearance between the features as per virtual component (VC)).
If P_1 Geometrical Tolerance of a particular feature on Part 1.
P_2 Geometrical Tolerance assigned of corresponding mating feature on Part 2.
C_e Working clearance between the said features as per the VC.

$$\text{Then; } C_{MINI} = P_1 + P_2 + C_e \tag{1}$$

2.3 Study of Distribution of A_{mini} Using a Case Study

An assembly that has features, to which tolerances of linear dimensions and concentricity are assigned, is selected as a case study and dealt. The sub-assembly shown in Fig. 1 depicts an assembly of stepped shaft inside a stepped hole. The axes of ϕ 40 shaft (A) and Hole (B) are datum for the $\phi 16$ hole and shaft for concentricity. Both the assembly and part drawings are shown in Fig. 1.

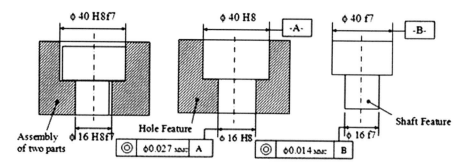

Fig. 1 Assembly of a stepped shaft in a stepped hole

As per Theory of Concentricity Tolerance [7], the sum of A_{mini} value of fit 40 H8f7 (0.025) and the of A_{mini} value of fit 16 H8f7 (0.016) has to be distributed to hole and shaft features as their concentricity tolerance based on functional requirements. As the process capability of shaft making machines and processes are better than hole making machines and processes, a two third to one third ratio is followed for assigning the concentricity tolerance to hole and shaft feature. (This is only an arbitrary ratio, but in real practice, the actual process capabilities of the concerned machines and processes shall be used for deciding the ratio)

Therefore; *Concentricity tolerance for Hole* $= (0.025 + 0.016) X 2/3 = 0.027$
Concentricity tolerance for Shaft $= (0.025 + 0.016) X 1/3 = 0.014$

2.4 Worst Case Analysis Using Virtual Condition

If we check for the Virtual sizes of the mating features of $\phi 16$ mm fit we find that;

Virtual size of the 16 mm Hole = min. size of hole – geometric tolerance = $16 - 0.027 = 15.973$

Virtual size of the 16 mm Shaft = max. size of shaft + geometric tolerance = $15.984 + 0.014 = 15.998$

The VC of the ϕ 16 mm hole feature and shaft feature are shown in the Fig. 2.

Therefore in the worst case scenario, since the virtual size of shaft is larger than the hole, it seems the assembly with positive clearance is not possible.

But as per Eq. (1) we find that;

$C_{MINI} = P_1 + P_2 + C_e = 0.027 + 0.014 + (15.973 - 15.998) = 0.027 + 0.014 - 0.025 = 0.016$

Even though based on the worst case, we arrive at a conclusion that the assembly is not possible, we find that the assembly is still possible based on Eq. (1) because there exist a minimum clearance of 0.016 mm between the mating features of ϕ 16 mm hole and shaft.

Method of Distribution of A_{mini}:

Hence, it is inferred that the ratio of distribution of A_{mini} and the type of geometry influences the VC. So, a very careful analysis is required to decide the ratio in

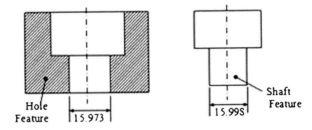

Fig. 2 Virtual component of ϕ 16 mm hole and shaft feature

which the A_{mini} value is to be distributed between the shaft and hole features for their geometric tolerances.

Also the assembly of the features has to be ensured by both the Worst Case Analysis and Eq. (1)

3 An Integrated Methodology for Geometric Tolerance Analysis

The following integrated methodology/procedure is suggested for Geometric Tolerance Analysis of mechanical assemblies.

Make an exhaustive analysis of the assemblies and identify the linkages for which assignment of geometric tolerances is essential. Based on the functional requirement of the mechanism/linkage, establish the basic sizes and tolerances either using limits and fits approach or functional dimensioning approach or using an automated method of tolerance allocation or analysis.

Establish the types of geometrical form or position features that have to be controlled in the relevant linkage based on its functional requirements. Analyze the geometrical requirements in the component parts of the linkage and group the appropriate features. For assigning values for geometric tolerances to component parts of the linkages, distribute the minimum allowance value (A_{mini}) between the component parts of the linkages to the relevant Geometrical Tolerance of the shaft and hole type features in an appropriate ratio as discussed in Article 2.3.

Upon specifying the types and values, check the coherence and completeness of tolerances assigned and the feasibility of assembly using a graphic analysis method by developing the geometrical reference frames (GRF) and tolerance diagrams (TD). Superimpose the GRF and TD of one group of features belonging to one part on the GRF and TD of the corresponding group in the mating part to check the feasibility of assembly.

Check the coherence and completeness of tolerances assigned and the feasibility of assembly using a graphic analysis method using VC. Superimpose the VC Diagram of one group of features belonging to one part on the VC Diagrams of the corresponding group in the mating part to check the feasibility of assembly. Ensure the assembly of features analytically by checking with Eq. (1).

4 Verification of Suggested Integrated Methodology Using a Case Study

An assembly shown in Fig. 3 is considered for verification of the suggested integrated methodology for Geometric Tolerance Analysis and Value Specification.

Assumption:

It is assumed that the types of geometrical form or position features that have to be controlled in this case study are already established as position, perpendicularity and symmetry for functional features such as spigot Y, gap W, and holes Z on Part 2 and their corresponding features T, V, and S on Part 1 as shown in Fig. 3. No other functional Feature is considered for the analysis.

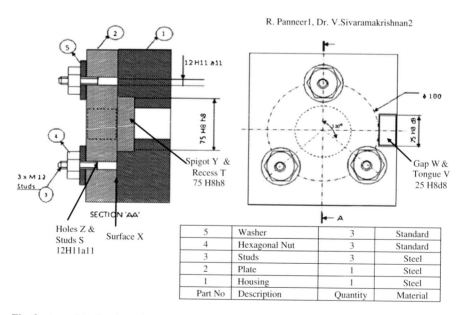

Fig. 3 Assembly drawing of the case study

An Integrated Methodology for Geometric Tolerance Analysis

4.1 Grouping of Features and Assignment of Basic Sizes and Tolerances

Based on the above information further tolerance analysis is carried out to suggest values for those features [7, 8]. Figure 3 shows the sub-assembly of two component Parts 1 and 2. Part 2 is located by the spigot Y, gap W, and surface X with corresponding features T, V, and U on Part 1. These constitute the Group 1 features. Parts 2 and 1 are secured together by the studs S through the holes Z. The holes Z with datum group 1 form Group 2 in Part 2 and studs S with datum group 1 form Group 2 in Part 1.

Table 1 Basic size, tolerances, and allowances of the mating features [8]

Mating features	Basic size	Hole feature dimension	Tol. on hole feature	Shaft feature dimension	Tol. on shaft feature	Max allowa. (A_{maxi})	Min allowa. (A_{mini})
'Y & T' (spigot and recess)	75	+ 0.046 0	0.046	0 −0.046	0.046	0.092	0
'W & V' (gap and tongue)	25	75 + 0.033 0	0.003	75 −0.065 −0.098	0.033	0.131	0.065
'Z & S' (holes and studs)	12	25 + 0.110 0 12	0.110	25 −0.290 −0.400 12	0.110	0.510	0.290

Table 2 Geometric groups and tolerances of features in Part 2

Group No.	Feature Letter and name	Number of features	Geometrical tolerance Type	Symbol	Magnitude (half of A_{mini})	Datum feature MMC
1	W (gap)	1	Symmetry and perpendicularity	⟂	0.035 (nearly half of $A_{mini} = 0.065$)	Surface 'X'
	Y (spigot)	1	Position and perpendicularity	⊕⟂	ϕ 0 (because $A_{mini} = 0$)	
2	Z (holes)	3	Position and perpendicularity	⊕⟂	ϕ 0.150 (nearly half of $A_{mini} = 0.290$)	Group 1 (W & Y)

Table 3 Geometric groups and tolerances of features in Part 1

Group No.	Feature		Geometrical tolerance			Datum feature MMC
	Letter and name	Number of features	Type	Symbol	Magnitude (half of A_{mini})	
1	V (tongue)	1	Symmetry perpendicularity		0.030 (nearly half of $A_{mini} = 0.065$)	Surface 'U'
	T (recess)	1	Position perpendicularity		$\phi\, 0$ ($A_{mini} = 0$)	
2	S (studs)	3	Position perpendicularity		$\phi\, 0.140$ (half of $A_{mini} = 0.290$)	Group 1 (V & T)

Based on the functional requirement of this mechanism/linkage, the basic sizes and tolerances are assigned to the fits between **'Y & T'**, **'W & V'**, **'Z & S'** as mentioned in Table 1. Hole basis system [8] is followed in assigning tolerances for linear dimensions in the linkages. Based on the method suggested vide Article 2.3 for assigning values for geometric tolerances, the A_{mini} values between these fits is distributed between the shaft and hole type features. In this case study, the A_{mini} value is distributed in the ratio of almost 1:1 to the hole and shaft features (an arbitrary ratio chosen for the case study).

But as stated earlier, the actual ratio of distribution can be decided based on the manufacturing process capability of the particular feature.

4.2 Arithmetical Analysis for Checking the Coherence and Completeness of Tolerances Assigned

The grouping of features, values assigned for symmetry, position, and perpendicularity tolerances of the features by distributing A_{mini} value to shaft and hole features etc., are projected in Tables 1, 2, and 3. (All sizes are given in mm). As can be seen in the following analysis, arithmetically it is established there is coherence and completeness of tolerances assigned and feasibility exists in assembling the corresponding mating features.

4.3 Graphical Analysis for Coherence and Completeness of Tolerances Assigned Using GRF and TD

Upon specifying the types and values, and after arithmetically verifying the coherence and completeness of the Geometrical Tolerances values assigned they have to be verified graphically [7]. GRF and TD are concepts that are useful in analyzing the Geometrical Tolerances graphically by establishing true geometric positions of the features and their tolerance zones. The usefulness of Graphical Analysis of the Geometrical Tolerances can be explained with the help of the case study as explained below.

4.3.1 Graphical Analysis using GRF: Representation of Axes and Median Planes of the Features Using GRF

GRF are diagrams showing the true geometric positions of all features in one positional, concentric or symmetrical group. Figure 4 shows the graphical representation of Group 2 feature of Parts 1 and 2 with its datum features drawn in a chosen scale. The figure itself is sufficed to explain the method of generating GRF.

4.3.2 Graphical Analysis using TD: Representation of tolerance zones of the features using TD

TD and GRF with the tolerance zone superimposed upon it. The MMC TD for Parts 1 and 2 are shown in Fig. 5. The figure itself is sufficed to explain the method of generating TD. The TD of one group of features belonging to one part can be superimposed on the TD of the corresponding group in the mating part to check and ensure the coherence and completeness of the tolerances specified.

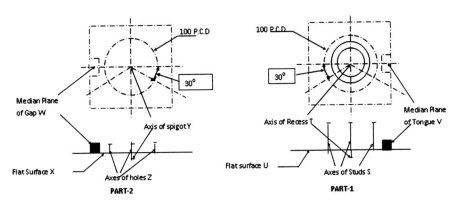

Fig. 4 Geometrical reference frames for parts 1 and 2

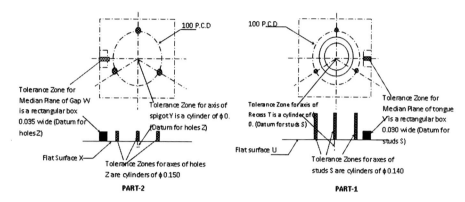

Fig. 5 MMC tolerance diagrams for parts 1 and 2

4.4 Graphical Analysis for Checking the Feasibility of Assembly in Worst Case Using VC

After ensuring the coherence and completeness of Geometrical Tolerances assigned on individual features such as the hole feature and shaft feature using arithmetical and graphical analysis (Using GRF and TD), the feasibility of their assembly has to be ensured. VC is very useful to check and ensure the assembly of the features. It is a diagram, which shows graphically, the virtual sizes of all features in one group. Virtual size of Hole feature is the minimum size of the Hole minus the geometry tolerance (e.g., position, concentricity, symmetry, straightness etc.,) at MMC. In practice, the virtual size of a finished hole or a gap is its actual measured size minus twice its measured error from true position.

Virtual size of a shaft feature is the maximum size of the shaft plus the geometry tolerance (e.g., position, concentricity, symmetry, straightness etc.,) at MMC. In practice, the virtual size of a finished shaft or a tongue is its actual measured size plus twice its' measured error from true position.

The usefulness of VC in ensuring the assembly can be conveniently explained with the help of the same case study. The virtual sizes of the hole and shaft features in the Parts 1 and 2 are shown in Tables 4 and 5. Based on these tables and graphical illustration of VC shown in Fig. 6, we can understand that the virtual sizes of corresponding features in both Parts 1 and 2 are the same and hence we

Table 4 Virtual size of gap W, hole Z, and recess T (hole features)

Feature	Minimum size of the feature	Geometry tolerance	Virtual size of the feature (min. size − geometry tolerance)
Gap W	25.000	0.035	24.965
Holes Z	12.000	0.150	11.850
Recess T	75.000	0.000	75.000

An Integrated Methodology for Geometric Tolerance Analysis

Table 5 Virtual size of tongue V, studs S, and spigot Y (shaft features)

Feature	Maximum size of the feature	Geometry tolerance	Virtual size of the feature (max. size + geometry tolerance)
Tongue V	24.935	0.030	24.965
Studs S	11.710	0.140	11.850
Spigot Y	75.000	0.000	75.000

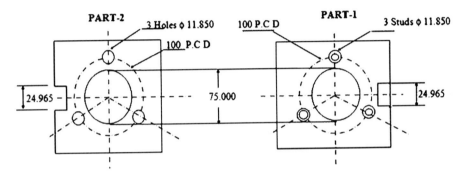

Fig. 6 Virtual component for parts 2 and 1

can ensure that these components will assemble and hence the tolerances and allowances are satisfactory.

4.5 Analytical Analysis Using Equation (1) for Checking the Feasibility of Assembly in Worst Case

After ensuring the coherence and completeness and the feasibility of the assembly using GRF and TD, the feasibility of the assembly has to be ensured analytically using Eq. (1) [$C_{MINI} = P1 + P2 + C_e$]. The C_{MINI} values derived analytically using Eq. (1) are shown in Table 6. Based on Table 6, we can establish that the C_{MINI} values of corresponding features in both Parts 1 and 2 are either zero or

Table 6 Verification with respect to Eq. (1)

Pair of features	P1 (geo. tol. of hole)	P2 (geo. tol. of shaft)	Working clearance C_e	$C_{MINI} = P1 + P2 + C_e$
W & V	0.035	0.030	24.965–24.965 = 0	= 0.035 + 0.030 + 0 = 0.065
Z & S	0.150	0.140	11.850–11.850 = 0	= 0.150 + 0.140 + 0 = 0.290
T & Y	0	0	75.000–75.000 = 0	= 0 + 0 + 0 = 0

positive values and hence we can ensure that these components will assemble and hence the tolerances and allowances are satisfactory.

5 Comments

Based on these methods further studies can be made and software packages can be developed for automatic generation of GRF, TD, and VC diagrams. Such a computer aided verification of shaft and hole feature dimensions will ensure assembly quickly which will result in economic benefits and energy savings.

6 Conclusions

Geometrical Tolerance Values for related features in a linkage can be assigned based on the minimum clearance value that exists in the linkage.

The concepts of GRF and TD are useful to check the coherence and completeness of the Geometrical Tolerances and hence can be included as additional features in design software.

The concept of VC is useful to ensure the assembly of the linkage. By superimposing the VC of related features, the virtual allowance can be obtained. Such a feature can be incorporated in design software so that at the design stage itself the assembly can be ensured at an expected quality level.

References

1. Rober cvet ko (1997) Solving tolerance constraints based on TTRS concept, Brigham Young University, USA
2. Clement A, Desrochers A, Riviere A (1997) Theory and practice of 3-D tolerancing for assembly. In: Proceeding of CIRP international working seminar on computer-aided tolerancing, Penn State University, 16–17 May 1991
3. Ward K (1993) Integrating geometric form variations into tolerance analysis of 3-D assemblies. Brigham Young University, USA
4. Dabling JG (2001) Incorporating geometric feature variation with kinematic tolerance analysis of 3D assemblies. Brigham Young University, USA
5. Wick C, Veilieux RF (1987) Quality control and assembly, Vol 4. Society of manufacturing Engineers, USA
6. Dipl.-Ing. Hochmuth R et al (2002) An approach to a general view on tolerances in mechanical engineering, University Erlangen-Nuremberg, Germany
7. Gladman CA (1994) Manual on geometric analysis of engineering designs, Australia
8. PSG College of Technology (2007) Design data, data book of engineering, India

Complexity on Parallel Machine Scheduling: A Review

D. K. Behera

Abstract The intended audience is scheduling practitioners and theoreticians as well as beginners in the field of scheduling. The purpose of this paper is to review the area of parallel machine scheduling (PMS) on issues of complexity. A critical review of the methods employed and applications developed in this relatively new area are presented and notable successes are highlighted. The PMS algorithms are discussed. We have given up-to-date information on polynomially type of problems based on non-preemptive criteria. It is shown that parallel machine makespan-minimization problem is NP-hard even for the two-machine problem. Moreover, the two-machine problem can be solved by the pseudo polynomial algorithm.

Keywords Parallel machine • Scheduling • Non-preemptive • Makespan • Polynomial • Complexity

1 Introduction

In the twenty-first century, the scheduling research area has made extraordinary advances in the development of techniques that enable improved solutions to practical problems [1, 2]. Notwithstanding the strengths of current techniques, the problems being addressed by current scheduling methods are generally NP-hard and solved only approximately [3]; there is scope for improvement in techniques for accommodating different classes of constraints and for optimizing under different sets of objective criteria [4–7]. The running time or time complexity of an algorithm expresses the total number of elementary operations, such as additions, multiplications, and comparisons, for each possible problem instance as a function of the size of the instance. The input size of a typical scheduling problem is bounded by the number of jobs n, the number of machines m and the number of bits to represent

---

D. K. Behera (✉)
Mechanical Engineering Department, Jadavpur University, Kolkata 700032, India
e-mail: dkb_igit@rediffmail.com

S. Sathiyamoorthy et al. (eds.), *Emerging Trends in Science,*
Engineering and Technology, Lecture Notes in Mechanical Engineering,
DOI: 10.1007/978-81-322-1007-8_34, © Springer India 2012

the largest integer (the processing time, tardiness, the due date etc.,). An algorithm is said to be polynomial or a polynomial-time algorithm, if its running time is bounded by a polynomial in input size [3–7]. The most real-world problems are difficult to solve to optimality [3, 7–9]. So, Polynomial-time algorithms (PTA) was introduced by Cobham in year 1964 in deterministic machine models and later by Edmonds in 1965 saying that polynomial time represents efficient computation. An algorithm with rational input is said to run in polynomial time if there is an integer say k such that it runs in O (n^k) times where n is the given input size, and all numbers in intermediate computations can be stored with O (n^k) bits. We term it as a linear-time algorithm when the value of k becomes unit. PTA are persistently called "efficient" or "good". This big O notation is used to classify algorithms by how they respond (based on processing time requirements) to changes in input factor or size. Big O notation has utility when efficiency is looked into for analyzing algorithms. The number of hierarchy depends on the particulars of the machine model on which the algorithm runs, but different types of machines typically vary by only a constant factor in the number of hierarchy needed to execute an algorithm. In parallel machine scheduling (PMS), the relationships between time and space being the criteria of analysis of complexity, it is important to study for deterministic and non-deterministic problems [1–8]. Although traditional techniques such as complete enumeration, dynamic programming, integer programming, and branch and bound were used to find the optimal solutions for small- and medium-sized problems, they do not provide efficient solutions for the problems with large size [10, 11] (Table 1).

2 Notation and Classification

The use of $\alpha|\beta|\gamma$ notation given by Graham et al. [1, 2, 4, 8] for scheduling problems, where α is the machining environment, β is the set of restrictions, and γ is the objective function. Say, $\alpha = 1$ which denotes a single machine, while $\alpha = P$ is a parallel machine environment. For γ, C_j is the total completion time objective. Parallel Machines (PM):

Table 1 The time complexity of different types of problem seen in the literature

Sublinear	O(1)	Constant-time
	O(log log n)	Double logarithmic
	O(log n)	Logarithmic
	O(log$^k n$)	Polylogarithmic; K *is a constant*
	O(n^a)	a < 1 is a constant; e.g.,O $(\sqrt{n})$ for a = 1/2
	O($n/$log$^k n$)	k is constant
Linear	O(n)	
Super linear	O(n log$^k n$)	
	O(n^c)	Polynomial; $c > 1$ is a constant; e.g., O $(n\sqrt{n})$ for c = 3/2
	O(2^n)	Exponential
	O(2^{2^n})	Double exponential

Complexity on Parallel Machine Scheduling: A Review   375

means more than one machine is performing the same function. Table 2 gives the general notation/parameters considered in any scheduling problem. The PM can be:

- Identical: all machines have the same speed factors, and they can process all the jobs.
- Uniform: parallel machine system with different speed factor, and each job has a single operation.
- Unrelated: there is no relation between machines.

In a Parallel Machine Environment we consider a simple case say $Pm|rj$, $Mj|wjTj$ which denotes a system with m machines in parallel. Job j arrives at release date rj and has to leave by the due date dj. Job j may be processed only on

Table 2 Notation/parameters for scheduling

Data	n	Number of jobs
	$p_i\ (p_{i;j})$	Processing time of job i (on machine j)
	d_i	Due date of job i
	s_i	Desired starting time of job i
	r_i	Release date of job i
Variables	C_i	completion time of job i
	E_i	Earliness of job i: $E_i = \max(0;\ d_i - C_i)$
	L_i	Lateness of job i: $L_i = C_i - d_i$
	T_i	Tardiness of job i: $T_i = \max(0;\ C_i - d_i)$
	U_i	Flag of tardiness for job i: $U_i = 1$ if i is tardy and 0 otherwise
Constraints	Permu	In a flow shop problem the job sequence is the same for each machine
	Pmtn	Jobs can be interrupted and resumed later
	Nmit	No machine idle times are allowed
	Ssd	Sequence dependent setup times occur between jobs
Criteria	$f_{max}/C_{max}/L_{max}/L_{min}/T_{max}/E_{max}/$ $\overline{C}\ \overline{C}\ w)\ /\overline{T}/\overline{T}w/\overline{E}\ /\overline{E}w/\overline{U}\ \overline{U}w$	General maximum function strictly increasing with the completion times/maximum completion of jobs: $C_{max} = \max i = 1::n(C_i)/$ maximum lateness of jobs: $L_{max} = \max i = 1::n(L_i)/$minimum lateness of jobs: $L_{min} = \min i = 1::n(L_i)/$ maximum tardiness of jobs: $T_{max} = \max i = 1::n(Ti)/$maximum earliness of jobs: $E_{max} = \max i = 1::n(E_i)/$sum of completion times: $C = Pn/(Ci)$ (weighted sum)/sum of tardiness: $T = Pn/(T_i)$ (weighted sum)/sum of earliness: $E = Pn/(E_i)$ (weighted sum)/number of late jobs: $U = Pn(U_i)$ (weighted sum)

one of the machines belonging to the subset Mj. If job j is not completed in time a penalty $wjTj$ is incurred.

A Complexity Hierarchy may be in following order as per nature of problem.

1. $1||C$max,
2. $P2||C$max,
3. $F2||C$max,
4. $Jm||C$max,
5. $FFc||C$max.
6. $1||L$max,
7. $1|prmp|L$max,
8. $1|rj|L$max.
9. $1|rj, prmp|L$max,
10. $Pm || L$max.

One standard approach for designing polynomial time approximation algorithms for a (difficult, NP-hard) optimization problem P is stated as follows:

(a) Relax some of the constraints of the hard problem P to get an easier problem P' (the so-called relaxation).
(b) Work out (in polynomial time) an optimal solution S_t for this easier relaxed problem p'.
(c) Translate (in polynomial time) the solution S_t into an approximate solution S for the original problem P.
(d) Analyze the quality of solution S for P by comparing its cost to the cost of solution S' for P'.

3 Scheduling Algorithms

Scheduling theory is concerned with the optimal allocation of scarce resources to activities over. Time horizon [4–6].The practice of this field dates to the first time two humans contended for a shared resource and developed a plan to share it without bloodshed. Algorithm may be defined as a succession of operations producing a solution to a problem through data manipulation. These data can be constants, or variables, or both kinds which can be arranged into data structures. Algorithms can be viewed as: precise type and approximate type. Precise analysis is quite tedious and at times unattainable to perform.

Thus scheduling algorithm arises [10, 11]. It is classified based on

1. Basic
   (a) as soon as possible
   (b) as late as possible
2. Time constrained
   (a) force directed
   (b) integer linear programming

(c) iterative refinement
3. Resource constrained
   (a) List based
   (b) static lists
4. Miscellaneous
   (a) Simulated annealing (SA)
   (b) path based

Figure 3.4 gives details of type of problem in a more elaborate way. Further heuristics can be classified into three types [12]. They are

- Index-development based on dispatching rules etc.
- Solution-construction like NEH.
- Solution-improvement (metaheuristics such as tabu search, SA etc).

Unfortunately, a simple, accurate, and time-invariant cost model for parallel machines does not exist The LPT, MULTIFIT, COMBINE, LISTFIT heuristics can also be applied in PMS for solving problems [1, 2, 13–16, 19–34] (Fig. 1).

An exact solution can be found by diverse methods of reduced enumeration, typically by a branch-and-bound algorithm. It is doubtful that an exact solution can be found by a polynomial-time algorithm. An algorithm is called an approximation algorithm if it is possible to found analytically how close the generated solution is to the optimum (either in the worst-case or on usual). The performance of a heuristic algorithm is usually analyzed experimentally, all the way through a number of runs using either generated instances or known benchmark instances.

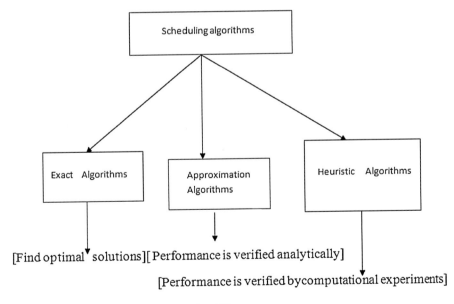

Fig. 1 The classification of scheduling algorithims

We define a ρ approximation algorithm to be an algorithm that runs in polynomial time and delivers a solution of value at most ρ times the optimum for any instance of the problem, i.e., $\frac{F(SA)}{F(SOPT)} \leq \rho$. The value of ρ is called a worst-case ratio bound. OPT stands for optimum value (Fig. 2).

In Fig. 3 P is polynomial time complexity problem and NP-hard belongs to non-deterministic polynomial. NP-Complete problems are the hardest problems in NP and P is subsets of NP.

As incase of PMS problem which is considered as hard optimization problems, finding this optimal solution is too hard because of the following reasons:

- Even with the best programming language available.
- Even with the fastest modern computer available.

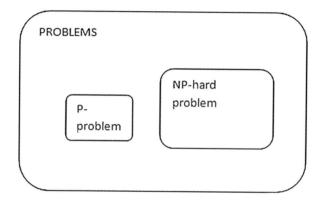

Fig. 2 Different types of problems as observed in scheduling

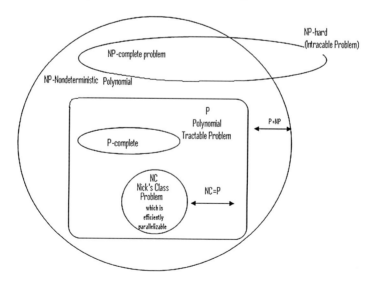

Fig. 3 A typical view of complexity classes and their relationships

Complexity on Parallel Machine Scheduling: A Review

- Even with the best programmer in the world.
- Even with the best and latest operating system.
- Even more years in the future.

The time complexity of an algorithm is the number of steps performed by this algorithm. For instance, our enumeration algorithm for the $P||C_{max}$ has time complexity big $O(m^n)$, since it evaluate m^n solutions. In any parallel identical machine scheduling denoted as $P||C_{max}$ the following parameters are looked into.

Given data/information   for each job, its duration
Constraint   perform all jobs
Decision   assign jobs to machines
Objective   end the last job as early as possible

4 Literature Review

Classical PMS considers a series of identical machines with a number of jobs and diverse processing times [1, 8, 10–34]. It assumes that the jobs are ready at time zero, and machines are endlessly available during the whole scheduling horizon. The simplest makespan problem arises in classical PMS when jobs are sequence independent and preemption is allowed. When preemption is permitted, the processing of a job can be interrupted and the remaining processing can be completed later, possibly on a different machine. When preemption of the jobs is permitted on all machines, the minimum makespan is obtained by: $M = \max\left[\sum_{j=1}^{n} p_j / m, \max_j \{p_j\}\right]$ where n is the number of jobs, p_j is the processing time of task j, and m is number of machines

It is shown that parallel machine makespan-minimization problem is NP-hard [1, 2, 6, 10] so far for the two-machine scheduling problem. Moreover, the two machine problem can be solved by the pseudo polynomial algorithm but solving problems with more than two machines is very tough and it becomes a Non-deterministic Polynomial-time hard problem which is NP-hard. Using some heuristics for generating one or more near-optimal individuals in the initial step can get better the last solutions obtained by meta-heuristic algorithms. Different criterion can be used for evaluating the efficiency of scheduling algorithms, the most important of which are makespan and flowtime [23]. Many researchers studied PMS problems in past. Cheng and Sin [8] and later Mokotoff [1] surveyed a PMS problem and Allahverdi et al. [13, 31] investigated a comprehensive review of setup time research for scheduling problems classifying into batch, non-batch, sequence independent, and sequence-dependent setup. Potts and Kovalyov [14] reviewed the literature on family scheduling models with single-machine, shop problems, and parallel machine. Brono et al. [17] proved that even a two-machine system for finding the weighted sum of flow times with an unequally weighted set of jobs is NP-hardness [19, 20, 27, 34]. A comparative analysis of PMS studied by Behera and Laha [18] indicates Listfit is better than all other algorithms.

5 Conclusions and Future Research

From the extensive literature review presented here, it can be concluded that interest in the area of PMS is growing. More direct search methods need to be explored for suitability to simulation optimization problems in PMS algorithms.

In this work, we consider a comprehensive survey of the PMS problems which is one of the most common and thoroughly studied problems in the scheduling literature. The papers surveyed include exact as well as heuristic techniques for many different multi-objective approaches. In numerous papers, SA is compared with Tabu search on scheduling problems and SA is observed to perform better than Tabu Search. SA forces the designer to either spend too much time or incur losses on the quality of solutions in scheduling problems. Different types of methods such as LPT, MULTIFIT, LISTFIT is studied in the literature [1, 2, 8, 13, 18, 19, 31]. Research in PMS will continue and is promising and there is scope for improvement.

References

1. Mokotoff E (2001) Parallel machine scheduling problems: a survey. Asia-Pacific J Oper Res 18:193–242
2. Baker KR, Trietsch D (2009) Principles of sequencing and scheduling. Wiley, New york
3. Garey MR, Johnson DS (1979) Computers and intractability: a guide to the theory of NP completeness. Freeman, San Francisco
4. Graham RL, Lawler EL, Lenstra JK, Kan AHGR (1979) Optimization and approximation in deterministic sequencing and scheduling: a survey. Ann Discrete Math 5:287–326
5. Karp RM (1972) Reducibility among combinatorial problems: complexity of computer computations. Plenum Press, New York, pp 85–103
6. Lawler EL, Lenstra JK, Kan AHGR, Shmoys DB (1993) Sequencing and scheduling: algorithms and complexity. Handbooks in operations research and management science 4, logistics of production and inventory, North Holland, Amsterdam, pp 445–524
7. Papadimitriou CH (1994) Computational complexity, Addison-Wesley, Boston
8. Cheng T, Sin C (1990) A state-of-the-art review of parallel-machine scheduling research. Eur J Oper Res 47:271–292
9. Filippi C, Romanin-Jacur G (2009) Exact and approximate algorithms for high-multiplicity parallel machine scheduling. J Sched 12:529–541
10. Levner E (2007) Multiprocessor scheduling: theory and applications. Itech Education and Publishing, Vienna, p 436. ISBN 978-3-902613-02-8
11. Koulamas C (1993) Total tardiness problem: review and extensions. Oper Res 42:1025–1041
12. Chen CL (2009) A bottleneck-based heuristic for minimizing makespan in a flexible flow line with unrelated parallel machines. Comput Oper Res 35:3073–3081
13. Allahverdi A, Gupta JND, Aldowaisan T (1999) A review of scheduling research involving setup considerations. Omega 27:219–239
14. Potts CN, Kovalyov MY (2000) Scheduling with batching: a review. Eur J Oper Res 120:228–249
15. Lee H, Guignard M (1996) Hybrid bounding procedure for the workload allocation problem on parallel unrelated machines with setups. J Oper Res Soc 47:1247–1261
16. Weng MX, Lu J, Ren H (2001) Unrelated parallel machine scheduling with setup consideration and a total weighted completion time objective. Int J Prod Econ 70:215–226

17. Bruno J, Sethi R (1978) Task sequencing in a batch environment with setup times. Found Control Eng 3:105–117
18. Behera DK, Laha D (2012) Comparison of heuristics for identical parallel machine scheduling. Adv Mater Res 488–489:1708–1712
19. Radhakrishnan S, Ventura JA (2000) Simulated annealing for parallel machine scheduling with earliness/tardiness penalties and sequence-dependent set-up times. Int J Prod Res 38:2233–2252
20. McNaughton R (1959) Scheduling with deadlines and loss function. Manage Sci 6:1–12
21. Kim DW, Kim KH, Jang W, Frank Chen F (2002) Unrelated parallel machine scheduling with setup times using simulated annealing. Rob Comput Integr Manufact 18:223–231
22. Koulamas C (1997) Decomposition and hybrid simulated annealing heuristics for the parallel-machine total tardiness problem. Nav Res Logistics 44:105–125
23. Izakian H, Abraham A, Snášel V Comparison of heuristics for scheduling independent tasks on heterogeneous distributed environments, collected from internet
24. Armentano VA, Yamashita DS (2000) Tabu search for scheduling on identical parallel machines to minimize mean tardiness. J Intell Manufact 11:453–460
25. Park MW, Kim YD (1997) Search heuristics for a parallel machine scheduling problem with ready times and due dates. Comput Ind Eng 33:793–796
26. Li X, Yalaoui F, Amodeo L, Chehade H (2002) Metaheuristics and exact methods to solve a multiobjective parallel machines scheduling problem. J Intell Manuf 23(4):1179–1194
27. Brucker P, Sotskov YN (2007) Complexity of shop-scheduling problems with fixed number of jobs: a survey. Math Meth Oper Res 65:461–481
28. Johnson DS, Aragon CR, Mageoch LA, Schevon C (1989) Optimization by simulated annealing: an experimental evaluation; part 1, graph partitioning. Oper Res 37:865–892
29. Ghirardi M, Potts CN (2005) Makespan minimization for scheduling unrelated parallel machines: a recovering beam search approach. Eur J Oper Res 165(2):457–467
30. Martello S, Soumis F, Toth P (1997) Exact and approximation algorithms for makespan minimization on unrelated parallel machines. Discrete Appl Math 75:169–188
31. Allahverdi A, Ng CT, Cheng TCE, Kovalyov MY (2007) A survey of scheduling problems with setup times or costs, Eur J Oper Res
32. Chen CL, Chen CL (2009) Hybrid metaheuristics for unrelated parallel machine scheduling with sequence-dependent setup times. Int J Adv Manufact Technol 43(1-2):161–169
33. Tavakkoli-Moghaddam R, Mehdizadeh E (2007) A new ILP model for identical parallel-machine scheduling with family setup times minimizing the total weighted flow time by a genetic algorithm, IJE Transactions a: basics, vol 20(2)
34. Pinedo ML (2008) Scheduling—theory, algorithms, and systems. Prentice–Hall, Englewood Cliffs

Design of Optimal Maintenance Strategy

A. Sarkar and D. K. Behera

Abstract In the Gas Turbine Power Plant industry, the effective maintenance strategy will solve the fast cost-effective responses to unpredictable and ever-changing plant shutdowns. Continued pressure on companies to reduce costs and improve customer satisfaction has resulted in increasingly detailed examinations of maintenance strategies. A judiciously selected maintenance program can provide significant advantages in relation to quality, safety, availability, and cost reduction in these plants. A wrong decision can lead to the failure of the components and consequent economic losses. Therefore, justification of any given maintenance strategy within an organization must consider cost and other related criteria viz availability, criticality of machines (CM), maintenance support facility (MSF). The purpose of this study is to select the apposite maintenance strategy based on these criteria to improve the availability of power plant. The analytical hierarchy technique is used to come to the appropriate selection of maintenance strategy.

Keywords Maintenance strategy • Analytical hierarchy technique • Costs • Availability

1 Introduction

There are various types of maintenance programs that power plants can use by selecting a right maintenance strategy [1]. Hence the decision of determining the strategy for power plant machinery maintenance is most significant. There is a dearth of development and applicability of an Analytical Hierarchy process (AHP) in selecting maintenance strategy for power plants [2]. In this section the selection of maintenance strategy for power plants has therefore been presented using the

---

A. Sarkar (✉)
Department of Mechanical Engineering, N.I.T.Agartala, Agartala 799055, India
e-mail: sarkarasis6@gmail.com

D. K. Behera
Department of Mechanical Engineering, I.G.I.T, Sarang, Dhenkanal 759146, India
e-mail: dkb_igit@rediffmail.com

S. Sathiyamoorthy et al. (eds.), *Emerging Trends in Science,*
Engineering and Technology, Lecture Notes in Mechanical Engineering,
DOI: 10.1007/978-81-322-1007-8_35, © Springer India 2012

analytical hierarchical process. Decisions associated with determining appropriate machine maintenance require the integration of (1) commercial (2) engineering (3) operational (4) safety, health, and environmental (SHE) issues. The important factors identified in maintenance strategies are criticality of machines (CM), Cost (C), availability (A), reliability (R),maintainability (M), maintenance support facility (MSF), SHE, condition monitoring facility, ease of maintenance [3, 4]. A set of questionnaires are designed and a survey was made from the respondents about their opinions. A five-point scale is used for each of the factor/criteria relating to the equipment selection problem (Fig. 1) and the respondents were asked to rate [5, 6].

2 Literature Review

A manufacturer's overall business success will be determined by the following integrated dimensions of performance: fitness of the manufacturer's products for use and their producibility. History has left the development and management of availability performance with great room for advancement. In this respect, design of the best maintenance strategy will play a great role. Researcher Richard G Lamb has given importance of Availability engineering and management for manufacturing plant performance in the year 1964. The researcher Suleyman Ozekci pointed out that the deterioration and failure process depends on the environment so that the failure rate at any time is a function of the prevailing environmental state. This necessitates the use of the intrinsic age of a device, rather than the real age, in reliability and maintenance problems n view of the fact that choosing of the most suitable maintenance strategy for different equipment is a crucial decision for managers, a large number of studies have been devoted to this field of research. In the literature, Murthy and Asgharizadeh (1999) [7] recommended a methodology based on game theory for selection of maintenance strategy for the companies which outsource the maintenance operations. Almeida and Bohoris (1995) [8] discussed a brief review of different decision theory concepts along with their applicability in the choosing of the most appropriate maintenance strategy. Bevilacqua and Braglia (2000) [9] proposed a multi criteria decision making (MCDM) method based on AHP. In this research sufficient attributes have been considered in the form of a

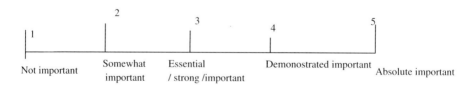

Five Scale for Questionnarie

Fig. 1 Five scale for questionnaires

crisp MCDM method. Bertolini and Bevilacqua (2006) [10] selected maintenance strategy for a set of centrifugal pumps used in an Italian oil refinery by using a hybrid method of goal programming and AHP. Al-Najjar and Alsyouf (2003) [11] proposed a combination of fuzzy inference system (FIS) and simple additive weighting (SAW), with considering a few failure causes as attributes, to make the optimum decision about maintenance strategy. Azadivar and Shu (1999) [12] proposed a new method considering 16 different characteristic parameters as criteria for each class of systems in a just-in-time environment. Gaonkar et al. (2008) [13] and Wang et al. (2007) [14] present a fuzzy AHP approach to model the uncertainty in the choosing process of the optimum maintenance strategy.

3 Methodology Used

A five-point scale is used for each of the factor/criteria relating to the equipment selection problem (Fig. 1):

(1) Respondents were asked to rate each factor how they liked each using the above scale.
(2) Mean value of each factor/criteria is calculated as follow:
(3) The mean value of each factor $= \sum x_i f_i / N$, where x_i number of respondents corresponding to rate associated with a particular factor, f_i corresponding value/rate of that particular factor, N—Total no of respondents. The results of survey are given in Fig. 2.
(4) Cut-off value is determined taking the mean of highest (4.8) and the lowest (1.2) mean rating value of all factors considered in survey.

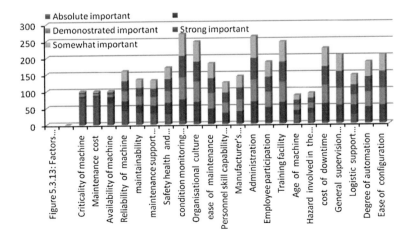

Fig. 2 Survey results of factors

(5) Mean rating value of the factors above the cut-off value (Fig. 2) are considered as important criteria or factors for selection of maintenance strategy. The following factors/criteria are found important. CM, COST(C), Availability (A), Reliability (R), Maintainability (M), MSF, SHE.

4 Case Study

4.1 Selection of Maintenance Strategy

For thispurpose of selection of appropriate maintenance for power plant machinery in plant five alternative maintenance strategies such as $M1$, $M2$, $M3$, $M4$, and $M5$ have been taken [15]. Five maintenance alternatives (M, $i = 1, 2,..., 5$) are then evaluated by AHP [16] model in terms of the criteria. Figure 3 presents the systematic procedure to understand the hierarchy for problems. The types of maintenance alternatives are: corrective maintenance, preventive maintenance (PM), total productive maintenance (TPM), reliability centered maintenance (RCM), Terotechnology.

5 Results

Table 1 presents the pair wise comparison matrix of Criteria and Fig. 4 presents the importance.

Table 2 also gives the overall rating of each maintenance. It is seen from Table 2 and Fig. 5 that maintenance alternative $M4$ (with rating of 0.3588) is the

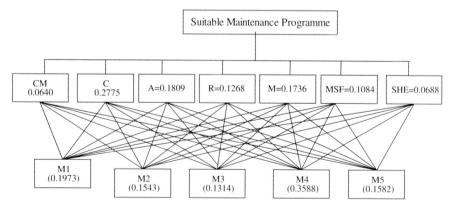

AHP model for machine criticality

Fig. 3 AHP model for by availability (relative priority 0.1809)

Design of Optimal Maintenance Strategy

Table 1 Pair wise comparison matrix of criteria

	CM	C	A	R	M	MSF	SHE	Relative priority
CM	1	1/5	1/2	1	1/3	1/2	1/2	0.0640
C	5	1	3	2	1	2	4	0.2775
A	2	1/3	1	1	2	3	3	0.1809
R	1	½	1	1	1	1	2	0.1268
M	3	1	½	1	1	2	3	0.1736
MSF	2	½	1/3	1	½	1	2	0.1084
SHE	2	¼	1/3	1/2	1/3	1/2	1	0.0688

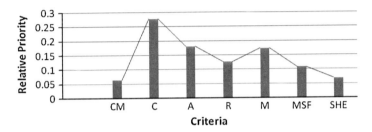

Fig. 4 The importance of maintenance criteria

Table 2 Overall rating of maintenance alternatives

	Priority	M_1	M_2	M_3	M_4	M_5
CM	0.064	0.1325	0.0449	0.5028	0.2522	0.0676
C	0.2775	0.0760	0.0493	0.1355	0.4982	0.2410
A	0.1809	0.2683	0.0422	0.1594	0.4308	0.0993
R	0.1268	0.4755	0.1421	0.0735	0.26487	0.0441
M	0.1736	0.1563	0.4299	0.0746	0.2913	0.0479
MSF	0.1084	0.1185	0.0754	0.0488	0.2810	0.4763
SHE	0.0688	0.2744	0.4258	0.0763	0.1731	0.0504
Overall		0.1973	0.1543	0.1314	0.3588	0.1582

most preferred one considering above seven criteria followed by maintenance $M1$, $M2$, $M5$, $M3$. That means RCM > Corrective maintenance > PM > Terro technology > TPM. That is, RCM is the best maintenance under the following criteria.

6 Concluding Remarks

Categorization of maintenance strategy for selection of appropriate maintenance decision making is very important, AHP is sound surrogate to make it possible to rank maintenance strategy. The AHP model is also unique in determining the

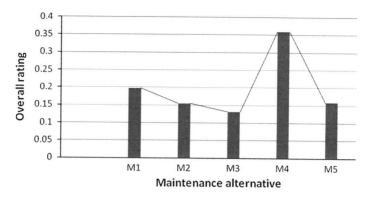

Fig. 5 Overall rating of maintenance alternative

machine criticality. Though the suggested methodology is applied to determine the best maintenance strategy and power plant machinery criticality determination for formulation of maintenance decision making, there is nothing unique about the environment and hence this approach can be generalized to any industry or plant. Similarly, we can determine the critical components of the particular machine for the purpose of maintenance decision making. By the exercise carried out in this paper it is found that reliability centered maintenance is the best maintenance strategy.

References

1. Carter ADS (1986) Mechanical reliability, 2nd edn. Macmillan, London
2. Samanta B, Sarkar B, Mukherjee SK (2002) An application of AHP for selection of maintenance personnel in mine. J Mines Met Fuels Appeared
3. Lamb GR (1995) Availability engineering and management for manufacturing plant performance, Prentice hall Inc
4. Ozekici S (1995) Optimal maintenance policies in random environments. Eur J Oper Res 82:283–294
5. Sastry GVYN (1997) Maintenance management of heavy earth moving machinery. Indian Min Eng J 36(8):27–28
6. Schottl A (1996) A reliability model of a system with dependent components. IEEE Trans Reliab 45(2):267–271
7. Murthy DNP, Asgharizadeh E (1999) Optimal decision making in a maintenance service operation. Eur J Oper Res 116:259–273. doi: 10.1016/S0377-2217(98)90202-8
8. Almeida AT, Bohoris GA (1995) Decision theory in maintenance decision making. Qual Maintenance Eng 1(1):39–45. doi:10.1108/13552519510083138
9. Bevilacqua M, Braglia M (2000) The analytic hierarchy process applied to maintenance strategy selection. Reliab Eng Syst Saf 70(1):71–83. doi: 10.1016/S0951-8320(00)00047-8
10. Bertolini M, Bevilacqua M (2006) A combined goal programming–AHP approach to maintenance selection problem. Reliab Eng Syst Saf 91(7):839–848. doi:10.1016/j.ress.2005.08.006
11. Al-Najjar B, Alsyouf I (2003) Selecting the most efficient maintenance approach using fuzzy multiple criteria decision making. Int J Prod Econ 84(1):85–100. doi: 10.1016/S0925-5273(02)00380-8

12. Azadivar F, Shu V (1999) Maintenance policy selection for JIT production systems. Int J Prod Res 37(16):3725–3738. doi: 10.1080/00207549919001
13. Gaonkar RSP, Verma AK, Srividya A (2008) Exploring fuzzy set concept in priority theory for maintenance strategy selection problem. Int J Appl Manage Technol 6(3):131–142
14. Wang L, Chu J, Wu J (2007) Selection of optimum maintenance strategies based on a fuzzy analytic hierarchy process Int J Prod Econ 107(1):151–163. doi: 10.1016/j.ijpe.2006.08.005
15. Taha HA (1977) Operation research, an introduction, 6th edn. Prentice hall
16. Tam CY, Tyummala VMR (2001) An application of the AHP in vendor selection of a telecommunication system, omega. Int J Manage Sci 29:171–182

Effect of Dummy Machines in the Gupta's Heuristics for Permutation Flow Shop Problems

A. Baskar and M. Anthony Xavior

Abstract Optimizing the Make Span is one of the important performance parameters in a flow shop scheduling problem (FSP). The problem is NP hard. In a general Permutation FSP, the total number of possible sequences is ($n!$). The job for a shop floor supervisor is that he has to schedule the jobs such that the total processing time is minimum. In the shop floor, due to limited computing facilities and capabilities, still the classical heuristics are popular as they are simple and easy to compute manually for smaller problems. However, in many cases, we get only one sequence. This paper deals with improving the make span in an FSP using the concept of Dummy Machine in Gupta's Heuristics. In such cases, we can get two more sequences having make spans that may be optimal/near optimal. The effect of dummy machine is analyzed using the well-known Taillard benchmark problems.

Keywords Flow shop • Scheduling • Make span • Heuristics • Dummy machine

1 Introduction

In an FSP, with m—machine permutation is considered in which, each job i, $i = 1$, 2,...., n needs to be processed on each machine j, $j = 1$, 2,..., m, in that order during an uninterrupted processing time $t_{ij} \geq 0$. The requirement is to find a processing order of the n jobs, the same for each machine, such that the make span is minimized. If the number of machines, $m = 2$, the popular Johnson's algorithm [1] generates an optimal solution. If $m = 3$ then, extended Johnson's algorithm can be used subject to the condition of minimum machining time of any job in machine I

A. Baskar (✉)
SMBS, VIT University, Vellore 632014, Tamil Nadu, India
e-mail: a.baaskar@gmail.com

M. Anthony Xavior
SMBS, VIT University, Vellore 632014, Tamil Nadu, India

S. Sathiyamoorthy et al. (eds.), *Emerging Trends in Science,*
Engineering and Technology, Lecture Notes in Mechanical Engineering,
DOI: 10.1007/978-81-322-1007-8_36, © Springer India 2012

(OR) minimum machining time of any job in machine III $<$ maximum machining time of any job in machine II.

For $m > 2$, various Classical Heuristic procedures have been suggested over the years. Palmer [2] proposed the slope index (SI) method. Gupta [3] proposed the functional algorithm. Both CDS and RA heuristics proposed by Campell et al. and Dannenbring [4, 5] respectively are based on Johnson's algorithm for the 2 machine problem. An improved iterated greedy algorithm (IIGA) was proposed by Quan-Ke Pan et al. [6] to solve the no-wait flow shop scheduling problem (FSP) with the objective to minimize the make span. A discrete firefly meta-heuristic with local search for make span minimization in permutation FSP was presented by Mohammad Kazem Sayadia et al. [7]. The main contribution of their work is to concern time-dependent weights which results in a more realistic insight for decision makers via considering time value of money in long scheduling problems.

Computational intelligence heuristics like simulated annealing, ant-colony optimization, differential evolution, particle swarm optimization, and fuzzy approaches have in recent years emerged as the most promising research directions for single-objective and multi-objective flow shop and job shop scheduling. An improved heuristic for permutation flow shop scheduling was proposed by Uday Kumar Chakraborty and Dipak Laha [8]. Dipak Laha and Uday Kumar Chakraborty proposed an efficient hybrid heuristic for make span minimization in permutation flow shop scheduling and a new constructive heuristic, based on the principle of job insertion, for minimizing make span [9, 10]. It may be noted that the Heuristic procedures are approximate methods and if more accuracy is required, Evolutionary algorithms or Meta Heuristics can be used which require more computing capabilities, time, and relevant software.

2 Proposed Concept of Dummy Machine

The authors have proposed the concept of Dummy Machines that can be used with any classical heuristics for computing many more additional sequences having improved make spans towards the bounds. The dummy machines have processing time of zeros for all jobs. This machine is introduced initially before the first machine and a sequence having optimal or near optimal make span will be determined using the original algorithm. Secondly, the dummy machine with zero processing times will be introduced as the last machine and another sequence having optimal or near optimal make span will be determined using the original algorithm. Altogether, three sequences will be computed and they will be evaluated for the optimality in terms of make span requirement. All the jobs have to pass through the dummy machines invariably. By the introduction of dummy machines, the number of machines is increased by one and also the position of some machines; as a result, the rank value/weight of different processing times get changed resulting in different index values. This gives different sequences which are computed for the make span.

Effect of Dummy Machines in the Gupta's Heuristics

Procedure:

Step 1: Let '*n*' number of jobs to be processed in '*m*' processing centers. It is assumed that all jobs are available at zero processing time. At any time, each machine can process at most one job and each job can be processed on at most one machine. The order of processing remains the same.

Step 2: Using the Gupta's Algorithm, the sequence with Optimal make span to be computed.

Step 3: Now, the dummy machine is introduced as the first machine and the total number of machines is now $(n + 1)$.

Table 1 Make spans for Taillard problems for 20 jobs

Seed	Lower bound	Make span	New make span	Movement towards LB %	Count (max. 2)
5 machines, 20 jobs					
873654221	1,232	1,409	1,409	0.0	0
379008056	1,290	1,424	1,424	0.0	0
1866992158	1,073	1,255	1,255	0.0	0
216771124	1,268	1,485	1,485	0.0	0
495070989	1,198	1,367	1,367	0.0	0
402959317	1,180	1,387	1,387	0.0	0
1369363414	1,226	1,403	1,403	0.0	0
2021925980	1,170	1,395	1,395	0.0	0
573109518	1,206	1,360	1,360	0.0	0
88325120	1,082	1,196	1,196	0.0	0
5 machines, 20 jobs					
587595453	1,448	1,829	1,829	0.0	0
1401007982	1,479	2,021	1,958	4.26	1
873136276	1,407	1,773	1,773	0.0	0
268827376	1,308	1,678	1,678	0.0	0
1634173168	1,325	1,781	1,715	4.98	1
691823909	1,290	1,813	1,808	0.39	1
73807235	1,388	1,826	1,649	12.75	1
1273398721	1,363	2,031	2,019	0.88	1
2065119309	1,472	1,831	1,831	0.0	0
1672900551	1,356	2,010	1,961	3.61	2
5 machines, 20 jobs					
479340445	1,911	2,833	2,721	6.86	1
268827376	1,711	2,564	2,535	1.69	1
1958948863	1,844	2,977	2,699	15.08	2
918272953	1,810	2,603	2,603	0.0	0
555010963	1,899	2,733	2,733	0.0	0
2010851491	1,875	2,707	2,660	2.51	2
1519833303	1,875	2,670	2,670	0.0	0
1748670931	1,880	2,523	2,523	0.0	0
1923497586	1,840	2,583	2,583	0.0	0
1829909967	1,900	2,707	2,627	4.21	2

Step 4: Using the Parent Algorithm, the sequence and its corresponding make span to be computed for the new problem.

Note: Gupta's algorithm reduces the problem into a 2 machines, n jobs problem and the Johnson's algorithm is used to find the sequence and make span to be computed using the original data.

Step 5: Applying the concept, the Dummy machine to be introduced as the last machine and steps 1–4 are to be repeated to obtain other sequence and make span. Among the three, the lowest make span can be selected.

Table 2 Make spans for Taillard problems for 50 jobs

Seed	Lower bound	Make span	New make span	Movement towards LB %	Count (max. 2)
5 machines, 50 jobs					
1328042058	2,712	2,920	2,920	0.0	0
200382020	2,808	3,032	3,032	0.0	0
496319842	2,596	3,034	3,034	0.0	0
1203030903	2,740	3,156	3,042	4.16	2
1730708564	2,837	3,188	3,064	4.37	1
450926852	2,793	3,154	3,154	0.0	0
1303135678	2,689	2,969	2,969	0.0	0
1273398721	2,667	3,236	3,179	2.14	2
587288402	2,527	3,255	2,780	18.80	2
248421594	2,776	3,167	3,167	0.0	0
10 machines, 50 jobs					
1958948863	2,907	3,660	3,661	1.69	2
575633267	2,821	3,645	3,619	0.92	1
655816003	2,801	3,659	3,534	4.46	1
1977864101	2,968	3,707	3,707	0.0	0
93805469	2,908	3,664	3,664	0.0	0
1803345551	2,941	3,584	3,584	0.0	0
49612559	3,062	3,806	3,806	0.0	0
1899802599	2,959	3,758	3,758	0.0	0
2013025619	2,795	3,548	3,424	4.44	1
578962478	3,046	3,964	3,912	1.71	2
20 machines, 50 jobs					
1539989115	3,480	4,759	4,759	0.0	0
691823909	3,424	4,398	4,398	0.0	0
655816003	3,351	4,471	4,471	0.0	0
1315102446	3,336	4,776	4,742	1.02	1
1949668355	3,313	4,642	4,642	0.0	0
1923497586	3,460	4,505	4,505	0.0	0
1805594913	3,427	4,758	4,758	0.0	0
1861070898	3,383	4,554	4,554	0.0	0
715643788	3,457	4,470	4,470	0.0	0
464843328	3,438	4,549	4,549	0.0	0

3 Effect of Dummy Machine in the Gupta's Heuristics

To demonstrate the new procedure, Taillard bench mark problems [11] have been used. Taillard had proposed a set of bench mark problems for permutation FSP starting from 5 machines, 20 jobs up to 20 machines, 500 jobs, a total of 120 problems which are being used by many Researchers to compare and validate the results with other known Algorithms. In Taillard's problems, the processing times vary from 1 to 99 time units and they are generated using a random number generator for different seeds. In this paper, those problems are used in the Gupta's

Table 3 Make spans for Taillard problems for 100 jobs

Seed	Lower bound	Make span	New make span	Movement towards LB %	Count (max. 2)
5 machines, *100 jobs*					
896678084	5,437	5,592	5,592	0.0	0
1179439976	5,208	5,657	5,657	0.0	0
1122278347	5,130	5,619	5,619	0.0	0
416756875	4,963	5,286	5,286	0.0	0
267829958	5,195	5,623	5,623	0.0	0
1835213917	5,063	5,259	5,259	0.0	0
1328833962	5,198	5,557	5,557	0.0	0
1418570761	5,038	5,509	5,509	0.0	0
161033112	5,385	5,821	5,821	0.0	0
304212574	5,272	5,740	5,735	0.09	1
10 machines, *100 jobs*					
1539989115	5,759	6,858	6,638	3.82	2
655816003	5,345	6,284	6,184	1.87	1
960914243	5,623	6,609	6,581	0.50	1
1915696806	5,732	6,783	6,783	0.0	0
2013025619	5,431	6,436	6,436	0.0	0
1168140026	5,246	6,138	6,138	0.0	0
1923497586	5,523	6,456	6,456	0.0	0
167698528	5,556	6,602	6,554	0.86	1
1528387973	5,779	6,356	6,356	0.0	0
993794175	5,830	6,852	6,506	5.93	2
20 machines, *100 jobs*					
450926852	5,851	7,586	7,586	0.0	0
1462772409	6,099	7,709	7,607	1.67	1
1021685265	6,099	7,481	7,481	0.0	0
83696007	6,072	7,895	7,540	5.85	2
508154254	6,009	7,657	7,657	0.0	0
1861070898	6,144	7,590	7,590	0.0	0
26482542	5,991	8,167	7,481	11.45	2
444956424	6,084	7,892	7,892	0.0	0
2115448041	5,979	7,604	7,604	0.0	0
118254244	6,298	7,965	7,711	4.03	1

Table 4 Make Spans for Taillard problems for 200 and 500 jobs

Seed	Lower bound	Make span	New make span	Movement towards LB %	Count (max. 2)
10 *machines*, 200 *jobs*					
471503978	10,816	12,079	12,062	0.16	1
1215892992	10,422	12,008	12,008	0.0	0
135346136	10,886	12,395	12,251	1.32	2
1602504050	10,794	11,812	11,812	0.0	0
160037322	10,437	11,634	11,634	0.0	0
551454346	10,255	11,892	11,730	1.58	1
519485142	10,761	12,390	12,299	0.85	1
383947510	10,663	11,791	11,791	0.0	0
1968171878	10,348	12,217	12,073	1.39	2
540872513	10,616	11,738	11,738	0.0	0
20 *machines*, 200 *jobs*					
2013025619	10,979	13,375	13,356	0.17	1
475051709	10,947	13,103	13,103	0.0	0
914834335	11,150	13,663	13,607	0.50	1
810642687	11,127	13,365	13,365	0.0	0
1019331795	11,132	12,983	12,983	0.0	0
2056065863	11,085	13,601	13,590	0.1	1
1342855162	11,194	13,079	13,079	0.0	0
1325809384	11,126	13,704	13,479	2.02	1
1988803007	10,965	13,318	13,300	0.16	1
765656702	11,122	13,609	13,500	0.98	1
20 *machines*, 500 *jobs*					
1368624604	25,922	30,529	30,167	1.40	2
450181436	26,353	30,743	30,660	0.31	1
1927888393	26,320	30,620	30,276	1.31	2
1759567256	26,424	30,400	30,400	0.0	0
606425239	26,181	29,892	29,892	0.0	0
19268348	26,401	29,874	29,874	0.0	0
1298201670	26,300	30,488	30,308	0.68	1
2041736264	26,429	30,296	30,296	0.0	0
379756761	25,891	29,147	29,147	0.0	0
28837162	26,315	30,325	30,258	0.25	1

heuristics and the make spans are found. Dummy machines are introduced and once again, by using the original algorithm, make spans are computed and analyzed for the improvement towards the lower bound. The results are tabulated in Tables 1, 2, 3, 4.

In these tables, the following terms are used:

Seed   Numbers used for random numbers generation, given in Taillard's tables.
LB   Lower Bound of the Make Span for a FSP.
Make Span Total completion time for all jobs using Gupta's algorithm.

New Make Span	New lowest Make Span after applying the Dummy Machines concept.
Movement towards LB	Percentage improvement towards Lower Bound.
Count	Number of additional sequences having Make Spans equal to or less than the make span obtained using original algorithm.

The entire process of generating the processing times, their corresponding lower bounds, make span computation using the algorithm depends on the random numbers. Hence, the generation of random numbers from the fed seed is more critical and due care is paid for this while writing the codes.

4 Conclusion

Tables 1, 2, 3, 4 show the output when the codes are generated and run in MATLAB. As all are aware of, the Heuristics methods are only approximate methods which produce varying results for different size and nature of problems. In this paper, the concept of Dummy machines is used with Gupta's Heuristic Algorithm and analyzed using Taillard's bench mark problems for permutation flow shop scheduling. MATLAB R 2008a version and a PC with 4 GB RAM and i5 Pentium processor is used for running the codes. A total of 47 problems out of 120, that is about 37 % cases, show significant improvement varying from 0.09–18.80 % with an average of 3.17 % which is reasonably good.

The authors tried this concept in other Classical Heuristics also. The dummy machines when used with Palmer SI algorithm produce improved results for 112 problems out of 120 problems, that is, in 93.3 % of the cases, the improvement varies from 0.03 up to 8.57 % with an average of 2.29 %. In case of Rapid Access algorithm, better results are obtained for 107 problems out of 120 problems, that is, in 89.17 % of the cases, the improvement varies from 0.13 up to 11.75 % with an average of 2.87 %. The dummy machines when used with CDS algorithm produce better results for 75 problems out of 120 problems which is comparatively less, that is, in 62.5 % of the cases only, the improvement varies from 0.01 up to 8.70 % with an average of 1.47 %. In these cases, we get $(m + 1)$ additional sequences against two in Gupta's and hence the results are better than Gupta's Algorithm.

The results are encouraging and the general conclusion is that we get many more sequences with optimal/near optimal make spans. We get a few more sequences with better make spans than the sequences obtained using the Parent Algorithm. The main advantage is that this can be effectively used even by the shop floor supervisor with minimum effort. As the classical algorithms, the procedure is so simple and for smaller problems, manual computations are possible with no errors. The authors have solved problems up to 10 jobs, 10 machines with little effort.

References

1. Johnson SM (1954) Optimal two and three machine production scheduling with set up times included. Naval Res, Log 1, No 1
2. Palmer DS (1965) Sequencing jobs through a multi-stage process in the minimum total time- a quick method of obtaining a near optimum. Oper Res 16:101–107
3. Gupta JND (1971) A functional heuristic algorithm for the flow shop scheduling problem. Oper Res 22:39–47
4. Campbell HG, Dudek RA, Smith ML (1970) A heuristic algorithm for the n job m machine sequencing problem. Manage Sci 16:630–637
5. Dannenbring DG (1977) An evaluation of flow-shop sequencing heuristics. Manage Sci 23:1174–1182
6. Pan Q-K, Wang L, Zhao B-H (2008) An improved iterated greedy algorithm for the no-wait flow shop scheduling problem with makes span criterion. Int J Adv Manuf Technol 38:778–786
7. Sayadia MK, Ramezaniana R, Ghaffari-Nasab N (2010) A discrete firefly meta-heuristic with local search for makes span minimization in permutation flow shop scheduling problems. Int J Ind Eng Comput 1:1–10
8. Chakraborty UK, Laha D (2007) An improved heuristic for permutation flow shop scheduling. Int J Inf Commun Technol 1(1)
9. Laha D, Chakraborty UK (2009) An efficient hybrid heuristic for make span minimization in permutation flow shop scheduling. Int J Adv Manuf Technol 44:555–569
10. Laha D, Chakraborty UK (2009) A constructive heuristic for minimizing make span in no-wait flow shop scheduling. Int J Adv Manuf Technol 41:97–109
11. Taillard E (1993) Bench marks for basic scheduling problems. Eur J Oper Res 64(2):278–285

Productivity Improvement in Railway Reservation Process: A Case Study

Krishna Kumar Singh Sengar and Apratul Chandra Shukla

Abstract As we all know the population of India reached 121 Crore (As per census report of 2011) last March. This has created a huge gap between supply and demand systems. A long queue of common people can be seen struggling for commodities and for paying electricity bills, telephone bills, municipal water supply bill, and in railway stations or bus stands. Everyday a lot of precious time is wasted in this process. In this paper, some effective tools and techniques related to work study have been used for identifying problems and giving alternative solution.

Keywords Ergonomics • Method study • Multiple activity charts • Workstudy

1 Introduction

In India thousands of passengers travel by trains to reach their destination. Most of them have a keen desire to get their berths reserved in advance to make their journey happy and comfortable. But it is unfortunate to say that they have to face several mental and physical problems in the reservation office for booking their berths. This research tries to determine unproductive time in offline railway reservation counters in Ujjain.

2 Objectives of Research

Following are the research objectives:

- To map the complete process of booking a Railway reservation ticket.
- To determine bottlenecks/unproductive activities in the present system.
- To suggest modifications in process/layout/tools/resources to improve the productivity and eventually reduce waiting time.

K. K. S. Sengar (✉) · A. C. Shukla
Department of Mechanical Engineering, Ujjain Engineering College, Ujjain, India
e-mail: krishna.sengar@gmail.com

A. C. Shukla
e-mail: apratulshukla@yahoo.com

S. Sathiyamoorthy et al. (eds.), *Emerging Trends in Science,*
Engineering and Technology, Lecture Notes in Mechanical Engineering,
DOI: 10.1007/978-81-322-1007-8_37, © Springer India 2012

399

3 Research Methodology

In the pursuit of better operational performance and profitability, organizations are looking for strategies to improve their operational performance and boost their profitability. As competition intensifies due to changes in the industry structure and emergence of new technologies, organizations are determined to reduce operational cost while enhancing their profitability, [1].

The effect of operational performance on profitability has been extensively discussed in both manufacturing and service organizations. From the quality management perspective, several studies show that certain practices such as employee training, employee involvement, and process improvement enhance profitability, [2, 3]. Empirical findings suggest a direct link between quality and profitability in manufacturing and service organizations, [4]. In case service organizations several studies have addressed the role of service quality, customer satisfaction, and business performance, [5].

3.1 Splitting the Process into Elements

In this research videotape has been recorded individual activities of three operators for a certain duration of time. This tape was further used for determining the number of activities performed by an operator in one cycle. By slowing these videos, we have calculated the average time taken by an individual operator for performing these activities in one cycle. We further analyzed pattern of hands usage by operator in individual activities. This motivated us to improve the present layout. The improved layout is discussed later in this section. One cycle is split into the following number of activities.

3.1.1 Form Receiving and Inspection

This is the first step for reservation and cancellation. During this activity, the operator stretches his hand towards window opening and takes form from customer (Fig. 1).

3.1.2 Data Entry to Computer

This is the second activity, which operators perform. During activity he/she starts filling information from Reservation Requisition Form to computer terminal

Fig. 1 Block diagram of activities in one cycle

3.1.3 Writing on Form

This is the third step taken by operator for proceeding further. During this element, the operator writes essential information on Reservation Requisition Form,i.e., PNR Number, Money Exchange Details, Number of Berth Booked, and Signature.

3.1.4 Transaction and Delivery

This is the final step taken by the operator for completing one cycle. During this cycle, exchange of money takes place along with delivery of ticket to the customer.

3.1.5 Delay

This is a non-value adding activity, which occurs occasionally. It mainly consists of rectification of mistakes in form, Intervention of third person between a cycle like peon, upper authority, breakdown of machine, waiting for confirmation from customer etc.

3.2 Multiple Activity Charts

The recorded video has been used for plotting MAC. We had focused involvement of three resources (i.e., Computer operator, Reservation requisition form, Computer and Printer) in the above four events. It helps us to determine relative idleness between resources and reason for this idleness.

3.3 Right/left Hand Utilization Pattern

The video footage is again used to determine the two-hand usage in the above-mentioned activities. This analysis helps us to reveal the work distribution between two hands. The table regarding two hand utilization patterns explains later in this paper.

3.4 String Diagram

By taking appropriate scale, the well-known concept of string diagram and plotted plan view of operator's workplace layout is used. The video for determining number of hand motion in existing workplace layout as well as distance traverse along with time taken for this movement in string diagram are again analyzed. The following hand movements in one cycle of reservation have been classified.

- Hand moves from central position to window opening for taking reservation requisition form.
- Putting reservation requisition aside over bunch of forms.
- Money taken from customer and move it to central position for counting.
- After counting money hands move from central position to drawer.
- Collecting change from drawer for ending transaction.
- Ticket tearing delivering to customer.
- After giving money to customer hands moves back to central position.

Two figures given below showing existing and proposed workplace layout (Fig. 2).

The two hands analysis of three operators shows us 76.35, 76.82, 120.67s usage of right hand compares with 60, 64.75, 117.3s of left hand. The new layout will design by considering this fact in mind and we also try to eliminate non value adding movements in railway reservation process. Figure 2 shows the existing and proposed layout. The calculations of existing and proposed layout are explained in data analysis section.

4 Data Analysis

The data collection work for this research started in mid of April, 2011, after getting permission for video recording from DRM office of Western Central Railway at Ratlam.

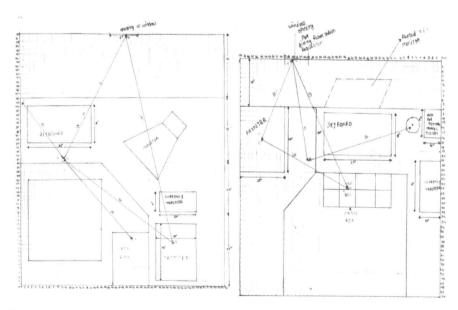

Fig. 2 String diagram of existing and improved layout *left* side existing and *right* side improved

Productivity Improvement in Railway Reservation Process

The video recording was performed on 25–27th, May 2011 by professional cameraman. The first 3 h video tape recorded bird eye view of all three windows. After that, camera was set to record 1 h individual working procedure of operators along with layout design of work station.

4.1 Average Time Taken by Three Operators with the Help of Multiple Activity Chart

Table 1 showing us the calculations for average time taken in each activity of one cycle by three operators. The name assigned to three operators is $R1$, $R2$, and $R3$ respectively.

4.2 Calculation of Right/left Hand Utilization Pattern

It is generally noticed that for booking a ticket for customer, sometimes both hands are used for operators simultaneously and few occasions' single hand has been used. Table 2 Given below explains us the hand usage pattern.

4.3 Calculation of Distance Traversed, Time Taken, and Speed of Hands in Existing Layout

A number of hands movement during one cycle has already being explained. During this movement hand traverse certain distance over string diagram. This distance traverse has been determined by hands with measuring scale. Later with the help video tape, average time taken in this movement has been calculated

Table 1 Average time taken by operator in four events

S. No.	Activities	Time taken by operators (time in seconds)			
		$R1$	$R2$	$R3$	Average time
1	Average time taken in form receiving and inspection	14.5	13.3	18.83	15.54
2	Average time taken in data feeding	34.66	39.30	56	43.32
3	Average time taken in filling details on reservation requisition.	21.66	24.3	43.66	29.88
4	Average time taken in transaction and delivery	21.08	17.61	20.5	19.73
5	Average time delay	12.66	20.3	32.33	21.76
Total		104.56	114.81	171.32	130.23

Table 2 Right/left hands utilization pattern for three operators

S.No.	Activities	Time taken (in seconds) Operator "R1"		Time taken (in seconds) Operator "R2"		Time taken (in seconds) Operator "R3"	
		RH	LH	RH	LH	RH	LH
1	Form taken and Inspection	10.83	6.75	10.61	8.84	13.5	14.5
2	Data Feeding	30	23.41	33.3	29.61	43.5	43.5
3	Writing on form	19.5	19.5	19.38	16	45	41.3
4	Transaction and Delivery	16	10.3	13.53	10.3	18.67	18
	Total	76.33	60	76.82	64.75	120.67	117.3

Total distance traversed by hands

$$D \text{ (existing)} = 72 + 36 + 72 + 28 + 86 + 36 + 88 = 418 \text{ inches}$$

Total time taken by clerk in above seven activities

$$T1 \text{ average} = 2 + 1 + 2 + 3 + 5 + 3 + 2 = 18 \text{ s.}$$

Speed of hands during all movements

$$\text{Speed } (S1) = \frac{418}{18} = 23.2 \text{ inches/s}$$

4.4 Calculation of Distance Traversed, Time Taken, and Speed of Hands in Improved Layout

We have incorporated various improvements in workplace layout and again calculate distance traverse, time taken and speed of hands.

Total distance traversed in new layout

$$D \text{ (improved)} = 42 + 56 + 60 + 60 + 24 + 19 = 261 \text{ inches.}$$

We already calculated the average speed of hands

$$S1 = 23.2 \text{ inches/s}$$

We determined the distance traversed by hands in new layout

$$D \text{ (improved)} = 261 \text{ inches}$$

We know the formula.

$$\text{Time} = \frac{\text{distance}}{\text{speed}} \tag{1}$$

Time taken by hands for traversing distance in new layout

$$T2av = \frac{261}{23.2} = 11.35 \text{ s}$$

4.5 Results

The above calculation clearly shows that, improved layout gives good results. It saves considerable amount of unproductive movements and time taken by hands. The results are given below.

Percentage of saving in hands movement in improved layout.

$$\text{Dsv.} = \frac{418 - 261}{418} \times 100 = 37.5\,\%$$

Percentage of saving in time taken in improved layout.

$$\text{Tsv.} = \frac{18 - 11.35}{18} \times 100 = 36.94 = 37\,\%$$

Percentage of saving in floor area of improved layout.

$$\text{Area1 (existing layout)} = 62 \times 56 = 3,472 \text{ sq. inches.}$$

$$\text{Area2 (new layout)} = 50 \times 50 = 2,500 \text{ sq. inches.}$$

$$\text{Asv.} = \frac{3472 - 2500}{3472} \times 100 = 27.99 = 28\,\%.$$

5 Suggestions and Conclusion

5.1 Suggestions to Case Organization

- Man/Machine/Material charts shows that, man was nearly 100 % busy. There is no such type of conditions appeared, when man was idle and machine doing automatic work done. This can be overcome by dividing activities between two persons. This required practical implementation in order determine amount of time savings.
- The major proportion of delay time contains wrongly filled reservation requisition form. The duration of delay time by three operators is 12.66, 20.3, and 32.33 respectively. This problem can be short out by making requisition form simpler. The specimen of present form is given below. By careful examination of reservation requisition gives us following flaws.
  - Improper fonts and size of letter.
  - One side of form contains two languages like Gujrati/Hindi and next side Marathi/English.
  - Sometimes customer forgot to fill personnel details in reservation requisition

- This can be reduced, by increasing size of form and using standard fonts, size. The railway should put one Demo filled form along with instruction, how to fill a form. They also highlight a point where customer makes frequent mistakes e.g. name and address of applicant, mobile no., and signature and form not accepted without these details. In proposed proforma we used check box more often for simplifying form. We also remove some unnecessary information's to avoid confusion. The proposed proforma given below (Fig. 3).

- The faulty layout makes uneven distribution of work between two hands. This can be reduced by proposing new layout. The comparison between existing and new layout is already explained in data analysis section.
- The height of train's information charts should place at eyelevel to customer and it will mention up/down train's number along with other information's separately. The tempering of charts will be protected by putting it behind the glass or any transparent material. This will ease customers for getting appropriate information's from charts to protect clerks to random enquiry along with delay caused due to mistakes in reservation requisition form.
- The white keyboards should be replaced by black one. The no./symbol/alphabets is in white color. This will reduced the possibility of reflection from keyboard. We can also explore the possibilities of special designed keyboards having keys which are quite often used by clerks.
- The digital display showing current status to customer should be positioned at their eyelevel and it also being replaced by TFT-LCD (Thin Film Transistor-Liquid Crystal Display) which shows each and every details filled by clerk to that screen.
- The delay time caused due insufficient change can be resolve by using card transaction.

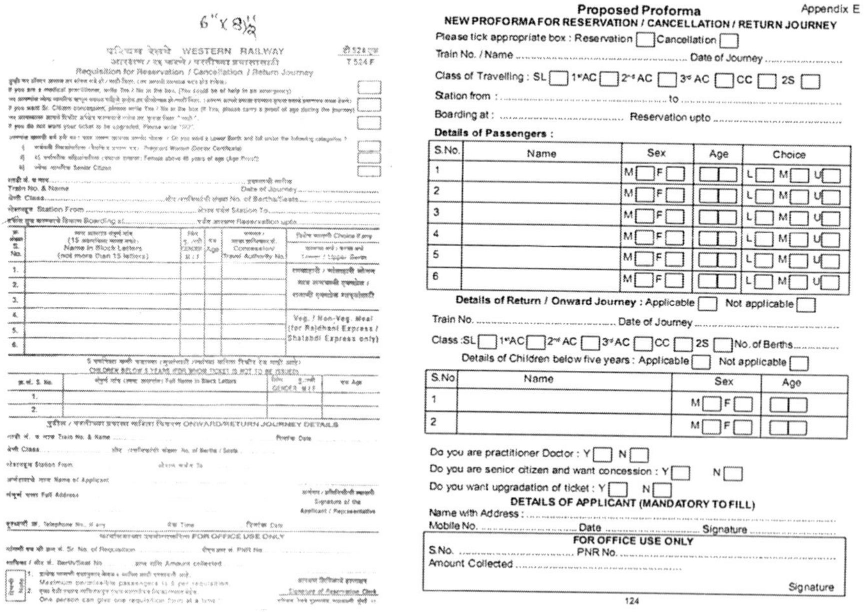

WESTERN RAILWAY

Proposed Proforma Appendix E

NEW PROFORMA FOR RESERVATION / CANCELLATION / RETURN JOURNEY

Please tick appropriate box : Reservation ☐ Cancellation ☐

Train No. / Name .... Date of Journey ....

Class of Travelling : SL ☐ 1st AC ☐ 2nd AC ☐ 3rd AC ☐ CC ☐ 2S ☐

Station from : .... to ....

Boarding at : .... Reservation upto ....

Details of Passengers :

S.No.	Name	Sex	Age	Choice
1		M ☐ F ☐	☐☐	L ☐ M ☐ U ☐
2		M ☐ F ☐	☐☐	L ☐ M ☐ U ☐
3		M ☐ F ☐	☐☐	L ☐ M ☐ U ☐
4		M ☐ F ☐	☐☐	L ☐ M ☐ U ☐
5		M ☐ F ☐	☐☐	L ☐ M ☐ U ☐
6		M ☐ F ☐	☐☐	L ☐ M ☐ U ☐

Details of Return / Onward Journey : Applicable ☐ Not applicable ☐

Train No. .... Date of Journey ....

Class :SL ☐ 1st AC ☐ 2nd AC ☐ 3rd AC ☐ CC ☐ 2S ☐ No. of Berths........

Details of Children below five years : Applicable ☐ Not applicable ☐

S.No	Name	Sex	Age
1		M ☐ F ☐	☐☐
2		M ☐ F ☐	☐☐

Do you are practitioner Doctor : Y ☐ N ☐

Do you are senior citizen and want concession : Y ☐ N ☐

Do you want upgradation of ticket : Y ☐ N ☐

DETAILS OF APPLICANT (MANDATORY TO FILL)

Name with Address : ....

Mobile No. .... Date .... Signature ....

FOR OFFICE USE ONLY

S.No. .... PNR No. ....

Amount Collected ....

Signature

124

Fig. 3 *Left* existing and *right* proposed proforma reservation requisition

5.2 Future Scope and Limitations

Each research is not a complete work in itself. There is always future scope to improve. This work is no exception. Today computer simulation makes us to derive the feasibility of tedious processes without being actual materialized along with minimal error in results. The various improvements those are suggested for layouts can be easily simulated with available software equipped with manikin module. It will help us to determine the feasibility of improvements. The computer simulation completely eliminates need of model formation to proposed layout. This simulation is beyond the scope of this paper.

The queuing analysis is initial stage of this research, which already done and beyond the scope of this paper.

The sample size taken for observing present system is small. The video recording captured 1 h. Activities of individual operators, but random arrival pattern of customers interrupted real time video recording. During the time of analysis, video slowing down to its usual speed have done with VLC player there is no special software available for this purpose.

5.3 Conclusion

Indian Railways is the backbone of the country's economy. Millions of people travel daily from one place to another. But getting reservation in trains is a tedious job through offline reservation counters. It consumes a lot of time and creates dissatisfied customers. The Railway Reservation system should be properly studied to protect losses to railway and win customer satisfaction. This project is an attempt to shed light on the present system and it also tried to find solutions to the present problems.

References

1. Neumann Patrick W, Dul jan Parast MM, Fini EH (2010) The effect of productivity and quality on profitability in US airline industry. Managing Serv Qual 20(5):458–474
2. Douglas TJ, Judge WQ Jr (2001) Total quality management implementation and competitive advantage: the role of structural control and exploration. Acad Manag J 44(1):158–169
3. Kaynak H (2003) The relationship between total quality management practices and their effects on firm performance. J Oper Manage 21(4):405–435
4. Buzzell R, Gale B (1987) The PIMS: linking strategy to performance. Free Press, New York
5. Zhao X, Yeung ACL, Lee TS (2004) Quality management and organizational context in selected service industries of China. J Oper Manage 22(6):575–587
6. Al-Hawari M, Hartley N, Ward T (2005) Measuring banks automated service quality: a confirmatory factor analysis approach. Mark Bull 16:1–19
7. Bitner MJ, Brown SW, Meuter ML (2000) Technology infusion in service encounters. J Acad Mark Sci 28(1):138–149

8. Bruke RR (2002) Technology and the customer interface: what customer want in the physical and virtual store. J Acad Mark Sci 30(4):411–432
9. Danaher PJ, Gallagher RW (1997) Modeling customer satisfaction in telecom New Zealand. Eur J Mark 31(2):122–133
10. Ganguli S, Roy SK (2010) Service quality dimensions of hybrid services. Managing Serv Qual 20(5):404–424
11. Gerald LB (2002) Auditing hospital queuing. Manag Auditing J 17(7):397–403
12. International Labour Office (2009) Introduction to work study, 3rd edn. Oxford and IBH, Geneva
13. Juga J, Juntunen J, Grant DB (2010) Service quality and its relation to satisfaction and loyalty in logistics outsourcing relationship. Managing Serv Qual 20(6):496–510
14. Charbes J, Chiapetta J (2011) HRM, ergonomics and work psychodynamics: a model and a research agenda. Humanomics 27(1):53–60
15. Keith AW, Benjamin TBC, Strenger M (2010) Achieving wait time reduction in the emergency department. Leadersh Health Serv 23(4):304–319
16. Kotler P (2010) Marketing management, 13th edn. Pearson Education Inc., pp 336–351
17. McDonnell J (2007) Music, scent and time preferences for waiting lines. Int J Bank Mark 25(4):223–237
18. Mittal KM (2010) Queuing analysis for outpatient and inpatient services: a case study. Manag Decis 48(3):419–439
19. Roger B (1998) Queues, customer characteristics and policies for managing waiting lines in supermarkets. Int J Retail Distrib Manage 26(2):78–87
20. Seth N, Desmukh SG (2005) Service quality model: a review. Int J Qual Reliab Manage 55(9):913–949
21. Seth N, Desmukh SG, Vrat P (2006) A conceptual model for quality of service in the supply chain. Int J Phys Distrib Logistics Manage 36(6):547–575
22. Towill DR (2009) Frank Gilberth and health care delivery method study driven learning. Int J Health Care Qual Assur 22(4):417–440
23. Yeung ACL, Cheng TC, Lai K (2006) An operational and institutional perspective on total quality management. Prod Oper Manage 15(1):156–170
24. Zeithaml VA, Berry LL, Parasuraman A (1996) The behavioral consequences of service quality. J Mark 60(2):31–46

Simulation Study of Servo Control of Pneumatic Positioning System Using Fuzzy PD Controller

D. Saravanakumar and B. Mohan

Abstract Servo pneumatic positioning system is a mechatronic approach that enables to use Pneumatic Cylinders as multi-position actuators. This technology can replace less efficient electromechanical systems in many applications. The nonlinear servo control for the pneumatic system is very important because the pneumatic system exhibits nonlinearities such as valve dead zone, compressibility of air, frictional changes, etc. The mathematical model of the servo pneumatic positioning system is derived for various physical laws and equations. The main parameters that are modeled are pressure dynamics, valve motion dynamics, mass flow rate, friction variation, and dead zone. This model is used to simulate the system behavior in Matlab-Simulink software. The paper also presents a Fuzzy PD controller design for the nonlinear servo positioning system. The designed controller performance is validated using simulation tests. From the results, it is observed that the developed controller exhibits good servo characteristics.

Keywords Mathematical modeling · Simulation · Fluid power systems · Mechatronics

1 Introduction

Pneumatic actuators are widely used in the field of automation, robots, and manufacturing. Traditionally, pneumatic actuators were used for motion between two hard stops. It is very useful for the manipulation and rapid motion of mechanical

---

D. Saravanakumar (✉)
Department of Production Technology, Madras Institute of Technology,
Anna University, Chennai, India
e-mail: saravanapoy@gmail.com

B. Mohan
Department of Mechanical Engineering, College of Engineering,
Anna University, Chennai, India
e-mail: mohan@mitindia.edu

S. Sathiyamoorthy et al. (eds.), *Emerging Trends in Science,*
Engineering and Technology, Lecture Notes in Mechanical Engineering,
DOI: 10.1007/978-81-322-1007-8_38, © Springer India 2012

objects and also in assembly, monitoring, packing, stacking, clamping, and fixing of various manufacturing products.

The advantages of Pneumatic drives are the attainment of high speeds in linear motion, without additional mechanical transmissions, considerable force development and prolonged intense operation without risk of overheating. Other advantages include high ceiling, high mechanical efficiency, long working life, and a broad working temperature range. They are also characterized by low operating costs, simple storage of the compressed air energy, and compatibility with portable power sources. Of comparable practical importance is the ability of pneumatic drives to operate in wet, dusty, and chemically aggressive atmospheres that pose the risk of fire or explosion, and in the presence of radiation and electromagnetic fields, as well as mechanical vibrations.

In order to expand the capabilities of equipment based on pneumatic drives, the trend is to employ electro pneumatic servo drives, characterized by higher speed and precision. One significant problem is that such drives are nonlinear: the pressure within the pneumatic cylinder, the frictional force, and the compressed air flow rates through the chokes of the pneumatic drive all vary in nonlinear fashion. The dead zone and time delay characteristics of the servo valve also adds to the complexity for designing controller for the system.

Mathematical modeling of the system becomes very important for development of controller for the system and also to optimize the parameters in the pneumatic system. Jun et al. in [1] has presented a model for dynamics of air. Gulati and Barth in [2] have presented a linear model for the servo pneumatic system. Lin-Chen et al. in [3] developed model for individual elements in the servo pneumatic system. Le et al. in [4] and Jin-feng et al. in [5] has developed a model for mass flow rate equation inside a valve with switching characteristics. Xin-too et al. in [6] formulated a model for rotary pneumatic actuator. Andrighetto et al. in [7] have developed a model for Dead zone in the servo valve. Kazerooni in [8] formulated a model for force created by the cylinder. Najafi et al. in [9] has developed a model for pneumatic cylinder with cushioning. Ke et al. in [10] has developed a model for the entire system based on the response of the system. Saleem et al. in [11] has presented an simulation environment for pneumatic systems. Rao et al. in [12] has developed a non-linear model for the pneumatic servo system based on the system response. Hildebrant et al. in [13] has formulated a dynamic relationship for the pressure in a chamber. Andrighetto et al. in [14] have developed a model for frictional forces inside the cylinder. No literature presents an exact model for servo pneumatic positioning system which is required for development of an accurate controller for the system. So a nonlinear mathematical model of the servo pneumatic positioning system is presented in this paper.

Development of accurate and fast controller for highly nonlinear system is very difficult using conventional controllers. Fuzzy Logic provides a solution for the control challenge as it works on knowledge base as manual control. Tain in [15] and Llangostera et al. in [16] has developed fuzzy based controller for the Pneumatic actuators. Chen et al. in [17] has described design of Adaptive fuzzy controller for the servo pneumatic process. Rezoney et al. in [18] has developed fuzzy based controller for robot manipulator using artificial pneumatic muscles. Takosoglu et al. in [19] has developed fuzzy based PD controller for servo pneumatic system and validated the performance of fuzzy controller. The paper also presents design of fuzzy PD controller for the servo pneumatic positioning system.

2 System Description

A schematic diagram of the pneumatic servo positioning system is shown in Fig. 1. The main components in the system are Pneumatic Cylinder, Proportional Servo Valve, Position Transducer, and an electronic control unit. The position of the pneumatic cylinder is controlled by the Proportional Servo valve which regulates the flow rate of the compressed air. The position transducer senses the current position value which is used by the controller to control the spool movement in the Proportional Servo Directional Control Valve.

The controller of pneumatic servo-system processes the position error $e(t) = y_o(t) - y(t)$ into voltage signal u(t) in the solenoid of proportional directional control valve controlling the slide of the rodless cylinder, where $y_o(t)$ is the input signal, and y(t) the signal generated by displacement transducer of the rodless cylinder's linear slide of $x(t)$ coordinate.

3 Mathematical Modeling

The systematic methodology of the pneumatic actuator nonlinear mathematical modeling is presented from applying physical laws and recent literature information. The full system constitutes a fifth order nonlinear dynamic model of the

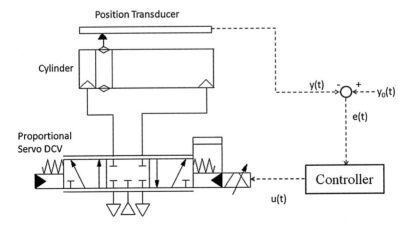

Fig. 1 Schematic of pneumatic servo positioning system

pneumatic positioning system and considers the nonlinearity of the dead zone, the mass flow rate, the pressure dynamics, and the motion equation, that includes the friction dynamics. The model given in the paper is based on [19].

In the modeling of the system, the following assumptions are made [1, 4, 12]:

- The gas is perfect.
- The pressures and temperature within each chamber are homogeneous.
- Kinetic and potential energy terms are negligible.
- Tube length can be ignored when air supply is very close to valve and cylinder.
- Valve dynamics is sufficiently faster than mechanical systems dynamics.
- Efficient area of valve will be used as the control input.

The equation of motion for the cylinder considering the friction effect is given in (1).

$$\ddot{x} = \frac{1}{M+m} \left[A(P_1 - P_2) - F_{\text{fric}} + mg \sin \theta \right] \tag{1}$$

where x is the piston position, $\dot{x}$ or v is velocity, $\ddot{x}$ is acceleration, M is internal mass of the cylinder and slide, m is external load, A is internal area of the cylinder chamber, P_1 and P_2 are the absolute pressure inside cylinder chambers, F_{fric} is the frictional force, g is acceleration due to gravity and θ is inflation angle of the piston.

The equation of motion for the spool of the proportional valve is given in (2).

$$\ddot{x}_r = \frac{1}{m_r} \left[\frac{k_m}{R_c} (u - nBl_c v_r) \left(1 - e^{\frac{R_c}{L}t} \right) - f_l v_r - k_s x_r \right] \tag{2}$$

where x_r and v_r are position and velocity of the valve spool, m_r is mass of the spool, k_m is solenoid coefficient, R_c is resistance of solenoid, u is the coil voltage, n is number of coil, B is the Magnetic field density, l_c is length of the coil, L is the inductance of the coil, f_l is coefficient of viscous friction, k_s is spool spring rate.

The equations pressure in the cylinder chambers is given in (3) and (4).

$$\dot{P}_1 = \frac{\kappa}{A(l_0 + x)} \left\{ R\dot{m}_1 - P_1 Av - \frac{\kappa - 1}{\kappa} \alpha \left[A_{10}(T_1 - T_a) + A(T_1 - T_2) \right] \right\} \tag{3}$$

$$\dot{P}_2 = \frac{\kappa}{A(l + l_0 - x)} \left\{ -R\dot{m}_2 + P_2 Av - \frac{\kappa - 1}{\kappa} \alpha \left[A_{20}(T_2 - T_a) + A(T_2 - T_1) \right] \right\} \tag{4}$$

where κ is a diabatic exponent, l_0 is length of dead zone of cylinder, l is full stroke length of the cylinder, R is specific gas constant, α is overall heat transfer coefficient, T_1 and T_2 are temperatures in cylinder chambers, T_a is ambient temperature, A_{10} and A_{20} are heat transfer surfaces, $\dot{m}_1$ and $\dot{m}_2$ are the mass flow rates.

The equations of mass flow rates through proportional control valve is given in (5) and (6).

$$\dot{m}_1 = \partial_0 x_r \sqrt{T_0} \left[\frac{C_{14} P_s w_{14}}{\sqrt{T_s}} - \frac{C_{45} P_1 w_{45}}{\sqrt{T_1}} \right] \tag{5}$$

$$\dot{m}_2 = \partial_0 x_r \sqrt{T_0} \left[\frac{C_{23} P_2 w_{23}}{\sqrt{T_2}} - \frac{C_{12} P_s w_{12}}{\sqrt{T_s}} \right] \tag{6}$$

where ∂_0 is air density, T_0 is ambient air temperature, P_s is supply air pressure, T_s is supply air temperature, C_{14}, C_{45}, C_{23} and C_{12} are sonic conductance consistent with the standard ISO 6358-1989 for critical pressure ratio, w_{14}, w_{45}, w_{23} and w_{12} are nonlinear flow function (sonic flow and subsonic flow) depending on the pressure ratio and on the critical pressure ratio, given by,

$$w_{14} = \begin{cases} 1 & \text{if } 0 \le \frac{P_1}{P_s} \le b_{12}, \\ \sqrt{1 - \left(\frac{\frac{P_1}{P_s} - b_{14}}{1 - b_{14}} \right)^2} & \text{if } b_{12} \le \frac{P_1}{P_s} \le 1 \end{cases} \tag{7}$$

$$w_{45} = \begin{cases} 1 & \text{if } 0 \le \frac{P_a}{P_2} \le b_{45}, \\ \sqrt{1 - \left(\frac{\frac{P_a}{P_2} - b_{45}}{1 - b_{45}} \right)^2} & \text{if } b_{45} \le \frac{P_a}{P_2} \le 1 \end{cases} \tag{8}$$

$$w_{12} = \begin{cases} 1 & \text{if } 0 \le \frac{P_1}{P_s} \le b_{12}, \\ \sqrt{1 - \left(\frac{\frac{P_1}{P_s} - b_{12}}{1 - b_{12}} \right)^2} & \text{if } b_{12} \le \frac{P_1}{P_s} \le 1 \end{cases} \tag{9}$$

$$w_{23} = \begin{cases} 1 & \text{if } 0 \le \frac{P_a}{P_2} \le b_{23}, \\ \sqrt{1 - \left(\frac{\frac{P_a}{P_2} - b_{23}}{1 - b_{23}} \right)^2} & \text{if } b_{23} \le \frac{P_a}{P_2} \le 1 \end{cases} \tag{10}$$

where P_a is atmosphere pressure and b_{14}, b_{45}, b_{12}, b_{23} are critical pressure ratios.

4 Fuzzy PD Controller Design

For accurate and fast control of the highly nonlinear servo pneumatic system, discrete fuzzy PD controller is designed. The controller has two inputs $e(k)$—position error and $\Delta e(k)$—change in position error and one output $u(k)$—voltage of the coil. The fuzzy PD controller is designed using fuzzy logic toolbox of Matlab-Simulink software. The Fuzzy controller works on the knowledge base containing IF–THEN rules for undetermined predicates and fuzzy control mechanism.

The error in the range of −2 to 2 and change in error in range of −0.2 to 0.2 are the inputs to the fuzzy controller. The output from the controller is in the range 0–5. All the three variables (error, change in error, output) has five triangular

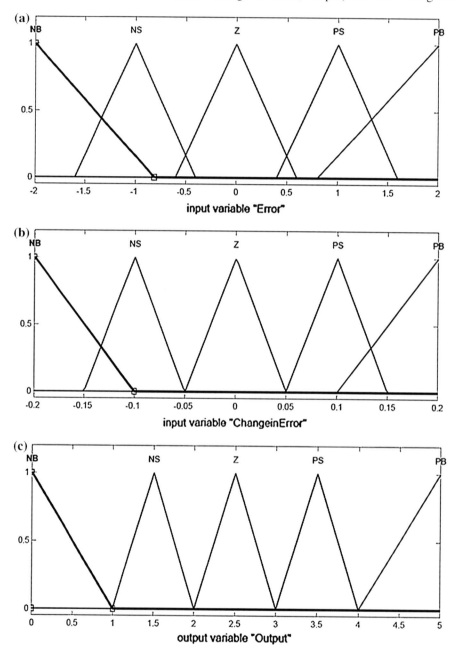

Fig. 2 Membership functions of variables error, change in error, and output

membership functions NB, NS, Z, PS, and PB. There range of all the membership functions of variables are shown in Fig. 2.

The fuzzy rule base consists of IF–THEN rules. The membership function for output is decided based on the rule base created which is shown in Table 1. The output control surface of the created fuzzy controller is shown in Fig. 3. It shows the three dimensional plot between input and output variables.

5 Simulation and Results

The simulation model of the servo pneumatic positioning system is created based on (1)–(10) using Matlab-Simulink software. The model of the system is shown in Fig. 4.

The simulation is conducted using the following values for the parameters.

- Pneumatic Unit: $p_z = 0.6$ MPa, $T_z = 298$ K.
- Normal air: $T_0 = 298$ K, $p_0 = 0.6$ MPa, $\rho 0 = 1.225$ kg/m^3, $R = 288$ Nm/KgK, $\varepsilon = 1.4$.

Table 1 Rule base for fuzzy PD controller

		Change in Error				
		NB	NS	Z	PS	PB
Error	NB	NB	NB	NB	NB	NB
	NS	NB	NS	NS	NS	PS
	Z	NB	NS	Z	PS	PB
	PS	NS	PS	PS	PB	PB
	PB	PB	PB	PB	PB	PB

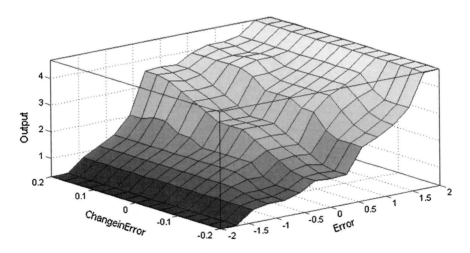

Fig. 3 Output control surface of fuzzy PD controller

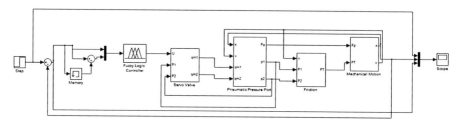

Fig. 4 Simulink block diagram model of servo pneumatic positioning system using fuzzy PD controller

- Proportional Directional Control valve: $C_{14} = C_{12} = C_{23} = C_{45} = 1.462 \times 10^{-8}$ m^{4}s/Kg, $b_{14} = b_{12} = b_{23} = b_{45} = 0.28$.
- Pneumatic Rodless Cylinder: $A = 49 \times 10^{-5}$ m^2, $l_0 = 0.02$ m, $l = 0.2$ m, $M = 1$ kg, $m = 3$ kg, $f_t = 250$ Ns/m, $k_p = 3$ N/Pa, $F_k = 100$ N, $F_{pr} = 200$ N, $v_k = 0.1$ m/s.

The Simulation of the servo pneumatic positioning system is conducted mainly to check the operation of the designed fuzzy PD controller in various system conditions. The developed controller must be tested for both disturbance avoidance and setpoint tracking performances. Step signal is used to test the disturbance avoidance property and Ramp signal is used to test the setpoint tracking property. Figure 5a shows position or displacement and position error obtained during a step

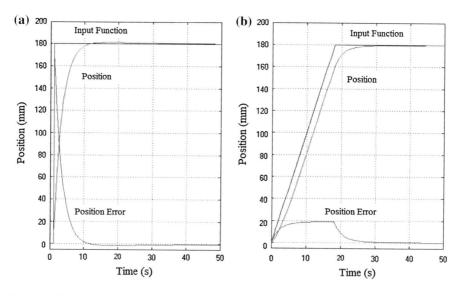

Fig. 5 a Simulation results of position and position error for step change in input signal from 0 to 180 mm at time = 1 s. **b** Simulation results of position and position error obtained when a ramp input signal from 0 to 180 mm with velocity 10 mm/s at time = 0 s

Simulation Study of Servo Control

Table 2 Performance indexes for fuzzy PD controller

	Settling time (s)	Overshoot (mm)	IAE	ISE	ITAE
Step input	28	1.37	9207.8	1614022.73	93208.2
Ramp input	32	0	567.76	6027.8	5328.7

change in input signal from 0 to 180 mm at time = 1s. Figure 5b shows position or displacement and position error obtained when a ramp input signal from 0 to 180 mm with velocity 10 mm/s at time = 0 s.

The Performance of the Fuzzy PD controller for the servo pneumatic positioning system is checked by means of standard performance index such as settling time, overshoot, positioning error, integral absolute error (IAE), integral time absolute error (ITAE) and integral square error (ISE). Table 2 shows the performance indexes of Fuzzy PD controller for step and ramp input signals.

6 Conclusion

The paper presents a nonlinear mathematical model for the servo pneumatic positioning system. The developed model includes the effect of friction, dead zone, mass flow rate variations, and pressure dynamics. The simulation model of the system is developed using Matlab-Simulink software. The paper also presents a nonlinear Fuzzy PD controller designed for fast and accurate control of the positioning system. The designed controller is tested for both disturbance avoidance and setpoint tracking capabilities. The performance indices and the response plots show that the developed controller has better servo performance.

References

1. Jun J, Kanako K, Kawamura S (2006) A cascaded feedback control for trajectory tracking of pneumatic drive systems. In: Proceedings of the SICE-ICASE international joint conference, pp 441–446
2. Gulati N, Barth EJ (2009) A globally stable, load-independent pressure observer for servo control of pneumatic actuators. IEEE/ASME Trans on Mechatron 14(3):295–306
3. Lin-Chen YY, Wang J, Wu QH (2003) A software tool for pneumatic actuator system simulation and design. J Comput Ind 51:73–88
4. Le MQ, Pham MT, Moreau R, Redarce T (2010) Comparison of a PWM and a hybrid force control for a pneumatic actuator using on/off solenoid valves. In: Proceedings of the international conference on advanced intelligent mechatronics, pp 1146–1151
5. Jin-ferg S, Xiao-xian Y, Ling-cong N (2010) Double-acting cylinder PWM pneumatic servo systems analysis method. In: Proceedings of the international conference on intelligent computation technology and automation, pp 153–156
6. Xin-tao M, Qing-jun Y, Jin-jun W, Gang B (2009) Control strategy for pneumatic rotary position servo systems based on feed forward compensation pole-placement self-tuning method. J Cent South univ Technol 16:608–613

7. Andrighetto PL, Valdiero AC, Bavaresco D (2008) Dead zone compensation in pneumatic servo systems. In: Proceedings of the ABCM Symposium series in mechatronics, pp 501–509
8. Kazerooni H (2005) Design and analysis of pneumatic force generators for mobile robotic systems. IEEE/ASME Trans Mechatron 10(4):411–418
9. Najafi F, Fathi M, Saadat M (2009) Dynamic modeling of servo pneumatic actuators with cushioning. Int J Adv Manuf Technol 42:757–765
10. Ke J, Wang J, Jia N, Yang L, Wu QH (2005) Energy efficient analysis and optimal control of servo pneumatic cylinders. In: Proceedings of the IEEE conference on control applications, pp 541–546
11. Saleem A, Wong CB, Pu J, Moore PR (2009) Mixed-reality environment for frictional parameters identification in servo-pneumatic system. J Simul Model Pract Theory 17:1575–1586
12. Rao Z, Bone GM (2008) Nonlinear modeling and control of servo pneumatic actuators. IEEE Trans Control Syst Technol 16(3):561–569
13. Hildebrandt A, Kharitonov A, Sawodny O, Gottert M, Hartmann A (2005) On the zero dynamic of servo pneumatic actuators and its usage for trajectory planning and control. In: Proceedings of the international conference on mechatronics and automation, pp 1241–1246
14. Andrighetto PL, Valdiero AC, Carlooto L (2006) Study of the friction behaviour in industrial pneumatic actuators. In: Proceedings of the ABCM Symposium series in mechatronics, pp 369–376
15. Tian J (2010) A system of position control for cylinder based on fuzzy control. In: Proceedings of the sixth international conference on natural computing, pp 52–56
16. Llangostera HM, Jimenez SH (2000) Control of a pneumatic servo system using fuzzy logic. In: Proceedings of the 1st FPNI-PhD Symposium, pp 189–201
17. Chen C, Chen P, Chen C (1993) A Pnneumatic model-following control system using a fuzzy adaptive controller. Automatica 29(4):1101–1105
18. Rezoney A, Boudowa S, Hamerlain F (2009) Fuzzy logic control for manipulated robot actuated by pneumatic artificial muscle. J Electr Syst (01):1-6
19. Takosoglu JE, Dindorf RF, Lashi PA (2009) Rapid prototyping of fuzzy controller pneumatic servo system. Int J Adv Manuf Tech 40:349–361

Part V
Propulsion Systems

Studies on Starting Thrust Oscillations in Dual-Thrust Solid Propellant Rocket Motors

S. Deepthi, S. K. Kumaresh, D. Aravind Kumar, J. Darshan Kumar and M. Arun

Abstract In this paper numerical studies have been carried out to examine the starting transient flow features of dual-thrust motors (DTMs) during the pre-ignition chamber conditions. Numerical computations have been carried out with the aid of a standard k-ω turbulence model. We concluded from the numerical results that in DTMs with highly loaded propellants the hot igniter gases can create pre-ignition thrust oscillations due to flow unsteadiness and recirculation. Under these conditions the convective flux to the surface of the propellant will be enhanced, which will create multiple-flame fronts leading to undesirable start-up transient. We also concluded that the prudent selection of the port geometry, without altering the propellant loading density, for damping the total temperature fluctuations within the motor is a meaningful objective for the suppression and control of instability and/or pressure/thrust oscillations often observed in solid propellant rockets with non-uniform port geometry.

Keywords Solid rocket motor · Starting transient · Internal flow choking · Thrust oscillations · Non-uniform port geometry

1 Introduction

The art of igniting practical rocket motors as it has evolved over the years may have been meeting the needs of this technology, essentially in the prediction of starting transient phase of operation of solid propellant rocket motors with

---

S. Deepthi (✉) · S. K. Kumaresh · D. Aravind Kumar · J. Darshan Kumar
Department of Aeronautical Engineering, Kumaraguru College of Technology, Coimbatore 641049, Tamil Nadu, India
e-mail: deepthi_shan2006@yahoo.co.in

M. Arun
Department of Mechanical Engineering, Kumaraguru College of Technology, Coimbatore 641049, Tamil Nadu, India

S. Sathiyamoorthy et al. (eds.), *Emerging Trends in Science, Engineering and Technology*, Lecture Notes in Mechanical Engineering, DOI: 10.1007/978-81-322-1007-8_39, © Springer India 2012

uniform port geometry, but the underlying physical processes remain undefined and poorly understood particularly in the case of dual-thrust motors (DTMs). This research topic, although interesting in its own right, has been motivated by several practical problems. The developments of large and more sophisticated solid propellant rocket motors, like PSLV third stage motor of ISRO, Titan, and Space shuttle's solid rocket motors of NASA, have emphasized the deficiencies on starting/ignition transient prediction [1]. These rocket motors do not lend themselves to the costly trial and error development techniques, and the radical differences in the size and design of these rocket motors defy extrapolation of the empirical knowledge gained in the development of previous, more conventional rocket motors.

Although many studies have been reported on modeling of DTMs starting transient, the simulation of the internal flow features of high-performance DTMs are still an active research topic [1–15] Of late Sanal Kumar et al. [1], reported the phenomena of internal flow choking in DTMs during the starting transient phase with different physical origin, which has received considerable attention in the scientific community. The motivation for the present study emanates from the desire to demonstrate the variations of starting transient flow features of DTMs due to its difference in grain orientation (i.e., horizontal and flip horizontal position). In this study two different high-velocity transient DTMs are selected, one with narrow head-end port and the other with wide head-end port but with same propellant loading density and grain configuration. It is known that the accurate evaluation of the internal flow features is inevitable for the exact starting transient prediction of dual-thrust motors. Note that in a rocket motor the mass flux is the largest at the throat and making the port size comparable to the throat cross section, the flux will be comparable to that at the throat. Under these conditions, the convective flux to the surface of the propellant is enhanced. The enhanced mass flux implies increase in the local Reynolds number of the flow. This implies a reduction in heat transfer film thickness and enhanced heat transfer to the propellant with consequent enhancement in the burn rate.

This paper addresses the preliminary design challenges associated with development of high-performance DTMs because of its large size, high length to diameter ratio, and complex geometry demanding thrust-time trace shape requirements. The DTMs with highly loaded propellants are selected for parametric analytical studies to examine the pre-ignition thrust oscillations due to high temperature igniter mass flow from non-uniform port geometry. Under these conditions flow unsteadiness and recirculation can occur and as a result the convective flux to the surface of the propellant will be enhanced, which will create multiple-flame fronts leading to undesirable start-up transient. Predicting transient unsteady flow in any DTM is a daunting task. In this paper an attempt has been made to predict the internal flow features during the pre-ignition chamber condition to examine the starting thrust oscillations for meeting all performance objectives while accommodating program development uncertainties.

2 Numerical Methodology

Numerical simulations have been carried out with the help of a two-dimensional standard k-omega model. This turbulence model is an empirical model based on model transport equations for the turbulence kinetic energy and a specific dissipation rate. This code solves standard k-omega turbulence equations with shear flow corrections using a coupled second order implicit unsteady formulation. In the numerical study, a fully implicit finite volume scheme of the compressible, Reynolds-Averaged, Navier-Stokes equations is employed. Compared to other available models this model could well predict the turbulence transition in duct flows and has been validated through benchmark solutions. Therefore, this model has been used for demonstrating the flow physics of DTMs. Compressible flows are typically characterized by the total pressure P_o and total temperature T_o of the flow. The viscosity is determined from the Sutherland formula.

Initial wall temperature, inlet total pressure, and temperature are specified. In the present compressible flow calculations, isentropic relations for an ideal gas are applied to relate total pressure, static pressure, and velocity at a pressure inlet boundary. Note that the motor exit geometry (nozzle) considered in this study is a short straight duct followed by the convergent duct. The Courant-Friedrichs-Lewy number is initially chosen as 3.0 in all of the computations. An algebraic grid-generation technique is employed to discretize the computational domain. A typical grid system in the computational region is selected after the detailed grid refinement exercises. The grids are clustered near the solid walls using suitable stretching functions. The motor geometric variables and material properties are known a priori. The grid systems in the computational region are shown in Figs. 1 and 2.

3 Results and Discussion

In this study high-performance dual-thrust solid propellant rocket motors with convergent and divergent port geometries are considered. The geometry and the baseline values are selected based on a typical solid propellant rocket motor. Initial and boundary conditions are the same in both cases.

Fig. 1 Grid system in dual-thrust motor with divergent port

Fig. 2 Grid system in dual-thrust motor with convergent port

Figure 3 shows the contours of velocity magnitude for the divergent and convergent port cases respectively. We observed that when time advances the flow field significantly altered in both cases. The recirculation bubble is discernable in the case of divergent port case and it further expands downstream when time advances. We have seen that at the identical inflow conditions the magnitude of the velocity fluctuation is more prominent for the rocket motor with divergent port than with the convergent port. Figure 4 shows the comparison of the axial temperature history of two different DTMs. It is evident from this figure that the steady state temperature will attain rapidly in the case of DTM with convergent port than with divergent port. We also observed that the temperature fluctuation is more prominent for divergent port case due to the flow unsteadiness associated with separation and reattachment. Figure 5 shows the formation of recirculation bubbles at the transition region of a DTM with divergent port. Note that the downstream of the separation point, the velocity u close to the surface becomes negative and so a region of reverse flow is established. Because of their ability to transfer

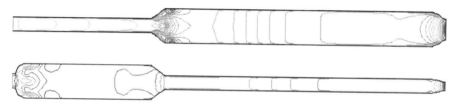

Fig. 3 Comparison of the contours of velocity magnitudes of DTMs, with horizontal and flip horizontal positions, at the same interval of time

Fig. 4 Comparison of the axial total temperature history of DTMs with convergent and divergent ports

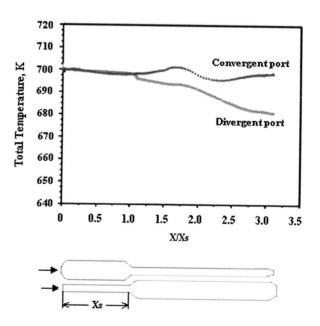

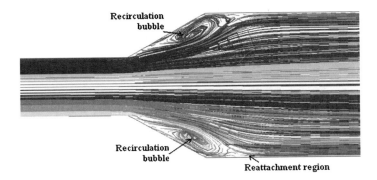

Fig. 5 Demonstrating the formation of recirculation bubbles, in enlarged view, at the transition region of a dual-thrust solid propellant rocket motor with divergent port

momentum laterally, turbulent flows are more able than laminar flows to negotiate regions of adverse pressure gradients.

Whether or not separation actually takes place, the general effect of the adverse pressure gradient is to give rise to a localized region of slow moving fluid stretching away from the wall. Because of the continuity condition, which can be applied over the whole cross-sectional area, the axial flow velocities must necessarily increase elsewhere in order to compensate for this effect. There is therefore a tendency for flows to become increasingly non-uniform whenever positive axial pressure gradients are encountered. As a result Mach number value will be higher at the transition region with low local pressure and relatively higher temperature. Later due to flow separation Mach number will be decreased, pressure will be increased without any significant reduction in temperature. It appears that temperature profile started decline after the pressure becomes steady state. These features can be discerned from Fig. 6.

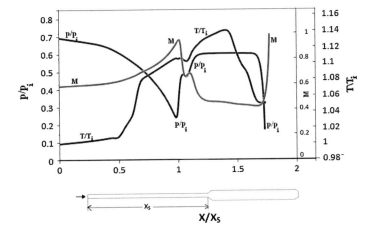

Fig. 6 Demonstrating the flow features, like pressure, temperature and Mach number, in non-dimensional form along the axis of a DTM with divergent port

Figure 7 shows the thrust transient at different axial locations of a DTM with divergent port. It can be seen from this figure that at the nozzle-end ($x/x_s = 1.7$) at the beginning slight thrust oscillation is noticed probably because of the unsteady transient flow. It is also observed that thrust differential is relatively large while comparing head-end ($x/x_s = 0$) and nozzle-end of the DTM. This will demand more powerful vectoring technique of the entire vehicle.

We have conjectured from these studies that the DTM designer can ascertain that there is a possibility of more thrust oscillations in motors with divergent port than with convergent port geometry. Although such oscillations do not peril the launcher operation, they induce some penalties to the overall performance. Hence, in the mission point of view, to avoid the unpredictable performance variation one has to recast the intrinsic shape of the grain geometry at the transition region. The suppression or the control of such oscillations is then a meaningful objective for any solid rocket motor designer. While comparing the characteristic curves, unlike the divergent case, the total temperature variation for the convergent port case is found amplifying along the axis during the initial period like a sinusoidal wave. Further the magnitude of the total temperature got diminished and reached the steady state value faster than the divergent port case presumably for readjusting it with the corresponding low velocity and pressure values, which has altered due to the port geometry. It was seen that the igniter temperature fluctuations are diminished rapidly in the case of convergent port case irrespective of the igniter total pressure. Note that the total temperature has to be readjusted with the corresponding velocity and the pressure values within the motor, which has altered due to the sudden variation of the port geometry. We conjectured that in DTMs with highly loaded propellants the mass flux of the hot gases moving past the burning surface is large. Under these conditions, the convective flux to the surface of the propellant will be enhanced, which in turn enhance the local Reynolds number.

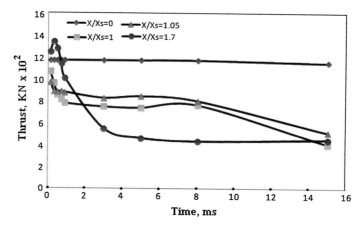

Fig. 7 Demonstrating the thrust transient of a DTM with divergent port

4 Concluding Remarks

Compressibility effects are encountered in gas flows at high velocity and/or in which there are large pressure variations. We observed that the igniter temperature fluctuations will be diminished rapidly and will reach the steady state value faster in the case of DTMs with convergent port than the divergent port irrespective of the igniter total pressure. From these studies one can deduce that the thrust/pressure oscillations, pressure-rise rate and the unexpected peak pressure often observed in solid rockets with non-uniform ports are presumably contributed due to the joint effects of the geometry dependent driving forces, transient burning and the chamber gas dynamics forces. We concluded that the prudent selection of the port geometry, without altering the propellant loading density, for damping the total temperature fluctuation within the motor is a meaningful objective for the suppression and control of instability and/or pressure/thrust oscillations often observed in DTMs with non-uniform port. This paper is a pointer towards meeting the high-performance rocket motors design challenges without altering the mission demanding thrust-time trace shape requirements.

Acknowledgments The authors thank Professor V.R. Sanal Kumar, Senior Member American Institute of Aeronautics and Astronautics (AIAA) and Editor International Journal of Aerospace Engineering for reviewing our work and further giving valuable guidance for improving the quality of this paper. The authors also thank the management of Kumaraguru College of Technology for extending their support for the successful completion of this paper.

References

1. Sanal Kumar VR, Raghunandan BN, Kawakami T, Kim HD, Setoguchi T, Raghunathan S (2008) Boundary layer effects on internal flow choking in dual-thrust solid rocket motors. J Propul Power 14(2)
2. Raghunandan BN, Sanal Kumar VR, Unnikrishnan C, Sanjeev C (2001) Flame spread with sudden expansions of ports of solid rockets. J Propul Power 17(1)
3. Peretz A, Kuo KK, Caveny LH, Summerfield M (1973) Starting transient of solid propellant rocket motors with high internal gas velocities. AIAA J. 11(12):1719–1729
4. Sanal Kumar VR, Raghunandan BN, Kim HD, Sameen A, Setoguchi T, Raghunathan S (2006) Starting transient flow phenomena in inert simulators of SRMs with divergent ports. AIAA J Propul Power 22(5):1138–1141
5. Sanal Kumar VR, Raghunandan BN, Kim HD, Sameen A, Setoguchi T, Raghunathan S (2006) Studies on internal flow choking in dual-thrust motors. AIAA J Spacecraft Rockets 43(5):1139–1143
6. Sanal Kumar VR, Kim HD, Raghunandan BN, Sameen A, Setoguchi T, Raghunathan S (2006) Fluid-throat induced shock waves during the ignition transient of solid rockets. AIAA J Spacecraft Rockets 43(1)
7. Raghunandan BN, Madhavan NS, Sanjeev C, Sanal Kumar VR (1996) Studies on flame spread with sudden expansions of ports of solid propellant rockets under elevated pressure. Defence Sci J 46(5):417–423
8. Sanal Kumar VR, Kim HD, Raghunandan BN, Setoguchi T, Raghunathan S (2005) Internal flow simulation of high-performance solid rockets using a k-ω turbulence model. Int J Therm Fluid Sci 14(2)

9. Ikawa H, Laspesa FS (1985) Ignition/duct overpressure induced by space shuttle solid rocket motor ignition. J Spacecraft Rockets 22:481
10. Salita M (2001) Modern SRM ignition transient modeling (part 1): introduction and physical models. AIAA paper, No AIAA-2001-3443
11. Alestra S, Terrasse I, Troclet B (2002) Identification of acoustic sources at launch vehicle lift-off using an inverse method. In: AIAA aerospace sciences meeting and exhibit, 40th, Reno, NV, AIAA-2002-926, 14–17 Jan 2002
12. Kumar M, Kuo KK (1984) Flame spreading and overall ignition transient. Prog Astronaut Aeronaut 90:305–360
13. Sanal Kumar VR (2003) Thermoviscoelastic characterization of a composite solid propellant using tubular test. J Propul Power 19(3):397–404
14. Sabnis JS, Gibeling HJ, McDonald H (1989) Navier-Stokes analysis of solid propellant rocket motor internal flows. J Propul Power 5:657–664
15. Blomshield FS, Mathes HB (1993) Pressure oscillations in post-challenger space shuttle redesigned solid rocket motors. J Propul Power 9:217–221

Vibration Analysis of a Constant Speed and Constant Pitch Wind Turbine

T. Sunder Selwyn and R. Kesavan

Abstract In order to increase the reliability of the rotating machineries like rotor, gear box, high speed shaft, low speed shaft, generator, and yaw of wind turbine (WT), it is important to monitor the vibration level. The key aim of the paper is to study the vibration characteristics experienced in the 400 kW WT at different load and operating conditions. The experimental vibration analysis of WT components is done and vibration characteristic curves are generated with the help of data acquisition software. These vibration characteristic curves are studied and observations made from them for their severity and effects are analyzed. Some indispensable recommendations have been given for vibration control.

1 Introduction

While fossil fuels are the main fuels for thermal power, there is fear that they will get exhausted eventually in this century. Therefore, we are in need to harvest any other renewable resources. In the continuous search of clean, safe, and renewable energy sources, wind power has emerged as one of the most attractive solutions. Wind turbine (WT) offers a cost-effective alternate renewal energy source. Modern WTs are of slender and elastic construction, above all the rotor blades and the tower. They are, therefore, structures which are extremely prone to vibration. In addition, there is no lack of excitations as the discussion of cyclically alternating rotor forces has shown. These forces can excite certain subsystems or even the entire turbine to vibrate dangerously. The vibration of WT components such as gearbox, generator, nacelle, high speed shaft, low speed shaft, tower, and nacelle

---

T. Sunder Selwyn (✉) · R. Kesavan
Deptartment of Production Technology, Madras Institute of Technology, Chennai, India
e-mail: sunder.selwyn@gmail.com

R. Kesavan
e-mail: kesavan@mitindia.edu

S. Sathiyamoorthy et al. (eds.), *Emerging Trends in Science,*
Engineering and Technology, Lecture Notes in Mechanical Engineering,
DOI: 10.1007/978-81-322-1007-8_40, © Springer India 2012

affect the dynamic stability of the complete system by causing excessive stress and fatigue. The main objectives of this paper are to analyze the vibration characteristic experienced in the gear box, bearing, yaw drive, low speed shaft, high speed shaft, and generator of a WT at different load and operating conditions. The rotor system, generator system, and gear box system which are connected in series, have less Failure Criticality Index but they do cause large downtime and replacement costs [1]. It is generally suggested that the manufacturer should provide a good logistical support and maintain spares of blade, gear, and generator at nearby sites. It is clearly understood that if the yaw system fails, then the electricity generation will be reduced as the yaw system is connected parallel to the entire system.

2 Literature Review

Li [2] stated that there are many types of techniques available for condition monitoring such as vibration analysis, oil analysis, thermography, strain measurements, visual inspection, and self diagnostic sensors but out of all those methods vibration analysis is most preferable in CM because the results obtained at the end of vibration analysis give a definite pattern of vibration spectrum of different parts. Yang et al. [3] investigated condition monitoring and fault diagnosis techniques used in this paper for a WT with a synchronous generator. First, a new condition monitoring technique was proposed based on the phenomena observed in a series of torque-speed experiments. Then, the proposed technique was verified by detecting generator winding and rotor imbalance faults, and investigated a WT mechanical fault by power signal analysis with the aid of the continuous wavelet transforms (CWT). They concluded from their investigations that it was possible, using wind velocity and power transducers, to develop a simple and cheap WT condition monitoring and fault diagnosis system, without resorting to costly vibro-acoustic transducers.

Quan et al. [4] profess "In order to study dynamic characteristics and vibration response of gear system of WT mathematical modeling of gear box by taking into account different parameters like moment of inertia, torsional damp coefficient, torsional stiffness coefficient and the mass of the system is done. The mathematical modeling provides clear information over governing equation of vibration for the system". Hau [5] articulates that the components for total vibration of WTs, the structural dynamic considerations, and the design parameters due to which vibration occur. McMillan and Ault [6] point up that the levels of benefits are dependent on a variety of factors including wind profile, typical downtime duration, and WT subcomponent replacement cost. Cruden et al. [7] avow that there are many methods of data acquisition available for collection of vibration data from WT such as SCADA, online data acquisition, two-way data acquisition, and one-way data acquisition. One-way data acquisition is suitable for experimental analysis since collection of data and report generation can be made according to our requirements.

The literature obtained for vibration in WT was a preliminary knowledge for the development of the experimental vibration analysis in WT components. The components of WT exhibit different vibration spectrum depending upon their dynamic characteristics. The vibration is the important problem in WT which has its major contribution in mechanical component failure. So vibration in WT components has to be continuously monitored for preventive maintenance. The DAS with suitable vibration meter is needed for continuous vibration measurement in WT components at different load and operating conditions. The main objective of vibration measurement is to find the root cause of vibration in individual components and vibration control by means of vibration isolation.

3 Scope

The scopes of the project are as follows:

1. To measure the vibration characteristics of the WTs within the wind speed of 3.3–23 m/s.
2. To record the vibration parameters such as velocity, acceleration, and displacement within the frequency range of 1–20 kHz.

4 Vibration Measurement Scheme

Vibration analysis is used to determine the operating and mechanical conditions of equipment. A major advantage is that vibration analysis can identify developing problems downtime. This can be achieved by conducting regular monitoring of machine either on continuous basis or at scheduled intervals.

All rotating machines produce vibrations that are a function of the machine dynamics such as the alignment and balance of the rotating part. Measuring the amplitude of vibration at certain frequencies can provide valuable information about the accuracy of shaft alignment and balance, the condition of bearings or gears, and the effect on the machine due to resonance from the housing and other structural elements. The basic vibration measurement scheme of WT with 250 kW is shown in Fig. 1. Its effectiveness relies heavily on someone detecting unusual noises or vibration levels. Ultimately, vibration analysis can be used as part of an overall program to significantly improve equipment reliability. This can include more precise alignment and balancing better quality installations and repairs, and continuously lowering the average vibration levels of equipment in the plant. The time taken (T) to complete one cycle of motion is known as the period of vibration or time period.

$$T = \frac{2\pi}{\omega} \tag{1}$$

where ω is called the circular frequency.

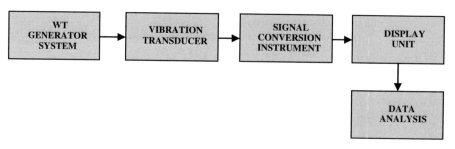

Fig. 1 Basic vibration measurement scheme

4.1 Frequency of Vibration

The number of cycles per unit time is called the frequency of oscillation or the frequency and is denoted by

$$f = \frac{1}{T} = \frac{\omega}{2\pi} \qquad (2)$$

Amplitude refers to the maximum value of a vibration. This value can be represented in terms of displacement, velocity, or acceleration.

5 Vibration Analysis of WT

The vibration analysis of the WT is used to diagnose the fault and prevent its occurrence. In WT, the rotor, gear box, high speed shaft, low speed shaft, generator, and yaw are rotating machineries. The flaw of the rotating machineries can cause a significant reduction in power generation and reliability. According to the ISO 10816-3 vibration severity chart, the velocity range up to 3.5 mm/s, the vibration for 400 kW WT is acceptable, and between 3.5 and 7 mm/s, the vibration is satisfactory. But beyond 7 mm/s, the vibration is unacceptable.

5.1 Vibration Analysis of Low Speed Shaft

The Fig. 2 shows the influence of vibration relating to velocity with time. The sudden spectrum rise in the later part of velocity graph depicts the alignment problems. The low speed frequently reaches an unacceptable range of vibration level. From the graph, it is evident that it reaches four times the intolerable range, within a minute. The high vibration in the low speed shaft is due to blade interference and defects in main bearing. The absurd vibrations of rotor system and low speed shaft are an effect of instability, poor structural framework, axial

Vibration Analysis of a Constant Speed

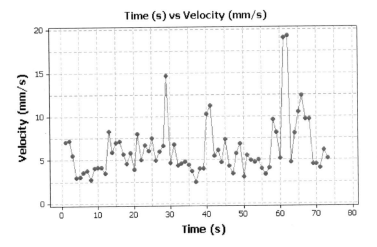

Fig. 2 Low speed shaft vibration relating to velocity with time

interference, and poor aerodynamic profile which increase the static divergence, stall, and flutter.

The continuous variation in acceleration graph at operation shown in Fig. 3 denotes mass imbalance, looseness, and increased rotor speed. The Fig. 4 shows that the displacement is varying at a moderate interval but jumps are high which denotes cyclic loading. The frequency spectrum represents that amplitude of vibration increases with increase in operating cycle. The main bearing of the rotor are subjected to abnormal wear, each time a WT is shut off or started. In these conditions, the entire weight of the rotating element rests directly on the lower half of

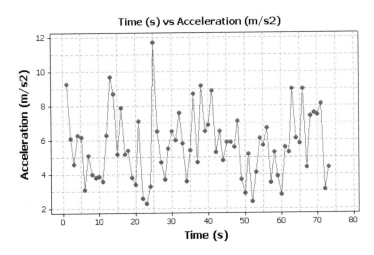

Fig. 3 Low speed shaft vibration relating to the acceleration with time

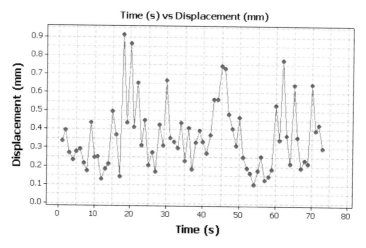

Fig. 4 Low speed shaft vibration relating to the displacement with time

the bearings which result bearing damage. The slackness of the main bearing at the end causes vibration (Fig. 5).

5.2 Vibration Analysis of Gear Box

The failure of WT gearbox is engendered due to many phenomenon such as amplified vibration, prolonged fatigue and stress, inevitable impact load, and increased

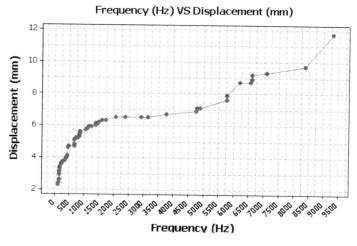

Fig. 5 Low speed shaft vibration relating to the frequency with displacement

Vibration Analysis of a Constant Speed 435

shaft speed. The vibration is augmented in gearbox owing to machine imbalance which is the effect of improper assembly and improper design.

The gear tooth damage is caused by abrasion, anisotropy, friction, and improper lubrication. The ill evitable fatigue is mainly because of excessive cyclic load which is a consequence of wind gusts and poor load distribution. The impact load has the main concern to increase vibration and reduce the lifetime of gearbox.

The Fig. 6 describes the vibration effect due to wind turbulence. The gearbox vibration is little and most probability within the limit. However, it reaches the maximum unacceptable range due to the impact of sudden load on the brake, uncertainty in the wind. The defects in the gear and misalignment of high speed and low speed shaft pull down the gear efficiency. The Fig. 7 represents the mass imbalance in a

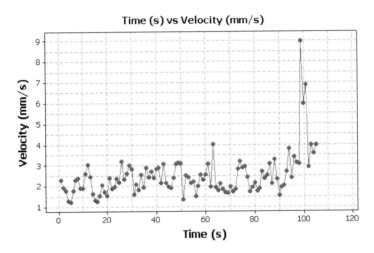

Fig. 6 Gear box vibration relating to the velocity with time

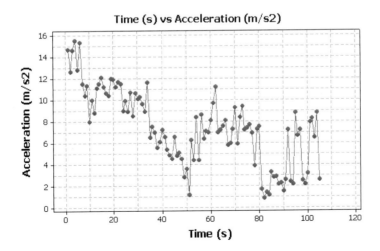

Fig. 7 Gear box vibration relating to the acceleration with time

minor level and friction. The maximum displacement for analyzed time is exposed in Fig. 8 and reveals high at initial state so vibration should be curtailed. The frequency spectrum delineates the effect of fluctuating load and increased wind speed (Fig. 9).

5.3 Vibration Analysis of High Speed Shaft

The shafts comprised with gear box are the low and high speed shafts, each of them has their own observable fact of failure. The damage of high speed shaft is owing to bent shaft, misaligned bearing, mass imbalance, and mechanical looseness.

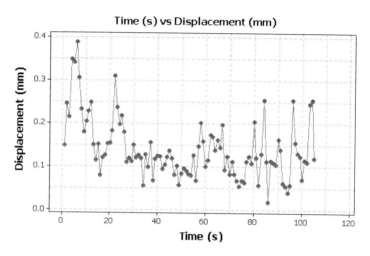

Fig. 8 Gear box vibration relating to the displacement with time

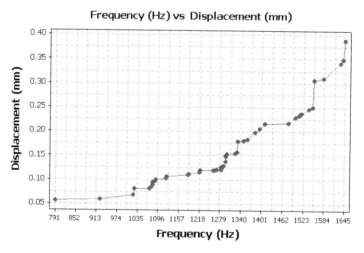

Fig. 9 Gear box vibration relating to the frequency with displacement

Vibration Analysis of a Constant Speed 437

The Fig. 10 makes known that the high speed shaft has extremely heavy vibration and the limit crosses its allowable value. The vibration in the high speed shaft causes damage in the gear box, coupling, and generator.

The extreme vibration is noted in the high speed shaft that causes frequent misalignment problem in the generator. The acceleration and displacement graphs shown in Figs. 11 and 12 are the evidences for that the level of vibration in the high speed shaft is terribly high. Displacement is the actual change in distance or position of an object relative to a reference point and is usually expressed in units of mm (Fig. 13).

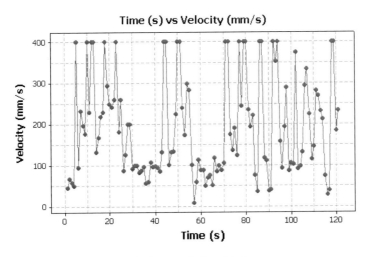

Fig. 10 High speed shaft vibration relating to velocity with time

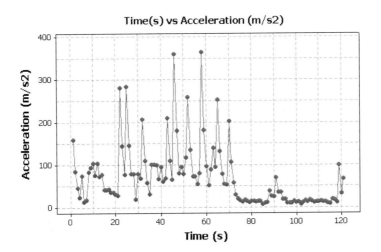

Fig. 11 High speed shaft vibration relating to the acceleration with time

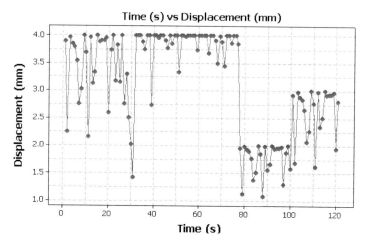

Fig. 12 High speed shaft vibration relating to displacement with time

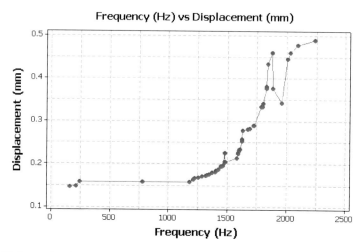

Fig. 13 High speed shaft vibration relating to frequency with displacement

5.4 Vibration Analysis of Generator

The failure of generator system is largely due to heavy vibration and electrical faults. The vibration has the effects in generator system such as soft foot, excessive brush wear, increase of electrodynamic fields, eddy current effect, and overhanging. The fault in rotor is engendered due to rotor lead failure and improper fixing of wedges and banding. The other nominal causes such as vibration, carbon brush worn out occur due to frequent transition from generator G1 to G2, improper fixing, and increased wind speed. The impact load caused due to radial preload and axial load damages bearing which in turn produces deformation in generator components.

The graph in Fig. 14 clearly shows that the rise in velocity in a rapid pattern denotes the increased speed and transition state. The key monitoring parameters for generators include bearings, casing and shaft, line frequency, and running speed (Fig. 15).

The generator rotors always seek the magnetic center of their casings. As a result, they tend to thrust in the axial direction. In almost all cases, this axial movement, or endplay, generates a vibration (Figs. 16, 17).

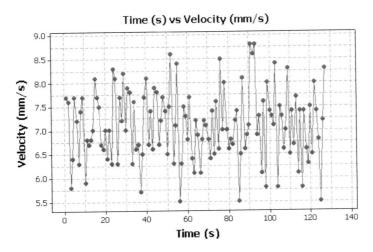

Fig. 14 Generator vibration relating to the velocity with time

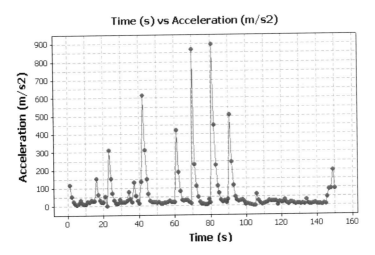

Fig. 15 Generator vibration relating to acceleration with time

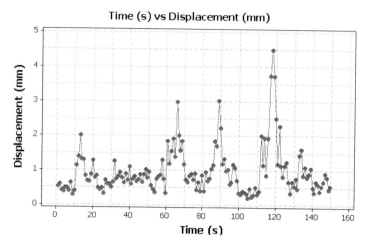

Fig. 16 Generator vibration relating to the displacement with time

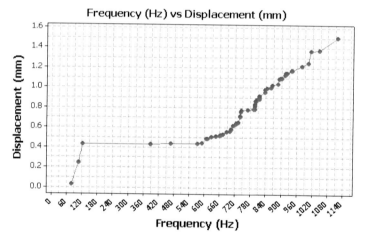

Fig. 17 Generator vibration relating to frequency with displacement

5.5 *Vibration Analysis of Yaw System*

The letdown of yaw system is induced due to heavy vibration, yaw motor worn out, yaw gear failure, and yaw brake failure. The profound vibration is engendered by wind turbulence, improper yaw braking, and yaw planetary failure that owe to the stress on gear tooth and friction.

The velocity graph of the yaw system shown in Fig. 18 depicts that the velocity rise is moderate and jump in velocity at shorter intervals throughout the curve is due to wind gusts.

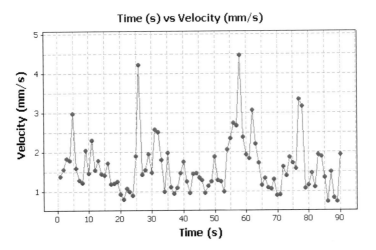

Fig. 18 Yaw vibration relating to the velocity with time

The yaw gear tooths are stressed by excessive cyclic loading and a structural dynamics. The abrasion of pinion and sleeve ring, the condomination of grease, and melting of the lubricant due to high temperature made high vibration in the planetary. The gust wind makes the yaw vibration. The single jump in acceleration shown in Fig. 19 at the later part of the curve represents fault in yaw planetary. The fluctuation in displacement at shorter intervals denote fault in sleeve ring. The frequency spectrum denotes increased vibration for analyzed condition (Figs. 20, 21).

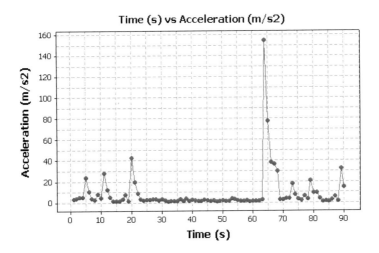

Fig. 19 Yaw vibration relating to acceleration with time

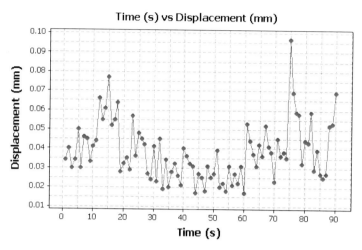

Fig. 20 Yaw vibration relating to displacement with time

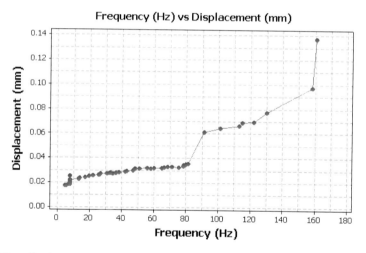

Fig. 21 Yaw vibration relating to the frequency with displacement

6 Conclusion

The vibration characteristic curves are generated with the help of vibration meter and data acquisition software. These vibration characteristic curves are computed and observations made from them for their severity and effects are analyzed. This paper concluded that large vibrations occurred in the high speed shaft, which reduce the reliability of the system. They require high speed shaft vibration monitoring measurement points in addition to standard casing measurement points. This requires the addition of permanently mounted proximity sensor, or displacement, transducers that can measure actual shaft deviation. The fluctuations in the wind, frequent grid off, and the transition of generator are the primary causes of vibration. The gear box vibration is also due to over loading, abrasion, and chemical attack. The generator is an overhanging structure and the misalignment of the high speed shaft makes the generator to vibrate.

References

1. Sunder Selwyn T, Kesavan R (2011) Computation of reliability and birnbaum importance of components of a wind turbine at high uncertain wind. Int J Comput Appl 32(4):42–49
2. Li Y (2011) Discussion on the principles of wind turbine condition monitoring system. Int Conf Mater Renew Energy Environ 1:621–624
3. Yang W, Tavner PJ, Crabtree CJ (2009) An intelligent approach to the condition monitoring of large scale wind turbines. In: European wind energy conference, Marseille, pp 1–8
4. Quan L, Yongqian L, Yongping Y (2010) Vibration response analysis of gear driven system of wind turbine. IEEE Int Conf Intell Comput Intell Syst 3:380–383
5. Hau E (2006) Wind turbines fundamentals technologies application economics, 2nd edn. Springer, Berlin
6. McMillan D, Ault GW (2008) Condition monitoring benefit for on shore wind turbines: sensitivity to operational parameters. IET Renew Power Gener 2(1):60–72
7. Cruden A, Booth C, Leithead W, Swiszcz G (1996) A data acquisition platform for the development of a wind turbine condition monitoring system. In: Proceedings of IEEE international symposium on product and quality and integrity, reliability and maintainability symposium, pp 30–36

Part VI
Aircraft Design and Manufacturing

Comparative Study Between Experimental Work and CFD Analysis in a Square Convergent Ribbed Duct

K. Sivakumar, E. Natarajan and N. Kulasekharan

Abstract In this present work, the local heat transfer and Nusselt number of developed turbulent flow in a convergent/divergent square duct have been investigated computationally. The angle of convergence of the duct is about $1°$. The computational analysis had been conducted within the range of Reynolds number from 10,000 to 77,000. The outcome of local heat transfer coefficient and Nusselt number of computational fluid dynamic analysis of the convergent/divergent duct is compared with the experimental data. There is no significant difference in Nusselt number value of CFD and experimental analysis, but the heat transfer coefficient for CFD is slightly (15 %) higher that the experimental work due to heat transfer losses and practical difficulties in fabrication work.

Keywords Rib turbulators · Turbulent flow · Aspect ratio · Heat transfer · Ribspacing

1 Introduction

In internal cooling of gas turbine blades, coolant air is circulated through serpentine passages and discharged through the bleed holes along the trailing edge of the blade (Fig. 1). With rotation, the flow is subjected to Coriolis forces induces secondary flow in planes perpendicular to the streamwise flow direction; this causes the migration of core fluid towards the trailing wall for radially outward flow. Thus, rotation destabilises the flow and enhances heat transfer along the trailing wall, while it stabilises the flow and reduces the heat transfer along the opposite wall. In addition, centrifugal buoyancy influences the radially outward flow along the heated walls and, at high rotation numbers, can have a significant effect on the surface heat transfer.

K. Sivakumar (✉)
Department of Mechanical Engineering, Valliammai Engineering College, Chennai, India

E. Natarajan
Institute for Energy Sutdies, Anna University, Chennai, India

N. Kulasekharan
Savitha Engineering College, Chennai, India

S. Sathiyamoorthy et al. (eds.), *Emerging Trends in Science, Engineering and Technology*, Lecture Notes in Mechanical Engineering, DOI: 10.1007/978-81-322-1007-8_41, © Springer India 2012

Fig. 1 Schematic sketch of a typical gas turbine blade

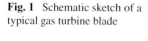
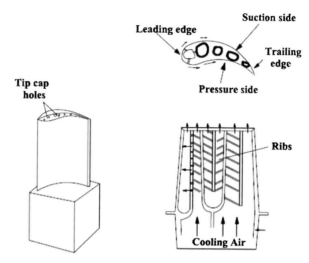

Turbulent heat transfer and fluid flow characteristics of air in rib-roughened tubes, annuli, ducts and between parallel plates have been studied extensively because of their important applications. Such a heat transfer enhancement method is widely used in cooling passages of gas turbine blade, compact heat exchangers, fuel elements in advanced gas-cooled reactor and electronic cooling devices and so on. Taking the cooling passages is usually approximated by a rectangular duct with a pair of opposite rib-roughened walls.

In the computational investigations reported in the literature, the internal cooling passages of a gas turbine have been modelled by either square or rectangular channels having two ways. As presented in Abuaf and Kercher [1], in typical airfoils with multipass cooling circuits, the cross-sectional area of radial passage usually varies along the passage from root to tip. In other words, the cooling passages are actually convergent to some extent. Such a geometric variation may induce substantial difference in both flow and heat transfer characteristics compared to those models with straight rectangular channel.

However, no significant study was found in the existing literature that deals with heat transfer in ribbed roughened convergent ducts.

A large number of experimental studies are reported in the literature on internal cooling in turbine blade passages, particularly for square coolant passages [2–7]. In a recent study, heat transfer in a rectangular channel with aspect ratio AR = 4:1 has been reported for smooth surfaces by Griffith et al. [8] for rib-roughened surfaces. Non-monotonic behaviour with respect to the rotation number was observed for the smooth channel while for the ribbed channel, the trailing surface heat transfer shows strong dependence on the rotation number. Data for 1:4 AR duct with rotation is similarly limited, with Agarwal et al. [9] recently reporting mass transfer data for smooth and ribbed ducts with rotation.

Most of the earlier computational studies on internal cooling passage of the blades have been restricted to three-dimensional steady RANS simulations [10–12]. The flow and heat transfer through a two pass smooth and 45° rib-roughened rectangular duct with an aspect ratio of 2 has been reported by Al-Qahtani et al. [13] using a Reynolds stress turbulence model. They have found reasonable match with experiments although in certain regions there are significant discrepancies. Wang et al. [14] found that for a smooth square duct, a mild streamwise variation of cross-mid-vertical sectional area may induce significant difference in the local and average heat transfer behaviours.

In the study, computational validation was conducted to find the developing heat transfer coefficient and Nusselt number for turbulent flow in a ribbed convergent/divergent ducts with uniform heat flux boundary condition, and also this value had been compared with the ribbed square convergent/divergent duct was conducted experimentally by Wang et al. [14] with identical mass flow condition.

2 Experimental Setup

A schematic of the experimental system as reported in [14] is shown in Fig. 2. This is an open test loop, with the room air being drawn by a 3.5 KW blower situated at the downstream end of the test loop. The forced air goes through an entrance section pre-plenum, test section after plenum and a 60 mm diameter, 1,440 mm length pipe equipped with a multiport averaging Pitot tube to measure the flow rate.

The ribs are manufactured as integral parts on two opposite walls of the duct. The convergent square ribbed duct has the largest cross-section of 58 × 58 mm^2

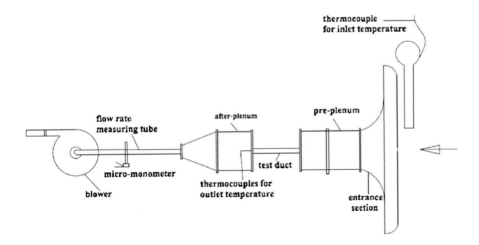

Fig. 2 Schematic diagram of test apparatus

and the smallest cross-section of 41 × 41 mm^2. Each test duct is 500 mm long. With e/Dm = 0.081 and p/e = 17.8, where Dm is the duct average value of hydraulic diameter. To get a detailed distribution of the local heat transfer coefficient, 120 thermocouples are imbedded in one of the ribbed walls along its centre line to measure.

3 Data Reduction

The local heat transfer coefficient was calculated from the total net heat transfer rate and the difference of the local wall temperature and the local bulk mean air temperature.

$$h_x = \frac{(Q - Q_{\text{loss}})}{A(T_{w,x} - T_{b,x})} \tag{1}$$

The local wall temperature used in Eq. (1) was read from the output of the thermocouple. The local bulk air temperature air was calculated by the following equation:

$$T_{b,x} = T_m + \frac{(Q - Q_{\text{loss}})A(x)}{Amc_p} \tag{2}$$

where $A(x)$ is the heat transfer surface area from the duct inlet to the position where the local heat transfer coefficient was determined. The heat loss to the environment Q_{loss} was estimated by heat conduction through the plastic foam and the two end losses. For most of the studied values, the ratio of Q_{loss}/Q was less than 5 %. This estimation was confirmed by the thermal energy balance between the fluid enthalpy increase and the total power input. In data reduction, Q_{loss} was determined from the measured outlet fluid temperature.

The local and the duct average Nusselt number were defined by

$$Nu_x = \frac{h_x D_m}{k} \tag{3}$$

$$Nu = \frac{(Q - Q_{\text{loss}})/D_m}{Ak(T_w - T_m)} \tag{4}$$

The characteristic length, the reference temperature and the average wall temperature were determined by

$$D_m = \frac{D_{h,\text{in}} - D_{h,\text{out}}}{2} \tag{5}$$

$$T_m = \frac{T_{b,\text{in}} + T_{b,\text{out}}}{2} \tag{6}$$

$$T_w = \frac{1}{A} \int_0^A T_{w,x} \, dA \tag{7}$$

As for most cases of the internal convection heat transfer, the fluid properties are evaluated at the mean temperature of the fluid in the duct. The Reynolds number was defined by

$$\mathrm{Re}_m = \frac{U_m D_m}{v} \tag{8}$$

where U_m is the duct mean cross-sectional average velocity. This is equal to the cross-section average velocity at the duct mid-section.

4 Numerical Solution Procedure

The computational model studied in the present study includes a segment of the inlet plenum and the test section. Inlet for the computational domain is the inlet to the domain, so that more realistic flow with entry effect can be simulated for the test section. The flow is assumed to leave the test section to ambient with a zero static pressure boundary condition at the outlet.

In this study, a commercial computational fluid dynamic analysis has been used to carry out the analysis and post processing. The code provides mesh flexibility and also includes several turbulence models. Computations are performed for turbulent flows depending on the Reynolds number. The energy equation is solved neglecting radiation. The renormalisation-group RNG $k-\varepsilon$ model is employed as the turbulence model for turbulent flow. The Reynolds averaged Navier-Stokes equations are solved numerically in conjunction with transport equations for turbulent kinetic energy and dissipation rate.

Full buoyancy effects are included in transport equations. Near wall regions are fully resolved for y^+ values between 0.1 and 0.5. Structured computational cells were created with a fine mesh near the plate walls and coarse mesh away from the plate walls with the help of bi-exponent ratio scheme in GAMBIT 2.4. Steady inline solver was used with second order upwinding scheme for convective terms in the mass, momentum, energy, and turbulence equations for both laminar and turbulent flows. For pressure discretisation the PRESTO scheme has been employed while for pressure–velocity coupling discretisation the SIMPLE algorithm has been used.

The general forms of the governing equations for this problem are:

$$\frac{\partial \rho u_j}{\partial x_j} = 0 \tag{9}$$

$$\frac{\partial}{\partial x_j}(\rho u_j u_i) = \frac{\partial \rho}{\partial x_j} + \frac{\partial}{\partial x_j}(\tau_{ij} - \overline{\rho u_i' u_j'}) \tag{10}$$

$$\frac{\partial}{\partial x_j}[u_i(\rho E + p)] = \frac{\partial}{\partial x_j}\left[\left(\lambda + \frac{C_p \mu_t}{pr_t}\right)\frac{\partial T}{\partial x_j} + u_i(t_{ij})_{eff}\right] \quad (11)$$

$$\overline{-\rho u'_i u'_j} = 2\mu_t - \frac{2}{3}\rho k \partial_{ij} \quad (12)$$

Continuity, momentum and energy equations are solved by the finite volume method with structured meshes and the coupling between the velocity and pressure given by the SIMPLE algorithm. The advection terms in the momentum, energy, mass transport, turbulent kinetic energy and specific dissipation equations were discretised using the second order upwind algorithm. Grid independence studies resulted in meshes having about 1, 50,000 elements for each case.

The convergence criteria were 10^{-4} for velocity, 10^{-6} for k and u and 10^{-8} for energy. The inlet boundary conditions were set as velocity inlet as the mass flow inlet with the outlet set as pressure outlet. The heat flux boundary condition was used in all the surfaces and no-slip.

5 Result and Discussion

The local Nusselt number for the orthogonal ribbed plate with respect to position of the ribbed duct for the Reynolds number value of 10,000 has been shown in Fig. 3. It illustrates that the computational fluid dynamics analysis of local Nusselt first increases before it sticks the first rib due to the entrance effect. In the entrance the local eddies are formed, due this the heat transfer area increases because of that local Nusselt number increases. After second and third ribs, the flow will be

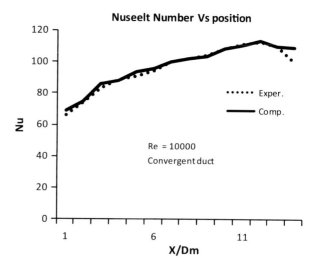

Fig. 3 Local Nusselt number for the 90 ribbed duct

steady and both experimental and CFD will coincide. At the end, Nusselt number value for experimental analysis will be reduced because of constant heat flux.

Figure 4 shows the local heat transfer coefficient for the orthogonal ribbed duct with respect to positions of ribs for the Reynolds number 10,000. The heat transfer coefficient for computational fluid dynamics analysis is 15 % higher than the experimental result. In experimental work, more heat loss will take place due to various factors as follows.

1. Convection heat loss throughout test section, but we assumed in CFD analysis, duct sections are perfectly insulated.
2. Heat loss due to radiation, but in CFD analysis, no radiation loss will be neglected.

5.1 Numerical Simulations for 90° Ribs

The design includes continuous 90° ribs are arranged as shown in Fig. 5. The channel was a convergent square ribbed duct that has the largest cross-section of 58 × 58 mm^2 and the smallest cross-section of 41 × 41 mm^2. Each test duct is 500 mm long. Grid independence studies resulted in meshes having about 150,000 elements for each case.

5.2 Velocity and Temperature

The velocity contours of a square 90° ribs on chosen planes are as shown in Fig. 6 are discussed in this section. A vertical and a horizontal plane passing

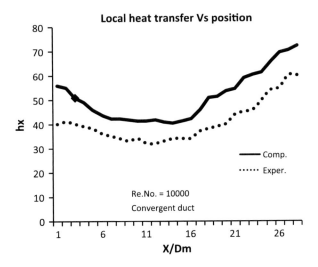

Fig. 4 Local heat transfer coefficient for the 90 ribbed duct

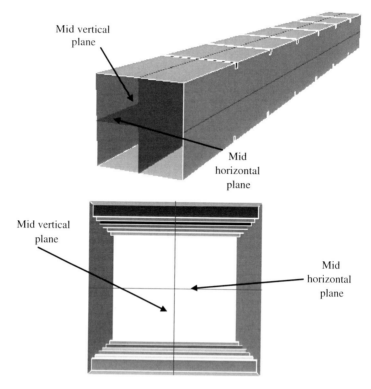

Fig. 5 Planes used for data extraction

through the mid of the channel are considered. Figure 6 shows the contours of velocity magnitude on these planes. Along these planes it can be observed that the velocity at inlet is lower than the outlet of the duct, due to the flow acceleration in the stream-wise direction. The naturally expected acceleration in a converged channel flow is further augmented by the flow blockage offered by the ribs on the opposing walls, which is evident from the contours on the mid-vertical plane.

The contours of temperature for the square convergent duct on the mid-horizontal plane and vertical planes are shown in Figs. 7 and 8. Near the ribs, the temperatures are higher than the core of the duct, because the cold air entering the channel inlet is less.

The contours of heat transfer coefficient on the side wall, bottom wall and rib surfaces are shown in Fig. 9. The effect of flow entering the test section can be observed from the peak value of local heat transfer coefficient 'h' on the side and bottom walls. The edges of the ribs show higher heat transfer coefficient values due to the edge effect and the flow separation from the rib front corners. View from the exit side of the test section shows the rib rear side 'h' values which are lower than that in the rib front side.

Comparative Study Between Experimental Work and CFD Analysis 455

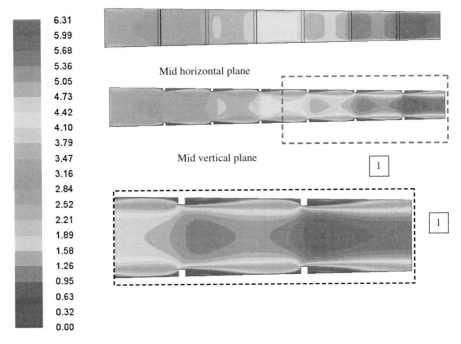

Fig. 6 Contours of velocity (m/s^2) magnitude on a mid-vertical plane

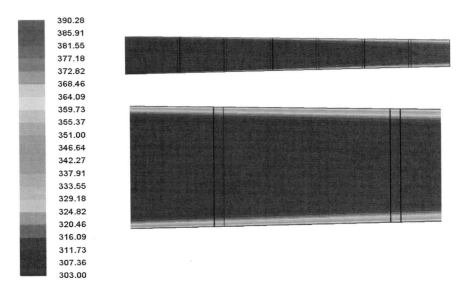

Fig. 7 Contours of temperature (K) on mid-horizontal plane

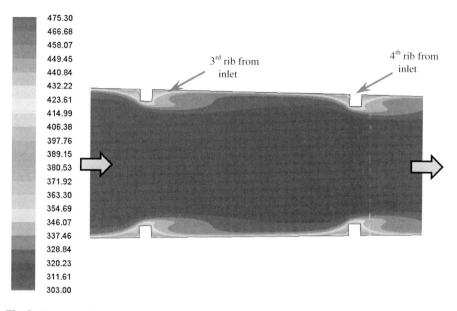

Fig. 8 Contours of temperature (K) on a vertical plane

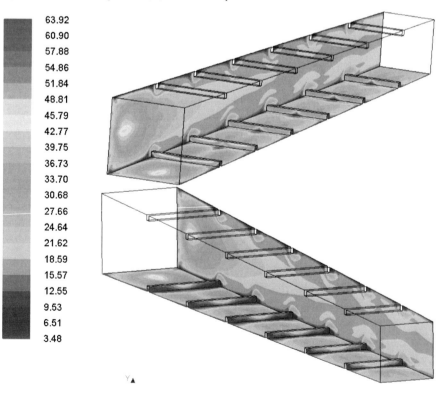

Fig. 9 Contours of heat transfer coefficient 'h' (W/m^2-K) on the side and bottom wall

6 Conclusions

The work reported here is a comparative study between computational fluid analyses of the convergent/divergent square duct with orthogonally and inline arranged ribbed duct with experimental work for same configuration of duct. The heat transfer coefficient for experimental work is 15 % lesser than the computational fluid dynamic analysis because of the convection and radiation heat loss and practical difficulties for fabrication work. And also in computational fluid dynamic, heat flux is assumed to be constant, but in practical case it is impossible to maintain. But several Nusselt number between experimental work and computational fluid dynamics had been same. This paper also described the different types of contours (velocity, temperature and heat transfer coefficient) with respected to rib positions.

References

1. Abuaf N, Kercher DM (1983) Heat transfer and turbulence in a turbine blade cooling circuit. ASME J Turbomach 116:169–177
2. Han JC, Park JS (1988) Developing heat transfer in rectangular channel with rib turbulators. Int J Heat Mass Transf 31(I):183–195
3. Han JC (1988) Heat transfer and friction characteristics in rectangular channels with rib turbulators. J. Heat Trasnf 110:321–328
4. Wagner JH, Johnson BV, Graziani RA, Yeh FC (1992) Heat transfer in rotation serpentine passages with trips normal to the flow. J Turbomach 114:847–857
5. Johnson BV, Wagner JH, Steduber GD, Yeh FC (1994) Heat transfer in rotating serpentine passages with trips skewed to the flow. J Turbomach 116:113–123
6. Johnson BV, Wagner JH, Steuber GD (1993) Effect of rotation on coolant passage heat transfer. NASA contractor report—4396, vol II
7. Chen Y, Nikitopoulos DE, Hibbs Acharya RS, Myrum TA (2000) Detailed mass transfer distribution in a ribbed coolant passage. Int J Heat Mass Transf 43:1479–1492
8. Griffith TS, Al-Hadhrami L, Han JC (2002) Heat transfer in rotating rectangular cooling channels with angled ribs. J Heat Transf 124:617–625
9. Agarwal P, Acharya S, Nikitopoulos DE (2003) Heat transfer in 1:4 rectangular passages with rotation. J. Turbomach 125:26–733
10. Rigby DL, Steinthorsson E, Ameri AA (1997) Numerical prediction of heat transfer in a channel with ribs and bleed. In: ASME paper 97-GT-431
11. Lacovides TB, Launder BE (1995) Developing buoyancy modified turbulent flow in ducts rotating in orthogonal mode. J. Turbomach 117:474–484
12. Lacovides H (1988) Computation of flow and heat transfer through rotating ribbed passage. Int J Heat Fluid Flow 19:393–400
13. Qantani MA, Chen HC, Han JC (2003) A numerical study of flow and heat transfer in rotating rectangular channels (AR = 4) with 45° rib turbulators by Reynolds stress turbulence model. J Heat Transf 125(1):19–26
14. Wang LH, Tao WQ, Wang QW, Wong TT (2001) Experimental study of developing turbulent flow and heat transfer in ribbed convergent/divergent square duct. Int J Heat Fluid Flow 22(6):603–613

Optimization of Rounded Spike on Hypersonic Forebody Reattachment

A. Sureshkumar, S. Nadaraja Pillai and P. Manikandan

Abstract Generally, the designing of supersonic and hypersonic vehicles has some major problems of reducing drag and aerodynamic heating which are considered as the important criteria. In hypersonic vehicles, the frontal area of blunt forebody nose is larger, so the drag and aerodynamic heating will be high. This blunt body generates more heat energy that can cause ablation to the vehicle materials and will affect the performance. Basically, blunt body produces more drag, so as to minimize drag and aerodynamic heating spike with aero disk was introduced. Presently, researchers introduced spike with aero disk in hypersonic blunt body; it is fitted at the front of the forebody stagnation point and it is projected upstream side. Normally, blunt body produces bow shock; this strong bow shock has changed into weak shock due to the aero disk with spike. So the aero disk creates an oblique shock wave and is useful to reduce drag. Normally, the high speed flows are best defined as the regime where certain physical flow phenomena become progressively more important as the Mach number is increased to higher values. The high temperature regions in the flow field around a blunt-nosed body and the massive amount of flow kinetic energy in a hypersonic free stream are converted into internal energy of the gas across the strong shock wave, hence reducing very high temperatures in the shock layer nearer the nose. The temperature in the shock layer of a high speed re-entry vehicle can be very high. The objective of this research is to design and optimize the model very effectively and to reduce drag, aerodynamic heating, and to delay the reattachment point on the forebody. In this research, an optimization of rounded aero disk with spike, the changing the aero disk diameter ratios along Y-directions 2r, 2.5r, 3r, 3.5r, and 4r at different ratios. It is found that the results detailed to minimize the drag, aerodynamic heating, and also delay reattachment point. It is found that the reattachment point changes on the forebody with respect to diameter of the rounded spike (aero disk). The model was created and simulated by using of CFD commercial software.

A. Sureshkumar (✉) · S. Nadaraja Pillai
Department of Aeronautical Engineering, GKM College of Engineering and Technology,
Chennai 600063, India
e-mail: suresh_aero12@yahoo.co.in

P. Manikandan
Department of Aerospace Engineering, Madras Institute of Technology, Chennai 600044, India

S. Sathiyamoorthy et al. (eds.), *Emerging Trends in Science,*
Engineering and Technology, Lecture Notes in Mechanical Engineering,
DOI: 10.1007/978-81-322-1007-8_42, © Springer India 2012

Keywords Aero disk · Blunt forebody · Spike · Shock waves · Aerodynamic heating

1 Introduction

The design and optimization of high speed vehicles are important criteria in this world. Now everyone is doing some research in this high speed flow problems. So, the high speed vehicles' major problems are high drag and aerodynamic heating [1]. Normally, the blunt body produces bow shock; this shock wave generates more drag and affects the vehicle thrust [2]. Previously, blunt body with spiked problems were solved in many shapes and sizes. The spiked blunt body hypersonic vehicles design performed well in this previous one. For further development, this research introduced spike with aero disk. This aero disk is very useful in reducing drag and aerodynamic heating. On one hand, it is desired to use a pointed spike with forebody in hypersonic vehicles to minimize the drag for less fuel consumption during the take-off (ascent) phase [3]. In the high speed regimes, a blunt forebody generates a strong bow shock ahead of it. This shock wave creates more excessive drag and aerodynamic heating attained by the downstream flow. This type of problem investigated by experimental agency in the different models were studied [4, 5]. Some researchers studied related with aerodynamic heating in the heat transfer for hypersonic flow problems. They were done with spiked fore bodies at Mach 6.8 [6]. In high speed flows, the spike tip feels some oscillations and vibrations [7]. Further development of spike with forebody needs to reduce drag and aerodynamic heating. The shape of the spike is cylindrical rod and it is fitted at the stagnation point on the forebody nose. The ultimate aim of the research is to reduce drag, aerodynamic heating, and delay the reattachment point on the forebody. So to further develop the needs to reduce drag and aerodynamic heating in the hypersonic flow problem, researchers have introduced hemispherical shape of aero disk in front of the spike. This hemispherical shape of the aero disk is to reduce drag efficiently compared with the previous one in hypersonic forebodes [8, 9] and [10]. Normally, flow separation and reattachment points are important criteria of this research, so as to delay the reattachment point on the forebody, get some amount of drag reduction and aerodynamic heating [11, 12]. The different shapes and sizes of spike configurations was modeled and solved in the high speed flows [13]. The design parameters were studied by changing the shape of aero disk diameter and spike length dimensions in the hypersonic vehicles to optimize the design of the spike with aero disk [14]. This model was created and analyzed by using Gambit and Fluent softwares. The aerodynamic heat reduction is not an easy one, but through analyzing we brought forth reduction of heat transfer. The objective of this research is to optimize the design and development of spike with aero disk forebody. It is used to minimize the drag, aerodynamic heating, and delay the reattachment point on the hypersonic forebody. Hence, the aero

Fig. 1 Aero disk-spike with blunt forebody

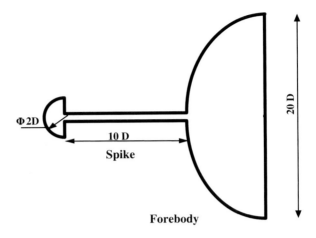

disk diameter changes along the Y-directions in front of blunt forebody. It is found that the reattachment point changes on the forebody with respect to diameter of the rounded spike (aero disk).

2 Modified Aero Spike

A hemispherical shape (Rounded spike) aero disk is used in this problem to change the aero disk diameters along the Y-direction at different ratios like 2r, 2.5r, 3r, 3.5r, and 4r. Where r is the radius of the aero disk, which is half of the diameter D as shown in Fig. 1. The entire radius is not changed as the traditional configuration. Only the aero disk frontal area faced by the flow is changed. For example, forebody diameter, $D = 50$ cm or 0.5 m, aerodisk diameter, d ranges from 0 to 0.4 D with a step 0.05 and along Y-axis varies from 2r to 4 r in steps of 0.5 r. Forebody diameter, $D = 0.5$ m, aerodisk diameter, $d = 0.025$ m, and radius, $r = 0.0125$ m. Overall spike to aerodisk length, $L = 0.25$ m (constant value for all cases), spike diameter, $l_d = d / 10$ (constant value for all cases). The aero disk diameter changes along the Y-axis only either in positive or negative directions, not in X-axis. Figure 2 shows the model made for the flow analysis for this problem. Aerodisk with spike model for case I with 2r is shown in Fig. 3 is full domain with unstructured triangular mesh for the flow analysis. The closed view of the spike with aerodisk body is shown in Fig. 4.

Since it is a compressible flow problem, the analysis is considered as density based and an inviscid. This problem is mainly related with the flow analysis of 2 dimensional case.

The model was considered as wall and domain considered as the pressure farfield. The defined boundary values along the X-directions are Mach number 3, Input pressure 1 atmosphere, operating conditions 0 atmosphere, and Courant number is 0.1.

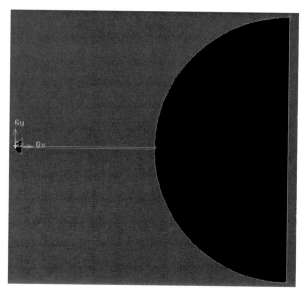

Fig. 2 Aero disk-spike with blunt body

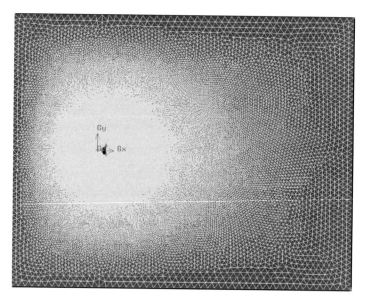

Fig. 3 Meshed geometry model

Optimization of Rounded Spike on Hypersonic Forebody Reattachment 463

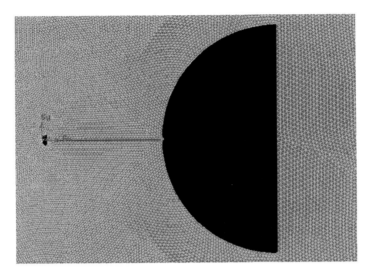

Fig. 4 Closed view of the meshed model

3 Results and Discussion

The rounded aero disk spike with forebody models having different dimensions are shown in Figs. 5, 9, and 13 are the pressure distributions over it. The formation of the shockwave is also evident from the Figure.

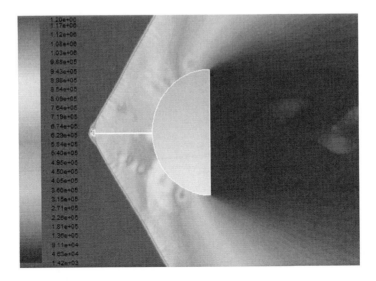

Fig. 5 Static pressure distribution for case 2r

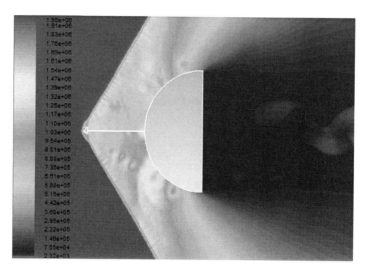

Fig. 6 Coefficient of pressure for case 2r

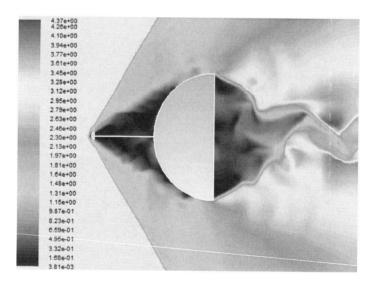

Fig. 7 Velocity Mach number for case 2r

It is seen that the pressure change before and after the shock shows the oblique shock characteristics. The coefficient of pressure changes also studied in this research at different r values. In Figs. 6, 7, 10, 11, 14, and 15 show the velocity distribution based on the Mach number. Here the reattachment of the flow from the aero disk to the forebody is of the research interest. At the lower r value, the reattachment point is delayed and for higher r values it is reversed. It is also evident from the vector plot shown in Figs. 8, 12, 16, 17, 18, and 19. Even though various models made with various geometries of r value, few are shown.

All the above-mentioned graphs represent the variations of the coefficient of pressure studied for cases 2r, 2.5r, 3r, 3.5r, and 4r. All the aerodynamic flow properties are studied in this research and reattachment point is also studied by the simulation.

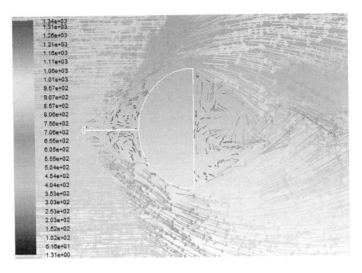

Fig. 8 Velocity vector for case 2r

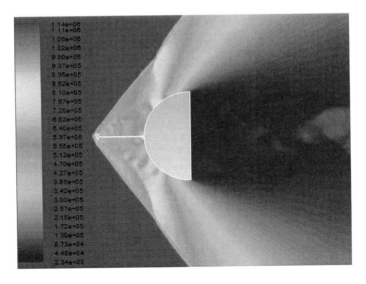

Fig. 9 Static pressure distribution for case 3r

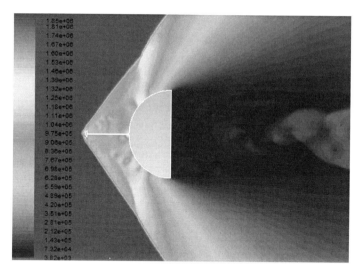

Fig. 10 Coefficient of pressure for case 3r

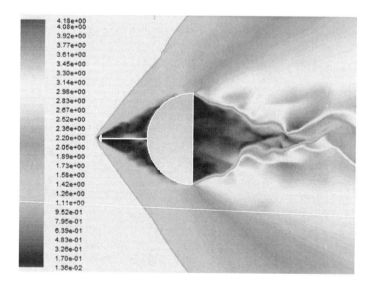

Fig. 11 Velocity Mach number for case 3r

Optimization of Rounded Spike on Hypersonic Forebody Reattachment 467

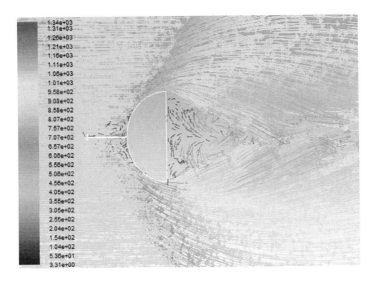

Fig. 12 Velocity vector for case 3r

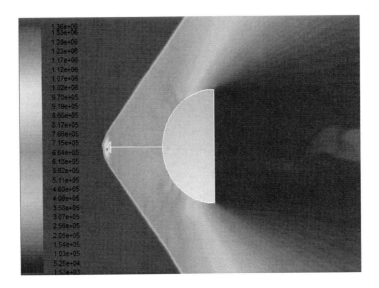

Fig. 13 Static pressure distribution for case 4r

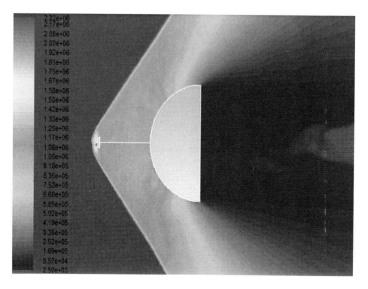

Fig. 14 Coefficient of pressure for case 4r

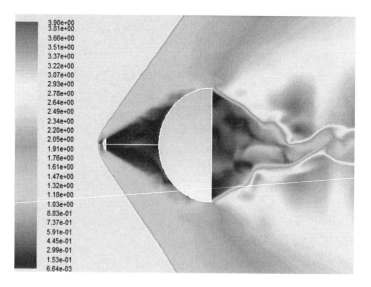

Fig. 15 Velocity Mach number for case 4r

Optimization of Rounded Spike on Hypersonic Forebody Reattachment

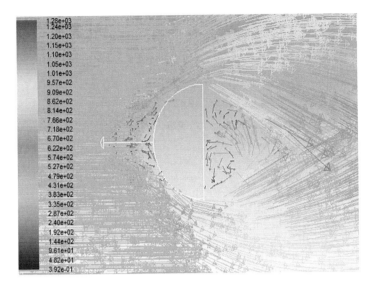

Fig. 16 Velocity vector for case 4r

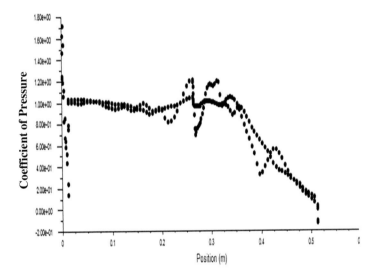

Fig. 17 Coefficient of pressure for case 2r

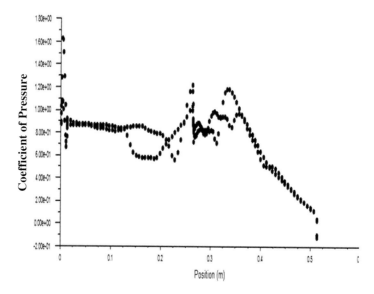

Fig. 18 Coefficient of pressure for case 3r

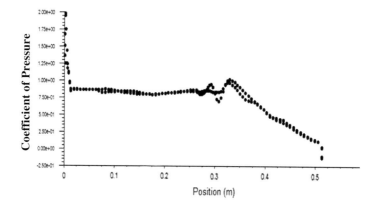

Fig. 19 Coefficient of pressure for case 4r

4 Conclusion

Hypersonic forebody spike with aerodisk has been made and analyzed for various configurations starting from 2r to 4r in steps of 0.5r. The results are discussed for the pressure and velocity distributions and also for coefficient of pressure over the forebody. The velocity distributions are shown in Mach number plot. Also the contour plot is shown inorder to understand the reattachment point on the forebody.

The goal of this research is to understand how much delay in reattachment for various configuration of r value. It is found from the results that the lower configuration r value shows delayed reattachment point on the forebody. The significance of such aerodisk configuration on reattachement is inferred. This research is presently extended to further configurations.

References

1. Anderson JD (2007) Fundamentals of aerodynamics, 4th edn. McGraw-Hill, USA
2. Anderson JD (2000) Hypersonic and high temperature gas dynamics. McGraw Hill
3. Maull DJ (1960) Hypersonic flow over axially symmetric spiked bodies. J Fluid Mech 8(4):584–592
4. Menezes V, Saravanan S, Jagadeesh G, Reddy KPJ (2003) Experimental investigations of hypersonic flow over highly blunted cones with aerospikes. AIAA J 41(10):1955–1966
5. Karlovskii VN, Sakharov VI (1986) Numerical investigation of supersonic flow past blunt bodies with protruding spikes. Fluid Dyn 21(3):437–445
6. Mehta RC (2000) Numerical heat transfer study over spiked blunt bodies at Mach 6.8. J Spacecraft Rockets 37(5):700–703
7. Calarese W, Hankey W (1985) Modes of shock-wave oscillations on spike-tipped bodies. AIAA J 23(2):185–192
8. Ahmed M, Qin N (2010) Drag reduction using aero disks for hypersonic hemispherical bodies. J Spacecraft and Rockets 47(1):62–80
9. Bogdonoff SM, Vas IE (1959) Preliminary investigations of spiked bodies at hyper- sonic speeds. J Aerosp Sci 26(2):65–74
10. Album HH (1968) Regarding the utility of spiked blunt bodies. J Aerosp Sci 5(1):425–449
11. Thurman WE (1964) A flow-separation spike for hypersonic control of a hemisphere cylinder. AIAA J 2(1):159–161
12. Ahmed M, Qin N (2011) Numerical investigation of aero heating characteristics of spiked blunt bodies at Mach 6 flight conditions. Aeronaut J 115:377–386
13. Roveda RB (2009) CFD study of spiked blunt body configurations. AIAA paper
14. Ahmed M, Qin N Surrogate-based multi-objective aerodynamic design optimization of hypersonic spiked bodies. In: Proceedings of the 28th Inetrnational symbosium of shick waves, Manchester;17–22 July 2011

Thermo Mechanical Analysis and Design of Rotating Disks

Priyambada Nayak and Kashi Nath Saha

Abstract This study proposes a numerical method for evaluation of thermo-elastic stresses and deformation states of rotating varying thickness disks, either solid or annular. The solutions are obtained under plane stress assumption and up to the elastic limit of the material by using minimization of total potential energy principle. The solution algorithm is implemented with the help of MATLAB computational simulation software. The results are validated with those of other researchers for appropriate values of system parameters and very good agreement is observed. Some new results are furnished in the form of limit angular speed of rotating disks under thermo-mechanical loading. The results considering density variation with temperature are also presented but they have not much effect on limit angular speed for different disk geometries and temperature distribution profiles.

Keywords Rotating disks · Variational method · Thermo-elastic stresses

1 Introduction and Literature Review

Rotating disks are widely used in many engineering applications and primarily they need to be designed for approximate uniform stress distributions. Such disks are often subjected to thermal field resulting in variable material properties. Theoretical studies pertaining to elastic and elastic–plastic stress analysis of rotating disk made of isotropic material were reported in textbooks by Calladine [1], Timoshenko and Goodier [2] and Srinath [3]. Manson [4] presented a finite difference solution of the equilibrium and compatibility equations for elastic stresses in a symmetrical disk, and subsequently he reported a simplified method for determining the disk profile under the combination of centrifugal and thermal loading [5]. Various numerical

---

P. Nayak (✉) · K. N. Saha
Mechanical Engineering Department, Jadavpur University, Kolkata 700032, India
e-mail: priyambada53@gmail.com

K. N. Saha
e-mail: kashinathsaha@gmail.com

S. Sathiyamoorthy et al. (eds.), *Emerging Trends in Science,*
Engineering and Technology, Lecture Notes in Mechanical Engineering,
DOI: 10.1007/978-81-322-1007-8_43, © Springer India 2012

schemes such as truncated Taylor's expansion method [6], perturbation method using power series solution method [7], etc., have been used to investigate elastic–plastic deformations of rotating disks.

Millenson and Manson [8] presented an analytical–numerical method which accounts variations of material elastic and thermal-physical characteristics, profile geometry and density, as well as thermal load, plastic flow and creep. Vivio and Vullo [9] investigated the elastic stress states of rotating disks having radial density variation subjected to thermal load based on two independent integrals of the hyper geometric differential equation. They also investigated variable thickness rotating solid and annular disks using homotopy perturbation method (HPM) and Adomian's decomposition method (ADM). Kordkheili and Naghdabadi [10] proposed a semi-analytical thermo-elasticity solution for hollow and solid rotating axisymmetric disks made of functionally graded materials with power-law distribution for the thermo-mechanical properties of the constituent components.

Afsar and Go [11] analyzed thermo-elastic field in a thin circular rotating disk of functionally graded material based on the two dimensional thermo-elastic theories. The axi-symmetric problem, formulated in terms of a second order ordinary differential equation, was solved by finite element method. Peng and Li [12] investigated the steady thermal stresses in a functionally graded annular rotating disk by numerically solving the Freedholm integral equation. Callioglu et al. [13] investigated the stress state of functionally graded rotating annular disks subjected to parabolic temperature distributions by using infinitesimal deformation theory of elasticity.

The present study proposes a numerical method for evaluation of thermo-elastic stresses and deformation states of rotating varying thickness disks, either solid or annular. The variational method proposed by Bhowmick et al. [14] is used to investigate the stress and deformation states of rotating disks. The solution of the governing partial differential equation is obtained by assuming series approximation of the radial displacement field following Galerkin's principle. The solutions are obtained under plane stress assumption and up to the elastic limit of the material by using minimization of total potential energy principle. A thorough literature review in this area reveals that analysis of similar problems are carried out mostly through analytical methods and no dedicated work on limit angular speed of rotating disks under thermo-mechanical loading has been carried out. The solution algorithm is implemented with the help of MATLAB computational simulation software. The results are validated with those of other researchers for appropriate values of system parameters and very good agreement is observed. The effect of disk geometry and temperature field variation on the performance of rotating disks is considered and normalized value of limit angular speed is furnished.

2 Mathematical Formulation

Radial displacement occurs in a disk due to the centrifugal load as well as due to thermal expansion. The solution for the displacement field is obtained from the minimization of total potential energy principle $\delta (U+V) = 0$, where U and V

Thermo Mechanical Analysis and Design of Rotating Disks

are the strain energy and potential energy respectively. It is assumed that the disk is symmetric with respect to the mid-plane, and a state of plane stress ($\sigma_z = 0$) persists in the loaded disk. Using Hooke's law for plane stress case, the stress–strain relations are given by $\epsilon_r = (\sigma_r - \upsilon\sigma_\theta)/E$ and $\epsilon_r = (\sigma_r - \upsilon\sigma_r)E$ and $\epsilon_\theta = (\sigma_\theta - \upsilon\sigma_r)/E$. When the disk material is subjected to a temperature field, it experiences a strain arising due to uniform expansion proportional to the temperature rise T, and the linear constitutive thermoelastic equations take the form $\epsilon_r = \frac{1}{E}(\sigma_r - \upsilon\sigma_\theta) + \alpha T$ and $\epsilon_\theta = \frac{1}{E}(\sigma_\theta - \upsilon\sigma_r) + \alpha T$. The inverse relations are given by, $\sigma_r = \frac{E}{(1-\upsilon^2)}[\epsilon_r + \upsilon\epsilon_\theta - (1 + \upsilon)\alpha T]$ and $\sigma_\theta = \frac{E}{(1-\upsilon^2)}[\epsilon_\theta + \upsilon\epsilon_r - (1 + \upsilon)\alpha T]$ where E, α and υ are respectively, elastic modulus, thermal expansion coefficient, and Poisson's ratio. It is assumed that the disk material properties (E, α and υ) remain constant in the variable thermal field of the disk. A list of notations for the variables used is provided as an appendix to this paper.

The strain energy of the disk comes from the stress and strain field, and expressed as, $U = \frac{1}{2}\int\limits_{Vol} (\sigma\ \epsilon)\,dv = \int\limits_{Vol}(\sigma_\theta\ \epsilon_\theta + \sigma_r\ \epsilon_r)\,dv$. Using the strain–displacement relations for an axi-symmetric body having axi-symmetric deformations, given by $\epsilon_r = \frac{du}{dr}$ and $\epsilon_\theta = \frac{u}{r}$ we get,

$$U = \frac{\pi E}{1 - \upsilon^2} \int\limits_a^b \left\{ \frac{u^2}{r} + 2\upsilon u \frac{du}{dr} + r\left(\frac{du}{dr}\right)^2 - (1 + \upsilon)\alpha T r\left(\frac{u}{r} + \frac{du}{dr}\right) \right\} h\,dr \quad (1)$$

The potential energy V arising from centrifugal force is given as follows:

$$V = -\int\limits_{vol} u\,(\omega^2 r)\,dm = -2\pi\omega^2 \int\limits_a^b \rho u r^2 h\,dr \quad (2)$$

Substituting the expressions of U and V in the energy principle $\delta\,(U+V) = 0$, the governing equilibrium equation becomes

$$\delta\frac{\pi E}{1 - \upsilon^2} \left\{ \int\limits_a^b \left(\frac{u^2}{r} + 2\upsilon u\frac{du}{dr} + r\left(\frac{du}{dr}\right)^2\right) h\,dr \right\}$$

$$- 2\pi\omega^2 \int\limits_a^b \rho u r^2 h\,dr - \frac{2\pi E}{(1 - \upsilon^2)}(1 + \upsilon)\alpha T \int\limits_a^b r\left(\frac{u}{r} + \frac{du}{dr}\right) h\,dr = 0 \quad (3)$$

Equation (3) is expressed in normalized co-ordinate to facilitate the numerical computation work and the governing equation takes the form

$$\delta \left[\begin{array}{c} \dfrac{E\bar{r}}{(1-v^2)} \int_0^1 \left\{ \dfrac{u^2}{(\bar{r}\xi+a)} + \dfrac{2vu}{\bar{r}}\left(\dfrac{du}{d\xi}\right) + \dfrac{(\bar{r}\xi+a)}{\bar{r}^2}\left(\dfrac{du}{d\xi}\right)^2 \right\} h\,d\xi \\[4mm] -2\omega^2\bar{r}\int_0^1 \rho(\bar{r}\xi+a)^2 u h\,d\xi - \\[4mm] \dfrac{2E(1+v)\alpha T\bar{r}}{(1-v^2)} \int_0^1 (\bar{r}\xi+a)\left(\dfrac{u}{(\bar{r}\xi+a)} + \dfrac{du}{\bar{r}d\xi}\right) h\,d\xi \end{array} \right] = 0 \qquad (4)$$

In Eq. (4), the non-dimensional radial co-ordinate, $\xi = \frac{(r-a)}{\bar{r}}$ where $\bar{r} = (b-a)$. The magnitude of the displacement field is dictated by external load and it is also governed by the boundary conditions of the disk. The displacement functions $u(\xi)$ in Eq. (4), can be approximated by a linear combination of sets of orthogonal coordinate functions as,

$$u\,(\xi) \cong \sum c_i \phi_1, \; i = 1, 2, \ldots, n \qquad (5)$$

The set of orthogonal functions ϕ_i are developed through Gram-Schmidt scheme, in which a starting function is used to generate the higher order orthogonal functions. To satisfy the boundary conditions of a solid disk we require, $u|_{(0)} = 0$ and $\sigma_r|_{(b)} = 0$. The displacement function that satisfies above boundary conditions is given below:

$$\phi_0\,(r) = \frac{\alpha}{r}\left[(1+v)\int_a^r T\,(r)\,r\,dr + \frac{(1-v)\,r^2}{b^2}\int_a^b T\,(r)\,r\,dr \right] \qquad (6)$$

For an annular disk the boundary conditions are $\sigma_r|_{(a)} = 0$ and $\sigma_r|_{(b)} = 0$, the corresponding displacement function is given by

$$\phi_0\,(r) = \frac{\alpha}{r}\left[(1+v)\int_a^r T\,(r)\,r\,dr + \frac{(1-v)\,r^2 + (1-v)\,a^2}{b^2+a^2}\int_a^b T\,(r)\,r\,dr \right] \; (7)$$

The function $\phi_0\,(r)$ is normalized and starting with it an orthogonal set of polynomial functions, following the Gram–Schmidt scheme, is generated. Now, substituting the series approximation of $u(\xi)$ in Eq. (4) and replacing the operator 'δ' by $\partial/\partial c_j, j = 1, 2, \ldots, n$, according to Galerkin's error minimization principle, we obtain the governing differential equation in matrix form, as given below

$$\frac{E\bar{r}}{1-v^2} \sum_{i=1}^n \sum_{j=1}^n c_i \int_0^1 \left\{ \frac{\phi_i \phi_j}{\bar{r}\xi+a} + \frac{v}{r}\left(\phi_i'\phi_j + \phi_i\phi_j'\right) + \frac{(\bar{r}\xi+a)}{\bar{r}^2}\phi_i'\phi_j' \right\} h\,d\xi$$

$$= \omega^2\bar{r} \sum_{j=1}^n \int_0^1 \rho\left\{(\bar{r}\xi+a)^2\,\phi_j\right\} h\,d\xi + \frac{E\alpha T}{(1-v)} \sum_{j=1}^n \int_0^1 \left\{\phi_j\bar{r} + (\bar{r}\xi+a)\,\phi_j'\right\} h\,d\xi \qquad (8)$$

where $()'$ indicates differentiation with respect to normalized coordinate ξ. Solution of Eq. (8) yields the solution vector $\{c_i\}$, obtained through a single step matrix inversion process. The numerical integration and differentiation and all other associated mathematical operations are carried out in the computational platform of MATLAB software.

3 Results and Discussions

The numerical values of the system parameters considered in the study are, $E = 210$ GPa, $\upsilon = 0.3$, $\rho = 7{,}800$ kg/m^3, $\alpha = 0.21 \times 10^{-6}\,°C^{-1}$ and $\sigma_0 = 350$ MPa. The analysis is carried out for four different disk geometries, uniform, taper, exponential, and parabolically varying thickness. For an exponentially varying disk, the thickness is given by $h(\xi) = h_0 \exp[-n(\xi)^k]$, whereas for a parabolically varying disk the expression becomes, $h(\xi) = [1 - n(\xi)^k]$. Uniform thickness and linearly varying thickness (taper) are obtained by setting $n = 0$ and $k = 1$ respectively, in the expression of parabolic thickness variation. The profiles considered in this study are derived by using constant volume criteria, which would help to characterize the performance of the disk in normalized plane. The thickness of the uniform disk is taken as 5 % of its outer radius and corresponding to its volume, other disk profiles are calculated. It is further assumed that for varying thickness disks, the tip thickness is 1 % of outer radius.

The dimensionless angular speed $\omega_y b\sqrt{\rho/\sigma_0}$ corresponding to the onset of yielding is defined as normalized limit angular speed and has been considered as one of the most important design parameters. The analysis is carried out for four temperature distribution profiles: uniform, parabolic, exponential and linear, as shown in Fig. 1 a–b. For each temperature distribution profile, two different

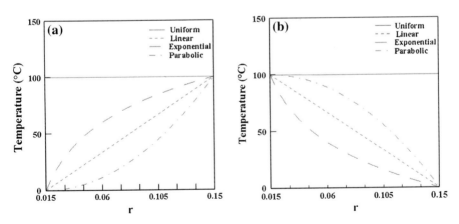

Fig. 1 Temperature distributions profiles for **a** Specified outer surface temperature (T_b), and **b** Specified inner surface temperature (T_a)

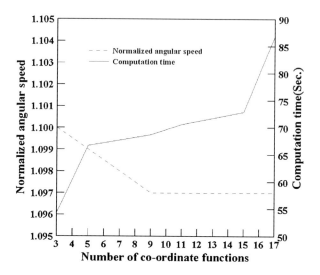

Fig. 2 Variation of normalized angular speed and computational time with number of co-ordinate functions

temperature boundary conditions are assumed, (1) $T_a = 0$, T_b specified and (2) T_a specified, $T_b = 0$. In Fig. 1, the boundary temperature is taken as 100 °C.

A convergence study for the adequate number of functions is carried out on normalized angular speed for a uniform disk and is reported in Fig. 2. The figure shows that a good convergence is achieved with nine functions and hence the subsequent computation for various parameter variations is carried out with 11 coordinate functions. The computational time, as shown in the figure is also a governing criterion for selection of the number of functions. All the functions are denoted numerically by using 24 Gauss points.

The validation of the present numerical scheme is made with the results of Vivio and Vullo [9] and is presented in Fig. 3 a–e. Figure 3 a–c shows the comparison for displacement field and radial and tangential stresses for rotating taper disks of constant density whereas in Fig. 3 (d, e) the disk is subjected to density variation. In all the cases a fairly good agreement is obtained. The details of disk geometries and other parameter values for this comparative study are not reported here as they may be found elsewhere [9].

For the prescribed disk geometries and temperature distributions, variations in normalized speed with outer surface temperature T_b are presented in Fig. 4 a–d. The inner surface temperature T_a is set to 0 °C. It is observed that with increase in outer surface temperature T_b, the normalized speed decreases for all type of disk geometries, and amongst the different geometries normalized speed is maximum for exponential disk.

The results are also presented for variation in inner surface temperature, which is shown in Fig. 5 a–d. In this case the outer surface temperature T_b is set to 0 °C. It is observed that the effect of disk geometry and temperature profile on the result with increase in inner surface temperature T_a remains almost identical with little quantitative change.

Thermo Mechanical Analysis and Design of Rotating Disks 479

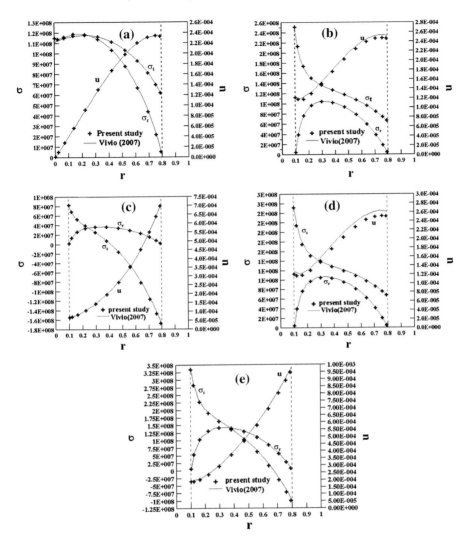

Fig. 3 Comparison of displacement (u), radial (σ_r) and tangential stresses (σ_t) for **a** Solid taper disk. **b** Annular taper disk. **c** Annular taper disk subjected to temperature gradient. **d** Annular taper disk with density variation and **e** Annular taper disk subjected to both density and temperature gradient

The effects of temperature distributions on normalized speed are shown in Figs. 4 and 5 for different disk geometries. The results are shown once again in Fig. 6 a, b for particular thickness profiles to highlight the influence of temperature distribution. In Fig. 6a effect of temperature distributions on normalized speed is shown for a uniform disk with variation in outer surface temperature and in Fig. 6b the case of an exponential disk with variation in inner surface temperature is considered.

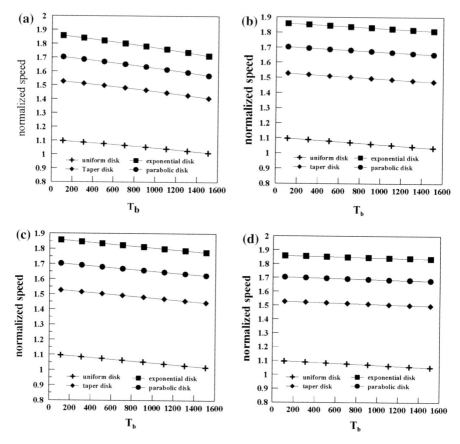

Fig. 4 Variation of normalized speed with outer surface temperature (T_b) for different temperature distributions **a** Uniform. **b** Linear. **c** Exponential and **d** Parabolic temperature

In the figures it is seen that the initial value of normalized speed is same for all temperature distributions and it decreases with increase in the surface temperature of the other boundary. The normalized speed decreases for all the temperature distributions for uniform disk geometry but the effect of uniform temperature is most significant. On the other hand for an exponential disk the uniform temperature distribution gets upper hand with relative changes in other temperature distributions.

The effect of density variation with temperature for the prescribed disk geometries and temperature distribution are depicted in Fig. 7 a–d. The inner surface temperature T_a is set to 0 °C and variation of normalized speed with increasing outer surface temperature (T_b) is plotted. The variation of density is governed by the relation $\rho(r) = \rho_0 \left[1 - \{\alpha T(r)\}^3\right]$, and for the specified value of thermal expansion coefficient, the effect is found to be small.

Thermo Mechanical Analysis and Design of Rotating Disks

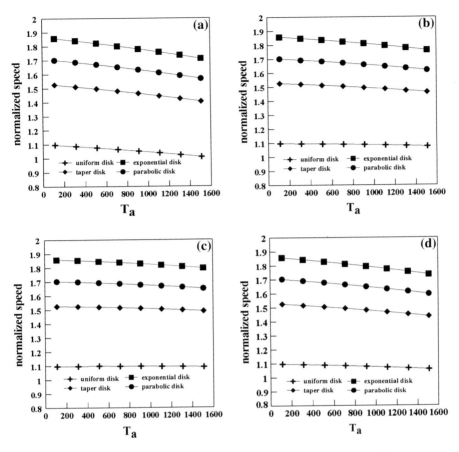

Fig. 5 Variation of normalized speed with inner surface temperature (T_a) for different temperature distributions: **a** Uniform. **b** Linear. **c** Exponential and **d** Parabolic temperature

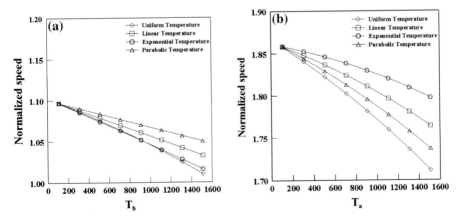

Fig. 6 Effect of temperature distributions on normalized speed for particular disk geometries: **a** Uniform disk with variation in outer surface temperature and **b** Exponential disk with variation in inner surface temperature

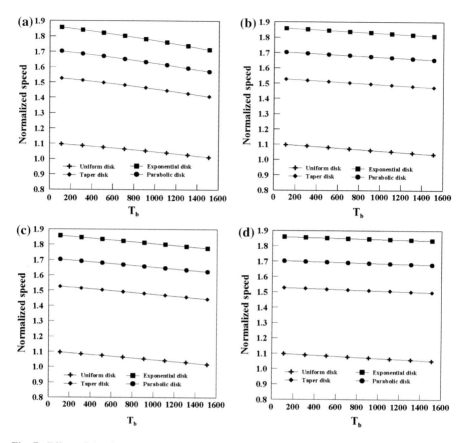

Fig. 7 Effect of density variation on normalized speed for different temperature distributions of outer surface (T_b): **a** Uniform. **b** Linear. **c** Exponential and **d** Parabolic temperature

4 Conclusion

This study deals with thermo-elastic stresses and deformation states of rotating solid or annular disks of varying thickness. A numerical method based on variational principle has been proposed to investigate the stress and deformation states of rotating disks taking the radial displacement field as unknown variable. The analysis is carried out for combined thermal load and inertia effects. Assuming a series approximation following Galerkin's principle, the solution of the governing partial differential equation is obtained. The solution algorithm is implemented with the help of MATLAB computational simulation software. The effect of disk geometry and temperature field variation on the performance of rotating disks is considered and normalized value of limit angular speed is furnished. These results are furnished graphically as design monograms which might prove helpful for the practicing engineers.

Appendix

a, b	Inner and outer radii of disk
c_i	The vector of unknown coefficients
h_0	Thickness at the root of the disk
h	Thickness at any radius r
n, k	Parameters controlling the thickness variation of the disk
r, θ, z	Radial, tangential and axial directions
u	Displacement field of the disk
U	Strain energy of the disk
V	Work potential of the disk
δ	Variational operator
$\rho, \upsilon, \alpha, E$	Density, Poisson's ratio, coefficient of thermal expansion and elasticity modulus of the disk material
ω	Dimensional angular speed of the disk
ω_y	Dimensional value of limit angular speed
T	Temperature at any radius r
T_a, T_b	Inner and outer surface temperature of the disk
$\epsilon_r, \epsilon_\theta$	Strains in radial and tangential direction
σ_r, σ_θ	Radial and tangential stresses
σ_{VM}	Von-Mises stress
σ_0	Yield stress of the disk material
$\emptyset_i$	Set of orthogonal functions used to approximate displacement field
$\emptyset_0$	Start function of orthogonal set of functions
ξ	Normalized co-ordinate in radial direction

References

1. Calladine CR (1969) Engineering plasticity. Pergamon Press, Oxford
2. Timoshenko S, Goodier JN (1970) Theory of elasticity. McGraw-Hill, New York
3. Srinath L (2001) Advanced strength of materials. Tata McGraw-Hill, New York
4. Manson SS (1947) The determination of elastic stresses in gas turbine disks. NACA Rep 871:241–251
5. Manson SS (1950) Direct method of design and stress analysis of rotating disks with temperature gradient. NACA Rep 952:103–116
6. Sterner SC, Saigal S, Kistler W, Dietrich DE (1994) A unified numerical approach for the analysis of rotating disks including turbine rotors. Int J Solids Struct 31(2):269–277
7. You LH, Zhang JJ (1999) Elastic-plastic stresses in a rotating solid disk. Int J Mech Sci 41(3):269–282
8. Milenson MB, Manson SS (1948) Determination of stresses in gas turbine disks subjected to plastic flow and creep. NACA Rep 906:277–292
9. Vivio F, Vullo V (2007) Elastic stress analysis of rotating converging conical disks subjected to thermal load and having variable density along the radius. Int J Solids Struct 44:7767–7784

10. Kordkheli HSA, Naghabadi R (2007) Thermoelastic analysis of a functionally graded rotating disk. Compos Struct 79:508–516
11. Afsar AM, Go J (2010) Finite element analysis of thermoelastic field in a rotating FGM circular disk. Appl Math Model 34:3309–3320
12. Peng XL, Li XF (2010) Thermal stresses in rotating functionally graded hollow cylinder disks. Compos Struct 92:1896–1904
13. Callioglu H, Demir E, Sayer M (2011) Thermal stress analysis of functionally graded rotating discs. Sci Res Essays 6(16):3437–3446
14. Bhowmick S, Misra D, Saha KN (2008) Approximate solution of limit angular speed for externally loaded rotating solid disk. Int J Mech Sci 50:163–174

Part VII
Recent Trends in Civil Engineering

A Study on Radioactivity Levels in Sedimentary and Igneous Rocks Used as Building Materials in and Around Tiruchirappalli District, Tamil Nadu, India

P. Shahul Hameed, G. Sankaran Pillai, G. Satheeshkumar and K. Jeevarenuka

Abstract The paper presents the results of the study on the activity concentrations of primordial radionuclides (238 U, 232 Th and 40 K) in the rock samples collected from eight sedimentary rocks and six igneous rocks which supply stones for construction of buildings in Tiruchirappalli district. In sedimentary rocks, the mean activity concentrations of 238 U, 232 Th and 40 K were found to be 5.4, 12.4, and 372.8 Bq kg^{-1}, respectively. On the other hand, in igneous rocks the mean activity concentrations of 238 U, 232 Th and 40 K were distinctly higher and found to be 15.5, 135 and 859.4 Bq kg^{-1}, respectively. The mean radium equivalent activity (Raeq) recorded in both sedimentary (32.8 Bq kg^{-1}) and igneous rocks (278 Bq kg^{-1}) was well within the limit prescribed for dwellings (370 Bq kg^{-1}) except Narthamalai ($S13$) (689.3 Bq kg^{-1}). The mean absorbed dose rate from igneous rock (124.5 nGy h^{-1}) exceeded the prescribed limit of 55 nGy h^{-1}. The mean annual effective dose from the sedimentary (0.089 mSv y^{-1}) and igneous rocks (0.63 mSv $y-1$) did not exceed the prescribed limit (1 mSv y^{-1}) except the igneous rock from Narthamalai ($S13$) (1.48 mSv y^{-1}). The study concludes that the sedimentary and the igneous rocks analysed were radiologically safe when used as building materials except igneous rock from Narthamalai ($S13$).

Keywords Primordials • Building materials • Radium equivalent • Absorbed dose rate • Annual effective dose rate

---

P. Shahul Hameed (✉) · G. Sankaran Pillai · G. Satheeshkumar
Environmental Research Centre, J. J. College of Engineering and Technology,
Tiruchirappalli 620009, Tamil Nadu, India
e-mail: drps_zo@yahoo.co.in

K. Jeevarenuka
Dept of Civil Engineering, Shivani Engineering College, Tiruchirappalli 620009,
Tamil Nadu, India

S. Sathiyamoorthy et al. (eds.), *Emerging Trends in Science,*
Engineering and Technology, Lecture Notes in Mechanical Engineering,
DOI: 10.1007/978-81-322-1007-8_44, © Springer India 2012

1 Introduction

Human beings are continuously exposed in their homes and places of work to ionise radiation and external gamma radiations [1]. The building materials derived from rocks and soil contain the natural radionuclides such as uranium (^{238}U) and thorium (^{232}Th) and their daughter products and singly occurring potassium (^{40}K). The internal radiation exposure, affecting respiratory tract, is due to radon and radon decay products which emanate from building materials [2]. Knowledge of radioactivity present in building materials enables one to assess any possible radiological risk to human health [3]. In order to assess the possible radiological hazards to human health, it is important to study the radioactivity levels emitted by the building materials. The worldwide average specific activity of ^{238}U, ^{232}Th and ^{40}K in the earth's crust is estimated at 32, 45 and 412 Bq kg^{-1}, respectively UNSCEAR [4]. The main objective of the present study is to identify and determine natural radionuclide activity in different rocks (Sedimentary and Igneous rocks) which are used as building materials in and around Tiruchirappalli district and to evaluate the specific activity concentrations, absorbed and annual effective dose rate to human. This study would also be useful for establishing a baseline data on primordial radionuclides present in different rock samples from various quarries located in Tiruchirappalli district.

2 Materials and Methods

2.1 Study Area

Tiruchirappalli city is located in central part of Tamil Nadu, South India (10° 48′ N: Lat. 78° 42′ E: Long.) on the northern bank of river Cauvery. It is the fourth largest city in Tamil Nadu and spread in an area of 5,114 km^2. The population density of the city is 3,601 persons per km^2. Tiruchirappalli district is naturally endowed with rich building material resources such as sand, stones, granites, cements and bricks which are avidly utilised by several adjacent districts also and the granites and cement are exported to other countries as well. In general, rocks are classified into three types namely igneous, metamorphic and sedimentary based on their formation from earth curst. However, igneous and sedimentary rocks are commonly found in Tiruchirappalli district.

2.2 Sample Collections

The rock samples were collected from 14 stone quarries spread over various locations of Tiruchirappalli district Fig. 1. About 2 kg of the rock sample was collected from each quarry. The solid matrix of the samples was powdered and sieved through 500 μm mesh. The samples were air-dried for several days to remove moisture and unwanted foreign particles. The samples obtained were oven dried

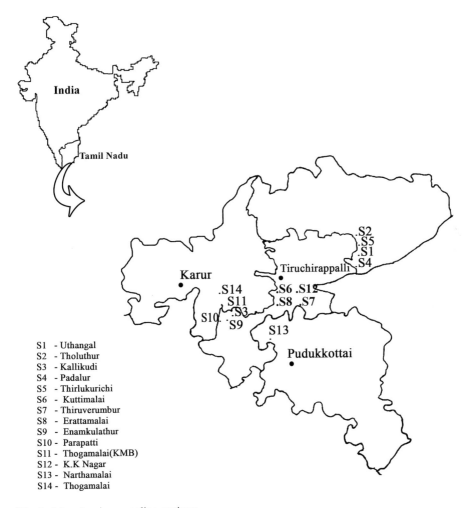

Fig. 1 Map showing sampling stations

at 110 °C until they reached a constant weight. Each sample was placed in 250 ml plastic Marinelli beaker having diameter of ~6 cm and height of ~9.5 cm. The container was sealed hermetically and externally using adhesive tape and kept aside for about a month to ensure secular equilibrium between radium and its daughter products before being taken for gamma ray spectrometric analysis [5].

2.3 Specific Activity Measurement

For the present study ORTEC make (USA) digiBASE 3″ × 3″ NaI (Tl) detector (Model: 905-4) was used for spectrometric analysis of the samples. The detector is coupled with 14 pin photomultiplier tube (PMT). The PMT is attached to a base which has inbuilt HV, preamplifier, digital pulse shaping unit and 8 k MCA. The base powered by a PC via USB connection. The MAESTRO-32 spectral analysis software is used for spectral analysis. The detector assembly is mounted inside a massive lead shielding of 10 cm thick to reduce the background radiation. In order to reduce the contribution of X-ray fluorescence, the inner surfaces of the lead shield is provided graded lining with copper sheet (0.8 mm thick) and cadmium sheet (1.5 mm thick). The standard reference materials RGU-1 (Uranium ore), RGTh-1 (Thorium ore), and RGK-1 (Potassium Sulphate) having the activity of 1,065, 1,608 and 4,810 Bq, respectively, procured from International Atomic Energy Agency (IAEA) were used for efficiency calibration of the system. These standards were obtained from Radiological Safety Division, Indira Gandhi Centre for Atomic Research, Kalpakkam. The sample container was placed on the top of the detector for counting. The same geometry was used to determine peak area of samples and references. The detector has an energy resolution of 57.6 keV for the 1173.5 keV of ^{60}Co gamma lines. The background and samples were counted for a period of 20,000 s. Background measurements were subtracted in order to get net counts for the sample. The concentration of ^{40}K was measured directly by its own gamma ray peak at 1,461 keV, whereas ^{238}U and ^{232}Th were estimated with the help of their gamma emitting daughter products. i.e. ^{214}Bi (1,764 keV) and ^{208}Tl (2614.6 keV), respectively. The below detectable limits (BDL) were 2.03 Bq kg^{-1} for ^{214}Bi, 4.7 Bq kg^{-1} for ^{208}Tl and 18.9 Bq kg^{-1} for ^{40}K and they were calculated using the following formula [6].

3 Results and Discussions

3.1 Activity Concentration

The data on activity concentrations of primordial radionuclides in the rock samples collected from 14 quarries spread over Tiruchirappalli district were presented in Table 1. Out of the 14 quarries sampled 8 are sedimentary rocks (stone) and 6 are igneous rocks (granite). It is evident from Table 1 that in sedimentary rocks, the activity concentrations of ^{238}U varied from BDL to 6.9 (Kuttimalai) Bq kg^{-1} with the mean value of 5.36 Bq kg^{-1} and ^{232}Th from BDL to 12.4 (Kuttimalai) Bq kg^{-1} with the mean value of 12.4 and ^{40}K from 70.6 (Uthangal) to 912.2 Bq kg^{-1} (Kuttimalai) with the mean value of 372.8. On the other hand, in igneous rock the concentrations of ^{238}U and ^{232}Th were found to be distinctly higher than that in sedimentary rock. In igneous rock the activity concentration of ^{238}U fluctuated from 5.5 (Enamkulathur) to 33.1 Bq kg^{-1} (Narthamalai) with the mean value of 15.5 Bq kg^{-1} and ^{232}Th from 38.9 (Enamkulathur) to 410 Bq kg^{-1} (Narthamalai) with the mean value of 135 Bq kg^{-1} and ^{40}K from 739.5 (Enamkulathur) to 995.3 Bq kg^{-1} (Thogamalai) with the mean value of 859.4 Bq kg^{-1}. Out of eight sedimentary rocks, the concentrations of ^{238}U in five stations were recorded BDL and only three stations ($S6$, $S7$ and $S8$) recorded the

A Study on Radioactivity Levels in Sedimentary and Igneous Rocks

Table 1 Specific activity concentration of ^{238}U, ^{232}Th and ^{40}K and radium equivalent (Ra_{eq}) in rock samples from Tiruchirappalli District

Sampling station	Station code	Type of rock	Specific activity concentration (Bq kg^{-1})			Ra_{eq} (Bq kg^{-1})
			^{238}U	^{232}Th	^{40}K	
Uthangal	S1	Sedimentary	BDL	BDL	70.6 ± 22.5	5.4
Tholuthur	S2	Sedimentary	BDL	BDL	613.6 ± 25.9	47.2
Kallikudi	S3	Sedimentary	BDL	BDL	137.5 ± 29.0	10.5
Padalur	S4	Sedimentary	BDL	BDL	176.6 ± 29.6	13.6
Thirlukurichi	S5	Sedimentary	BDL	BDL	92.5 ± 22.5	7.09
Kuttimalai	S6	Sedimentary	6.9 ± 2.7	12.4 ± 5.9	912.2 ± 29.0	94
Thiruverumbur	S7	Sedimentary	4.5 ± 2.9	BDL	165.7 ± 27.1	17.3
Erattamalai	S8	Sedimentary	4.7 ± 3.2	BDL	814.7 ± 35.5	67.4
Range			BDL–6.9	BDL–12.4	70.6–912.2	5.4–94.0
Mean			5.4	12.4	372.8	32.8
Enamkulathur	S9	Igneous	5.5 ± 3.0	38.9 ± 6.5	739.5 ± 29.7	118
Parapatti	S10	Igneous	23.5 ± 3.5	129.4 ± 8.0	763.4 ± 31.3	267.3
Thogamalai (KMB)	S11	Igneous	7.7 ± 3.4	115 ± 8.0	995.3 ± 33.5	249
K.K.Nagar	S12	Igneous	11.5 ± 2.7	40.1 ± 6.2	950.5 ± 29.3	142
Narthamalai	S13	Igneous	33.1 ± 4.2	410 ± 10.7	909.6 ± 33.9	689.3
TAMIN (Thogamalai)	S14	Igneous	12.2 ± 3.0 77 ± 8	798.3 ± 34.5	183.8	
Range			5.5–33.1	38.9–410	739.5–995.3	118–689.3
Mean			15.5	135	859.4	278

*BDL = Below Detectable Level

values of 6.9 ± 2.7, 4.5 ± 2.9 and 4.7 ± 3.2 Bq kg^{-1}, respectively, which are above detectable limit. On the other hand, the concentration of ^{232}Th was recorded only in one station ($S6$) having the value of 12.4 ± 5.9 Bq kg^{-1} and all the remaining five stations registered the value of BDL. However, the concentration of ^{40}K was recorded in all six stations. Unlike sedimentary rocks concentrations of ^{238}U, ^{232}Th and ^{40}K were distinctly higher in igneous rock. The activity of the radionuclides was calculated following the method described by [7]. The variation of natural radioactivity levels at different sampling sites is due to the variation of concentrations of ^{238}U, ^{232}Th and ^{40}K in the geological formation [8]. The radionuclides linked minerals such as zircon, iron oxides and fluorite play an important role in controlling the distribution of uranium and thorium. Zircon usually contains uranium and thorium concentrations which ranged from 0.01 to 0.19 % and 1 to 2 %, respectively [9]. Uranium in iron oxides is first trapped by adsorption [10]. The high uranium content in the mineralised granite and pegmatite is attributed to the ability of iron oxide in them on adsorbing uranium. The abundance of potassium to some extent is proportional to silica content of the rocks. Since, a higher silica content of granite is associated with higher value of ^{40}K [11]. It is observed that high percentage of U_3O_8, ThO_2 and K_2O in granite was responsible for higher activity concentration. Zircon typically contains 2–2000 g t^{-1} of ^{232}Th and 5–4000 g t^{-1} of ^{238}U [12]. A good co-relation exists between ^{232}Th and ^{238}U in igneous rocks ($r^2 = 0.80$), where as a poor correlation is observed between ^{232}Th and ^{40}K ($r^2 = 0.080$). The mean activity concentrations of ^{238}U, ^{232}Th and ^{40}K in sedimentary rocks are well within the average values for India namely, 14.8, 18.3 and 370 Bq kg^{-1}, respectively [13]. In contrast the mean activity concentration of ^{238}U, ^{232}Th and ^{40}K in igneous rocks is exceeding the India limit.

3.2 Radium Equivalent Activity (Ra_{eq})

Consequently, the distribution of primordials in rocks and soils is not uniform. Hence, the total exposure to radiation from the primordials has been defined in terms of radium equivalent (Ra_{eq}) in Bq kg^{-1} in order to compare the specific activity of materials containing variable amounts of ^{226}Ra, ^{232}Th and ^{40}K [14–16]. The radium equivalent is calculated on the assumption that 370 Bq kg^{-1} of ^{226}Ra or 260 Bq kg^{-1} of ^{232}Th or 4810 Bq kg^{-1} of ^{40}K produce the same gamma dose rate. Therefore, Ra_{eq} of any sample can be calculated using the following formula [17].

$$Ra_{eq} \left(Bq\, kg^{-1} \right) = A_{Ra} + (A_{Th} \times -1.43) + (A_K \times 1 - 0.077) \qquad (1)$$

where, A_{Ra}, A_{Th} and A_K are the specific activities of ^{238}U, ^{232} Th and ^{40}K. The safe value of Ra_{eq} in building materials must be less than 370 Bq kg^{-1} in order to limit the annual effective dose to 1 mSv for the general public [16]. Table 1 also provides the Ra_{eq} values which are a measure of total gamma radiation in a given sample. In sedimentary rocks, Ra_{eq} ranged from 5.4 to 94 Bq kg^{-1} with the mean value of 32.8 Bq kg^{-1} while in igneous rocks Ra_{eq} ranged from 118 (Enamlulathur) to 689 Bq kg^{-1} (Narthamalai) with the mean value of 278 Bq kg^{-1}. It has been recommended [18] to use the construction materials with an average Ra_{eq} value less than 370 Bq kg^{-1} for dwellings. The results of the present study indicate that the sedimentary rocks do not pose a significant radiological hazard when used for construction of buildings. The results also indicated that radium, thorium and potassium are not uniformly distributed in rocks from which building materials are derived [19]. Slunga et al, 1988 reported that the radioactivity varies often greatly, over a short distance. The variation between the concentrations of the radionuclides in the present study area depends on the type of geological formation of rocks. In sedimentary rocks, the mean activity concentration of ^{238}U, ^{232}Th and ^{40}K is much lower than the world average values reported by UNSCEAR [4]. In contrast, in igneous rock, the mean activity concentration of ^{232}Th and ^{40}K is much higher than the world average, whereas the concentration of ^{238}U is well within the limit.

The mean Ra_{eq} values are convenient to compare the total gamma activity of any given samples. As such, the maximum Ra_{eq} value of 289 Bq kg^{-1} was reported from Germany [20] and minimum value of 52 Bq kg^{-1} in Kurdistan

A Study on Radioactivity Levels in Sedimentary and Igneous Rocks 493

(Iraq) [21]. However, in the present study, the mean Ra_{eq} value for the igneous rock samples was found to be 278 Bq kg^{-1} which is well below the recommended limit of 370 Bq kg^{-1}. Since mean values of Ra_{eq} for sedimentary and igneous rocks analysed in the present study are well below the recommended level and they do not pose any radiological hazard when used as building materials.

3.3 The Absorbed Dose Rate in Air from Rock Samples

The absorbed dose rate(D) due to gamma radiations in air at 1 m above the ground surface for the uniform distribution of the naturally occurring radio nuclides (^{226}Ra, ^{232}Th and ^{40}K) were calculated based on guidelines provided by [22, 23]. It is assumed that the contributions from other radionuclides were insignificant. Therefore, Absorbed dose (D) can be calculated using the following formula

$$D\ (\mathrm{nGy}\ h^{-1}) = 0.462 A_{Ra} + 0.621 A_{Th} + 0.0417 A_K \qquad (2)$$

where A_{Ra}, A_{Th} and A_K are the specific activities of ^{238}U, ^{232}Th and ^{40}K. The results are given in Table 2. From the result it can be noticed that igneous rock (granites) exceed the safety limit of 55 nGy h^{-1} as proposed by UNSECAR [4]. While in sedimentary rock the dose rate well within the safety limit.

Table 2 The absorbed dose rate and annual effective dose rate

Sample code	Absorbed dose rate (D) (nGy h^{-1})	Annual effective dose rate	
		E_{in} (mSv y^{-1})	E_{out} (mSv y^{-1})
Sedimentary rock			
S1	3.07	0.015	0.004
S2	26.9	0.132	0.033
S3	6.7	0.033	0.008
S4	7.8	0.038	0.010
S5	6.6	0.032	0.008
S6	48.7	0.238	0.059
S7	9.5	0.047	0.012
S8	36.9	0.181	0.045
Range	3.07–48.7	0.015–0.238	0.044–0.059
Mean	18.27	0.089	0.022
Igneous rock			
S9	56.8	0.278	0.069
S10	120.8	0.592	0.148
S11	114.5	0.692	0.140
S12	69.1	0.339	0.084
S13	300.8	1.475	0.360
S14		0.419	0.104
Range	56.8–300.8	0.278–1.475	0.069–0.360
Mean	124.5	0.63	0.15

3.4 Estimation of Annual Effective Dose

The absorbed dose rate is converted into annual effective dose by a conversion factor (0.7 SvG y^{-1}) and indoor occupancy factor of 0.8 as proposed by UNSCEAR [4]. The conversion formula for annual effective dose is given below.

$$E_{in}(\text{Indoor effective dose} = D_\gamma \left(\text{nGy } h^{-1}\right) \times 10^{-6} \times 8760 h/y \times 0.8 \times 0.7 \text{ SvG } y^{-1})$$

(3)

$$E_{out}(\text{Outdoor effective dose} = D_\gamma \left(\text{nGy } h^{-1}\right) \times 10^{-6} \times 8760 h/y \times 0.2 \times 0.7 \text{ SvG } y^{-1})$$

(4)

The data on indoor and outdoor annual effective dose from sedimentary and igneous rocks sample analysed the present study are presented in Table 2. The mean indoor annual effective dose rate for the sedimentary rocks was 0.089 mSv y^{-1} and igneous rocks 0.63 mSv y^{-1}. Similarly, the mean outdoor annual effective dose rate for the sedimentary rocks was 0.022 mSv y^{-1} and igneous rocks 0.15 mSv y^{-1}. The data also indicated that these values are well within the safety limit (1 mSv y^{-1}) as proposed by UNSCEAR [4].

4 Conclusion

A preliminary database on the natural radioactivity levels in sedimentary and igneous rocks of Tiruchirappalli district is generated. The distribution of primordials in this study region is not uniform. The high radioactivity levels are found in granite type of rocks (igneous). All other radiological parameters like, radium equivalent, absorbed dose rate and effective dose are derived. Except the granites from Narthamalai ($S13$) in all other rocks the values are within the recommended limit which means that the building materials do not pose any radiological hazards.

Acknowledgments The authors thankfully acknowledge Atomic Energy Regulatory Board, Govt. of India, Mumbai for funding (Project No: AERB/CSRP/45/05/2010) and Shri. R. Mathiyarasu and Shri. S. Balasundar, Radiation Safety Division, Indira Gandhi Centre for Atomic Research, Kalpakkam for technical support and Prof. K. Ponnusamy, Chairman, J. J College of Engineering and Technology for providing facilities and support.

References

1. Damla N, Cevik U, Kobya AI, Celik A (2011) Yildirim: assessment of natural radioactivity and mass attenuation coefficients of brick and roofing tile used in Turkey. Radiat Meas 46:701–708
2. European Commission (EC): radiation protection 112 (1999) Radiological protection principles concerning the natural radioactivity of building materials directorate-general environment, nuclear safety and civil protection
3. Kumar A, Kumar M, Singh B, Singh S (2003) Natural activities of ^{238}U, ^{232}Th and ^{40}K in some India building materials. Radiat Meas 36:465–469

4. United Nations Scientific Committee on the Effect of Atomic Radiation (UNSCEAR). Report to the general assembly (2008) Annex B: exposures of the public and workers from various sources of radiation
5. Ramasamy V, Murugesan S, Mullainathan S (2004) Gamma ray spectrometric analysis of primordial radionuclides in sediments of Cauvery River in Tamilnadu. India Ecologica 2:83–88
6. Kpelgo DO, Lawluvi H, Faanu A, Awudu AR, Deatanyah P, Wotorchi SG, Arwui CC, Emi-Reynolds G, Darko EO (2011) Natural radioactivity and its associated radiological hazards in Ghanaian cement. Res J Environ Earth Sci 3(2):160–166
7. Sannappa J, Chandrashekara MS, Sathish LA, Paramesh L, Venkataramaiah P (2003) Study of background radiation dose in Mysore city, Karnataka state. India Radiat Meas 37:55–65
8. El-Arabi AM (2007) ^{226}Ra, ^{232}Th and ^{40}K concentrations in igneous rocks from Eastern desert, Egypt and its radiological implications. Radiat Meas 42:94–100
9. Cuney M, LeFort P, Wangeg Z (1987) Geology of granites and their metallogenetic relations. Science Press, Moscow, pp 852–873
10. Speer J, Solberg T, Becker S (1981) Petrography of the Uranium-bearing minerals of the Liberty Hill Pluton. South Carolina: phase assemblages and migration of uranium in granitoid rocks. Ecom Geol 76:162–175
11. Sannappa J, Ningappa C, Prakash Narasimha KN (2010) Natural radioactivity levels in granite regions of Karnataka State. Indian J Pure Appl Phys 48:817–819
12. Deer WA, Howie RA, Zussman J (1997) Rock forming minerals: orthosilicates. Geol Soc, UK, p 918
13. Mishra UC, Sadasivan S (1971) Natural radioactivity levels in Indian soils. J Sci Ind Res 30:59–62
14. OCED: exposure to radiation from the natural radioactivity in building materials (1979) Reported by a group of experts of the OCED, nuclear energy agency, Paris
15. Ionising radiation sources and biological effects (1982) United Nations scientific committee on the effects of atomic radiation (UNSCEAR), A/37/45, New York
16. Beretka J, Mathew PJ (1985) Natural radioactivity of Australian building materials, industrial wastes and by-products. Health Phys 48:87–95
17. Sroor A, Afifi S, Abdel-Haleem A, Abdel-Sammad M (2002) Environmental pollutant isotope measurements and natural radioactivity assessment for North Tushki area, South Western desert. Egypt Appl Radiat Isot 7:1–10
18. Somlai J, Horvath M, Kanyar B, Lendvai Z, Nementh CS (1998) Radiation hazard of coal slags as building material in Tatabanya town (Hungary). Health Phys 75(6):648–651
19. Slunga E (1988) Radon classification of building ground. Radiat Prot Dosim 24(114):39–42
20. Ahmed NK, Abbady A, El-Arabi AM, Michel R, El-Kamel AH, Abbady AGE (2006) Comparative study of the natural radioactivity of some selected rocks from Egypt and Germany. Indian J Pure Appl Phys 44:209–215
21. Ahmed A, Hussein I (2011) Natural radioactivity measurements of Basalt Rocks in Sidakan District Northeastern of Kurdistan region-Iraq. World Acad Sci, Eng Technol 74:895–905
22. Sources and effects of ionising radiation (2000) United Nations scientific committee on the effect of atomic radiation (UNSCEAR), New York
23. Beck HL, Decompo J, Gologak J (1972) In Situ Ge (Li) and NaI (Tl) gamma ray spectrometry. Health and safety Laboratory, Reported by HASL, AEC, New York, p 258

An Experimental Approach on Enhancing the Concrete Strength and its Resisting Property by Using Nano Silica Fly Ash

Yuvaraj Shanmuga Sundaram, Dinesh Nagarajan and Suji Mohankumar

Abstract This paper deals with the study of Nanotechnology experimentation in Civil Engineering which includes the development, advantages and limitations of Nano concreting technologies. For reducing carbon emission during cement manufacturing fly ash is used as a replacement in ordinary Portland cement which is termed as Portland pozzolana cement (PPC), this inclusion relatively increases the workability and the corrosion resisting capacity in concrete, but this replacement of fly ash in the ordinary Portland cement deviates the concrete strength consequently. Therefore, here we added Nano silica as an additive to fill up the deviation, and it is possible because the silica (S) in the sand reacts with calcium hydrate (CH) in the cement at Nano scale to form C–S–H bond as it improves the strengthening factor of concrete, which are in turn helpful in achieving high compressive strength even in early days. This process proved that the increase in strength may have a possibility of turning the concrete less alkaline because as the concentration of CH crystals is reduced the alkalinity of concrete will be reduced which can cause corrosion in reinforcement bars, Hence by the addition of Nanosilica, a significantly improved corrosion resistant property was identified in our experimental research. Also, the performance of reinforced Nano concrete and the fly ash added RC Beam Column joints were casted and their flexure strength results were compared with one another and their test results are presented in this paper.

Y. S. Sundaram (✉)
Research Scholar, Karpagam University, Coimbatore, India
e-mail: sanyuvayad@gmail.com

D. Nagarajan
PG Student, Karpagam University, Coimbatore, India
e-mail: dinukanch@gmail.com

S. Mohankumar
Dean/Civil Engineering RVS Faculty of Engineering, Coimbatore, India
e-mail: sujimohan2002@gmail.com

S. Sathiyamoorthy et al. (eds.), *Emerging Trends in Science,*
Engineering and Technology, Lecture Notes in Mechanical Engineering,
DOI: 10.1007/978-81-322-1007-8_45, © Springer India 2012

1 Introduction

Concrete is at something of a crossroads: there are many prospects and some risks involved in it. Regarding the opportunities concrete comes into a practice in the implementation by various researches, areas to innovate for engineers, use of different material which ensures the manufacturers to learn and to produce the concepts of new ingredients to be use which focus on the changes that are required to champion concrete and maintain its dominance within the global construction industry. Many researches has shown that a state-of-the-art process for high-performance cement adds a new changes to 'modern' cement technology, similarly it's the time to understand about what is "NANO TECHNOLOGY" and their development and use in construction industry as many innovations in concrete techniques are adopted. As concrete is most usable material in construction industry it's been requiring to improve its quality. The main aim of this research is to outline promising research areas in the field of Nano technologies in the Construction world.

1.1 What is Nano Technology?

Nanotechnology is defined as fabrication of devices or materials with atomic or molecular scale precision Nanotechnology is usually associated with study of materials of micro size i.e. one billionth of a meter (a Nanometer) or 10^{-9} m [1].

1.2 Definition of Nano Concrete

For discussions presented in this research, Nano concrete is projected as concrete made with Portland cement particles that are less than 500 Nano-meters as the cementing agent. Normally, cement particle sizes ranges from a few Nano-meters to a maximum of about 100 μm, at this point the average particle size of micro cement is reduced to 5 μm as an order of magnitude reduction is needed to produce Nano-cement [1]. The SEM image of the Nano silica we had taken for our investigation is shown in Fig. 1.

1.3 Nanoscale Concrete

Cement Reactivity at Nano Scale

We need to better know how to control the setting time of concrete. The evolution of the hydrogen profile shows the timing of the surface layer's breakdown [2].

An Experimental Approach on Enhancing the Concrete Strength

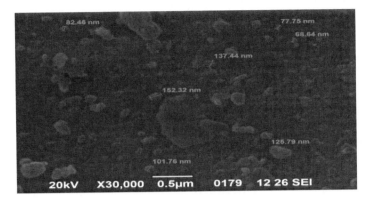

Fig. 1 The SEM image of nano silica

This information can be used to study the concrete setting process as a function of time, cement chemistry and temperature. For example, researchers used Nuclear Resonance Reaction Analysis (NRRA) to determine that in cement hydrating at 30 °C, the breakdown occurs for about 1.5 h. The surface disintegration then releases accumulated silicate into the surrounding solution, as it reacts with calcium ions to form a calcium-silicate hydrate gel, which acts as a binding agent with cement grains together and sets the concrete [4].

2 Objectives of this Investigation

To determine the improvement of Corrosion resistance and concrete Strength by using Nano silica in concrete:
  A. The objective of this investigation is

1. Studies on the Corrosion resisting property and Flexural strength of concrete were made separately by partially replacing of cement with fly ash and find out the optimum replacement percentage of fly ash to the weight of cement.
2. Different proportions of replacing of fly ash with cement for studies are 0, 10, 20, 30, 40, 50, 60, 70 and 80 %.
3. Same proportions of replacing of cement with fly ash with addition of Nano silica at a rate 2.5 % to the weight of cement are done and their results are studied.
4. A comparative analysis of Corrosion resistance of partially replaced fly ash (10, 20, 30, 40, 50, 60, 70 and 80 %) Concrete with the Nano concrete were carried out.
5. A comparative analysis of Flexural strength of partially replaced fly ash (20, 40 and 60 %) Concrete with the Nano concrete were carried out.
6. Then a comparative analysis Fly ash concrete and Nano Silica Fly ash concrete were listed out.

3 Materials and Methods

3.1 An Investigation of Nano Silica in Cement Hydration Process

With the introduction of nano technology, materials have been urbanized that can be applied to high-performance concrete design mix. Nano silica get reacts with calcium hydroxide (CH) to increase the strength carrying structure of cement: calcium silica hydrate (CSH) [5–8]. In this paper, interaction has been developed to distinguish the benefits when using unusual sizes of nano silica in cement. An experimental analysis was carried out to determine the performance of Nano silica. From these experiments the heat of hydration of several cement design mix was calculated. Finally, the compressive strength was determined for each cement paste. During these experiments it was found that as the silica particles decreased in size and their size distribution broadened, the CSHs became more inflexible; this increased the compressive strength.

3.2 Specimen Casting and Curing

- Initially cement, fly ash and Nano silica are taken and mixed for 1 min and after that the fine aggregate was also included and mixed and coarse aggregate was mixed thoroughly in dry state and cement were mixed for 1 min.
- Mixing with Super plasticizer is done with water as its being added within 2 min and the concrete was allowed to mix for 3 min overall.
- Compaction process is done by using table vibrator and the specimens are cured for 7, 14 and 28 days respectively.

4 Corrosion Test Setup

The test was carried out on 150×300 mm size concrete cylinders with a rod of height 450 mm in the center throughout the specimen. The test specimens are marked and removed from the molds after 24–48 h from casting depending upon the percentage of fly ash and submerged in clean fresh water for curing. The impressed current technique, which is commonly used for accelerating reinforcement corrosion in concrete specimen, was used for testing. The specimen was placed in water bath containing calculated quantity of dissolved sodium chloride (table salt) to act as electrolyte. Calculated voltage of current was passed through the concrete specimen till the concrete cracks. The percentage weight loss in rebars and the width of cracks in concrete were studied (Fig. 2).

Fig. 2 Impressed current test setup

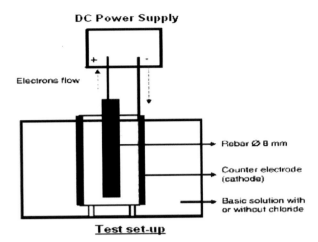

4.1 Results Attained for Corrosion Test

The testing was carried out using the Test Set Up and the graphs for the comparison of Mass of Actual Rust, Degree of Induced Corrosion and Crack Width between the Fly Ash added concrete and nS+Fly ash added concrete are plotted (Graphs. 1, 2).

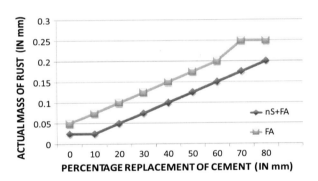

Graph. 1 Degree of induced corrosion comparison result for (M_{40}) 28 day

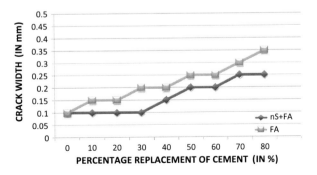

Graph. 2 Crack width comparison result for (M_{40}) 28 days

5 Flexure Test

The beam specimen casted for determining the flexure strength of concrete is 15 × 15 × 75 cm. The bed of machine should be provided with two steel rollers of 38 mm diameter on which the specimen is supported. Rollers are placed at center to center distance of 60 cm.

The test specimen is casted and cured for 28 days and tested for maximum load. The Flexural strength or Modulus of rupture (fb) is calculated using the formula,

$$fb = 3Pa/bd2$$

Where,

P maximum load,
a distance of loading from support,
b Breadth of specimen,
d Depth of specimen.

Table 1 Flexural strength attained

Specimen	Average load in (kN)	Flexural strength $\sigma = 3FL/2BD^2$ (kN/m2)
CS	17.74	3.94
F_{20}	18.09	4.02
F_{40}	15.48	3.44
F_{60}	11.48	2.55
NC	33.81	7.5
FN_{20}	22.14	5.76
FN_{40}	20.68	4.92
FN_{60}	17.82	3.96

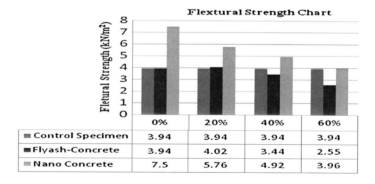

Graph. 3 Flexural strength chart

5.1 Results and Discussions

For the sample under three-point bending setup the following results are made:

At the outside of the bend (convex face) the stress will be at its maximum tensile value. These inner and outer edges of the beam or rod are known as the 'extreme fibers' (Table 1, Graph. 3).

6 Conclusion

From the above experiments and results we safely conclude the points as follow.

1. Nano concrete could control the carbon dioxide emission from the earth which is shown by using fly ash concrete products instead of cement concrete.
2. Thus the Nano particles which is in the form of silica can easily react with cement particles which are normally in Nano scale initiate the CSH reaction and hence its tend to accelerate the flexural strength of concrete.
3. It is found that the corrosion resisting property of the nS added concrete is comparatively higher than ordinary fly ash concrete.
4. The corrosion resistance of optimum percentage replacement of fly ash is higher in nano concrete than the ordinary fly ash concrete.
5. From the flexure graph, it is obvious that within FN_{20} and FN_{40}, the efficiency of concrete is maximum.
6. For concrete structures construction for petrol tanks, bunkers and silos, oil well, this type of special concrete can be preferred to influence more strength and performance.

References

1. Balaguru PN (2005), Nanotechnology and concrete: background, opportunities and challenges. In: Proceedings of the International conference—application of technology in concrete design, Scotland, UK, pp. 113–122
2. Boresi AP, Chong KP, Saigal S (2002) Approximate solution methods in engineering mechanics. Wiley, New York, pp 265–280
3. Balaguru P, Shah SP (1992) Fiber reinforced cement composites. McGraw-Hill, New York 530
4. Srivastava D, Wei C, Cho K (2003) Nanomechanics of carbon nanotubes and composites. Appl Mech Rev 56:215–230
5. Li G (2004) Properties of high-volume fly ash concrete incorporating Nano-SiO2. Cem Concr Res 34:1043–1049
6. Elmenshawi A, Brown T (2010) Hysteretic energy and damping capacity of flexural elements constructed with different concrete strengths. Eng Struct from the j Elsevier 32: pp 297–305
7. Mukherjee A, Joshi M FRPC reinforced concrete beam-column joints under cyclic excitation. Department of civil engineering, Indian Institute of Technology, Mumbai
8. Bjornstrom J, Martinelli A, Matic A, Borjesson L, Panas I (2004) Accelerating effects of colloidal nano-silica for beneficial calcium–silicate–hydrate formation in cement. Chem Phys Lett 392:242–248

Numerical Analysis and Study of Nanomaterials for Repair in Distressed Irrigation Structures

J. Nirmala, G. Dhanalakshmi and A. Rajaraman

Abstract In irrigation structures, seepage through the body wall and foundation play a major role for the assessment of failures in the structures. To design any kind of structure for the proper irrigation, the effect due to seepage should be considered widely. Many methods have been suggested for the prediction of seepage. Visualization of seepage route gives more impact than the numerical solution for that particular problem which can be achieved by MATLAB and GUI. Concrete resurfacer, Quick Setting cement, anchoring cement, hydraulic water stop cement, concrete bonding adhesive and concrete acrylic fortifier are some of the materials used to repair the concrete in any mass structure like dams, weirs and barrages. But, in practical, it is very difficult to extract the efficiency utmost from the materials used for repairing. So, Engineers must alter the method of transferring the repair method to the advanced pattern to enhance the behavior of the structure. In this view, nanomaterials for repairing are considered in this study. Nanomaterials are very small particles of material either by themselves or by their manipulation to create new large scale materials. The size of the particles, though, is very important because at the length scale of the nanometer, 10–9 m, the properties of the material actually become changed and enhanced. The precise size at which these changes are manifested varies between materials, but is usually in the order of 100 nm or less.

Keywords Distress • Finite element method • Graphical visualization • Nanomaterials

---

J. Nirmala (✉)
Department of Civil Engineering, Parisutham Institute of Technology and Science, Thanjavur, Tamil Nadu, India
e-mail: nimmijegan@gmail.com

G. Dhanalakshmi
Department of Civil Engineering, Oxford Engineering College, Trichy, Tamil Nadu, India
e-mail: g_dhanalakshmi2003@yahoo.co.in

A. Rajaraman
Former Director/Structural Engineering Research Center, Chennai, Tamil Nadu, India
e-mail: arraman_2000@yahoo.com

S. Sathiyamoorthy et al. (eds.), *Emerging Trends in Science, Engineering and Technology*, Lecture Notes in Mechanical Engineering, DOI: 10.1007/978-81-322-1007-8_46, © Springer India 2012

1 Need for Numerical Analysis and Finite Element Method

Numerical analysis is essential for scientific computing in the Engineering field. Numerical analysis is the study of algorithms for solving the problems of continuous mathematics, by which we mean problems involving real or complex variables. Numerical analysis is built on a strong foundation which is the mathematical subject of approximation theory. Finite Element Methods (FEMs) divides the problem of interest into a mesh of geometric shapes called finite elements. The potential within an element is described by a function that depends on its values at the cell corners and parameters defining the state of the element. Several such cells are assembled to solve the entire problem. A total energy associated with the mesh configuration is found as part of the calculation and this is minimized by adjusting the parameters specifying the elements. Williams (1995) stated that the solution can be refined by subdividing the regions of the mesh that contribute most to the total "energy" of the solution [1]. General purpose programs to perform these calculations are fairly complicated but fortunately very efficient commercial packages are available.

2 Numerical Solutions to Differential Equations

In Science and Engineering Solving differential equations is a fundamental problem. The seepage is analyzed only by the Laplace's equation $d^2\varphi/dx^2 + d^2\varphi/dy^2 = 0$ plus some boundary conditions. Sometimes we can find closed-form solutions using calculus. However, in general we must route to numerical solutions. In Differential equation all dependent variables are a function of a single independent variable where as in partial differential equation (PDE) in which all dependent variables are a function of several independent variables. Seepage analysis of irrigation structures works mainly on the basis of two dimensional Laplace equation. The PDE Toolbox provides a powerful and flexible environment for the study and solution of PDEs in two space dimensions and time. The equations are discretized by the FEM.

3 MATLAB

The objectives of the PDE Toolbox are to

Define a PDE problem, with 2-D regions, boundary conditions, and PDE coefficients.

Numerically solve the PDE problem, i.e., generate unstructured meshes, discretize the equations, and produce an approximation to the solution.
Visualize the results.

The graphical user interface (GUI), which is a self-contained graphical environment for PDE solving. For common applications we can use the specific physical terms rather than abstract coefficients. Using PDE Tool requires no knowledge of the mathematics behind the PDE, the numerical schemes or MATLAB. Advanced applications are also possible by defining the domain geometry, boundary conditions, and mesh description to the MATLAB workspace. With the available applications mode in MATLAB generic scalar mode is used for the numerical solution to be found out. Entire domain is treated as solid geometry model.

The basic equation of the PDE Toolbox is which we shall refer to as the Elliptic equation, regardless of whether its coefficients and boundary conditions make the PDE problem elliptic in the mathematical sense. Generalized Neumann and mixed boundary conditions are evaluated and the equation $H \times u = R$ represents the Dirichlet type boundary conditions.

Figure 1, shows the structural domain taken for the numerical analysis for the distress. Plot ratio determines the relative size of X and Y axes for the width and height of the domain.

The structural domain is constructed by the traditional concrete. The failure is found out at the junction of upstream and downstream structure. Quantitative study for the flaw was conducted in the structure for various parameters was carried out. The size of the flaw is varied for new depth and width. Following results show the seepage path through the domain various depth of the flaw in terms of the height of the domain on upstream side.

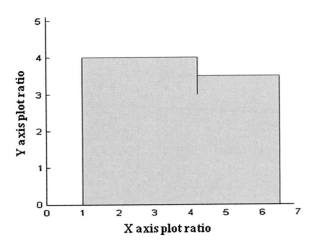

Fig. 1 Structural outline of the domain

4 Results and Discussions

The following results show the path of the seepage water in a generic scalar mode and it has been indicated as the red color arrow heads. The scalar factor for the pressure distribution is also well emphasized. It is clearly seen that the distribution factor is even for all the three cases. The contours were shown as black color curves and the values are useful for the geometrical parameters of the domain.

The above shown results are varied for repeated refinement or meshing pattern permitted in MATLAB. Initial mesh, refine, Jiggle are the three meshing pattern taken for the study and incorporated in the coding. Pattern refining is repeated for three times. The values of the contour in the maximum and minimum level was identified and the graphical representation was drawn for both cases. The entire numerical values have been normalized and found that initial mesh and two times repeating the refine mesh give the convergence of the result.

From Figs. 2, 3 and 4, it was found that the scalar property shown in the color bar remains the same in all the cases. So, the size of the failure is negligence to the bearing resistance of the structure. In the finalized selected meshing pattern, the depth of the crack is considered as constant and width of the flaw was varied. Maximum and minimum values of contour were noted down and the following graphs were drawn.

Width of the crack gives uniform maximum contour value as represented in the Fig. 5. But, the minimum value of contour varies for the various depth of the

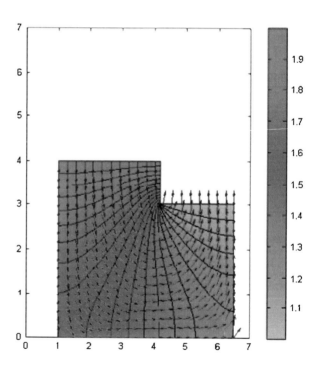

Fig. 2 Seepage route through the domain while the flaw is 0.005 h deep

Numerical Analysis and Study of Nanomaterials 509

Fig. 3 Seepage route through the domain while the flaw is 0.01 h deep

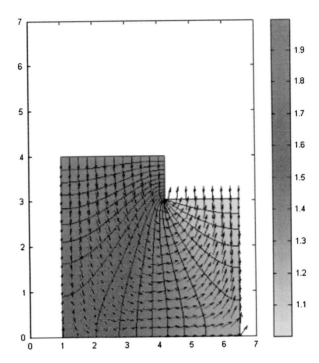

Fig. 4 Seepage route through the domain while the flaw is 0.1 h deep

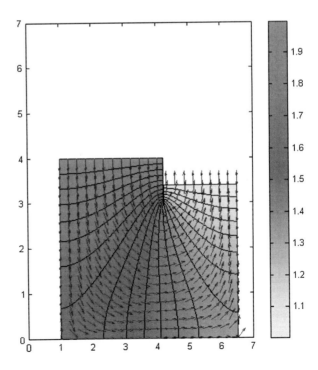

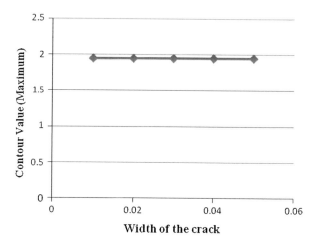

Fig. 5 Maximum contour value for various size of crack

flaw as shown in the Fig. 6. The scalar property shown by the contour in the plot is useful to predict the general geometrical changes of the flaw considered in the domain.

5 Need and Features of Nanomaterials

The notion of size as an independent degree of freedom which can be manipulated independent of composition, temperature and pressure to yield materials that possess new properties not exhibited by their conventional counterparts, is only very recently being realized from a commercial perspective. When materials possesses size features that are on the order of a few billionths of a meter, those materials

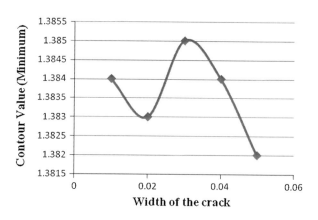

Fig. 6 Minimum contour value for various size of crack

often exhibit new properties not found in their ordinary material counterparts and those properties can be modified independently of the materials composition.

The trick is to produce nanomaterials with tightly controlled size and size distribution so that the size dependent properties emerge and are distinguishable. Coppa (1995) [2] illustrated that Nanomaterials have a wide particle size distribution may exhibit unique properties, but those properties are a statistical result of the ensemble of individual sizes present in the distribution. The compressive strength of the nano concrete by adding the super plasticizers is more than 50 % of ordinary concrete in 1, 7, 28 days. Figure 7 shows the comparison of nanomaterials with traditional plain concrete. Sobolev et al. [3] examined and found the application of Gaja a superplasticizer containing nano-SiO_2 particles, at a dosage of 1.3 % provides nearly twofold increase in concrete compressive strength at the age of 7 and 28 days. The early strength of the concrete with Gaja was 68.2 MPa, which is approximately three times higher than that of conventional concrete.

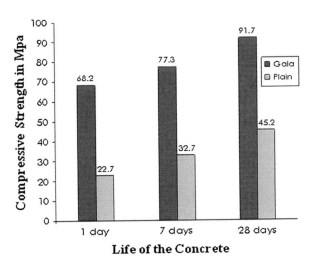

Fig. 7 Compressive strength of concrete with Gaja [3]

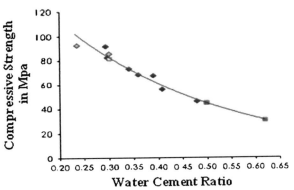

Fig. 8 Compressive strength of concrete with Gaja

Fig. 9 Compressive strength of nano silica

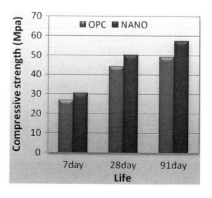

Figure 8 shows the 28th day compressive strength of the concrete made with Gaja had demonstrated the dependence on W/C.

6 Compressive Strength of Nanosilica

Khanzadi et al. [4] proved that Fig. 4 show the compressive strength of the specimens at 7th, 28th and 91st days. It can be seen that, when nano particles in a small amounts are added, the compressive strengths of concrete can be enhanced. This result is a result of increasing the bound strength of cement paste-aggregate interface by means of the filing effect of nano silica particles.

7 Conclusion

The created domain will undergo coarser meshing to analyze the parameters and the results will be improved in terms of compressive strength of the material. The manual method was overwhelmed by the numerical analysis for all the triangular nodes considered for refinement. The total energy associated in the mesh configuration will trim the results for numerical solution. As the nanomaterials are of high resistive strength as shown in the Fig. 9, it can fill up the flaw and increase the structural behavior of the domain. Nano silica will be studied thoroughly and the applications can be proved earlier by the future research.

References

1. Williams CDH (1996) Numerical solutions to Laplace's equation. 2 (19950226/1)
2. Coppa NV (1995) Nanomaterials—nanofacts. Nanomaterials Company, 1995

3. Sobolev K, Vivian IF (2005) Ferrada-Gutiérrez M Nanotechnology of concrete. Universidad Autonoma de Nuevo Leon, Mexico, pp. 9–16
4. Khanzadi M, Tadayon M, Sepehri H, Sepehri M (2010) Influence of nano-silica particles on mechanical properties and permeability of concrete

Part VIII
Geotechnical Engineering

Russell Projection and Mercator Projection by Harmonic Equations Comparison Study

Mohammed S. Akresh and Ali E. Said

Abstract One of the main key points in mapping is the projection system used. The accuracy depends on many things like data source and accuracy, projection system, and the way of mapping. Many problems appeared in using the traditional projection system (distance distortion, scale factor, ..., etc.), which can affect the spatial accuracy of the produced map. A new method has been implemented for map projection based on harmonic equations. It is different in the way of creation the zones. Main latitude and central longitude play the main steps in the projection system, also, the isometric coordinates and front and back algorithms up to 12th edge of series (1, 2, 5, 6). This paper presenting the expansion of the algorithms used in projection of Russell to 12th edge of series and compares between Russell and Mercator projections using harmonic equations in three regions distributed from the equator to the parallel 62° N. Also, distance distortions in the local system in the main projections of Russell and Mercator projections have been studied. The results show that Russell projection has less distortion in curvature near the equator than Mercator projection. The distance distortions in local system with the new method are less than in the traditional methods. The new method proved that it is better than the traditional methods in mapping.

Keywords Mercator projection · Russell projection · Algorithms · System · Distortion

1 Introduction

One of the main key points in mapping is the projection system used. The accuracy depends on many things like data source and accuracy, projection system, and the way of mapping. Many problems appeared in using the traditional projection system

---

M. S. Akresh (✉) · A. E. Said
Civil Engineering Department, Faculty of Engineering, Tripoli University, Tripoli, Libya
e-mail: Sab20084@mail.ru; makresh@yahoo.com

A. E. Said
e-mail: ali_said_ly@yahoo.com

S. Sathiyamoorthy et al. (eds.), *Emerging Trends in Science,*
Engineering and Technology, Lecture Notes in Mechanical Engineering,
DOI: 10.1007/978-81-322-1007-8_47, © Springer India 2012

(distance distortion, scale factor, ..., etc.), which can affect the spatial accuracy of the produced map.

New methodology in map projections use harmonic equations, all projections "Mercator, Lambert and Russell" have relationship among them, this relationship in standard parallel and center meridian; this method can cover big area [1–3]. Three different projections in coefficients, where each one has special coefficients for direct method, while indirect method all of them have conformal coefficients.

New coefficients for direct method in Russell projection to 12th edge of series were obtained, and comparison between projections Russell [4, 5] and Mercator has been made in different parallels from the equator to the parallel 62`north beside distortion of curve latitudes and transformations coordinates.

2 Methodology

The methodology used in this study based on analysis of data in Russell and Mercator projection from the side of distortion in curvature and distances using local systems compared with reverse solution of geodetic problems to find the distances on the ellipsoid, and define the zone width in two projection systems in form of spatial accuracy.

Starting from the standard parallel 0° equator, three probabilities for projections (Russell, Mercator) have been made. First 26°*26° (26° in direct latitudes and 26° longitude), second 24°*24°, third 22°*22° (Fig. 1).

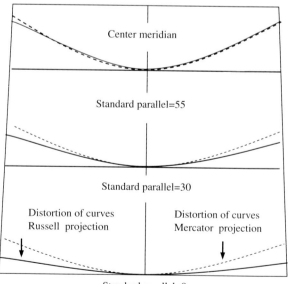

Fig. 1 Distortion of curves on different of parallels

2.1 Results

Tables (1, 2, and 3) shows the results obtained.

Table 1 shows the best results for Russell projection with high accuracy for zone 22, while Mercator projection gives medium accuracy for zone 22; also distortion of curves in Russell projection are less to Mercator projection.

Table 2 shows the best results for Russell projection with high accuracy for zone 20° while projection Mercator gives medium accuracy for zone 20°; also distortion of curves in Russell projection are less or equal to Mercator projection.

Table 3 shows good results for projections of Russell and Mercator with good accuracy for zone 11 × 14, also distortion of curves in Russell and Mercator projections are the same.

3 Local System Coordinates for City of Tripoli

The system coordinates using harmonic equation have good results for all zone, also give good result for the measured distance; Table 4 shows the results of five points tested in Russell and Mercator projection for decrease of distances

Table 1 Results of analyses between Mercator projection and Russell on equator

Degree of zone	22°	24°	26°
Projection Russell m = 1.0000000 *standard parallel* 0			
B	11°00'00.0000" N	12°00'00.0000" N	13°00'00.0000" N
L	11°00'00.0000" E	12°00'00.0000" E	13°00'00.0000" E
X meter	1,231,474.0186 m	1346571.6985 m	1,462,490.6601 m
Y meter	1,216,894.4811 m	1325891.8507 m	1,434,444.6608 m
dB	−0.00002" = 0.0005 m	−0.00006" = 0.0018 m	−0.00018" = 0.005 m
dL	−0.00002" = 0.0005 m	−0.00006" = 0.0018 m	−0.00016" = 0.005 m
Distortion of curves	11,282.008 m	14,646.159 m	18,619.711 m
Projection Mercator m = 1.00000000 *standard parallel* 0			
B	11°00'00.0000" N	12°00'00.0000" N	13°00'00.0000" N
L	11°00'00.0000" E	12°00'00.0000" E	13°00'00.0000" E
X meter	1,238,816.9139 m	1,356,044.119 m	1,474,448.2952 m
Y meter	1,209,112.3953 m	1,315,700.6935 m	1,421,366.5553 m
dB	+0.00015" = 0.003 m	+0.00043" = 0.013 m	+0.00125" = 0.0385 m
dL	+0.00011" = 0.003 m	+0.00033" = 0.010 m	+0.00095" = 0.0029 m
Distortion of curves	22,350.912 m	28,959.690 m	36,737.573 m

Table 2 Results of analyses between Mercator projection and Russell on standard parallel 30

Degree of zone	20	22	24
Projection Russell m = 1.0000000 standard parallel 30			
B	40°00'00.0000" N	41°00'00.0000" N	42°00'00.0000" N
L	10°00'00.0000" E	11°00'00.0000" E	12°00'00.0000" E
X meter	4,475,682.6590 m	4,596,780.4011 m	4,718,926.8526 m
Y meter	860,462.5904 m	933,975.0976 m	1,004,973.6954 m
dB	0.00003"	0.00009"	−0.00027"
	= 0.001 m	= 0.0023 m	= 0.008 m
dL	−0.00003"	−0.00013"	−0.00043"
	= 0.001 m	= 0.0039 m	= 0.013 m
Distortion of curves	43,338.512 m	52,456.843 m	62,416.450 m
Projection Mercator m = 1.00000000 standard parallel 30			
B	40°00'00.0000" N	41°00'00.0000" N	42°00'00.0000" N
L	10°00'00.0000" E	11°00'00.0000" E	12°00'00.0000" E
X meter	4,477,739.9099 m	4,599,296.0392 m	4,721,899.242 m
Y meter	854,684.6785 m	226,262.5637 m	994,938.8673 m
dB	−0.00016"	−0.00048"	−0.00122"
	= 0.005 m	= 0.015 m	= 0.038 m
dL	+0.00028"	+0.00103"	+0.00338"
	= 0.008 m	= 0.031 m	= 0.100 m
Distortion of curves	48,210.854 m	58,722.686 m	70,262.217 m

Table 3 Results of analyses between Mercator projection and Russell on standard parallel 55

Degree of zone	11 × 14	12 × 14	14 × 14
Projection Russell m = 1.0000000 standard parallel 55			
B	60°00'00.0000" N	61°00'00.0000" N	62°00'00.0000" N
L	7°00'00.0000" E	7°00'00.0000" E	7°00'00.0000" E
X meter	6,674,603.3602 m	6,785,801.5279 m	6,897,107.3519 m
Y meter	390,789.3242 m	379,241.3981 m	367,613.4522 m
dB	−0.00000"	0.00012"	+0.00066"
	= .000 m	= 0.0039 m	= 0.02 m
dL	+0.000063"	+0.00017"	−0.00054"
	= 0.002	= .005 m	= 0.016 m
Distortion of curves	20,177.386 m	19,697.379 m	19,206.201 m
Projection mercator m = 1.00000000 standard parallel 55			
B	60°00'00.0000" N	61°00'00.0000" N	62°00'00.0000" N
L	7°00'00.0000" E	7°00'00.0000" E	7°00'00.0000" E
X meter	6,674,749.3888 m	6,785,739.6435 m	6,896,721.9708 m
Y meter	390,112.7461 m	378,251.7287 m	366,276.3130 m
dB	+0.00007"	−0.00016"	−0.00089
	= 0.002 m	= 0.005 m	= 0.026 m
dL	−0.00008"	−0.00022"	+0.00068"
	= 0.002 m	= 0.007 m	= 0.021 m
Distortion of curves	20,676.570 m	20,246.101 m	19,791.076 m

Russell Projection and Mercator Projection by Harmonic Equations

Table 4 Points of study in city Tripoli

Points	B	L
A	32°52'00"	13°07'00"
B	32°53'00"	13°15'00"
C	32°48'30"	13°17'00"
D	32°48'00"	13°06'00"
E	32°50'30"	13°11'30"

Table 5 Local system coordinates for Mercator and Russell and distances measurements

	X meters	Y meters	Value of scale factor	Distance for plane (m)	Distance by sampson (m)	Distance by geodetic problems
Local system coordinates for Mercator $X_0 = 2,876,834.5725$ m, $Y_0 = 0$ m factor scale $m_{local} = 0.9984122$						
A	3,643,434.433	−362,992.825	1.00003926	S_{A-C}	S_{A-C}	S_{A-C}
C	3,636,407.915	−347,632.209	0.99990447	16,891.432	16,891.915	16,891.915
B	3,644,830.008	−350,456.345	0.99992879	S_{B-D}	S_{B-D}	S_{B-D}
D	3,636,103.194	−364,824.846	1.00005576	16,811.042	16,811.179	16,811.179
E	3,640,407.838	−356,077.979	0.99997786			
Local system coordinates for Russell $X_0 = 2,876,834.5725$ m, $Y_0 = 0$ m factor scale $m_{local} = 0.995641396$						
A	3,641,606.434	−363,195.082	1.00005729	S_{A-C}	S_{A-C}	S_{A-C}
C	3,634,630.989	−347,810.855	0.99992386	16,891.75	16,891.916	16,891.915
B	3,643,044.232	−350,662.709	1.00001504	S_{B-D}	S_{B-D}	S_{B-D}
D	3,634,269.139	−365,002.050	0.9999969	16,811.275	16,811.179	16,811.179
E	3,638,603.167	−356,269.914	0.99999807			

distortion "standard parallel 26° N and center meridian 17° E factor scale for general zone 0.99611165".

Table 5 Illustrate a good result for Russell projection compared with geodetic problems.

4 Conclusions

Map projection using harmonic equation is the suitable way in GIS, where all projections have connection among them, also it can make local systems within them. From the results obtained for projections of Russell and Mercator, may conclude that:
Russell projection shows better performance than of Mercator projection from equator to standard parallel 30'N or S, where Russell has big zone and less distortion curves.

Russell projection and Mercator projection near 55′N or S gave conformal results. Transformation of rectangular coordinate to geodetic coordinate in Russell projection is better than in Mercator projection.

New coefficients in Russell projection to edges 12 in series were obtained.

The Russell projection is the adequate projection for the city of Tripoli.

Because of the flexibility of local projection system proposed, it is very useful in GIS applications.

References

1. Akresh MS (2010) The constructing and development of scientific–fundamentals of technologies and transformation technology of systems coordinates Geographic information system "GIS" for the study area of Libya. PSU, Novopolotsk, p 131
2. Bugayevskiy LM, Snyder JP (1995) Map projections: a reference manual. Taylor Francis, London, p 328
3. Padshyvalau UP (1998) The theoretical background of formation coordinate environment for GIS. PSU, Novopolotsk, p 125
4. Akresh MS (2009) New method in map projection indirect coefficients. In: Proceedings of 12th AGILE international conference on geographic information science and ISPRS Hannover workshop 2009 high-resolution Earth imaging for geospatial information. Hannover of Germany, 2–5 June 2009. First paper in CD www.agile-online.org
5. Akresh M.S (2008) Work and calculation indirect coefficients in mercator projection for Libya. a12 Vestnk Magazine of Polotsk State University, Novopolotsk, pp 136–140

Stable Isotopic Analysis Using Mass Spectrometry and Laser Based Techniques: A Review

M. Someshwar Rao

Abstract Stable isotope ratios of hydrogen and oxygen are widely used in hydrology, biology, chemistry, environmental sciences, food and drug authentication, forensic science, geochemistry, geology, ecology, oceanography, and paleoclimatology. In water cycle processes, it is used in getting the information on groundwater circulation, discharge components in rivers, lake hydrology, sources of moisture in precipitation, recycling and transport of atmospheric moisture, glaciology at local, regional and continental scales. The isotope ratio mass spectrometer (IRMS) invented in 1940s led to the measurement of isotopes of water. The technological advancement led to the development of new range of IRMS that functions as programed operation provides high throughput with high precision and accuracy and uses microliters of sample analysis. However, their high cost and required operational skills have a wider limited use of these systems and applications of isotopes in hydrological science by researchers in general. Over the last one decade, less expensive and easy-to-operate spectroscopic methods for isotope analysis using lasers have emerged. The present paper provides insight into this advancement along with performance of these systems.

Keywords Mass spectrometry · Stable isotopes · Laser · International atomic energy agency

1 Introduction

Isotopes can be classified in two important categories: (1) stable isotopes and (2) unstable isotopes (also called radioactive isotopes). Stable isotopes are the atoms that do not decay with time. On the other hand, radioactive isotopes disintegrate by giving out alpha (α), beta (β) particles and/or gamma (γ) radiation, etc., and transform into another type of atom.

M. Someshwar Rao (✉)
National Institute of Hydrology, Roorkee, India
e-mail: 65somesh@gmail.com

S. Sathiyamoorthy et al. (eds.), *Emerging Trends in Science, Engineering and Technology*, Lecture Notes in Mechanical Engineering, DOI: 10.1007/978-81-322-1007-8_48, © Springer India 2012

The mass spectrometer is used for detection of stable isotopes that do not decay with time. It is not used for detection of unstable isotopes (also called radioactive isotopes) that decay and disintegrate by giving out alpha (α), beta (β) particles and gamma (γ) radiation, etc., and transform into daughter nucleus. If the daughter nucleus if radioactive, the process of disintegration with emission of radiation continues till it reaches to stable nuclear form and the process of disintegration stops. The radioactive isotopes are usually measured through detection of their emitted radiation or by counting parent or daughter isotopes using Accelerator Mass Spectrometer. The present paper limits its scope mainly in the measurement of stable isotope concentration.

2 Units of Measurement

To avoid system dependent and procedure dependent biases, measured isotope abundance ratios are compared to those of a standard material of known isotope composition using delta (δ) notation. The δ provides deviation of the isotope abundance ratio of the sample from that of a standard and is expressed in parts per thousand or per mill (%). Positive δ values indicate that the sample is enriched in the heavier isotope of the element with respect to the standard material. Negative δ value represents samples that are depleted in the heavier isotopes in comparison to the standard.

$$\delta^i E = \frac{\left(^i R_{\mathrm{sa}} - {}^i R_{\mathrm{ref}}\right)}{^i R_{\mathrm{ref}}}$$

with i denoting the mass number of the heavy isotope of element E, R_{sa} is the respective isotope number ratio of a sample, and R_{ref} is that of the reference material. The δ values are usually given in percentage (per mill i.e., parts per thousands) after multiplying the values by 1,000. For instance, in the case of oxygen,

$$\delta^{18} O = \frac{\left(^{18}O/^{16}O\right)_{\mathrm{sample}} - \left(^{18}O/^{16}O\right)_{\mathrm{standard}}}{\left(^{18}O/^{16}O\right)_{\mathrm{standard}}} \times 1000\,\%$$

The advantage of δ notation is that it does not require prior knowledge of isotope relations in a reference material. Recently, the δs are given the unit name Urey[1] with symbol Ur [1]. The per mill (%) then becomes mUr.

The δ notation can be converted into atomic fraction notation φ as;

$$\varphi\left(^i E\right) = \frac{x\left(^i E\right)}{x\left(^i E\right)_{\mathrm{reference}}} - 1$$

---

[1] The unit name *Urey* is given in the honor of Professor Harold. Urey of University of Chicago who won Nobel Prize in Chemistry in 1934 for his discovery of deuterium.

with x (^iE) denoting the atomic fraction: x (^iE) = N (^iE)/ΣN (^jE), where the summation is over all stable isotopes of element E.

The stable isotopes are usually expressed isotopes. Two fundamental processes explain the stable isotope deltas measured in most terrestrial systems: isotopic fractionation and isotope mixing. Isotopic fractionation is the result of equilibrium or kinetic physicochemical processes that fractionate isotopes because of small differences in physical or chemical properties of molecular species having different isotopes.

3 Stable Isotope Ratio Analysis Using Isotope Ratio Mass Spectrometry

In isotope ratio mass spectrometry, element isotopes ratios are determined very accurately and precisely. Typically, single focusing magnetic sector mass spectrometers with fixed multiple detectors (one per isotope) are used. The *molecular mass* of a sample is resolved according to its mass/charge ratios (m/q) by utilizing combination of electric and magnetic fields.

History of stable isotope measurement dates back to 1912 when Sir J. J. Thomson and his research assistant F. W. Aston developed mass spectrograph and separated isotopes of neon (^{20}Ne and ^{22}Ne). Over the past century, several advancements that have modified the early spectrograph to make it suitable to analyze microsized samples, multi-isotope detection, online sample preparation, gas/liquid inlet adoptability, compound specific analysis, high accuracy, precision and throughput results, automation in system operation, data analysis features, etc. However, the basic units of mass spectrometer still remain the same i.e., an inlet system, an ion source, an analyzer for ion separation, and a detector for ion registration.

Principles of IRMS is discussed in several reviews (see for example [2]) Schematic arrangement of a typical IRMS is shown in the Fig. 1. In the *ionization chamber* of the *inlet unit* of the IRMS, *electron impact source* ionizes sample. A small magnetic field (~500 gauss) provided in the same direction as the electron beam confines the ionization region to a diameter of <1 mm and helps increase ionization efficiency. The electron emission current typically around 1 mA in an ionization region with density of gas at 10^{-7} mbar generates an ion current of 8×10^{-10} A.

The electrons that do not participate in ionization hit the electron collector or "trap" constituting the *electron trap current*. The feedback mechanism from electron trap current regulates the ion current improving *S/N* ratio.

The ions are extracted from the ionization chamber by a combination of drawout plates and a *repeller plate* or pusher under the influence of positive accelerating voltage V. The acceleration of ions improves sensitivity, resolution, and peak shape as all the ions acquire same kinetic energy according to the equation:

$$E_{\text{kinetic}} = \frac{1}{2}mv^2 = qV$$

where, v = velocity of ion.

Fig. 1 Schematic illustration of magnetic sector mass spectrometer

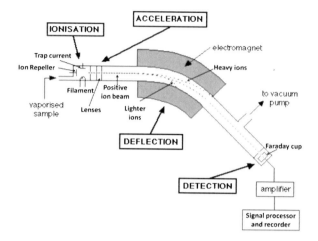

The beam is then focussed by series of lenses at the beam defining slit.

Subsequently, the ions are allowed to pass through a magnetic field perpendicular to the direction of electric field. A magnetic sector acts like an optical prism in geometrical optics: Monoenergetic ions of different mass are dispersed through the magnetic field. The path of the ions becomes circular with radius r maintained due to balance between centripetal force and Lorentz force given by the equation:

$$r = \frac{1}{B}\sqrt{\frac{2mV}{q}}$$

r radius in the ionic path in cm
B applied magnetic field in gauss
V potential difference in volts
e charge on the ions measured in terms of the number of electrons removed or added on ionization
m Atomic masses (e.g., oxygen $m = 16$)

Since q is invariably unity (1^+), in the mass spectrograph, ions of different mass gets collected at different positions according to the radius of curvature they follow.

In the magnetic sector, lighter ions follow a path with a small radius than heavier ions. The pole faces of the magnet are designed in such a way that the ion beam gets focussed both along y and z planes (Fig. 2). This increase in dispersion improves the abundance sensitivity and provides for higher transmission efficiency. Usually, laminated magnets are used to reduce hysteresis effect during magnetic scanning. Magnetic field is usually delivered by an electromagnet stabilized with a Hall probe. In practice, out of about 2000 molecules that enters into the mass spectrometer, one molecule as an ion reaches the detector.

Stable Isotopic Analysis

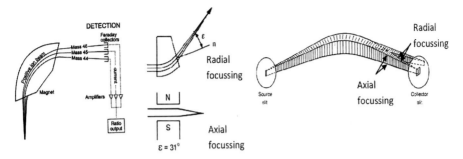

Fig. 2 Schematics of mass collection with universal triple collector Faraday cups. Also curved pole pieces enabling radial and axial focussing of ion beam are shown

In double focussing mass spectrometers, an electrostatic energy analyzer is additionally provided (Fig. 3). The energy filter bends lower energy ions more strongly than higher energy ions. The magnetic sector filters ions according to m/e and electrostatic filter according to their energies. Addition of electrostatic filter increases the resolving power of the instrument tremendously. This is especially useful in hydrogen isotope analysis in the presence of helium contamination.

A difficulty in measuring D/H isotope ratios is that, along with H_2^+ and HD^+ formation in the ion source, H_3^+ is produced as a byproduct of ion–molecule collisions. The amount of H_3^+ formed is directly proportional to the number of H_2 molecule and H^+ ions.

An isotope ratio mass spectrometer (IRMS) commonly has two or more ion collectors (usually, Faraday Cups) positioned to simultaneously measure the isotopic ion currents of interest, which are typically only a few mass units apart (see Fig. 4). The objective of the ion collection is to measure two or more ion currents that impinge on the collecting system. The ion current is fed through a highly

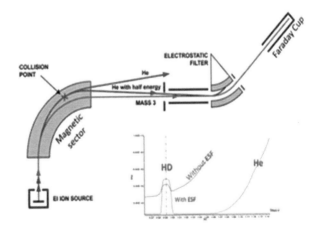

Fig. 3 Separating HD and He using electrostatic filter (*Courtesy*: Micromass instruments) collector along the focal plane of mass spectrometer for collecting major and minor beams

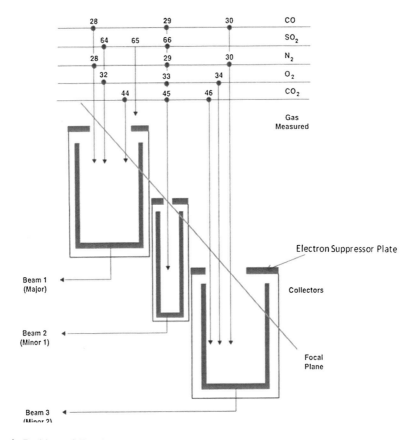

Fig. 4 Position of Faraday cups of universal triple collector along the focal plane of mass spectrometer

stable ohmic resistor of resistance 10^{8-11} Ω. The voltage produced across the resistors is simultaneously measured to compute the observed isotope ratios.

An electrode with a negative voltage located near the entrance to the Faraday collector helps to prevent secondary electrons from escaping from the cup. The cups are usually deeper and coated with graphite to reduce the probability of sputtering. Multiple cups (collection) offer simultaneous analysis of isotopic peaks of interest. Recent systems have come up with up to 10 Faraday cup arrangements. When the ion beam is scanned across the exit slit of the mass spectrometer, the ion current should be flat top over the collector width and falling to background outside the collector.

The interior of the mass spectrometer is kept at extremely low pressure (less than 10^{-7} Torr) using turbo molecular pumps or ion getter pumps and cryopumps to remove neutral species from the ionization region in the source and to minimize the number of collisions between ions and residual molecules along the flight path. These interactions would reduce the ions to lose variable amounts of kinetic energy and defocus the ion beams. A mass spectrometer must possess high

Abundance Sensitivity which is defined as the intensity of a large peak divided by intensity of the background one mass unit lower. If, for example, 107 counts are collected at mass position M and one count at mass position $M-1$, then the abundance sensitivity is 107 (or sometimes referred to as the reciprocal 10^{-7}).

The specification of IRMS is judged with respect to its (1) sensitivity of detection, (2) peak resolving efficiency, (3) linear relation between the ion current intensities and the measured ratios, (4) memory between subsequent introductions of sample and reference gases in the mass spectrometer, (5) chemical inertness of the hot filament, and (6) stability of ion currents over a time range much longer than required for the measurement of a single sample.

According to the inlet system, the gas source stable isotope ratio mass spectrometers come in two designs: (1) Dual inlet and (2) Continuous flow.

3.1 Dual Inlet System

In dual inlet (DI) system, the inlet is alternately switched for analysis between standard and reference gas (Fig. 5). While one gas flows to the vacuum chamber, the other is directed to a vacuum waste pump so that flow through the capillaries is never interrupted. Because both the sample and reference gases are measured under identical conditions, the method is very accurate and precise.

3.2 Continuous Flow

In the continuous flow technique, solids, liquids, and gases are reduced to simple molecules, for example, organic compounds are combusted to CO_2, H_2O, and N_2

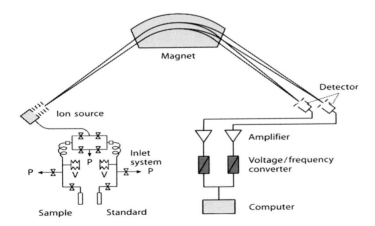

Fig. 5 Schematic representation of a DI-IRMS. P denotes pumping system, V denotes a variable volume bellow

in an elemental analyzer (EA). The different molecules are separated by passing through gas chromatography (GC) column and the clean and pure gas is introduced into the ion source in a stream of helium (Fig. 6). According to the interface these systems are termed as EA-IRMS, GC-IRMS, liquid chromatography (LC)-IRMS, etc., EA-IRMS provides bulk sample isotope analysis (BSIA) while GC-IRMS provides component specific isotope analysis. By measuring isotopic ratio of desired molecules source of origin of these isotopes and the processes occurred during their synthesis is determined.

4 Optical Techniques of Stable Isotope Analysis

IRMS is the conventional method for measuring isotope ratios and has benefited from over 40 years of research and development. Unfortunately, IRMS is incompatible with a condensable gas or a sticky molecule such as water. Therefore, in general, chemical preparation of the sample is required to transfer the water isotope ratio of interest to a molecule that is more easily analyzed. For the ^{18}O and ^{2}H analysis of water involves several hours of equilibration at high accurate temperature stabilization. IRMS instrumentation is expensive, voluminous, and heavy, confined to a laboratory setting, and usually requires a skilled operator. All or most of these issues may be addressed by optical measurement techniques. Optical isotope ratio instrumentation is cheaper, more compact, and easier to operate (airborne, or at a remote location, if needed) than IRMS instrumentation. In general, due to the extreme selectivity possible with infrared spectroscopy, chemical sample pre-treatment is usually not required. The optical measurement is nondestructive, fast, can be made online, even real-time, measurements can be made.

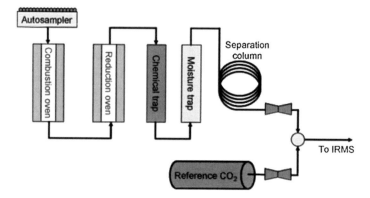

Fig. 6 Sample introduction to CF-IRMS using an autosampler, elemental analyzer (*EA*) and GC separator

Most molecular species exhibit strong absorption bands near and mid-infrared spectrum. For example, IR absorption region of water is shown in the Fig. 7. This feature, combined with the use of diode lasers, enables the resolution of single spectral lines of a molecule and to discriminate between different isotopologues. And, this by comparing with reference material can be used to enable isotopic composition of sample.

5 Cavity Ring Down Spectroscopy (CRDS)

Optical technique of isotope ratio measurements utilizes laser absorption within a cavity of high-reflectivity mirrors (reflectance >99.999 %). The system comprises optical cavity, built of at least two highly reflective mirrors in which photons propagate for a prolonged time. When a laser is allowed to illuminate the optical cavity, the laser resonates in the cavity (Fig. 8). It rings back and forth many times between the two mirrors making an effective absorption path length of several kilometers. The time behavior of the light intensity inside the cavity is monitored by the small fraction of light that is transmitted through the other mirror and detected by a fast photodetector (Fig. 9). The intensity of the light pulse inside the cavity decreases by a fixed percentage during each round trip due to absorption

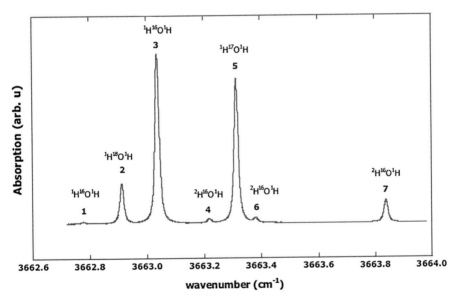

Fig. 7 Absorption spectra of water near IR at 2.73 μm (near 3663 cm^{-1}) range. The intensities are for a natural water sample with abundances: 0.998, 0.00199, 0.00038, and 0.0003 for $^1H^{16}O^1H$, $^1H^{18}O^1H$, $^1H^{17}O^1H$, and $^2H^{16}O^1H$, respectively

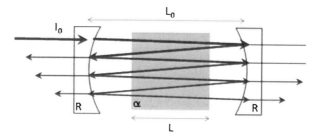

Fig. 8 Schematic of a CRDS. Depending whether the input intensity is pulsed or continuous, the observer looking at the exit mirror senses intensity decay or continuous absolute intensity related to the losses inside the cavity, respectively. R high reflective mirrors separated by the distance L_0 I_0: intensity of laser beam falling into the cavity α Absorption coefficient of the resonator

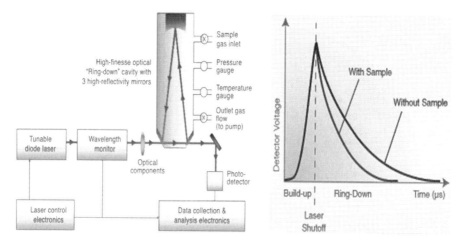

Fig. 9 Schematic of wavelength scanned CRDS (M/s Picarro) instrument and the cavity ring down spectrum

and scattering by the medium within the cell and reflectivity losses. As a result, measured intensity outside the cavity shows an exponentially decay (extinction) according to the relation.

$$I(t) = I_0 e^{-\frac{t}{\tau}}$$

The decay constant τ is called ring down time and is dependent on the loss mechanisms within the cavity. If something that absorbs light is placed in the cavity, the amount of light decreases faster—it makes fewer bounces before it is all gone.

A CRDS measures how long it takes for the light to decay to 1/e of its initial intensity, and the "ring down time" can be used to calculate the concentration of the absorbing substance in the gas mixture in the cavity. Also, because no laser

Stable Isotopic Analysis

light is entering the cavity during the measurement of the decay time curve, the technique is almost undisturbed to laser amplitude noise. An absorbing species in the cavity will increase losses according to Beer–Lambert law. Assuming the sample gas fills the entire cavity,

$$\tau = \frac{n}{c} \times \frac{l}{1 - R + X + \alpha l}$$

where α is the absorption coefficient for a specific analyte concentration, n is the index of refraction within the cavity, c is the speed of light in vacuum, l is the mirror separation, R is the mirror reflectivity, and X takes into account other miscellaneous optical losses.

From the relation, molar absorptivity, ε and the analyte concentration, C, can be determined from the ratio of both ring down times.

$$\frac{\tau}{\tau_0} = \frac{\alpha l}{1 - R} = \frac{\varepsilon l C}{2.303 \, (1 - R)}$$

Thus, the absolute abundance of individual molecules can be quantified through the amount of absorbance at a specific wavelength. For water isotope analysis, wavelength of laser scanning is done over H_2O spectral features of interest and measuring ring down time (optical loss) and laser wavelength, an optical spectrum is generated. The concentration of each isotopologue is then proportional to the area under each measured isotopologue spectral feature. In CRDS, measurement is done either on rate of absorption of a light pulse or magnitude of absorption of continuous laser light.

The effective path length of light inside the cavity depends upon the reflectivity (R) of the mirrors and is given by:

$$L_{\text{eff}} = L / (1 - R) \, ;$$

The cavity length is usually about 30 cm and $R = 0.99999$. This makes the effective path length about 30 km.

The CRDS analysis is done in two modes (1) time dependent decay of output light (Fig. 9) and (2) absorption mode (Fig. 10). In time dependent decay mode also called wavelength scan (WS-CRDS) mode, a continuous wave (CW) laser is allowed to pass into the cavity containing sample gas. When the photodetector signal reaches a threshold level (in a few tens of microseconds), the CW laser is abruptly turned off. The light inside the cavity exponentially decays to zero is monitored. The cavity is then filled with the gas species to be analyzed. The gas species accelerates the ring down time compared to a cavity without any gas.

In absorption technique, an absorption spectrum is obtained by rapidly sweeping the laser over the spectral region of interest and recording the detector output. The technique is capable of measurement of mole fraction down to the parts per trillion level.

In the absorption mode of operation, the cavity is pumped by a continuous light source. This is also named as cavity enhanced absorption spectroscopy (CEAS).

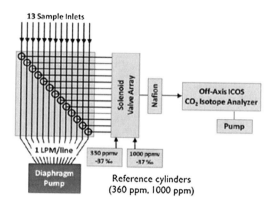

Fig. 10 A block diagram of multiline isotope analysis using Off Axis integrated cavity output spectrometry (*ICOS of* M/s LGR) and cavity-enhanced absorption spectra (*CEAS*) of 12 and 13 CO_2

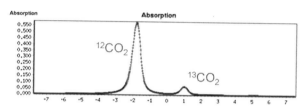

In CEAS, the total time-integrated intensity is measured. There is no need to switch the laser output on and off, and no fast detector and data acquisition system are needed, greatly reducing the complexity of the technique. Applications of laser-based isotopic technique and operation procedure of the system are discussed in detail elsewhere (see for example, [3, 4]).

6 Interlaboratory Comparison Exercise for δ^2H and δ^{18}O Analysis of Water Samples

The calibration and performance check of mass spectrometer is usually done using reference materials (RMs) of known isotopic composition distributed by national institute of standards and technology (NIST) USA (previous name: national bureau of standards-NBS) on behalf of the international atomic energy agency (IAEA), Vienna, Austria. Each RM unit consists of 20 mL water in a sealed glass ampoule. Since only very limited quantities of these materials exist, the distribution is limited to one unit of each per 3-year period of time. The isotopic values of reference standards are conventionally given with respect to the Vienna standard mean ocean water (VSMOW). The isotopic composition of VSMOW (RM8535) is estimated as [5]:

$$^2H/^1H = \left(1.5575 \times 10^{-4}\right); \quad ^{18}O/^{16}O = \left(2.0052 \times 10^{-3}\right); \quad ^{17}O/^{16}O = 3.79 \times 10^{-4}$$

Stable Isotopic Analysis

Similar to VSMOW, for calibration of water with depleted isotopic composition NIST introduced two depleted isotopic composition waters [6]:

Standard light antarctic precipitation (Water)-SLAP (RM 8536). It has isotopic composition:

$$^2H/^1H = 89 \times 10^{-6};\ ^2H = -428.0\ \% \text{ and } ^{18}O/^{16}O = 18.93 \times 10^{-4};\ ^{18}O = -55.5\%$$

Greenland ice sheet precipitation (Water)-GISP (RM 8537). Its isotopic composition is: $\delta^2H = -189.5\ \%$ and $\delta^{18}O = -24.78\ \%$.

Since the primary reference got exhausted, recently, NIST prepared standards VSMOW2 (RM 8535a) and SLAP2 (RM8537a) with the same primary standards isotopic composition. The reference materials are intended to be used to calibrate/check internal laboratory standards on VSMOW-SLAP scale.

For quality assurance IAEA conducts interlaboratory comparison exercises. In the past, it has conducted four such exercises: 1995, 1999, 2002, and recently in 2011. Under this, IAEA circulates samples without giving isotopic composition. After the exercise, IAEA publishes the report providing values measured at different laboratories and by IAEA. Such exercise also helps to precision and accuracy of new instruments and methodologies. In 2011, this exercise was made with an objective to compare the laser-based instruments with conventional mass spectrometers in addition to the normal interlaboratory calibration. The detail final report of this exercise has recently been released [7]. Under this exercise, waters labeled as OH-13, OH-14, OH-15 and OH-16 were distributed globally. 137 laboratories from 53 countries including three from India (IITM, Pune, PRL, Ahmedabad and NIH, Roorke) submitted their results. The comparison of results of national institute of hydrology (NIH), Roorkee (Lab code 153) with IAEA is given in the Table 1. Values as obtained from various laboratories and their comparison with IAEA for two waters IAEA-OH-13 and IAEA-OH-14 is also shown. The measured value at NIH is shown in the figure by arrow. The results in the Table 1 and Figs. 11 and 12 assure quality performance of the systems at NIH, Roorkee. The exercise demonstrated that the new range of laser-based equipment measures isotopic composition of water with same accuracy and precision. Within the country, in 2010 an interlaboratory comparison exercise was performed between five laboratories with an objective to exchange isotopic data and samples on a countrywide program. The samples were prepared and distributed by Physical Research Laboratory, Ahmedabad to NIH, Roorkee, IIT Kharagpur, CWRDM,

Table 1 IAEA samples and their measured average values at NIH during international interlaboratory comparison exercise, 2011

IAEA sample	IAEA value (%)	1σ (%)	NIH value	1σ (%)
OH-13	−0.96	0.04	−0.98	0.07
OH-14	−5.60	0.05	−5.52	0.08
OH-15	−9.41	0.04	−9.39	0.06
OH-16	−15.43	0.04	−15.40	0.04

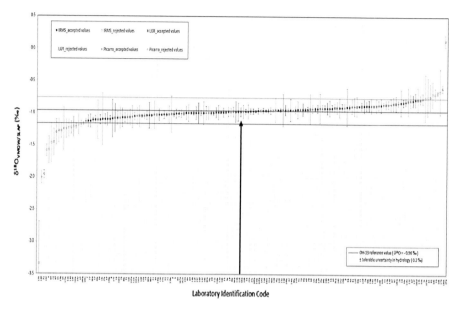

Fig. 11 Plot for $\delta^{18}O$ values of sample IAEA-OH-13. In the figure, arrow indicates NIH lab code and its data (*Reproduced from the IAEA report*: [6])

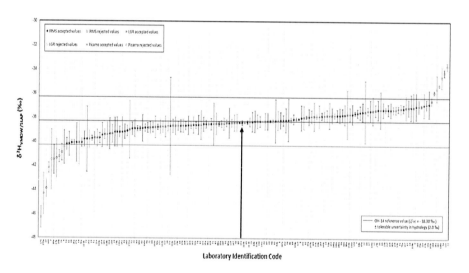

Fig. 12 Plot for δD values of sample IAEA-OH-14. In the figure, *arrow* indicates NIH lab code and its data (*Reproduced from the IAEA report*: [6])

Kozikode and NGRI, Hyderabad. The same samples were also measured at IAEA, Vienna. The interlaboratory comparison of samples is shown in the Fig. 13 clearly demonstrates high performance of the Indian systems.

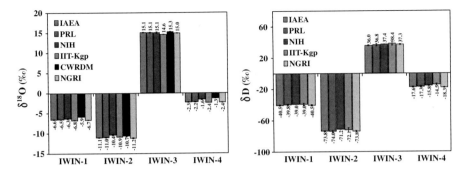

Fig. 13 Interlaboratory comparison of water for isotopic analysis in Indian laboratories

7 Conclusion

Stable isotope analysis of water helps tracing hydrological cycle and its pathways. The instrumental advancement led to Dual Inlet and Continuous Flow type mass spectrometers that have ability to analysis micrograms of samples, analysis of more than one type of isotope, automation, increased throughput and compound specific analysis. The resolution of these systems is quiet enough to resolve isotopic difference for their application in different hydroclimatic regime. The addition of laser-based equipment in the last decade brought new series of low-cost portable instruments with better efficiencies than the conventional mass spectrometer. The efforts taken by IAEA in developing standards such as VSMOW, SLAP, GISP, etc., and through interlaboratory comparison led to development of good laboratory practices and in analyzing instrumental performance. The interlaboratory comparison exercise indicates fairly good acceptance of results produced by conventional mass spectrometers and laser-based techniques at global scale. Also, these have demonstrated fairly good quality assurance of isotope laboratories working in India. In India, the potential of the technique has already been demonstrated through wide range of applications to groundwater and surface water system, oceanography and in climate studies (see the Refs. [8–22] for various Indian case studies). The laser-based isotope analyzers have started getting into country and this will help widening the scope of isotope technology to other branches of science and engineering in addition to the hydrology.

References

1. Brands WA, Coplen TB (2012) Stable isotope deltas: tiny, yet robust signatures in nature. Isot Environ Health Stud 1–17
2. Platzner IT (1999) Modern isotope ratio mass spectrometry. Wiley, Chichester
3. Groot PAD (2008) Handbook of stable isotope analytical techniques, vol 1. Elsevier, Amsterdam

4. IAEA (2009) Laser spectroscopic analysis of liquid water samples for stable hydrogen and oxygen isotopes, Training course series no. 35. IAEA, Vienna
5. De Wit JC, van der Straaten CM, Mook WG (1980) Determination of the Absolute hydrogen isotopic ratio of VSMOW and SLAP. Geostand Newslett 4(1):33–36
6. NIST (2005) Report of investigation: reference material 8535, 8536, 8537
7. Ahmad M, Aggarwal P, van Duren M, Poltenstein L, Araguas L, Kurttas T, Wassenaar LI (2012) Fourth interlaboratory comparison exercise for δ^2H and $\delta^{18}O$ analysis of water samples isotope. Final report, IAEA, Vienna
8. Bhattacharya SK, Gupta SK, Krishnamurthy RV (1985) Oxygen and hydrogen isotopic ratios in groundwaters and river waters from India. Proc Indian Acad Sci (Earth Planet Sci) 94(3):283–295
9. Kumar B, Rao MS, Gupta AK, Purushottaman P (2011) Groundwater management in a coastal aquifer in krishna river delta, south India using isotopic approach. Curr Sci 100:1032–1043
10. Datta PS, Bhattacharya SK, Tyagi SK (1996) ^{18}O studies on recharge of phreatic aquifers and groundwater flow paths of mixing in the Delhi area. J Hydrol 176:25–36
11. Datta PS, Deb DL, Tyagi SK (1996) Stable isotope investigations on the process controlling fluoride contamination in groundwater. J Contam Hydrol 24 (1):85–95
12. Krishnamurthy RV, De Niro MJ, Pant RK (1982) Isotope evidence for Pleistocene climatic changes in Kashmir, India. Nature 298(5875):640–641
13. Narayana Rao T, Radhakrishna B, Srivastava R, Mohana Satyanarayana T, Narayana Rao D, Ramesh R (2008) Inferring microphysical processes occurring in mesoscale convective systems from radar measurements and isotopic analysis. Geophys Res Lett 35:L09813. doi: 10.1029/2008GL033495
14. Navada SV, Rao SM (1991) Study of Ganga river-groundwater interaction using environmental oxygen-18. Isotopenpraxis 27:380–384
15. Navada SV, Nair AR, Rao SM, Paliwall BL, Doshi CS (1993) Groundwater recharge studies in arid region of Jalore, Rajasthan using isotope techniques. J Arid Environ 24:125–133
16. Navada SV, Suman S, Kulkarni UP (1993) Application of environmental ^2H and ^{18}O in Tapoban and Badrinath geothermal areas, Chamoli District, Uttar Pradesh. In: Proceedings of the 6th national symposium on mass spectrometry, Dehradun, pp 498–500
17. Ramesh R, Jani RA, Bhushan R (1993) Stable isotopic evidence for the origin of salt lakes in the Thar desert. J Arid Environ 25:117–123
18. Ramesh R, Bhattacharya SK, Gopalan K (1986) Climatic correlations in the stable isotope records of silver fir (Abies pindrow) trees from Kashmir, India. Earth Planet Sci Lett 79:66–74
19. Saravana Kumar U, Noble J, Navada SV, Rao SM, Nachiappan RmP, Kumar B, Murthy JSR (2001) Environmental isotope study on hydrodynamics of lake Naini, Uttar Pradesh, India. Hydrol Process 15:425–439
20. Shivanna K, Tirumalesh K, Noble J, Joseph TB, Singh G, Joshi AP, Khati VS (2008) Isotopic techniques to identify recharge areas of springs for rainwater harvesting in the mountain region of Gaucher area, Chamoli District, Uttarakhand. Curr Sci 94(2):1003–1011
21. Srivastava R, Ramesh R, Satya Prakash, Anilkumar N, Sudhakar M (2007) Spatial variation of oxygen isotopes and salinity in the Southern Indian. Ocean Geophys Res Lett 34. doi: 10.1029/2007GL031790
22. Yadava MG, Ramesh R (2004) Past monsoon rainfall variation in peninsular India recorded in a 331 year-old speleothem. Holocene 14:517–524

Part IX
Waste Management

An Analytical Study of Physicochemical Characteristics of Groundwater in Agra Region

Chandel Prerna, Piyus Kumar Pathak, Deepak Singh and Amitabh Kumar Srivastava

Abstract The Physicochemical parameters were studied for groundwater samples collected from different places in Agra region. The main purpose of this study was to find the quality of water for drinking, irrigation, and domestic purpose from selected location. There are seven locations where groundwater samples were collected and examined in Agra region. The present investigation is focused on the determination of physicochemical parameters such as temperature, turbidity, pH, total hardness, alkalinity, total dissolved solids, chlorides, fluoride, dissolved oxygen, calcium hardness, magnesium hardness, and selected heavy metals (Fe, Cr, Cd, and Pb). Most of the parameters were found within permissible limits of above standards. All these water samples have very low concentration of Pb and Cr. The concentration of Cd and Fe was observed higher than the permissible limit of WHO and BIS. In the light of analysis, we can conclude that all these water samples require some treatment before their use for drinking purpose.

Keywords Physico-chemical parameters · Groundwater · WHO · BIS · Heavy metals · Agra

1 Introduction

Groundwater occurs as a part of the hydrological transformations of permeable structured zones of the rocks gravel and sand. Groundwater can be obtained from aquifers to hypopheric zones. Fractured crystalline bedrocks are excellent sources of potable water in many parts of the world. The groundwater passing through soil dissolves various salts and causing direct threat to the life of human beings and other living beings. These substances may be the important components of the

---

C. Prerna (✉) · D. Singh · A. K. Srivastava
Bundelkhand Institute of Engineering and Technology, Jhansi 284001, Uttar Pradesh, India
e-mail: prerna_chandel@yahoo.com

P. K. Pathak
Sunder Deep Group of Institutions, Ghaziabad 201010, India

S. Sathiyamoorthy et al. (eds.), *Emerging Trends in Science, Engineering and Technology*, Lecture Notes in Mechanical Engineering, DOI: 10.1007/978-81-322-1007-8_49, © Springer India 2012

human environment. Some of them may be either beneficial or toxic depending on their concentrations. The mobilization of various toxics in an environment may be hazardous to human health [13]. Groundwater satisfies the domestic, agricultural, and industrial need of the people. In today's world, the demand of water is swiftly increasing due to substantial increase in population, industrialization, and urbanization. This demand is fulfilled by surface water and groundwater. Both the water resources largely bank on ice melting and rainfall. In this scenario, to provide safe drinking water is a very big accountability for the governments.

Today, a big part of the population does not have pure water to drink. Easily and regularly available clean drinking water is still a harsh task to achieve not only in deserts but also in most of the megacities and small towns.

Groundwater is an excellent reservoir of water but as rivers, lakes, and streams are influenced by natural and human factors, groundwater is also facing the same situation around the world. Human activities, hydrological aspects, and characteristics of recharged water affect the quality of groundwater. As groundwater is used in high extent, some troubles are created such as water logging, land subsidence, lowering of water table, sea water, and intrusion in coastal aquifers and deterioration in water quality [1]. Groundwater is a very sensitive topic which has significance not only at local level, but at global level also [2].

Groundwater quality is slowly but surely declining everywhere. Groundwater pollution is intrinsically difficult to detect, since problem may well be concealed below the surface and monitoring is costly, time consuming, and somewhat hit-or-miss by nature. Many times, the contamination is not detected until obnoxious substances actually appear in water used, by which time the pollution has often dispersed over a large area.

Keeping this in view, it was thought worthwhile to determine of Agra groundwater to establish the baseline for the estimation of future contamination by external agents. It is important because groundwater is consumed by about 80 % of population of Agra city.

2 Experimental Section

2.1 Study Area

The metropolitan city of Agra is one of the important industrial towns of north central India. It is situated about 200 km southeast of Delhi. The metropolitan city of Agra occupies an area of about 140 km^2 and lies between $27°8'-27°14'$ N latitude and $77°57'-78°04'$ E longitude. It is the 22nd largest town in India (population wise) and the 3rd largest town in Uttar Pradesh after Kanpur and Lucknow. The urban area of Agra is divided into Nagar Mahapalika (renamed as Municipal Corporation in 1994), Agra Cantonment Area and the Dayalbagh and Swamibagh Panchayat. The municipal area is further divided into three parts, viz., the main city, the Trans Yamuna and the Taj Ganj.

2.2 Water Sampling

In the present investigation, Groundwater samples from all seven regions of Agra city were collected each during premonsoon (May 2011) and postmonsoon (December 2011) seasons from various abstraction sources at various depths covering extensively populated area, commercial, industrial, agricultural, and residential colonies so as to obtain a good aerial and vertical representation and preserved by adding an appropriate reagents as and when required. The hand pumps were continuously pumped prior to the sampling, to ensure that groundwater to be sampled was representative of groundwater aquifer.

The Physicochemical analysis samples were transferred in clean 2 L plastic bottles previously rinsed with sample water. The water samples for trace element analysis were collected in acid-leached polyethylene bottles and preserved by adding ultra pure nitric acid (2 mL/L). The standard methods (APHA, 19th Edition) adopted for each parametric analysis of groundwater samples.

The samples for dissolved oxygen analysis were collected separately in 300 mL narrow mouth flat stopper (NMFS) BOD bottles and preserved on the spot by adding 2.0 mL each of alkali-iodide-azide reagent (700 g KOH and 150 g KI in 1 L distilled water, 10 g $NaNO_3$ dissolved in 40 mL distilled water was added) and Manganese Sulfate solution (400 g $MnSO_4$ $2H_2O$ in distilled water filtered and diluted to 1 L with distilled water) by a pipette dipping below the water surface to check the activities of microorganisms which can directly influence dissolved oxygen present. The stopper was fitted in bottle's mouth and mixed thoroughly. All the

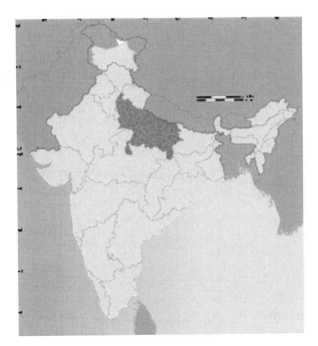

Fig. 1 Location of Uttar Pradesh

Fig. 2 Location of Agra in Uttar Pradesh

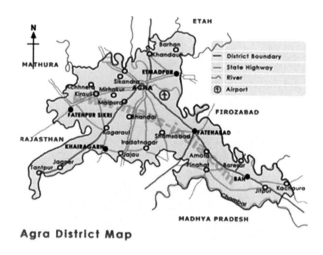

Agra District Map

samples thus collected were brought and kept at 4 °C temperature in refrigerator in the laboratory till analysis was completed. The water samples are chemically analyzed [3] (Figs. 1, 2).

3 Methodology

The pH and color was measured by pH Meter [4] and Hach color spectrometer. Total hardness, calcium, magnesium were measured by EDTA titration methods [5]. Total Alkalinity was determined by silver nitrate titrimetric methods using potassium chromate as an indicator [6]. Sulfate was determined by nephelometrically using ELICO-52 Nephelometer [7]. Fluoride content was measured by ELICO-52 Spectrometer. The Physicochemical analysis was carried out according to standard methods [8–10]. Iron, chromium, lead, and cadmium were determined by atomic absorption photometer [11]. Turbidity and TDS were observed with the help of water kit [12]. DO were determined by the standard methods of NEERI [13].

4 Results and Discussions

The temperature in the water from the study area out of seven locations of Agra city industrial area (Sikandara) shows maximum level whereas Ram Bagh has minimum level as shown in Fig. 3. In pH level of groundwater, the maximum level in the Kamla Nagar while the minimum pH content shows in the industrial

An Analytical Study of Physicochemical Characteristics

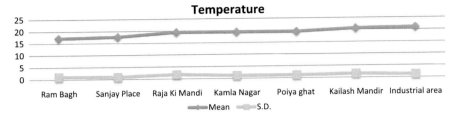

Fig. 3 Shows temperature value in Agra region

areas shown Fig. 4 which shows factors like photosynthesis, exposure of air, and disposal of industrial affect pH.

While DO content should be negligible according to standards, the presence of highest TDS content in the area Raja Ki Mandi and lowest level in industrial area is as shown in Fig. 5, which shows that water is not suitable for drinking purposes. The Highest chloride level present in the Raja Ki Mandi and industrial area shows minimum level in chloride content as shown in Fig. 6; higher the value of chloride can be attributed to be indicators of amount of industrial effluents being carried out in underground water. From the present studies, Raja Ki Mandi shows highest level in hardness whereas industrial area has lowest value as shown in Fig. 7 that indicates no suitable drinking water. Total hardness (Ca and Mg) contributing calcium and magnesium hardness so Raja Ki Mandi has highest level and

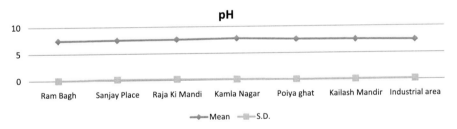

Fig. 4 Shows pH level in Agra region

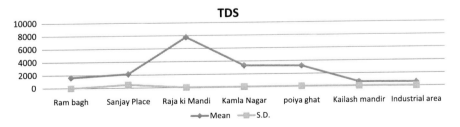

Fig. 5 Shows TDS value in Agra region

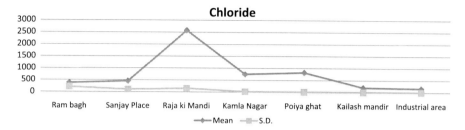

Fig. 6 Shows chloride values in Agra region

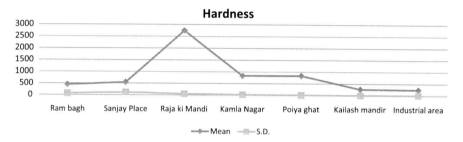

Fig. 7 Shows total hardness values in Agra region

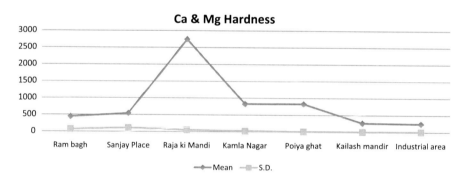

Fig. 8 Shows Ca and Mg hardness values in Agra region

industrial area has lowest level as shown in Fig. 8. Raja Ki Mandi has high turbidity whereas Sanjay place has low turbidity as shown in Fig. 9. Total Alkalinity value is higher in Raja Ki Mandi region which is not good for drinking purposes as shown in Fig. 10. From the studies, Fluoride content is present high in Kamla Nagar whereas low in industrial areas shown in Fig. 11 which shows that high value of fluoride has dental and skeletal problem in the study area.

An Analytical Study of Physicochemical Characteristics

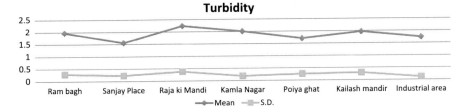

Fig. 9 Shows turbidity value in Agra region

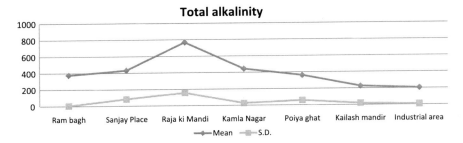

Fig. 10 Shows total alkalinity values in Agra region

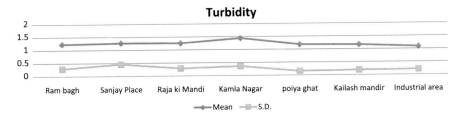

Fig. 11 Shows fluoride level in Agra region

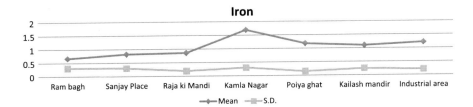

Fig. 12 Shows iron values in Agra region

In underground water scenario, Iron content is present highest in Kamla Nagar whereas lowest in Ram Bagh Fig. 12. No lead in groundwater at a detectable level in the instrument. Traces are likely to be present but at a concentration under

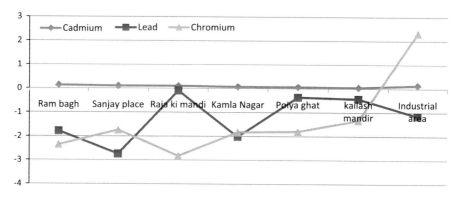

Fig. 13 Shows cadmium, lead, and chromium value in Agra region

the limit of detection. No chromium in groundwater is detectable in instrument. Therefore, it is predicted that traces are likely to be present but at a concentration under the limit of detection. Cadmium concentrations are high in industrial area whereas it is present low in Kamla Nagar that was shown in Fig. 13.

5 Conclusion

In the current study based on physicochemical parameters thus concluded that the sample location Kailash Mandir and industrial area (Sikandara) region—two location out of seven location in Agra—the water parameters are in the limit of BIS and WHO standards. Hence, these samples of water can be used for drinking, irrigation, fisheries, and cooking. Water sample location region mainly Raja Ki Mandi, Kamla Nagar, Poiya Ghat, Sanjay Place, and Ram Bagh are moderately higher in physicochemical parameters which may not much fit for drinking and cooking purposes. In general, this water may be boiling, cooled, filtered, and used for drinking purpose which are based on physicochemical studies with the water collected from the distance of 10–15 km in the Agra city.

References

1. Rastogi AK (1999) In Inland water resources-India. In: Durgaprasad MK, Pitchaiah SP (eds), vol 2. Discovery Publishing House, New Delhi, p 305
2. Shahbazi A, Esmaeili-sari A (2009) Groundwater quality assessment in north of Iran, a case study of the Mazandaron province. World Appl Sci J 5:92–97
3. Claessens L, Hopkinson Jr C, Rastetter E, Vallino J (2006) Effect of historical changes in land-use and climate on the water budget of an urbanizing watershed. Water Resour Res 42(3):W03426. doi:10.1029/2005WR004131
4. Yadav SS, Kumar R (2010) Ultra Chem 6(2):181–186

5. Kumar R, Yadav SS (2011) Int J Chem Sci 9(1):440–447
6. Yadav SS, Kumar R (2011) Adv Appl Sci Res 2(2):197–201
7. Manivasgam N (1984) Physico chemical examination of water. Pragati Publication, Meerut
8. Kumar R, Yadav SS (2011) Shodh Samiksha aur Mulyakan 2(22):19–20
9. American Public Health Association (1995) 19th edn. APHA, Washington
10. Shah M (2006) Poll Res 25(3):549–554
11. Bhutra MK, Soni A (2008) J Ind Council Chem 25(1):64–67
12. Kotaiah B, Kumaraswamy N (1994) Environmental Engineering laboratory Manual, 5th edn. Charotar Publishing House, India
13. Sandereson BB (1994) Manual of water and waste water analysis. NEERI, Nagpur

Experimental Studies on Decolourisation of Textile Effluent by Using Bioremediation Technique

S. Sathish and D. Joshua Amarnath

Abstract Biodegradation method has been proved to be very effective in treatment of this pollution source in an environmental friendly and cost competitive way. There is a need to develop alternative and cost-effective treatment processes for coloured effluents. The extent of decolourisation of direct orange dye by microbial degradation by employing the bacteria *Pseudomonas fluorescens* needs to be investigated in detail and an attempt was made to develop a cost-effective method for dye decolourisation using *P. fluorescens* bacteria. Decolourisation by bacterial strain *P. fluorescens* with respect to various nutritional sources, environmental parameters and its optimisation has been experimented in the present study.

Keywords Biodegradation · Decolourisation · Pseudomonas fluoresrcens

1 Introduction

The control of water pollution is becoming increasingly important these days. Release of dyes into environment constitutes only a small portion of water pollution. Currently, removal of dyes from effluents is by the physicochemical means. Such methods are often very costly and though the dyes are removed but accumulation of concentrated sludge creates a disposal problem. There is a need to find alternative methods of treatment that are effective in removing dyes from large volumes of effluents. At present, there is no satisfactory method to economically decolourise textile wastewater. Textile wastewater includes a large variety of dyes and chemicals additions that make the environmental challenge for textile industry not only as liquid waste but also in its chemical composition. Major pollutants in textile wastewaters are high-suspended solids, chemical oxygenated demand (COD), heat,

---

S. Sathish (✉) · D. Joshua Amarnath
Department of Chemical Engineering, Sathyabama Univeristy, Rajiv Gandhi Salai,
Jeepiaar Nagar, Chennai 600119, India
e-mail: sathi_5211@yahoo.com

S. Sathiyamoorthy et al. (eds.), *Emerging Trends in Science,*
Engineering and Technology, Lecture Notes in Mechanical Engineering,
DOI: 10.1007/978-81-322-1007-8_50, © Springer India 2012

colour, acidity, and other soluble substances. The removal of colour from textile industry and dyestuff manufacturing industry wastewaters represents a major environmental concern. In addition, only 47 % of 87 of dyestuff are biodegradable. It has been documented that residual colour is usually due to insoluble dyes which have low biodegradability as reactive blue, direct blue, and vat violet with COD/biochemical oxygen demand (BOD) ratio of 59.0, 17.7, and 10.8, respectively.

1.1 Textile Wastewater Characteristics

The common characteristics of textile wastewater are high COD, BOD, temperature, pH, suspended solids (SS), and dissolved solids (DS), solid materials, phenol, sulphur, and the colours caused by different dyes. Important pollutants in textile wastewater are especially the organics and then colour, toxic materials, inhibitor compounds, active substances, chlorine compounds, pH, salt, and dying substances.

1.2 Treatment of Textile Wastewaters

Colour is the first contaminant to be recognised in wastewater and has to be removed before discharging into water bodies. The presence of very small amounts of dyes in water (<1 ppm for some dyes) is highly visible and affects the aesthetic quality water transparency and gas solubility in lakes, rivers, and other water bodies. The removal of colour is often more important than the removal of the soluble colourless organic substances, which usually contribute to the major fraction of the BOD. Methods for the removal of BOD from most effluents are fairly well established. Dyes, however, are more difficult to treat because of their synthetic origin and complex aromatic molecular structures. A wide range of methods have been developed for the removal of synthetic dyes from waters and wastewaters to decrease their impact on the environment. They are divided in three major categories: physical, chemical, and biological.

1.3 Bioremediation

Bioremediation can be defined as any process that uses microorganisms or their enzymes to return the environment altered by contaminants to its original condition. It can also be described as a treatability technology that uses biological activity to reduce the concentration or toxicity of a pollutant.

Bioremediation process involves detoxification and mineralisation, where the waste is converted into inorganic compounds such as carbon dioxide, water, and methane. When compounds are persistent in the environment, their biodegradation

Experimental Studies on Decolourisation of Textile Effluent

often proceeds through multiple steps utilising different enzyme systems or different microbial populations. Contaminated wastewater, ground or surface waters, soils, sediments, and air where there has been either accidental or intentional release of pollutants or chemicals are the sites where bioremediation are employed. Effective Microorganism (EM) is the consortia of valuable and naturally occurring microorganisms which secretes organic acids and enzymes for utilisation and degradation of anthropogenic compounds. These days, microbes are collected from the waste water, residual sites, and distillery sludges which are believed to have the resistance against the hazardous compounds. This is particularly due to their tolerance capacity even at the higher concentrations of xenobionts.

Heavy metals and toxic organic pollutants which are believed to have resistance towards some of the microbes can be degraded using these tolerant microbes. Microbial consortium used in activated sludges and aerated lagoons are used recently for solid waste effluent removal. Biofilter technology is used to remove the hazardous chemicals and heavy metals from the effluents which contain these microbes capable of utilising the substrates rapidly due to its high surface to volume ratio and fixed cell nature. Microbial biodegradation is carried out by different organisms like Bacteria, Fungus, and Algae.

2 Microbial Bioremediation

The waste product of one process and using it as fuel or food for another process is one way to get biodegradation done; it makes intelligent use of resources decreasing the pollution and microbes does the same. They use these residual compounds as one of their substrate and grow on them, degrading or fragmenting them, which is highly valuable in case of bioremediation. *P. fluorescens* strain is reported to be a good candidate for remediation of some heavy metals and phenolics in heavily polluted sites. *Pseudomonas aeruginosa* is used for the reclamation of oil/metal contaminated soils by producing surfactants and tolerate to certain heavy metals and this strain is also used in decolourisation and degradation of dyes. *Pseudomonas* sp. has been characterised for complete and partial mineralisation of organophosphorus pesticides and fungicides, morpholine, and methyl parathion. *Pseudomonas* sp. is also involved in characteristic aromatic and aliphatic hydrocarbon degradation of oils. *Bacillus* sp. have been characterised and documented for their ability to degrade benzimidazole compounds. *Azotobacter* sp. as the biomatrix is used to remove the less toxic Cr^{3+} through biosorption.

2.1 Decolourisation Capability with Fungus and Bacterial Cultures

Bio-decolourisation of textile wastewater, as measured by the decrease in colour absorption, using two white-rot Basidiomycete fungi: *Phanerochaete chrysosporium* and *Tinctoria* sp., was reported as early as 1980. Both were clear examples of colour removal through microbial degradation of polymeric lignin molecules. Since then, the wood-rotting *P. chrysosporium* in particular has been the subject of intensive research related to the degradation of a wide range of recalcitrant xenobiotic compounds, including azo dyes. The mechanism of colour removal involves a lignin peroxidase and Mn-dependent peroxidase or lactase enzymes [1].

The decolourisation of three polymeric dyes by *P. chrysosporium* was confirmed by Glenn and their results suggested that the decolourisation was a secondary metabolic activity linked to the fungus' ligninolytic-degradation activity. The process, however, was slow and optimum decolourisation needed up to 8 days. *P. chrysosporium* was also shown to biodegrade the azo- and heterocyclic-dyes Orange II, Congo Red. The extent of colour removal varied depending on the dye complexity, nitrogen availability in the media, and ligninolytic activity in the culture.

At low nitrogen concentrations 90 % of the colour was removed within the initial 6 h, while when excess nitrogen was provided, up to 5 days were required to achieve 63–93 % decolourisation of the above-mentioned dyes. A series of other dyes were also tested and were decolourised by *P. chrysosporium* at various concentrations, particularly when veratryl alcohol was present in the medium. Veratryl alcohol is believed to stimulate the ligninase activity, which seems to be linked to decolourisation. Another white-rot fungus, *Pycnoporus cinnabarinus*, was also found to tolerate and decolourise high concentrations of dyes in a packed-bed reactor within 48–72 h incubation. An extracellular oxidase activity similar to the veratryl alcohol, which acts as an inducer for ligninolytic activity in *P. chrysosporium*, was observed in *P. cinnabarinus*, which confirmed its suitability for the degradation of dye effluent [2].

Enrichment procedures designed to obtain microbial agents suitable for decolourising dye-containing wastewater, resulted in the isolation of several strains of fungi capable of decolourisation. These included strains of *Myrothecium verrucaria* and of *Ganoderma* sp. Up to 99 % decolourisation was observed after 48 h incubation, which was mainly through adsorption to the fungus mycelium and was effective for a wide range of dyes. *M. verrucaria* was shown to have a very strong binding affinity to some azo dyes, which were recoverable by extraction with methanol, suggesting a hydrophobic–hydrophilic interaction in the dye binding mechanism.

De Angelis and Rodrigues tested colour removal of textile dyes using a *Candida* sp. yeast biomass (produced by the vast ethanol production industries in Brazil) and demonstrated 93–98 % decolourisation for several Procyon dyes at 100 mg/l. A strain of *Aspergillus sojae* B-10 was also shown to be able to decolourise the azo dyes, Congo Red and Sudan III in nitrogen-poor media after 3–5 days incubation. The decolourisation of molasses wastewater has also been investigated and several fungal cultures capable of decolourisation have been isolated. One culture possessing this capability was found to belong to the Basidiomycete group.

Recently, other facultative anaerobic fungi capable of growth on dyes as sole carbon sources have been reported. They, however, do not seem to be able to carry out decolourisation. They appear to cleave some of the bonds in these dyes to use as carbon sources, yet do not affect the chromophore centre of the dyes. This capability might be of significance when a consortium of microorganisms is employed

Experimental Studies on Decolourisation of Textile Effluent

in degrading dye-containing effluents when other decolourisers are present. Both types, the degrading and the decolourising microorganisms, would ultimately benefit from each other's activities to achieve more complete or faster biodegradation. Success in the decolourisation of paper-mill bleach-plant effluent has also been recently reported using unidentified marine fungi which produced the enzymes laccase, manganese peroxidase, and lignin peroxidase.

Numerous bacteria capable of dye decolourisation have been reported Actinomycetes, particularly *Streptomyces* sp. are known to produce extracellular peroxidases that have a role in the degradation of lignin and were also found effective in the degradation of dyes. Efforts to isolate bacterial cultures capable of degrading azo dyes started in the 1970s with reports of a *Bacillus subtilis*, then *Aeromonas hydrophila*, followed by a *Bacillus cereus*. Isolating such microorganisms proved to be a difficult task. Extended periods of adaptation in chemostat conditions were needed to isolate the first two *Pseudomonas* strains capable of dye decolourisation. An azoreductase enzyme was responsible for the initiation of the degradation of the Orange II dye by these strains and substituting any of the groups near the azo group's chemical structure hindered the degradation. Several other decolourising *Pseudomonas* and *Aeromonas* sp. were then reported by a Japanese group [3].

An alternation, however, from anaerobic to aerobic conditions was required to achieve complete degradation. This was necessary as different members of the consortium needed different conditions for optimum reaction and the main azobond cleavage needed the reductase enzymes, which are mainly functional under anaerobic conditions. In a review, described a host of bacterial cultures with capabilities to carry out decolourisation, including a *Rhodococcus* sp., *Bacillus cereus*, a *Plesiomonas* sp., and *Achromobacter* sp. compared the efficiency of a soil actinomycete culture, *Streptomyces chromofuscus*, to that of the previously described fungus *P. chrysosporium* and concluded that the soil bacterium could carry out decolourisation but to a lesser extent than the white-rot fungus. The presence of sulpho groups on the aromatic component of some dyes seemed to significantly inhibit the biodegradability of the sulfonated azo dyes by bacteria [4, 5].

2.2 Degradation of Direct Orange Dye

The UV and IR spectra after degradation of Direct Orange-102 by *P. fluorescens* showed that Direct Orange-102 degraded into another compound. The possible pathway of degradation is that Direct Orange-102 changed into an intermediate compound. This intermediate compound again changes into 3, 7-diamino-4 hydroxy-naphthalene-2 sulphonic acid sodium salt. This compound has changed again into 7-amino-3, 4-dihydroxy-naphthalene-2-sulphonic acid sodium salt or 3-amino-4-7-dihydroxynaphthalene-2 sulphonic acid sodium salt or 1, 3, 4, 5, 6, 7, 8, heptahydroxy naphthalene-2-sulphonic acid sodium salt. A perusal of the literature revealed that the end products obtained as a result of biodegradation of the dyes by the *P. fluorescens* were nontoxic in nature.

2.3 Methodology

Dye effluent was collected from a dyeing unit in Erode. The sample was collected in a brown bottle. Prior to the collection the sample bottle was rinsed thoroughly with the sample effluent. Then the sample was stored in refrigerator at 4 °C and used without any pretreatment. *The textile dye Direct Orange-102 used throughout the study.*

2.4 Microorganism and Culture Medium

(Table 1)

2.5 Decolourisation Experiment

Decolourisation of Direct Orange dye by *P. fluorescens* in the culture supernatants was determined using UV-Vis spectrophotometer. The effect of static and shaking condition of the bacterial culture on Direct Orange-102 were studied. 1 ml of bacterial cultures were transferred into separate 100 ml conical flask containing fresh nutrient medium containing Direct Orange-102 (250 mg/l) and were incubated at 30 °C, under static condition. One set of flask was incubated under agitation at 180 rpm and temperature of 30 °C while the second set was incubated under stationary condition at 30 °C for a period of 48 h. The un-inoculated dye Medium supplemented with respective dye served as the experimental. The control consisted flask without any microorganisms. All experiments were done in triplicates.

To determine the effect of pH on decolourisation, the full grown culture was inoculated in conical flasks containing 100 ml nutrient broth of varying pH (4–10) and was amended with 250 mg/l of Direct Orange-102. pH was adjusted using either $HCl(0.1\ M)$ or $Na_2CO_3(0.1\ M)$. In the similar fashion, the optimum temperature of dye decolourisation by selected bacterium was determined by evaluating the dye decolourisation at 20, 30, 37, 40, and 50 °C. After different time intervals aliquot (5 ml) of the culture media was withdrawn and supernatants obtained after centrifugation were used for analysis of decolourisation by UV-Vis double beam spectrophotometer (Table 2).

Table 1 Composition for nutrient medium (g/l) used for decolourisation studies

Nutrients	Composition
Sodium Chloride	5 g
Peptone	10 g
Beef extract	10 g
Distilled Water	1 l

Experimental Studies on Decolourisation of Textile Effluent

Table 2 Characteristics of textile wastewater used

Parameters	Values
pH	8.8
Biochemical oxygen demand (mg/l)	480
Chemical oxygen demand (mg/l)	1,826
Total suspended solids (mg/l)	15124.50
Total dissolved solids (mg/l)	3,860
Chloride (mg/l)	422
Total Kjeldahl nitrogen (mg/l)	75
Colour, absorbance (nm)	0.524
Cu^{2+} (mg/l)	38.5
Cl^- (mg/l)	1,820
Ca^{2+} (mg/l)	30
So_4^{2-} (mg/l)	245
Cd^{2+} (mg/l)	0.5
Fe^{2+} (mg/l)	6.8
Zn^{2+} (mg/l)	1.4
Mn^{2+} (mg/l)	4.8
Cr^{2+} (mg/l)	1.34
Ni^{2+} (mg/l)	0.65
Pb^{2+} (mg/l)	0.28

Various concentrations of dye (50, 100, 200, 300, 400, and 500 mg/l), and inoculum sizes of 10 % (1 ml), 20 % (2 ml), 30 % (3 ml), and 40 % (4 ml) were used to examine the effect of initial dye concentration and inoculums size on the decolourisation rate. *P. fluorescens* was cultivated for 48 h in conical flask containing 100 ml nutrient broth. Incubation was done at 30 °C.

3 Decolourisation Rate

Incubation of the dye Direct Orange-102 with *P. fluorescens* resulted in 60 % removal of colour during the first day itself. The decolourisation rate of dye by *P. fluorescens* at different dye concentration under static and agitated conditions shows that using 1 mg dye concentration, the percentage of decolourisation rate is 75 and 89 %, respectively for agitated and static conditions. Similarly using dye concentration at the levels of 4 and 5 mg, the percentage of dye decolourisation rate is 71, 74, 88, and 91 %, respectively, for the agitated and static conditions, respectively. It is found that static conditions are more efficient than the agitated conditions for the dye decolourisation.

The dye samples that show clear solution were from initial concentrations (1 and 2 mg), respectively. The results show that high rate of agitation decreases the bacterial growth and activities of some biologic substrates, such as enzymes, which play an important role in decolourising the dye.

4 Parameters Optimisation

The rate of colour removal increases with increasing temperature, within a defined range of 35–37 °C. The temperature required to produce the maximum rate of colour removal tends to correspond with the optimum cell culture growth temperature of 30–37 °C. Varying pH from 5 to 9 were experimented. Results clearly indicate that the optimum pH for colour removal is at slightly alkaline pH value of 8 and the rate of colour removal tends to decrease rapidly at strongly acid or strongly alkaline pH values.

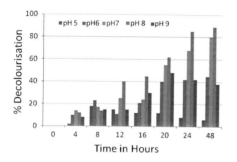
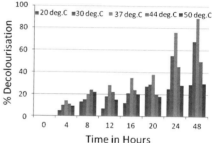

5 Result and Discussion

The COD values of composite wastewater are extremely high. In most cases, BOD/COD ratio of the composite textile wastewater is around 0.25 that implies that the wastewater contains large amount of non-biodegradable organic matter. The percentage decolourisation was found to be 80 % and the results of other parameter are tabulated in Table 3. Similarly, the dye decolourisation rate in static condition is more efficient than the agitated conditions. The optimum pH of the colour removal is slightly alkaline pH of 8, the rate of colour removal tends to

Table 3 Comparison of the physicochemical properties of the textile effluent treated with *P. fluorescens*

Physicochemical property	Untreated effluent	Effluent treated with *P. fluorescens*
pH	8.8	7.2
Decolourisation rate	0 % (Dark orange)	80 %
COD (mg/l)	750	108
Total suspended solids(mg/l)	386	91
Total dissolved solids (mg/l)	1,5124	2,000
Biochemical oxygen demand (mg/l)	480	44
Chloride (mg/l)	6,900	892

decrease rapidly at strongly acid or alkaline pH values. The optimum range of cell culture growth temperature was 30–37 °C. The decline in colour removal activity at higher temperatures is attributed to the loss of cell viability or to the denaturation of the azo reductase enzyme.

6 Conclusion

The biodecolourisation process studied presents a feasible and economical method of treating coloured effluents. Results show that textile wastewater having diverse characteristics could be decolourised effectively using *P. fluorescens*. The textile dye (Direct Orange) is degradable under aerobic conditions with a concerted effort of bacteria isolated from textile dye effluent. Nutrients (carbon and nitrogen sources) and physical parameters (pH, temperature and inoculum size) had significant effect on dye decolourisation. *P. fluorescens* showed highest decolourisation of Orange dye effectively during optimisation and *P. flourescens* showed consistent decolourisation of textile dye (Direct Orange) throughout the study.

Although decolourisation is a challenging process to both the textile industry and the wastewater treatment, the result of this findings and literature suggest a great potential for bacteria to be used to remove colour from dye wastewaters. Interestingly, the bacterial species *P. fluorescens* used in carrying out the decolourisation of Direct Orange dye in this study showed promising results. The ability of the strain to tolerate, decolourise dyes at high concentration gives it an advantage for treatment of textile industry wastewaters.

References

1. Fukuzumi T (1980) Microbial decolourization and defoaming of pulping waste liquors. In: Jurj TK, Higuchi T, Chang H (eds) Lignin biodegradation Microbiology, Chemistry and Potential Applications, CRC Press, Boca Raton pp 215–230
2. Brahimi-Horn MC, Lim KK, Liany SL, Mou DG (1992) Binding of textile azo dyes by Mirothecium verrucaria–Orange II, 10B (blue) and RS (red) azo dye uptake for textile wastewater decoluorization. J Ind Microbiol 10:31–36
3. Cripps C, Bumpus JA, Aust SD (1990) Biodegradation of azo and heterocyclic dyes by Phanerochaete chrysosporium. Appl Environ Microbiol 56:1114–1118
4. Eaton D, Chang HM, Kirk TK (1980) Fungal decolourization of kraft bleach plant effluent. Tappi J 63:103–109
5. Glenn JK, Gold MH (1983) Decolourization of several polymeric dyes by the lignin-degrading Basidiomycete Phanerochaete chrysosporium. Appl Environ Microbiol 45:1741–1747

Shade Improvement in Textiles Through Use of Water From RO-Treated Dye Effluent

S. Karthikeyan and R. I. Sathya

Abstract Wastewater treatment from textile industry is the burning alarm of recent years. The dye effluent consists of high total dissolved solids (TDS), total iron content (TIC) chlorides, hardness, and low pH. When this dye discharge gets recycled by reverse osmosis (RO) technology, the water becomes free from above particles and reused for dyeing. Apart from this the normal softened water also consists of TDS, TIC, chlorides and hardness, which prevent the dye penetration into the textile material. Hence, it incurs more salts and dye stuffs for improving the shade of the fabric. This paved the research between RO treated water dyed fabric and softened fresh water dyed fabric and its percentage of dye dispersion spectrum in woven textiles. The ultimate investigation discloses, RO treated water dyed fabric is apparently brighter and cost economic than softened fresh water and its dyed fabric. Moreover, the atmosphere is also free from effluent and greener.

Keywords Reverse osmosis treated water (RO water) • Soft water • Effluent • Chemicals • Dye penetration • Wastewater treatment • Woven textiles

1 Introduction

Effluent is a major concern in the textile industry, with the increase in stringent legislation. Dye wastewater usually consists of a number of contaminants, including acids, bases, dissolved solids, toxic compounds, and color [1]. The textile industry at the same time constantly seeks to preserve water that would be cost-effectively benefited from dye recovery and re-usage of water [2]. The effluent from the textile dyeing industries contains a variety of dyes, which are carcinogenic and mutagenic. The nature and extent of pollution depends on the biochemical oxygen demand (BOD), chemical oxygenated demand (COD), and dissolved oxygen (DO) in the aquatic bodies. The chemicals present in the dye effluent

---

S. Karthikeyan (✉) · R. I. Sathya
Department of Home Science (Textiles and Clothing), Gandhigram Rural Institute–Deemed University, Dindigul, India
e-mail: karthikeyanseenivasan@gmail.com

S. Sathiyamoorthy et al. (eds.), *Emerging Trends in Science, Engineering and Technology*, Lecture Notes in Mechanical Engineering, DOI: 10.1007/978-81-322-1007-8_51, © Springer India 2012

deplete DO and increase the BOD which causes heavy fish morality by interfering with respiratory physiology [3, 4].

Tirupur is one of the largest foreign exchange earning towns in India. Last year, the export turnover from the town was more than Rs 5,000 crore (Rs 50 billion). There are some 7,000 garment units in the town that provides employment opportunity to close to 1 million people. Membrane technology has wide range of applications in the textile industry. Various types of dyes and chemicals can be recovered from the textile effluent using this technology and a large proportion of wastewater can be reused. Membrane technology can be an efficient and cost-effective method for treating textile effluents. The conventional technologies for textile wastewater treatment are biological treatments (e.g., aerobic/anaerobic digestion, plant absorption, bio-scrubbers, bio-filtration etc.). Following chemicals treatment (e.g. neutralization, oxidation, reduction, catalytic oxidation, ion exchange etc.,) was given. Physical treatments such as (e.g. filtration, sedimentation, adsorption, and membrane etc.,) [5].

Membrane processes are microfiltration used to separate colloids from polymers with pores of 0.1 to 1 μ. Microfiltration membranes are made of several polymers including Poly (Ether Sulfone), Poly (Vinylidene Fluoride), and Poly (Sulfone). Ceramic, carbon, and sintered metal membranes have been employed where extraordinary chemical resistance or where high temperature operation is necessary. Ultrafiltration (UF) is used to separate polymers from salts and low molecular weight materials, with pores of 0.001 to 0.1 μ. Polymers are retained for reduced total oxygen demand (TOD), BOD, and COD. UF is an excellent mean to remove metal hydroxides, reducing the heavy metal content to 1 ppm or less BOD and COD.

UF is an excellent mean to remove metal hydroxides, reducing the heavy metal content to 1 ppm or less. Nano-Filtration (NF) has found wide application for water softening. It is also demonstrating ability to decolorized solutions. However, NF modules are extremely sensitive to fouling by colloidal material and polymers. For this reason extensive pretreatment is required. UF makes an excellent pretreatment substitute by eliminating the polymer addition, chlorine disinfectant and mixed-media per-filtration. Virtually, all NF and Reverse osmosis treated water (RO) membranes are thin film composite membranes. RO is used to remove almost everything from water mainly the valuable salt and organic compounds for re-usage. RO is very sensitive to fouling and the influent stream must be carefully pretreated. As a rule of thumb, the smaller the pore the higher the operating pressure [4–7, 10].

2 Materials and Methods

2.1 Selecting of Fabric

The fabric selected was cotton woven fabric (25 s count) which was selected for this study. The sample fabric is subjected to soft water in all wet processing operations. Another sample woven fabric is subjected to RO treated water in all wet processing operations. The basic processes consist of scouring, desizing, bleaching, dyeing, and finishing [8–10].

2.2 Selection of Dye and Dye Recipe

Reactive dye was selected for the study. The reactive dye is the most preferred dyes for dyeing the cotton fibers or fabric which has only low to moderate affinity for the fiber and exhausts procedures with these dyes require electrolytes to enhance dye-fiber interactions and lengthy after washing to achieve the required wet-fastness properties. Dyeing process is done with soft flow dyeing machine with the Dark blue Color.

Dark blue Dye Recipe:

Blue LSG:	0.38 %
Red LSB:	0.0082 %
Yellow LSR:	0.0016 %

The woven fabrics were wet processed and dyed with reverse osmosis water and the woven fabrics were wet processed and dyed with soft water.

3 Experiments and Testing Procedures

The woven fabrics wet processed and dyed with RO treated water and the fabrics wet processed and dyed with soft water tested for the following testing procedures.

- Bursting strength
- Fastness to sublimation
- Fastness to home laundering
- Fastness to rubbing
- Color fastness to perspiration
- Color fastness to light
- Color fastness to water
- Portability of the RO treated water and the soft water
- Spectrometer test for color matching shade depth

4 Results and Discussion

4.1 Table 1

Table 1 shows below the—pH Difference—RO treated water pH was found neutral and soft water pH as acidic. Total hardness—the total hardness found nil in RO treated water and 106 ppm in soft water total chloride—there was a huge

Table 1 Water parameters

Parameters	Soft water	RO treated water
pH (at 27°C)	5.7	7
Total hardness (as $CaCo_3$) (ppm)*	106	Nil
Total chlorides (as Cl^-) (ppm)	159.5	31.9
Total dissolved solids (TDS) (ppm)	541	247
Total iron content (TIC) (ppm)	0.43	0.14

*(ppm) parts per million

difference in ppm of chloride content between RO-treated water and soft water (Compared to RO-treated water the soft water shows a difference about 127.6 ppm which affects the dyeing process). Total sissolved solids—there was an indication of huge difference in ppm of Total dissolved solids (TDS) between RO treated water and soft water (compared to RO treated water the soft water shows a difference about 294 ppm which affects the dyeing process).The TDS make the shade duller and affect the auxiliary performances [6].

Total iron content (TIC), the results indicate a huge difference in ppm of TDS between RO treated water and soft water (compared to RO-treated water the soft water shows a difference about 0.29 ppm which also affects the dyeing process). RO-treated water has low concentration of chlorides, TDS and iron content. Whereas soft water has high total hardness (106 ppm as $CaCo_3$), total chloride (159.5 ppm), TDS (541 ppm), and TIC (0.43 ppm). Iron and manganese can affect the shade of dyed fabric or impart a yellowish tinge to bleached fabric.

4.2 Table 2

Table 2 shows the quality evaluation of dyed fabrics. The RO treated water and the soft water dyed fabric samples were tested for bursting strength, fastness to sublimation, fastness to home laundering, fastness to rubbing, color fastness to perspiration, color fastness to light, and color fastness to water. This is done to verify whether the RO treated water and soft water had made any significant comparative changes in the physical characteristics of both the fabrics after dyeing. But the test results between RO treated water dyed fabric and the soft water dyed fabric were more common in consideration with bursting strength, fastness to sublimation, fastness to home laundering, fastness to rubbing, color fastness to perspiration, color fastness to light, and color fastness to water.

4.3 Figure 1

Apart from the above results, there was a fruitful result of this research study has been noticed. In regards with the spectrometer test for color matching and shade depth. The RO treated water has a superior improvement in the color shade when compared to the soft water dyed fabric. i.e. The shade of the RO treated water

Shade Improvement in Textiles Through Use of Water

Table 2 Evaluation of dyed fabric

| | | Physical test results for dyed fabric | | | |
| | | Gray scale rating | | | |
S. No	Fabric testing	Soft water		RO water	
1	Bursting strength (PSI) (BS 4768 30 mm) average corrected bursting strength (PSI)* Average distension (mm)	130.6 11.8		133.4 11.20	
2	Color fastness to sublimation (IS 975 : 88) Gray scale rate staining on cotton/polyester	5/5		5/5	
3	Color fastness to home laundering accelerated at 40 ^{0}C (AATCC—61—1 A: 2003) AATCC-grey scale rating change in color Staining cotton	4.5 4.5		4.5 4.5	
4	Color fastness to rubbing (ISO 105 × 12: 2001) Grey rating stain on cotton (Dry/wet)	4–5/4		4–5/4	
5	Color fastness to perspiration (ISO 105 EO4: 1994) Grey scale rating (a) Change in color (b) Staining on cotton	Acid 4 5	Alkali 4–5 5	Acid 4 5	Alkali 4–5 5
6	Color fastness to light (Xenon Arc Lamp-AATCC-16 (OPTION 5): 2004) 10 AFU-gray scale rating change in color 20 AFU-gray scale rating change in color	4 2		3.5 2.5	
7	Color fastness to water (Grey scale rating ISO 105 E01: 1996) (a) Change in color (b) Staining on cotton	4–5 5		4–5 4–5	

*(PSI) Per Square Inch

dyed fabric is 100 % in its dye absorption. But in the case of the soft water dyed fabric the dye absorption level noticed in the spectrometer is only 94.41 % (Fig. 1)

K/S: Ratio of coefficient of absorption 'K' To Coefficient of scattering 'S'

L^*:   Light,

a^*:   Redder tone (positive value), Greener tone (negative value)

b^*:   Yellow tone (positive value), Blue tone (negative value)

DE*:   Color difference

DEcmc: Color difference equation (British Standard, AATCC, and ISO)

TL 84-10:   Light illuminant source

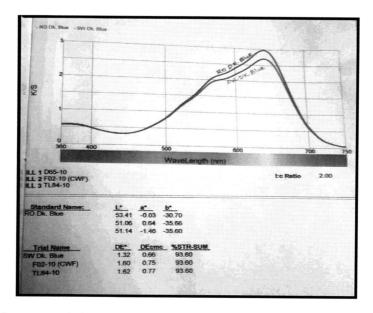

Fig. 1 Spectrometer shade percentage rating graph

F02-10(CWF): Light illuminant source
%STR-SUM: Percentage color shade difference between two samples

4.4 Table 3

The conservative estimate for putting up a basic workable model of RO membrane technology plant to achieve the salt removal/zero discharge stage comes to Rs 4–4.5 crore for a unit handling 10 lakh liters of effluent a day (Table 3).

Table 3 Cost analysis between RO water and soft water

S. No	Cost difference between Ro-water and soft water		
	Cost per liter	RO-Water (Paisa)	Softer water (Paisa)
1	Power	2.0	
2	Chemicals	2.5	
3	Man power	1.5	
4	Water and other	0.5	
5	Water procuring cost		4.5
6	Softening cost (Power and chemicals for controlling hardness and TDS mainly to remove magnesium and calcium).		0.9
	Total	6.5	5.4

In this connection, the cost analysis between RO treated water and soft water is worked. In the case of soft water the water procurement cost come around 4.5 Paisa/L and the softening cost reached to 0.9 Paisa/L. The total cost involved in soft water is 5.4 Paisa/L.

Beside this, the RO treatment for the effluent costs around 6.5 Paisa/L. This includes the power consumption (2.0 Paisa/L), chemicals (2.5 Paisa/L), manpower (1.5 Paisa/L), water and other (0.5 Paisa/L). Eventually, RO water process is cost-effective because it is only 1 paisa/L higher than the soft water.

5 Conclusion

The above research experimentally confirms the results that the fabric obtained after dyeing with RO treated water is greater than the soft water dyed fabric. In some tests, the result between RO treated water dyed fabric and soft water dyed fabric was similar. But the spectrometer test for color matching and shade depth provides a considerable result that the RO treated water dyed fabric is higher in shade than soft water dyed fabric. The RO-treated water dyed fabric is 6.41 % shade higher than that of a soft water dyed fabric, which is highly preferred by customers of the Indian apparel export like Nike, Cutter and Buck, Tommy, Hilfiger, Adidas, GAP, FILA, and Arrow [8]. In ecological point of view the industrial wastewater recycling is an urge, some of the advanced techniques like RO through membrane technology should be expanded to all dye houses to get the water which is free from Iron, TDS, chlorides, hardness and so on. To produce a good quality textile material after dyeing and also to keep the atmosphere greener, the only way is to use RO through membrane technology water for dyeing process which is cost-effective in the economical point of view also.

To put in a nutshell, the ultimate investigation concludes that RO water dyed fabric is better than soft water dyed fabric both in international customer and conservation point of view with social responsibility.

Acknowledgments First of all the author wishes to thank MS. Arora fashion ltd., Tirupur, for granting permission to undergo this research. He also extends his special thanks to his family members who have been the great inspiration at every stage of the research.

References

1. Chu W (2001) Dye removal from textile dye wastewater using recycled alum sludge. Water Res 35(13):3147–3152
2. Ramesh Kumar M, Saravanan K (2011) Textile dyeing industry waste water treatment with reverse osmosis membrane system. J Text Assoc 72(2):103–108
3. Muthu Kumar M, Sargunamani D, Senthilkumar M, Selvakumar N (2005) Studies on decoloration, toxicity and the possibility for recycling of acid dye effluents using ozone treatment. Dyes Pigm 64(I):39–44
4. Aslam MM, Baig MA (2004) Textile waste characterization and reduction of its COD and BOD oxidation. Electron J Environ Agric Food Chem ISSN:1579–437

5. Naveed S, Bhatti I, Ali K (2006) Membrane technology and its suitability for treatment of textile waste water in Pakistan. J Res (Sci) 17(3):155–164
6. Growther L, Meenakshi M (2009) Biotechnological approaches to combat textile effluents. Internet J Microbiol 7(1):384–470
7. Visvanathan C, Aim RB (2002) Membrane Bioreactor Applications in Wastewater Treatment http://www.faculty.ait.ac.th/visu/Prof visu's/chapters in books/11/MBR.text1.pdf
8. Chita Sivasankar A (2011) Dyeing stain on knit cities future. Indian Text J 28(12):16–21
9. Muthu Manickam M, Ganesh Prasad J (2005) Eco-friendly textiles.Text Indus Trade J 41–48
10. Satyanarayana YVV (2004) Total water management in cotton textile industry. Colourage 45(10):4–11

Part X
Physics

Fabrication and Electrochemical Properties of All Solid State Lithium Battery Based on PVA–PVP Polymer Blend Electrolyte

N. Rajeswari, S. Selvasekarapandian, J. Kawamura and S. R. S. Prabaharan

Abstract All Solid State Lithium Polymer Batteries (LPBs) have been fabricated by using solid polymer blend electrolyte (SPE) composed of biodegradable polymers, polyvinyl alcohol (PVA) and polyvinylpyrrolidone (PVP) blend with lithium perchlorate $LiClO_4$ at a different weight percent ratio. The effect of addition of lithium salt with blend polymers is investigated by XRD, FTIR, Electrochemical impedance (EIS), and Cyclic Voltammetry measurements as well as the charge–discharge performance. The room temperature electrolytic conductivity of the order of $10^{-4}\,S\,cm^{-1}$ has been achieved in the composition of 70PVA: 30PVP: 25 Wt % $LiClO_4$. The conductivity–temperature plot is found to follow the Arrhenius nature, which showed the decrease in activation energy with the increasing salt concentration. The electrolyte with the highest ion conductivity has been used in the fabrication of Lithium/Polymer battery with the configuration of Li (metal foil)/SPE/$LiCoO_2$. The galvanostatic charge/discharge performance is carried out from 3 to 4.6 V versus Li^+/Li. The electrochemical stability of the polymer blend electrolyte membrane has been found to be stable up to ~4.6 V versus Li/Li^+.

1 Introduction

Solid Polymer Electrolytes (SPEs) are promising materials as electrolytes in advanced electrochemical applications, particularly lithium ion battery. Lithium polymer battery technology is regarded as advanced batteries due to their distinct

---

N. Rajeswari (✉)
Department of Physics, Kalasalingam University, Krishnankoil 626126, Tamil Nadu, India
e-mail: rajikani.phy@gmail.com

S. Selvasekarapandian
Department of Nanosciences, Karunya University, Coimbatore 641114, Tamil Nadu, India

J. Kawamura
Institute of Multidisciplinary Research for Advanced Materials,
Tohoku Univeristy, Sendai, Japan

S. R. S. Prabaharan
Department of Electrical and Electronics Engineering, Manipal International University,
Petaling Jaya 47301, Malaysia

S. Sathiyamoorthy et al. (eds.), *Emerging Trends in Science,*
Engineering and Technology, Lecture Notes in Mechanical Engineering,
DOI: 10.1007/978-81-322-1007-8_52, © Springer India 2012

properties like moldability in various physical shapes, light weight, good electrode–electrolyte contact, and favorable mechanical properties [1, 2]. Replacement of the liquid electrolyte currently in use by the SPE yields several advantages including structural stability, low volatility, and the SPE provides liquid like medium for fast ion transport and thus are able to improve ion conductivity by an order of magnitude approaching that of liquid electrolytes at room temperature [3, 4].

Various methods have been employed to improve the Li^+ conductivity of polymer electrolytes which have led to the development of new routes of preparation of polymer electrolytes. For instance, cross-linking of two polymers, adding plasticizers to polymer electrolytes in the presence of inorganic inert fillers and also blending of two polymers is thought of as a new avenue to improve ion conductivities. Polymer blends are physical mixtures of structurally different polymers that interact through secondary forces and that are miscible to the molecular level. The most common interactions present in blends are hydrogen bonding, Ionic, and dipole interactions. The significant advantages of polymer blends are that the properties of the final product can be tailored to the requirement of applications, which cannot be achieved alone by one polymer. Generally, blending of two polymers not only results in the improvement of the mechanical strength but also helps increasing the conductivity by suppressing the crystallization of polymer chain [5].

Among polymers, PVA has excellent physical property such as mechanical strength, electrochemical stability, nontoxic, good film forming capability, and biocompatibility. It contains a hydroxyl group attached to methane carbons. These hydroxyl groups can be a source of hydrogen bonding. PVP is a vinyl polymer possessing planar and highly polar side groups due to the peptide bond in the lactam ring [6]. It is an amorphous polymer and possesses a high T_g because of the presence of the rigid pyrrolidone group, which can strongly draw a polar group and is known to form various complexes with other polymers. Therefore, a hydrogen-bonding interaction may take place between these two polymers [7, 8].

In the present study, prepared lithium ion conducting polymer blend electrolyte which is composed of PVA and PVP polymers and also examined the structural, vibrational, electrolytic conductivity, and electrochemical properties. Finally, Li-polymer battery performance using the new polymer blend electrolyte has been assembled $Li/[PVA-PVP + LiClO_4]/LiCoO_2$ and tested.

2 Experimental Technique

2.1 Materials Used

The polymer blend electrolyte has been prepared by solution casting technique. Polymers are PVA (PVA, Mw-1, 25,000, 88 % hydrolyzed, AR grade, s.d fine) and PVP (PVP, K90, Mw: 90,000, LR grade, s.d fine). Electrolyte salt is lithium perchlorate ($LiClO_4$, HIMEDIA) and the solvent is DMSO. These materials have been used as the raw materials in this study.

2.2 Film Casting

The appropriate quantities of PVA and PVP have been added to DMSO solvent and stirring the solution at room temperature to complete dissolution. Required quantity (5, 10, 15, 20, 25 and 30 Wt %) of lithium perchlorate is also dissolved in the above solution. The resultant mixture was stirred for 24 h to complete the homogeneous of the mixed solution. The final solution has been poured onto cleaned petri dishes and dried in oven at 70 °C for 6 days to ensure removal of the solvent traces. After drying, the films were peeled off from petri dishes and kept in desiccators until use. These polymer electrolytes have been characterized by different experimental techniques.

2.3 Characterization

The X-ray diffraction patterns of polymer electrolytes have been recorded at room temperature employing a PANalytical Xpert Pro with Cu Kα ($\lambda =$ 1. 5406 Å) as the source material. FTIR measurement has been performed with SHIMADZU-IR Affinity-1 spectrometer instrument in the wave number range of 400–4000 cm^{-1}. The Ionic conductivity study of the polymer electrolytes has been carried out in the temperature range of 303–373 K over a frequency range of 42 Hz–1 MHz using a computer interfaced HIOKI 3532 LCR meter. The Li-polymer cells were assembled by sandwiching the polymer blend electrolyte membrane between the cathode $LiCoO_2$ and lithium metal anode. Cathode is prepared by spreading a mixture of active electrode material, $LiCoO_2$, 85 Wt %, Conductive carbon (L6) 10 Wt %, binder (PVdF) 5 Wt %, and coated onto a special Al foil. The galvanostatic charge/discharge performance is carried out from 3 to 4.6 V versus Li^+/Li. (SP-150 Potentiostat/Galvanostat (Bio Logic Science Instruments) equipped with EC-Lab® Express software, Japan.

3 Results and Discussion

3.1 XRD Analysis

Figure 1 (a–e) shows that the X-ray diffraction pattern of pure 70PVA: 30PVP blend and 70PVA-30PVP blend with different weight percentage of $LiClO_4$. From the Fig 1, it can be clearly seen that the peak intensity is decreased and there is a increase of a full width half maximum of the PVA and PVP peaks due to the addition of lithium salts. The increased broadness and decreased intensity (indicating amorphous nature of the polymer electrolyte) have been found in the composition of 70PVA: 30PVP: 25 Wt %$LiClO_4$. DMSO plays an important role in decreasing the crystalline nature

Fig. 1 XRD pattern of (a) 70PVA: 30PVP (b) 70PVA: 30PVP: 5Wt % LiClO₄ (c) 70PVA: 30PVP: 15Wt % LiClO₄ (d) 70PVA: 30PVP: 25Wt % LiClO₄ (e) 70PVA: 30PVP: 35Wt % LiClO₄

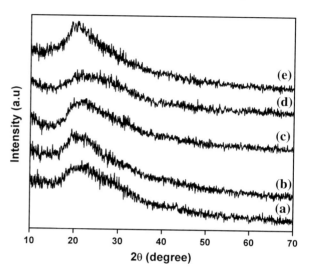

of the polymer electrolyte films [9]. None of the diffraction peaks relating to lithium perchlorate was observed indicating that lithium salt is completely dissociated.

3.2 FTIR Analysis

The FTIR spectroscopic technique has been used to identify the interactions in polymer blends. Fig 2 shows the IR spectra of 70PVA: 30PVP blend and 70PVA: 30PVP blend with different weight percentage of lithium perchlorate. From Fig 2, the band observed at 1607 cm^{-1} corresponds to the carbonyl stretching region of PVP and the hydroxyl stretching band of PVA is at 3313 cm^{-1}. These results revealed that a hydrogen-bonding interaction occurs between two polymeric components. The band corresponding to the methane group (CH$_2$) asymmetric stretching vibration occurs at about 2163 cm^{-1}. The bands at 1469, 1137, 943, and 856 cm^{-1} corresponding to CH$_2$ bending, C=O stretch, C–H, and OH bending, C=O symmetric stretching, and CH$_2$ rocking. The above-mentioned peaks get shifted and intensity varied when the salt doped with PVA-PVP blend. The peak around 640 cm^{-1} has indicated the presence of solvent (DMSO) in very small fraction and attached to the polymer [10].

3.3 Electrochemical Impedance Analysis

The most important property of a material for potential electrolyte applications is its ion conductivity. Figure 3 shows that the Nyquist plot for PVA: PVP blend with different weight percentage of lithium perchlorate. The plot consists of a

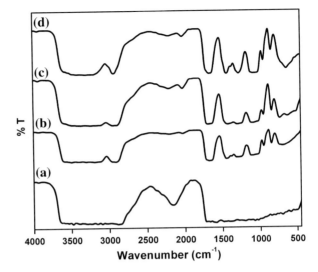

Fig. 2 FTIR analysis of (a) 70PVA: 30PVP (b) 70PVA: 30PVP: 20Wt % LiClO$_4$ (c)70PVA: 30PVP: 25Wt % LiClO$_4$ (d) 70PVA: 30PVP: 30Wt % LiClO$_4$

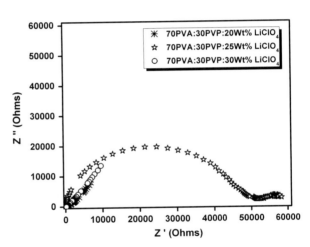

Fig. 3 The Nyquist plot for different weight percentage of LiClO$_4$ doped with 70PVA:30PVP blend at room temperature

semicircle in a high-frequency region, which corresponds to an equivalent circuit consisting of a parallel combination of a capacitance and a resistance (charge transfer, R_{ct}). This R_{ct} contains charge-transfer resistance as well as an electrical contribution based on a surface resistive layer formed on the electrode [11]. A tail-like flattened spike occurs at low-frequency region indicates the interfacial resistance at the electrolyte/SS electrode interface. The conductivity of the polymer complex films has been calculated from the bulk resistance (R_b) determined from the intersection of the high-frequency semicircle with the real axis in the complex impedance plots.

$$\sigma = t / AR_b$$

where t is the thickness of the film and A is the area of the sample.

Fig. 4 Arrhenius plot for PVA–PVP blend with different weight percentage of lithium perchlorate polymer electrolyte

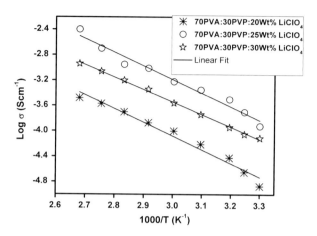

The highest value of σ obtained is 1.20×10^{-4} S cm^{-1} at 303 K for the composition of 70PVA: 30PVP: 25 Wt % LiClO$_4$. The ion conductivity initially increases with the salt concentration. However, after reaching a maximum value, it decreases because higher the salt concentrations lower the segmental motion of the polymeric chains.

Figure 4 shows that the conductivity increases linearly with increases temperature for all the composition. The regression values of the plots using a linear fit have been found to be close to unity suggesting that the temperature dependence Ionic conductivity for all the composition obeys Arrhenius relation.

$$\sigma = \sigma_o \exp(-E_a/KT)$$

where σ is related to the number of charge carriers in the films and E_a is related to the activation energy of ion transport associated with the configurational entropy of the polymer chains and K is the Boltzmann constant. The low activation energy has been observed in the high conductivity composition of 70PVA: 30PVP: 25 Wt % LiClO$_4$. The activation energy is found to be low on the order of 0.38 eV for higher conductivity polymer electrolyte.

3.4 Electrochemical Performance

The pouch type cells as shown in Fig. 5 was fabricated in argon filled glove box. The lithium metal foil (100 μm thick) was used anode and composite LiCoO$_2$ blended with a predetermined portion of polymer electrolyte in order to increase the mixed conductivity of the composite cathode. The surface electronic conductivity was enhanced by adding a conductive additive carbon. The slurry was spreaded over aluminum foil (100 μm thick) before being dried over vacuum over at ~90 °C. The cathode and lithium foil anode was separated by a microporous PP separator (Celgard 2400) and the electrolyte (1 M LiPF$_6$ in EC/DMC) was injected

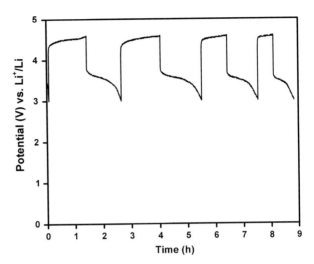

Fig. 5 Pouch type lithium polymer battery

into the pouch before sealed using a hand-operated heat sealer inside the glove box. The pouch was then vacuum sealed externally as an air tight barrier for any air and/or any atmospheric attack.

As-fabricated cell was allowed to equilibrate for 30 min and the ocv was measured to be 3.01 V. The cell was cycled galvanostatically at 10 μA and the electrochemical performance is shown in Fig. 6. As depicted, the charging curve (plateau potential) indicates that Li$^+$ deintercalation occurs at the cathode side from the LiCoO$_2$ according to the below reaction:

$$Li_{1-x}CoO_2 + xLi^+ + xe^- \rightarrow xLi^+ + xe^- + Li^0$$

The removed Li$^+$ during the charging process is deposited/plated onto Li anode electrochemically. Similarly, on discharge, a charge plateau occurs near ~3.6 V versus Li/Li$^+$ indicative of the intercalation of Li$^+$ into LiCoO$_2$. This process repeats as the cycle continues. This clearly indicates that the developed polymer facilitates the Li$^+$ transport across it between the cathode and the anode.

4 Conclusion

Polymer blend electrolytes consisting of PVA and PVP blend with different weight percentages of LiClO$_4$ have been prepared by solution casting method. The amorphous nature and polymer salt complex formation has been confirmed from the XRD and FTIR spectral studies. The room temperature electrolytic conductivity of the order of 10^{-4} S cm^{-1} with low activation energy 0.38 eV for the composition of 70PVA: 30PVP: 25 Wt % LiClO$_4$ has been observed from electrochemical impedance spectroscopy (EIS) analysis. The temperature dependence of Ionic conductivity of these electrolytes exhibited Arrhenius behavior. The developed SPE

Fig. 6 Typical galvanostatic cycling of the pouch type (see Fig. 5) cell for four charge/discharge cycles

has been employed and tested in a pouch-type cell against lithium metal and the cycling performance was evident that the SPE developed in this work acts as a electrolytic membrane across the cathode and anode allowing the Li$^+$ transport as electrolyte cum separator.

References

1. Gray FM (1991) Solid polymer electrolytes fundamentals and technological applications. VCH, New York
2. Maccallum JR, Vincent CA (eds) (1987 and 1989) Polymer electrolyte reviews, vol 1 and 2. Elsevier, London
3. Kelly IE, Owen JR, Steele BCH (1985) Poly (ethylene oxide) electrolytes for operation at near room temperature. J Power Sources 14:13
4. Koksbang R, Oslen II, Shackle D (1994) Review of hybrid polymer electrolytes and rechargeable lithium batteries. Solid State Ionics, 69:320
5. Michael MS, Prabaharan SRS (2004) Rechargeable lithium battery employing a new ambient temperature hybrid polymer electrolyte based on PVK + PVdF-HFP (copolymer). J Power Sources 136:408–415
6. De Queiroz AAA, Soares DTAW, Tizesniak P, Abraham GA (2001) Resistive-type humidity sensors based on PVP-Co and PVP-I$_2$ complexes. J Polym Sci B: Polym Phys 39:459
7. Feng H, Feng Z, Shen L (1993) A high resolution solid-state n.m.r and d.s.c study of miscibility and crystallization behavior of Poly (Vinyl alcohol)/Poly (N-Vinyl-2-pyrrolidone) blends. Polym 34:2516
8. Zhang X, Takegoshi K, Hikichi K (1992) High resolution solid-state 13C nuclear magnetic resonance study on Poly (Vinyl alcohol)/Poly (Vinyl-pyrrolidone) blends. Polym 33:712
9. Awadhia A, Agrawal SL (2007) Structural, thermal and electrical characterizations of PVA:DMSO:NH$_4$SCN gel electrolytes. Solid State Ionics 178:951–958
10. http://en.wikipedia.org/wiki/Dimethyl_sulfoxide_(data_page)
11. Sun HY, Takeda Y, Imanishi N, Yamamoto O, Sohn HJ (2000) Ferroelectric materials as a ceramic filler in solid composite polyethylene oxide base electrolytes. J Electrochem Soc 147:2462

Modeling Thermal Asymmetries in Honeycomb Double Exposure Solar Still

K. Shanmugasundaram and B. Janarthanan

Abstract In this paper, the modeling of honeycomb double exposure solar still has been analyzed include four elements, such as the glass cover, honeycomb structure, water mass, and bottom absorber. The energy balance equation has been written for the still parameters. The computer program has been written for the analytical solutions of water, glass, and bottom absorber temperature. To validate the analytical results, experimental observations for one of the typical days in February 2012 have been taken. The correlation coefficients have been found and their r^2 values for water and bottom absorber are presented.

Keywords Honeycomb solar still • Desalination • Correlation

1 Introduction

Supply of drinking water is a major problem in under developed as well as in some developing countries. Distillation method is a best solution for this problem. Distillation of brackish or saline water, wherever it is available, is a good method to obtain fresh water solar distillation uses the heat of the sun directly in a simple piece of equipment to purify water. The equipment commonly called a solar still John Ward [1]. The analytical and experimental study of a plastic solar water purifier has been done. The experimental study had shown that the sea water input with 35,000 ppm of totally dissolved solids is converted into portable with a totally dissolved solid of 1.2 ppm without the taste, smell, and color of plastic with distillate yield for up to 1,000 W/m^2 insolation and 35 °C ambient temperature Delyannis [2]. The analytical study of a historic background of desalination and renewable energies has been done. The historical path revealed the most important ideas and today on destination of sea and brackish water as well as renewable energy utilization with special

---

K. Shanmugasundaram (✉) · B. Janarthanan (✉)
Department of Physics, Karpagam University, Coimbatore, India
e-mail: sspriyashanmu2@gmail.com

B. Janarthanan
e-mail: bjanarthanan2002@yahoo.co.in

S. Sathiyamoorthy et al. (eds.), *Emerging Trends in Science,*
Engineering and Technology, Lecture Notes in Mechanical Engineering,
DOI: 10.1007/978-81-322-1007-8_53, © Springer India 2012

reference to use of solar energy for desalination Mathioulakis and Belessiotis [3]. The simple solar still has been designed and its incorporation in a multi-source and multi-use environment has been analyzed. It had shown that the efficiency and productivity increases in multi-source environment Singh and Tiwari [4]. The comparative study of monthly and annual performance of passive and active solar has been carried out and for different Indian climatic conditions. It has been concluded that the annual yield depends on water depth, inclination of glass cover for both active and passive solar still. The annual yield of active solar still for a given water depth increased linearly with the collector area.

Rajesh Tripathi and Tiwari [5], the condensing covers with characteristic dimension of 0.14 and 0.07 cm made of aluminum and copper has been designed to study the convective and evaporative heat transfer coefficients in the still. The study has been revealed that there is an increase of 15 and 7.5 % in the evaporative heat transfer coefficients due to the change in the size and material of the condensing cover Ghoneim [6]. The thermal performance of solar collector equipped with different arrangement of square honey comb material made with poly carbonate sheet of 10 mm cell size has been designed and analyzed. It is proved that the present of square type model honey comb suppress natural convection and increases the efficiency of the collector Shanmugan et al. [7]. The performance of acrylic mirror boosted solar distillation has been designed and analyzed. It has been concluded that the presence of booster mirror above the glass cover helps to increase the efficiency. The observed output of the still was 4.2L/m^2 at 890 W/m^2. Zhang Zhiqiang et al. [8] the thermal performance of solar air collector with transparent honey comb made of glass tube has been predicted. It has been proved that the aspect ratio L/P of the honey comb, size of the honeycomb, and optical properties of the material of the honeycomb influences the thermal performance of solar air collector. Ahmed et al, [9] the experimental and theoretical analysis of multistage solar distillation has been studied. It has been found that the internal pressure is directly proportional to the still productivity Bassel [10]. An attempt has been made to design, fabricate, and analyze hemispherical solar still. It has been found that the daily distilled water output from the still varied from 2.8 to 5.7 L/m^2 day. It is also found that the average efficiency of the still decreases by 8 % when the saline water depth increased by 50 %. The objective of this work, the modeling of honeycomb double exposure solar still has been analyzed include four elements, such as the glass cover, honeycomb structure, water mass, and bottom absorber. To validate the analytical results, experimental observations for one of the typical days in February 2012 have been taken. The correlation coefficients have been found and their r^2 values for water and bottom absorber are presented.

2 Construction of the Still

The base of the still is made of the 24 gage galvanized iron sheet. It is bent to have a base of length 100 cm and breadth 50 cm with aspect ratio L/W = 2.0, and practical evaporation area of 5,000 m^2. This 24-gage G.I. sheet was bent using

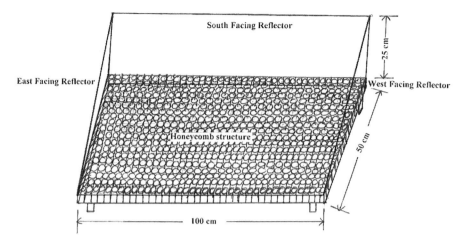

Fig. 1 Schematic diagram of the honeycomb double exposure

plate-bending machine. The sides of the base are bent upwards to a height of 8 cm, so that the vertical distance between the base and the cover glass becomes 8 cm. All the corners are joined using aluminum rivets. The upper glass cover is tilted at an angle 26°. The south facing, east facing, and west facing of the still covered with highly reflecting material from the outer surface that helps to gain additional energy through the reflection of sun rays to the still basin are closely packed with thin walled transparent honeycomb structure with small aspect ratio(L/D < 3). The schematic diagram of the honeycomb double exposure is shown in Fig. 1.

The presences of thin walled transparent honeycomb structure in the basin suppress the natural convection, obstruct the infrared radiation heat loss, and reduce conduction heat loss through the walls. The transparent honeycomb structure traps the total incident radiation inside the basin by means of total internal reflection.

3 Experimental Observations

The experimental observations have been measured for every 1 h interval from 9 am to 5 pm on 01.10.2010 are mentioned given below.

T_b: Temperature of the basin
T_w: Temperature of the water
T_g: Temperature of the glass cover
T_a: Temperature of the ambient air
H_s: Total solar insulation

To measure the temperature of the tilted cover glass, temperature of the water and temperature of the basin of the still, three thermocouples were used.

Fig. 2 Photograph of the honeycomb double exposure solar still

To measure the temperature of the glass cover, the first thermocouple is fixed at the middle of the glass cover-using adhesive. To measure the temperature of the water, the second thermocouple immersed in the water. To measure the temperature of the still basin, the third thermocouple placed at the middle part of the basin by the help of solder iron instrument. All the leads of the thermocouples are taken out of the still through small holes provided at the side of the still base. The photograph of the honeycomb double exposure solar still is depicted in Fig. 2.

4 Thermal Analysis

The condenser cover of the honeycomb double exposure solar still may be considered as a system having one thermally independent flat-plate. The high inclination of the cover toward the south is the characteristics of absorbed and transmitted energy. The latitude and time of the day are the parameters which play as important role in the performance of the condenser and the solar radiation reaching the components of the system have a distinct influence on the temperature of the system. Based on this fact, the analysis of the still includes four elements: the glass cover, honeycomb structure, water mass, and bottom absorber.

5 Energy Balance Equation

The time-dependent solar energy is transmitted through the transport cover to the water through the honeycomb structure and absorber bottom. The temperature of these elements varies as a result of the fraction of absorbed energy, heat losses, and thermal properties. The energy balance for the glass cover can be expresses as follows,

$$C_g \frac{dT_g}{dt} = H(t)\alpha_g A_g + h_1(T_w - T_g)A_g - h_2(T_g - T_a)A_g \qquad (1)$$

It is necessary to estimate the reflected energy from south, east, and west facing reflectors and also the transmittance of the glass tube honeycomb unit in the basin.

The energy input to the basin with honeycomb structure can be estimated by the equation derived by El-Swsify and Mettias (2002) and it is given by

$$I_H = \tau \tau_1 H(t) + \sum (I_r)_{S,E,W} \tag{2}$$

where,

$$(I_r)_{S,E,W} = (I_B)_e \sum f_{S,E,W} \tag{3}$$

The fraction of reflected solar energy form South East and West reflectors is given by El-Swiffy and Mettias [11].

$$f_s = Cos\gamma Cos\alpha S^{-1} \left(1 - \frac{1}{2}Tan \right. \tag{4}$$

$$f_E = 0.25 \sin^2 \gamma \frac{Cos^2 \alpha}{Sin\alpha} \tag{5}$$

$$f_W = 0.25 \sin^2 \gamma \frac{Cos^2 \alpha}{Sin\alpha} \tag{6}$$

The response of each glass tube in the closely packed honeycomb structure with water to the reflected rays from the reflectors in addition to the direct one falling on the still glass window remains the same. Consider a single glass tube in the honeycomb unit and beam of parallel rays irradiate into the glass tube openings at an arbitrary angle. The following assumptions have been made to simplify the analysis. The single glass tube in the honeycomb unit and the corresponding path of solar radiation is shown in the Fig. 3.

1. The solar rays from the reflectors in addition to the direct ray will be considered as parallel lines entering through the top of the glass tube and reflecting between the side walls.
2. The light entering the water surface and glass tube but transmitting to another glass tube will be converted into reflected light with in it.
3. The transmittance of the water surface and glass tube will not change with the incident angle.

When light transmitted through the water surface and glass tube, each time the intensity decays by a fraction τ, which is effective transmittance. Therefore, when the light goes from one glass tube into another, it passes through water and two glass walls, and its intensity decay by a fraction τ_e, which is effective transmittance. Therefore, when the light goes from one glass tube into another, it passes through water and two glass walls and its intensity decays to

$$\tau_e = \tau_2 \tau_3 \tau_4 = 0.64 \tag{7}$$

Fig. 3 Single glass tube in honeycomb unit

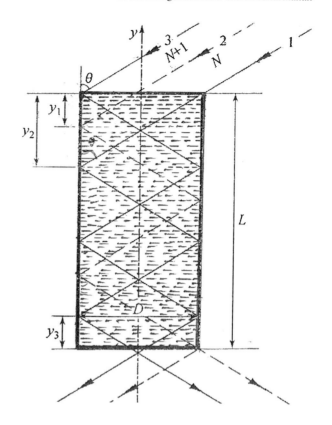

On the account of above transmitted light, the total fraction of reflected light inside the basin, or effective reflectivity will be

$$\rho_e = \tau_e + \rho = 0.64 + 0.05 = 0.69 \qquad (8)$$

With the above assumptions, consider the rays which incident on the glass tube at angle 'θ' and finally out of the glass tube as shown in the Fig. 3. Let the ray represented by line 1 is incident on a point in the wall after transmitted through the water with distance 'Y_α' from the opening of the glass tube. It can be derived that

$$Y_\alpha = \frac{D}{\tan\theta} \qquad (9)$$

If the height of the honeycomb is 'N' time the distance 'Y_α', thus length of the glass tube will be $L = N y_\alpha$. Thus, the light through the water and honeycomb will experience n times reflection before coming out of the glass tube.

The height of the tube is fixed, though the incident angle of the sunlight changes with time. Thus we have $L/Y = L/D \tan\theta$, which is not an integer. Among these light lines coming out of the glass tube, small fraction of lines experience one more reflection through the bottom wall section y_3. Since $Y_1 = Y_2$, the rays 2 and 3 must experience $N+1$ times reflection before emerging out of the tube. The fraction of rays which under goes $N + 1$ times reflection can be obtained as $X_1 = Y_1 / Y_2 = L - N Y_2 / Y_2 = L / D \tan \theta - N$. The fraction of rays having N times reflection can be written as $X_2 = 1 - X_1 = 1 + N - L / D \tan \theta$. The sum of above two fractions of rays gives the total transmittance of the glass tube with water. Therefore, the effective transmittance through the honeycomb unit with water can be expressed as a simple formula as

$$
\begin{aligned}
\tau_e &= \left[X_1 \rho_e^{N+1} + X_2 \rho_e^N \right] \\
&= \left[\left[\frac{L}{D} \tan \theta - N \right] \rho_e^{N+1} + \left[1 + N - \frac{L}{D} \tan \theta \right] \rho_e^N \right] \\
&= \left[1 + \left(N - \frac{L}{D} \tan \theta \right) (1 - \rho_e) \right] \rho_e^N
\end{aligned}
\tag{10}
$$

Based on the equation, we can obtain the effective transmittance of the honeycomb unit with water in the basin. The total radiation trapped within the basin can be estimated by using the effective transmittance of honeycomb with water and rays reflected from the east, west, and south facing reflector in addition to direct one falling from glass windows. Hence equation can be written as

$$
I_{Tot} = I_H \tau_e
\tag{11}
$$

For the water mass, the energy exchange is between the bottom absorber and the cover.

$$
C_w \frac{dT_w}{dt} = I_{tot} \alpha_w A_w + h_3 (T_b - T_w) A_b - h_1 (T_w - T_g) A_w
\tag{12}
$$

Finally, for the bottom absorber, the main energy source inside the thermal system can be written as

$$
C_b \frac{dT_b}{dt} = I_{tot} \alpha_b A_b - h_3 (T_b - T_w) A_b - h_4 (T_b - T_a) A_b
\tag{13}
$$

The solar energy that reaches each element of the system considered in the equation has been expressed through the fraction of total incident solar radiation. The fraction depends on the optical properties, such as transmittance, absorptance and reflectance, and on the distiller orientation.

The optical properties such as transmittance, absorptance, and reflectance for incident unpolarized radiation can be found by the expression given by Duffie and Beckmann (1991) [12].

$$
\tau = \frac{\tau_a}{2} \left[\frac{1 - r_\perp}{1 + r_\perp} \left(\frac{1 - r_\perp^2}{1 - (r_\perp \tau_a)^2} + \frac{1 - r_{\|\|}}{1 + r_{\|\|}} \left(\frac{1 - r_{\|\|}^2}{1 - (r_{\|\|} \tau_a)^2} \right) \right) \right]
\tag{14}
$$

$$
\alpha = \frac{1 - \tau_a}{2} \left[\left(\frac{1 - r_\perp}{1 - r_\perp \tau_a} + \frac{1 - r_{\|\|}}{1 - r_{\|\|} \tau_a} \right) \right]
\tag{15}
$$

$$\rho = \frac{1}{2}\left[r_\perp(1 + \tau_a r_\perp) + r_{||}(1 + \tau_a r_{||})\right] \tag{16}$$

where transmittance of the parallel and perpendicular components of unpolarized incoming radiation is

$$\tau_\perp = \frac{1 - r_\perp}{1 + r_\perp} \tag{17}$$

$$\tau_{||} = \frac{1 - r_{||}}{1 + r_{||}} \tag{18}$$

And reflection components of radiation can be obtained as

$$r_\perp = \frac{\sin^2(\theta_2 - \theta_1)}{\sin^2(\theta_2 + \theta_1)} \tag{19}$$

$$r_{||} = \frac{\tan^2(\theta_2 - \theta_1)}{\tan^2(\theta_2 + \theta_1)} \tag{20}$$

Transmittance with absorption losses for the covers can be estimated as

$$\tau_a = \exp\left(\frac{-kL}{Cos\theta_2}\right) \tag{21}$$

From Snell's law

$$\theta_2 = \sin^{-1}\left(\frac{n_2}{n_1}\sin\theta_1\right) \tag{22}$$

After rearranging the Eqs. (1), (12), and (13), equations can be written as

$$\frac{dT_g}{dt} + PT_g = Q \tag{23}$$

$$\frac{dT_w}{dt} + RT_w = S \tag{24}$$

$$\frac{dT_b}{dt} + MT_b = N \tag{25}$$

where

$$P = \frac{(h_1 + h_2)A_g}{C_g}$$

Modeling Thermal Asymmetries in Honeycomb Double Exposure Solar Still

$$Q = \frac{A_g}{C_g} \left[H(t)\alpha_g + h_1 T_w + h_2 T_a \right]$$

$$R = \frac{h_3 A_b + h_1 A_w}{C_w}$$

$$S = \frac{A_w(I_{tot}\alpha_w + h_1 T_g) + h_3 T_b A_b}{C_w}$$

$$M = \frac{h_3 A_b + h_4 A_b}{C_b}$$

$$N = \frac{A_b}{C_b}(I_{tot}\alpha_b + h_3 T_w + h_4 T_a)$$

After solving the Eqs. (23), (24), and (25), we get the solution for T_g, T_w and T_b as

$$T_g = T_{g0}e^{-Pt} + \frac{Q}{P}\left[1 - e^{-Pt}\right] \tag{26}$$

$$T_w = T_{w0}e^{-Rt} + \frac{S}{R}\left[1 - e^{-Rt}\right] \tag{27}$$

$$T_b = T_{b0}e^{-Mt} + \frac{N}{M}\left[1 - e^{-Mt}\right] \tag{28}$$

where T_{g0}, T_{w0} and T_{b0} are the initial glass, water mass, and absorber bottom temperature at the beginning of the experiment.

Analytical solutions for the water mass, glass cover, and bottom absorber have been used to calculate the temperature at different time intervals during the working hours of the system. Experiments have been conducted for many days and one of the typical days observations have been used for the calculation.

6 Results and Discussion

A program has been developed in BASIC programming language for the analytical solutions of water, glass, and bottom absorber temperature. To validate the analytical results, experimental observations for one of the typical days in February 2010 have been taken. The optical properties used in this model are

Parameter	Value
α_g	0.05
α_b	0.90

(continued)

Parameter	Value
α_w	0.88
k	32
C_w	4190 J/kgK
C_b	1465 J/kgK
C_g	820 J/kgK

The measured solar radiation and ambient temperature have been plotted with respect to the working hours of the device. Figures 4 and 5 show the variation of solar radiation and ambient temperature with respect to time for the typical days in the month of February 2010.

One of the typical days i.e., 6.02.2010 has been used for the calculation in the model. Graph has been drawn for the measured and calculated glass cover temperature on the typical day and is represented in the Fig. 6. It has been observed that

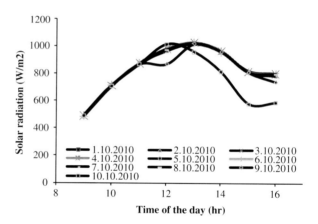

Fig. 4 Variation of solar radiation with time

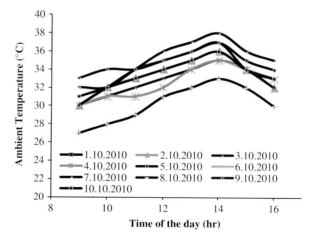

Fig. 5 Variation of ambient temperature with time

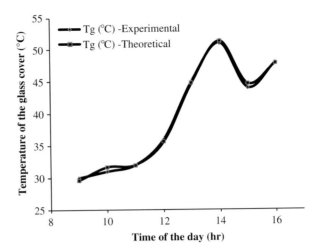

Fig. 6 Variation of modeled and observed glass cover temperature with time

the calculated values slightly underestimated with the experimental observations. Correlation coefficients have been obtained by plotting the experimental and modeled glass cover temperature and it is shown in the Fig. 7. It is observed that there are experimental asymmetries which slightly underestimated.

On the same day, experimental and modeled temperature of water and bottom absorber has been drawn. Figures 8 and 9 have shown the variation of experimental and modeled temperature of water and bottom absorber. It is observed that the calculated and observed values have same trend as experimental observation with much less error. The correlation coefficients have been found and their r^2 values for water and bottom absorber are presented. It is shown in the Figs. 10, 11.

From the graph it is clear that a large thermal inertia of the system is observed, since the maximum temperature of water and bottom absorber obtained at 13 h. Observed and calculated distillate yield is plotted with respect to time and it is shown in the Fig. 12. Correlation coefficient has been found for experimental and

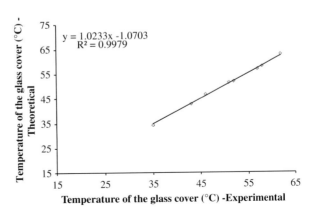

Fig. 7 Correlation model for glass cover temperature

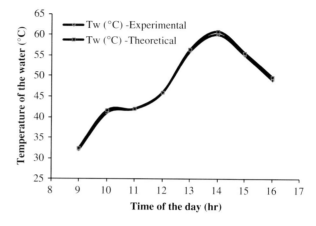

Fig. 8 Variation of experimental and modeled water temperature

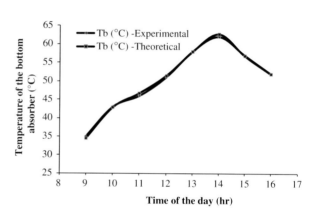

Fig. 9 Variation of experimental and modeled bottom absorber temperature

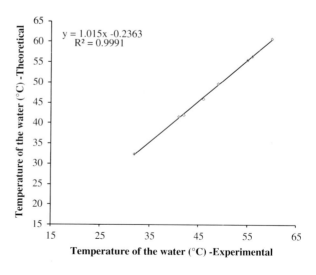

Fig. 10 Correlation model for water temperature

Fig. 11 Correlation model for bottom absorber temperature

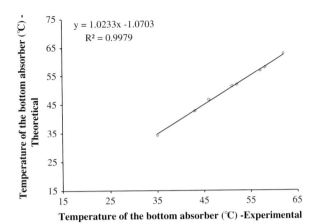

Fig. 12 Variation of experimental and modeled distillate yield with time

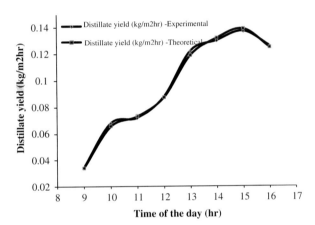

theoretical yield by plotting a graph shown in the Fig. 13. It has been observed that distillate yield depends on the water and bottom absorber temperature. There exists a good agreement between the experimental and theoretical results.

Among the correlation coefficients obtained for various parameters of the still, a strong thermal inertia can be shown for bottom absorber and water temperature. It correlates better. This is a result of the model dynamic effects governed by external reflectors such as ambient temperature and wind velocity. Moreover, the experimental instantaneous efficiency of the still is plotted with respect to time and it is shown in the Fig. 14.

7 Conclusion

A mathematical model to predict thermal asymmetries in specific honeycomb double exposure solar still is developed and validated. An approach has been carried out based on energy balance equations and a full set of equations, where

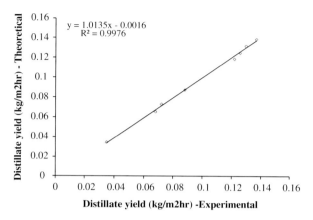

Fig. 13 Correlation model for distillate yield

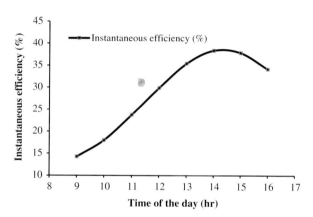

Fig. 14 Variation of experimental instantaneous efficiency with time

Table 1 Chemical analysis of the raw and distilled water

Nature of water	E. C. meg mhos/cc	Total hardness of caco$_3$ - mg/l	Calcium hardness Caco$_3$/l	Magnesium hardness Caco$_3$ mg/l
Raw water	1800	570	260	310
Distilled water	85	12	6	6
Percentage	4.7	2.1 %	2.3 %	2 %

the condenser is treated as a system composed of one element for the still. Predicted temperature of the glass cover slightly underestimated by the model, due to complexity of dynamical effects produced by external environmental factors. This can be extended to study the detailed thermal performance of the proposed still.

8 Chemical Analysis of the Result

The result of the chemical analysis of the distillate is given in Table 1. The chemical analysis has been done at Tamil Nadu Agriculture University Coimbatore.

From the table it is clear that the honey comb double exposure soar still reduced the present contaminants in the raw water and produced less contaminants distillate output.

Acknowledgments The authors would like to thank The University Grants Commission (UGC), the Government of India, New Delhi, for the award of Rajiv Gandhi National Fellowship (RGNF) for SC/ST (Junior Research Fellow).

References

1. Ward John (2003) A plastic solar water purifier with high output. Sol Energy 75:433–437
2. Delyannis E (2003) Historic background of desalination and renewable energies. Sol Energy 75:357–366
3. Mathioulakis E, Belessiotis V (2003) Integration of solar still in a multi-source, multi-use environment. Sol Energy 75:403–411
4. Singh HN, Tiwari GN (2004) Monthly performance of passive and active solar still for different Indian climatic condition. Desalination 168:145–150
5. Tripathi R, Tiwari GN (2004) Effect of size and material of a semi cylindrical condensing cover on heat and mass transfer for distillation. Desalination 166:231–241
6. Ghoneim AA (2005) Performance optimization of solar collector equipped with different arrangements of square-celled honeycomb. Int J Therm Sci 44:95–105
7. Shanmugan S, Rajamohan P, Mutharasu D (2008) Performance study on an acrylic mirror boosted solar distillation. Desalination 230:281–287
8. Zhiqiang Zhang, Ran Zuo, Ping Li, Wenjia Su (2009) Thermal performance of solar air collector with transparent honeycomb made of glass tube. Sci China Ser E: Technol Sci 52:1–7
9. Ahmed MI, Hrairi M, Ismail AF (2009) On the characteristics of multistage evacuated solar distillation. Renewable Energy 34:1471–1478
10. Ismail BI (2009) Design and performance of transportable hemispherical solar still. Renewable Energy 34:145–150
11. EI-Swify ME, Metias MZ (2002) Source Renewable Energy 26:1639–1650
12. Duffie JA, Beckman WA (1991) Solar engineering of thermal processes, Wiley Interscience

Optical Band Gap Studies on Dy^{3+} Doped Boro-Tellurite Glasses

K. Maheshvaran and K. Marimuthu

Abstract The Dy^{3+} doped boro-tellurite glasses were prepared following conventional melt quenching technique with the chemical composition (69–x) H_3BO_3 + $xTeO_2$ + $15Mg_2CO_3$ + $15K_2CO_3$ + $1Dy_2O_3$ (where x = 0, 10, 20, 30 and 40 wt%). The optical absorption spectra have been recorded at room temperature to explore the optical properties. The optical band gap energy has been calculated using Davis and Mott theory through direct allowed, indirect allowed, and indirect forbidden transitions. Through the optical absorption spectra, optical band gap (E_{opt}), band tailing parameter (B), and Urbach energy (ΔE) were calculated for all the prepared glasses and the results were reported. It is observed from the results that the band gap energy value decreases, whereas Urbach's energy increases with the increase in TeO_2 content.

Keywords Optical band gap · Band tailing parameter · Absorption edges · Urbach's energy

1 Introduction

Rare-earth-doped (RE) glasses were extensively studied due to their potential applications in the field of solid-state laser materials, optical memory devices, optoelectronics, magneto-optical devices, display devices, solar concentrators and phosphors, and so on. [1, 2]. Oxide-based glasses such as silicate, phosphate, borate, germinate, vanadate, and tellurite matrices are found to be promising materials for the photonic applications. Borate glasses are more suitable for RE ion doping due to their high transparency, low melting point, high thermal stability, RE ion solubility, easy preparation on large scale, shaping, and cost effective properties

---

K. Maheshvaran · K. Marimuthu (✉)
Department of Physics, Gandhigram Rural University, Gandhigram 624 302, India
e-mail: mari_ram2000@yahoo.com

S. Sathiyamoorthy et al. (eds.), *Emerging Trends in Science,*
Engineering and Technology, Lecture Notes in Mechanical Engineering,
DOI: 10.1007/978-81-322-1007-8_54, © Springer India 2012

[3]. The boron atom usually coordinates with four oxygen atoms forming $[BO_3]^{3-}$ or $[BO_4]^{5-}$ structural units [4]. Tellurite-based glasses exhibit excellent linear and nonlinear optical behavior due to their physical properties such as low melting temperature, high dielectric temperature, high dielectric constant, high refractive index, and large third order nonlinear susceptibility. The distinguished factor of these tellurite-based glasses is the atomic network arrangement which appears more open and very weak Te–O band as compared with the silica glasses. The Te–O band in tellurite glasses can be easily broken and this is an advantage for accommodating RE ions and heavy metal oxides [5]. The B_2O_3–TeO_2 glasses are also of interest to design various new optical devices. The studies made on these glasses have shown progressive changes in both the boron and tellurium coordination with the addition of other ions. The study on absorption spectra of solids provides essential information about the band structure and the energy gap. The technique is based on the principle that a photon with energy greater than the band gap energy will be absorbed. Analysis of the absorption spectra in the lower energy part gives information about the atomic vibrations, while that of the higher energy part of spectrum gives information about the electronic state in the atom [6]. When an electromagnetic wave interacts with a valence electron, both direct and indirect optical transitions occur across the energy gap. However, indirect transitions involve simultaneous interaction with lattice vibrations and the wave vector of the electron [7].

The dysprosium ion is one of the most efficient RE ions for mid–IR lasers (4.4 and 2.98 μm), telecommunication amplifiers, and Q–switching devices [2]. Emission spectrum of Dy^{3+} ion, exhibits $^4F_{9/2} \rightarrow {}^6H_{11/2}, {}^6H_{13/2}, {}^6H_{15/2}$ transitions in the visible region. Dy^{3+} ion is a good activator for the preparation of electron-trapping luminescent materials. Maria et al. reported the electron diffraction studies on the short-range order of TeO_2–B_2O_3 glasses [8].

The present work reports spectroscopic and band gap studies on Dy^{3+} doped boro-tellurite glasses. Davis and Mott theory has been used to analyze the fundamental absorption edges of the prepared glasses. Through the absorption spectra, absorption edges, optical band gap (E_{opt}), band tailing parameter (B), and Urbach energy (ΔE) have been calculated and reported for direct and indirect transitions.

2 Experimental

Dy^{3+} doped boro-tellurite glasses were prepared following conventional melt quenching technique. The chemicals used were H_3BO_3, TeO_2, Mg_2CO_3, K_2CO_3, and Dy_2O_3 of 99.99 % analytical grade purity. The chemical compositions and their codes are presented in Table 1.

Chemical composition of 10 g batches was thoroughly mixed and ground in an agate mortar to obtain homogeneous mixture. The ground chemicals were taken into a porcelain crucible and kept at a temperature of 900 °C in an electrical furnace for 30 min. The melt was stirred for homogeneous mixing and then poured onto a preheated brass plate and pressed by another brass plate to obtain uniform

Optical Band Gap Studies on Dy³⁺ Doped Boro-Tellurite Glasses

Table 1 Compositions of the Dy³⁺ doped the boro-tellurite glasses

Sl. no	Sample code	Composition
1	B0TD	$69B_2O_3 + 0TeO_2 + 15MgO + 15K_2O + 1Dy_2O_3$
2	B1TD	$59B_2O_3 + 10TeO_2 + 15\,MgO + 15K_2O + 1\,Dy_2O_3$
3	B2TD	$49B_2O_3 + 20TeO_2 + 15\,MgO + 15K_2O + 1\,Dy_2O_3$
4	B3TD	$39B_2O_3 + 30TeO_2 + 15\,MgO + 15K_2O + 1\,Dy_2O_3$
5	B4TD	$29B_2O_3 + 40TeO_2 + 15\,MgO + 15K_2O + 1\,Dy_2O_3$

thickness of 1.5 mm. The glass samples were annealed at 350 °C at 7 h to remove the strain and to improve the mechanical strength. The prepared glasses were well polished to obtain optical quality before further optical measurements.

X-ray diffraction pattern of the prepared glasses was used to confirm its amorphous nature. The optical absorption spectra were recorded in the wavelength 360–2,000 nm using CARY 500 spectrometer with a spectral resolution of ±0.1 nm. All these measurements were carried out at room temperature.

3 Results and Discussion

3.1 Absorption Spectra

The absorption spectra of the Dy³⁺ doped boro-tellurite glasses measured in the wavelength range 360–2,000 nm are quite similar and as a representative case absorption spectrum of the Dy³⁺: B0TD glass is shown in Fig. 1. The absorption spectra consist of 10 absorption bands located at around 5,922, 7,837, 9,158,

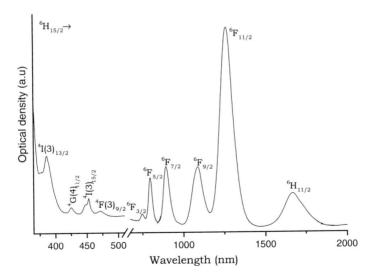

Fig. 1 Absorption spectra of the Dy³⁺: B0TD boro-tellurite glass

1,1124, 12,500, 13,387, 21,231, 22,173, 23,585, and 25,510 cm^{-1} corresponding to the transitions $^6H_{11/2}$, $^6F_{11/2}$, $^6F_{9/2}$, $^6F_{7/2}$, $^6F_{5/2}$, $^6F_{3/2}$, $^4F(3)_{9/2}$, $^4I(3)_{15/2}$, $^4G(4)_{11/2}$ and $^4I(3)_{13/2}$ originates from the ground state $^6H_{15/2}$. The absorption spectra of the Dy^{3+}: BXTD glasses resemble other Dy^{3+} doped glasses [3, 9].

The position and spectral intensity of certain energy level transitions of the RE ions are sensitive to the environment of the RE ion. They follow the selection rules $\Delta J \leq 2$, $\Delta L \leq 2$ and $\Delta S = 0$ and such transitions are referred as hypersensitive transitions [10]. For the Dy^{3+} ion, $^6H_{15/2} \rightarrow {}^6F_{11/2}$ is identified as the hypersensitive transition, and is found to be more intense than any other transitions. Because of the strong absorption of the host, the transitions that lie above $^4I(3)_{13/2}$ (26,000 cm^{-1}) could not be assigned. Analyzing the trends of the energy level positions of the Dy^{3+}: BXTD glasses, the order of magnitude is found to be Glasses > Aquo–ion > Crystalline hosts. Transition energy associated with the Dy^{3+}: BXTD glasses is found to be higher than the crystalline hosts and it may be due to the fact that the 4f orbitals of the Dy^{3+} ions interact relatively stronger in glass hosts than crystalline hosts.

The glass sample which starts to absorb completely at a characteristic wave length (λ_{edge}) in the ultraviolet region is defined as fundamental absorption edge and in the present work the value of λ_{edge}, is found to increase with increasing TeO$_2$ content. The fundamental absorption edge is found to be 350.62 nm for the Dy^{3+}: B0TD glass which is having zero wt% TeO$_2$ content. The value of the fundamental absorption edges found to increase with the increasing TeO$_2$ content. It is observed from Table 2 that the λ_{edge} is found to have blue shift for lower TeO$_2$ content and it makes considerable red shift with increasing TeO$_2$ content.

3.2 Optical Band Gap and Urbach's Energy Analysis

The study of optical absorption edge is useful to understand the optically induced transitions and the optical band gap of the glasses. The optical absorption spectra have been used to calculate optical band gap values. The absorption co-efficient $\alpha(v)$, near the edge of the spectrum is calculated using the following expression

$$\alpha(v) = (1/d) \ln\left(I_0/I_t\right) \tag{1}$$

Table 2 The fundamental absorption edge (λ_{edge}), optical band gap (E_{opt}), band tailing parameter (B), and Urbach energy (ΔE) corresponding to $r = 1/2$, 2, and 3 for the prepared Dy^{3+}:BXTD glasses

| Sample code | λ_{edge} (nm) | $r = 1/2$ | | $r = 2$ | | $r = 3$ | | |
		E_{opt} (eV)	B (cm^{-2}) eV)	E_{opt} (eV)	B (cm eV)$^{-1/2}$	E_{opt} (eV)	B (cm eV)$^{-2/3}$	ΔE (eV)
B0TD	350.62	3.631	833.033	3.443	11.157	3.471	8.229	0.297
B1TD	368.95	3.394	763.285	3.147	10.107	3.163	7.032	0.363
B2TD	388.49	3.251	711.007	2.993	9.005	2.997	6.067	0.438
B3TD	414.15	3.145	643.360	2.767	8.279	2.748	5.476	0.664
B4TD	485.01	2.955	596.261	2.602	7.771	2.618	4.849	0.728

where d is the thickness of the sample, I_0 and I_t are intensities of incident and transmitted radiations, respectively, and $\ln(I_0/I_t)$ corresponds to optical density (A). According to Mott and Davis theory [11], the absorption coefficient is calculated using the following equation $\alpha(\nu)h\nu = B(h\nu - E_{opt})^r$, where r can have values as 1/2, 1/3, 2, and 3 depending on the nature of the interband electronic transitions, such as direct allowed, direct forbidden, indirect allowed, and indirect forbidden transitions, respectively [1, 12–14]. B is the band tailing parameter, E_{opt} is the optical band gap energy and $h\nu$ is the incident photon energy.

The optical band gap values can be obtained by extrapolating the linear region of the $h\nu$ plots. Figure 2 shows the Tauc's plot of $(\alpha h\nu)^2$ as a function of $h\nu$ ($r = 1/2$) for direct allowed transitions of the prepared glasses. The direct band gap E_{opt} values were obtained by extrapolating the linear region of the curves to the zero absorption at which $(\alpha h\nu)^2 = 0$ and the results are presented in Table 2. The values of the band tailing parameter obtained from the slope of the curves of Fig. 2 are also presented in Table 2. Figures 3 and 4 exhibit the plots of $(\alpha h\nu)^{1/2}$ and $(\alpha h\nu)^{1/3}$ as a function of $h\nu$ corresponding to indirect allowed and forbidden transitions. The band gap values of the prepared glasses corresponding to indirect allowed ($r = 2$) and indirect forbidden transitions ($r = 3$) have been obtained from the linear region of the curves of Figs. 3 and 4, which were extrapolated to zero absorption and the results are presented in Table 2. It is observed from Table 2 that the direct band gap values are found to be higher than the indirect band gap values.

Both the direct and indirect band gap values are found to decrease with the increasing TeO_2 content, which indicates that the increasing TeO_2 content causes structural changes in the glass network. The band gap values of the TeO_2–B_2O_3 glasses reported by Halimah et al. [15] are found to be less than the Dy^{3+}: BXTD glasses whereas Yanmin et al. [13] reported that the B_2O_3–TeO_2–ZnO–Na_2O glasses posses higher band gap values than the prepared glasses.

The absorption region $\alpha(\nu)$ which lies between 10^2 and 10^4 cm^{-1} is defined as the Urbach's band tail region and is given by [14]

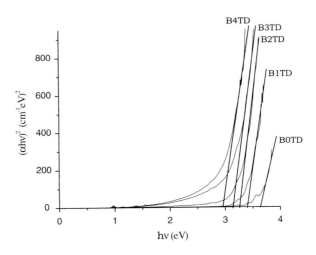

Fig. 2 Tauc's plot of $(\alpha h\nu)^2$ verses $h\nu$ for direct band gap measurements of Dy^{3+} doped boro-tellurite glasses

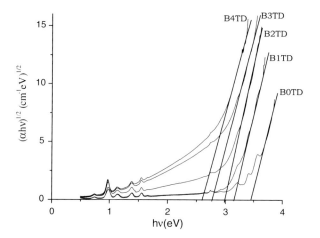

Fig. 3 Tauc's plot of $(\alpha h\nu)^{1/2}$ verses $h\nu$ for indirect band gap measurements of Dy^{3+} doped boro-tellurite glasses

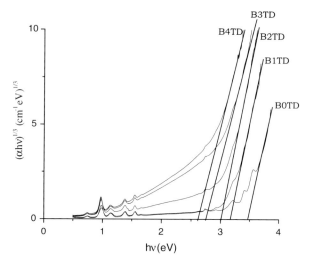

Fig. 4 Tauc's plot of $(\alpha h\nu)^{1/3}$ verses $h\nu$ for indirect forbidden band gap measurements of Dy^{3+} doped boro-tellurite glasses

$$\alpha(\nu) = \alpha_0 \exp\left(h\nu/\Delta E\right) \qquad (2)$$

where α_0 is a constant and ΔE is the Urbach energy corresponds to the optical transitions between localized tail states adjacent to the valance band and the extended states in the conduction band, lying above the mobility edge [16]. The Urbach energy is calculated from the reciprocal of the slope of the plots between ln $\alpha(\nu)$ and $h\nu$ and is shown in Fig. 5. The absorption region $\alpha(\nu) \leq 10^2$ cm^{-1} involves low-energy absorption and is attributed to the optical transitions between localized states. Figure 6 shows the optical band gap (E_{opt}) and Urbach's energy values with varying TeO_2 content. It is observed from Fig. 6 that the values of the Urbach's energy increases with the increase in TeO_2 content. The Urbach energy (ΔE) values

Fig. 5 Urbach's plot of Dy^{3+} doped boro-tellurite glasses

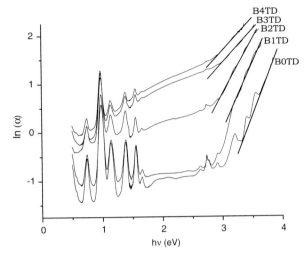

Fig. 6 Variation of optical band gap and Urbach's energy with TeO_2 content for the Dy^{3+} doped boro-tellurite glasses

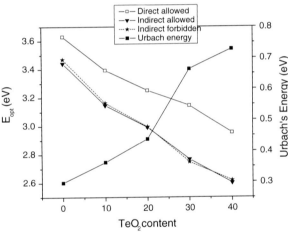

of the prepared glasses lie between 0.297 and 0.728 eV are in agreement with the reported Urbach energy (ΔE) values by Halimah et al. [15] and Ali [1].

4 Conclusion

Dy^{3+} doped boro-tellurite glasses were prepared and their optical behavior have been studied and reported. It is observed from the absorption spectra that the energy level positions of the prepared glasses follow the trend as Glasses > Aquoion > Crystalline hosts. The optical band gap (E_{opt}) for the direct allowed, indirect allowed, and indirect forbidden transitions has been studied and reported. The direct band gap values are found to be higher than the indirect band gap values. It is observed from the studies that the direct and indirect band gap values are found to decrease with the increasing TeO_2 content which causes the structural changes in the glass network. The Urbach energy (ΔE) values of the prepared glasses increase with the increase in TeO_2 content similar to the reported studies.

References

1. Ali AA (2009) J Lumin 129:1314–1319
2. Dwivedi Y, Rai SB (2009) Opt Mater 31: 1472–1477
3. Karunakaran RT, Marimuthu K, Surendra Babu S, Arumugam S (2010) J Lumin 130:1067–1072
4. Becker P (1998) Adv Mater 10:979–992
5. Murali A, Sreekanth chakradhar RP, Lakshmana Rao J (2005) Physica B 364:142–149
6. Ardelean I, Filip S (2005) J Optoelectr Adv Mater 7:745–752
7. Hosono H, Ikuta Y (2000) Nucl Instr Methods Phys Res B 166:691–697
8. Bursukova MA, Kashchieva EP, Dimitriev YB (1995) J Non-Cryst Solids 192: 40–44
9. Babu P, Jayasankar CK (2000) Opt Mater 15:65–79
10. Thulasiramudu A, Buddhudu S (2007) Spectrochim Acta Part A 67:802–807
11. Mott NF, Davis EA (1979) Electronic processes in non-cryst mater, 2nd edn. Clarendon Press, Oxford
12. Tauc J (1974) Amorphous and liquid semiconductor. Plenum Press, New York
13. Yanmin Y, Baojiu C, Cheng W, Guozhong R, Xiaojun W (2007) J Rare Earths 25:31–35
14. Urbach F (1953) Phys Rev 92:1324–1324
15. Halimah MK, Daud WM, Sidek HAA, Zainal AT, Zainal H, Hassan J (2005) Amer J Appl Sci 63 ISSN, 1546–9239
16. Shindhu S, Sanghi S, Agarwal A, Seth VP, Kishore N (2005) Mat Chem Phys 90:83–89

Preparation and Characterization of Cobalt Doped Mn-Zn Ferrites

M. Bhuvaneswari and S. Sendhilnathan

Abstract $Co_{(1-x)}Mn_yZn_yFe_2O_4$ (x = 0.9,0.5, 0.1 and y = 0.45, 0.25, 0.05.) nanoparticles of average crystalline size of 61 nm were prepared by chemical co-precipitation method. XRD, FTIR, SEM were utilized in order to study the effect of variation in the cobalt substitution and its impact on size and associated water content. The average crystalline size (D_{aveXR}) of the particles was found to be decreased from 85 to 41 nm with the increase in cobalt concentration. Fourier transform infrared spectroscopy (FTIR) spectra of the $Co_{(1-x)}Mn_yZn_yFe_2O_4$ in the range 400–4000 cm^{-1} were reported. The spinel structure and the crystalline water adsorption of cobalt doped Mn-Zn nanoparticles were studied by using FTIR.

Keywords Ferrites • Spinel • Co-precipitation • Crystalline size

1 Introduction

Magnetic nanoparticles are of great technological importance because of their use in magnetic fluid, information storage system, medical diagnostics, and so on. Various preparation techniques have been used for the synthesis of fine particles of ferrites, which exhibit novel properties when compared to their properties in the bulk. Non-conventional methods such as co-precipitation, thermal decomposition, sol-gel and hydrothermal methods have been widely used. Ultra

---

M. Bhuvaneswari (✉)
Department of Physics, Mount Zion College of Engineering and Technology,
Pudukkottai, India
e-mail: bhuvanjothi@gmail.com

S. Sendhilnathan
Department of Physics, Anna University of Technology,Tiruchirappalli, Pattukkottai
Campus, Rajamadam, Pattukkottai, India
e-mail: sendhil29@yahoo.co.in

S. Sathiyamoorthy et al. (eds.), *Emerging Trends in Science,
Engineering and Technology*, Lecture Notes in Mechanical Engineering,
DOI: 10.1007/978-81-322-1007-8_55, © Springer India 2012

fine ferrite particles can be prepared by chemical co-precipitation method. Auzans et al. [1, 2] have studied the preparation and properties of Mn-Zn ferrite nano particles, which were, used in ionic and surfacted ferrofluids with different degrees of Zn substitution prepared by co-precipitation method. Chandana Rath et al. [3] have reported the dependence on cation distribution of crystallite size, lattice parameter and magnetic properties in nano size Mn-Zn ferrite for different degrees of inversion of Zn substitution prepared by hydrothermal precipitation method.

The use of Mn-Zn ferrite for the preparation of temperature sensitive magnetic fluid by co-precipitation method has already been studied [4–6]. $Co_{0.2}Zn_{0.8}Fe_2O_4$ fine particles have been prepared by chemical co-precipitation method followed by sintering [7]. Control of crystalline size in the nanometer range by the variation of synthesis condition is always a difficult task. It becomes mandatory in the case of ferrofluid preparation using co-precipitation method. In order to prepare ferrofluid having such fine particles, specific size restriction is imposed considering the stability criteria. $Co_{(1-x)}Mn_yZn_yFe_2O_4$ (x = 0.9,0.5,0.1 and y = 0.45, 0.25, 0.05.) substituted ferrites were prepared by co-precipitation method have not yet been fully studied like Mn-Zn substituted ferrites. In this paper we report preparation of $Me_{1-x}Mn_yZn_yFe_2O_4$ fine particles, where Me = Co^{2+} with x = 0.9, 0.5, 0.1. Average crystalline size 61 nm by chemical co-precipitation method and the consequent change in their lattice parameter, crystalline size and associated water content due to increase in cobalt concentration were reported.

2 Synthesis and Characterization of $Co_{(1-x)}Mn_yZn_yFe_2O_4$ of Nanoparticles

2.1 Synthesis of Cobalt Doped Mn-Zn Ferrites

The cobalt doped ferrite nanoparticles synthesized by co-precipitation depends mostly on parameters such as reaction temperature, pH of the suspension, initial molar concentration etc. [4].Ultra fine particles of $Co_{(1-x)}Mn_yZn_yFe_2O_4$ (x = 0.9, 0.5, 0.1 and y = 0.45, 0.25, 0.05.) were prepared by co-precipitating aqueous solutions of $CoCl_2$, $MnCl_2$, $ZnCl_2$ and $FeCl_3$ mixtures respectively in alkaline medium. The mixed solution of $CoCl_2$, $MnCl_2$, $ZnCl_2$ and $FeCl_3$ in their respective stoichiometry (100 ml of 0.1 M $CoCl_2$, 100 ml of 0.45 M $ZnCl_2$, 100 ml of 0.45 M $MnCl_2$ and 100 ml of. 2 M $FeCl_3$ in the case of $Co_{0.1}Mn_{0.45}Zn_{0.45}Fe_2O_4$ and similarly for the other values of x) was prepared and kept at 333 K (60 °C).

This mixture was added to the boiling solution of NaOH (0.63 M dissolved in 1200 ml of distilled water) within 10 s under constant stirring. Nano ferrites are formed by conversion of metal salts into hydroxides, which take place immediately, followed by transformation of hydroxides into ferrites. The solutions were maintained at 358 K (85 °C) for 1 h. This duration was sufficient for the

Preparation and Characterization of Cobalt Doped Mn-Zn Ferrites

transformation of hydroxides into spinel ferrite (dehydration and atomic rearrangement involved in the conversion of intermediate hydroxide phase into ferrite) [4]. Sufficient amount of fine particles were collected at this stage by using magnetic separation. These particles were washed several times with distilled water followed by acetone and dried at room temperature.

2.2 XRD

The X-ray diffraction (XRD) patterns of the samples were recorded on a BRUKER-binary V2 (.RAW) powder diffractometer using Cu $K_\alpha (\lambda = 1.54060$ A°) radiation. Slow scans of the selected diffraction peaks were carried out in step mode (step size 0.02°, measurement time 5 s, measurement temperature 323 K (25 °C), standard: Si powder). The crystalline size of the nanocrystalline samples was measured using Debye- Scherrer formula,

$$D_{XRD} = \frac{0.89\lambda}{\beta \cos\theta}. \tag{1}$$

2.3 FTIR

FTIR spectra were recorded for the dried samples of $Co_{(1-x)}Mn_yZn_yFe_2O_4$ (x = 0.9, 0.5, 0.1 and y = 0.45, 0.25, 0.05.) with **Nexus 670** (range 400–4000 cm^{-1}) spectrometer. The dried samples were mixed with KBr and spectra were measured according to transmittance method.

3 Results and Discussions

Generally, XRD can be used to characterize the crystallinity of nanoparticles, and it gives average diameters of all the nanoparticles. The precipitated fine particles were characterized by XRD for structural determination and estimation of crystallite size. XRD patterns were analyzed and indexed using JCPDS. All experimental peaks were matched with (JCPDS #653111) the theoretically generated one and indexed. Analysis of the diffraction pattern using powder-X software [8] confirms the formation of cubic spinel structure for all the samples. All the compositions had a spinel structure. The XRD pattern for $Co_{(1-x)}Mn_yZn_yFe_2O_4$ (x = 0.9, 0.5, 0.1 and y = 0.45, 0.25, 0.05.) shown Fig. 1.

The broad XRD lines indicate that the particles are of nanosize range. The particle size was found to decrease from 85 to 41 nm with the increase in cobalt concentration. The crystallite size (D_{XRD}) was estimated by the Debye-Scherrer formula [9] using the full width at half maximum value of the respective indexed peaks. Though all the samples were prepared under identical condition, the

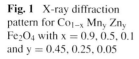

Fig. 1 X-ray diffraction pattern for $Co_{1-x}Mn_yZn_yFe_2O_4$ with x = 0.9, 0.5, 0.1 and y = 0.45, 0.25, 0.05

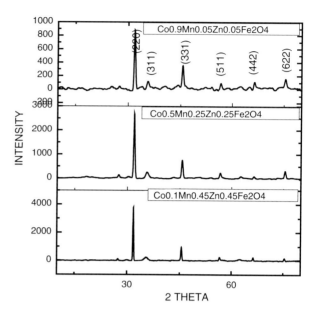

Table 1 The average crystalline size (D_{aveXR}) for $Co_{(1-x)}Mn_yZn_yFe_2O_4$

Composition	2θ	FWHM	SIZE 'D' nm	Ave. SIZE nm
$Co_{0.1}Mn_{0.45}Zn_{0.45}Fe_2O_4$	31.7	0.09	85.19	76.90
	45.5	0.12	71.08	
	56.5	0.12	74.43	
$Co_{0.5}Mn_{0.25}Zn_{0.25}Fe_2O_4$	31.9	0.14	56.81	64.69
	45.6	0.11	74.60	
	56.6	0.14	62.05	
$Co_{0.9}Mn_{0.05}Zn_{0.05}Fe_2O_4$	31.9	0.19	41.57	42.07
	45.6	0.19	43.36	
	56.6	0.21	41.27	
				61.15

crystallite size was not the same for all concentrations (Table 1). This was probably due to the preparation condition followed here, which gave rise to different rate of ferrite formation for different concentrations of cobalt, favoring the variation of crystalline size.

The variation of average crystalline size with the cobalt concentration is given in Fig. 2. Ferrofluids can be conveniently prepared by making use of particles in this size range. The average crystallite size (D_{aveXR}) for $Co_{(1-x)}Mn_yZn_yFe_2O_4$ (x = 0.9, 0.5, 0.1 and y = 0.45, 0.25, 0.05.) is shown in Table 1. The particle size was confirmed by SEM data (Fig. 3).

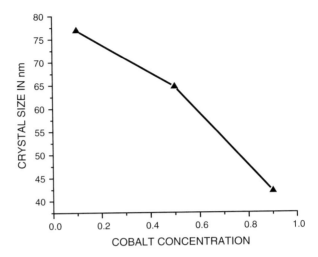

Fig. 2 Cobalt concentration versus particle size

3.1 SEM

3.2 FTIR

From the FTIR spectra for Fe$_3$O$_4$ and for Co$_{(1-x)}$Mn$_y$Zn$_y$Fe$_2$O$_4$ (x = 0.9, 0.5, 0.1 and y = 0.45, 0.25, 0.05.), the spectral similarities were observed. The main transmittance frequencies observed in the region 400–4000 cm^{-1} of the FTIR spectra for Co$_{(1-x)}$Mn$_y$Zn$_y$Fe$_2$O$_4$ (x = 0.9, 0.5, 0.1 and y = 0.45, 0.25, 0.05.) are summarized in Table 2.

Fig. 3 SEM images of cobalt doped ferrites

Table 2 The transmittance frequencies for Co$_{(1-x)}$Mn$_y$Zn$_y$Fe$_2$O$_4$

Sample	IR absorption bands cm^{-1}			
	v_1	v_2	v_3	v_4
Co$_{0.1}$Mn$_{0.45}$Zn$_{0.45}$Fe$_2$O$_4$	3460.66	1640.02	988.21	584.74
Co$_{0.5}$Mn$_{0.25}$Zn$_{0.25}$Fe$_2$O$_4$	3492.59	1646.12	987.29	584.03
Co$_{0.9}$Mn$_{0.05}$Zn$_{0.05}$Fe$_2$O$_4$	3031.56	1509.94	986.49	583.66

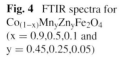

Fig. 4 FTIR spectra for Co$_{(1-x)}$Mn$_y$Zn$_y$Fe$_2$O$_4$ (x = 0.9,0.5,0.1 and y = 0.45,0.25,0.05)

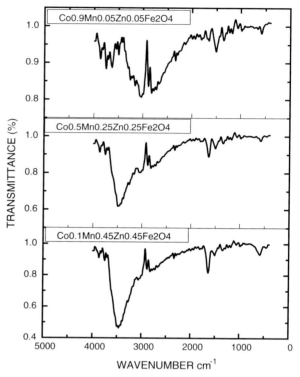

The broad feature between 3460.66 and 3031.56 cm^{-1} is due to O–H stretch (v_1), which corresponds to the hydroxyl groups attached by the hydrogen bonds to the iron oxide surface and the water molecules chemically adsorbed to the magnetic particle surface (associated water content) [8]. From these results, it appears that the hydroxyl groups are retained in the samples during the preparation of the uncoated Co$_{(1-x)}$Mn$_y$Zn$_y$Fe$_2$O$_4$. Ghose et al. [7] have reported that the presence of some hydroxyl ions are completely removed when the sample is sintered at temperatures ≥973 K. The O–H in-plane (v_2) and out-of-plane (v_3) bonds appear at 1640.02–1509.94 cm^{-1} and 988.21–986.49 cm^1, respectively. The spectrum of the uncoated sample Co$_{(1-x)}$Mn$_y$Zn$_y$Fe$_2$O$_4$ shows a strong band at v_4 (635.57–573.51 cm^{-1}) due to Fe$_3$O$_4$ [10]. The transmittance waveband at v_4 (584.74–583.66 cm^{-1}), which corresponds to the metal-oxygen bonds are considered as the confirmation for the ferrite formation. This is in good agreement with Zins et al. [1, 11–14] (Fig. 4).

4 Conclusion

The preparation technique of nano particles has a definite impact on the control of particle size and alteration of magnetic properties. The estimated cations from the product are in comparison with the initial substitution degree, indicating that the preparation procedure favors the formation of only ferrites. The formation of $Co_{(1-x)}Mn_yZn_yFe_2O_4$ (x = 0.9, 0.5, 0.1 and y = 0.45, 0.25, 0.05.) was confirmed by the X-ray diffraction. The average crystallite size (D_{aveXR}) decreased when the partial substitution of cobalt increased. The Cobalt doped Mn-Zn ferrite particles can be used to prepare ferro fluids with higher magnetization. FTIR was used to confirm the formation of Fe–O bonds and presence of the associated water content in the samples.

Acknowledgments One of the authors Dr. S. Sendhilnathan gratefully acknowledges DST (Ref No.SR/FTP/PS -59/2008) for the financial assistance received through the project.

References

1. Auzans E, Zins D, Blums E, Massart R (1999) Synthesis and properties of Mn-Zn Ferrite ferrofluids. J Mater Sci 34:1253
2. Auzans E, Zins D, Blums E, Massart R (1999) Magn. Gidrodinamika, Mn-Zn nanoparticles for ferrofluid preparation, Science direct 36:78
3. Chandana Rath S, Anand RP, Das KK, Sahu SD, Kulkarni SD, Date SK, Mishra NC (2002) Dependence on cation distribution of particle size, lattice parameter and magnetic properties in nanosized Mn-Zn ferrite. J Appl Phys 91(4):2211
4. Jeyadevan B, Chinnasamy CN, Shinoda K, Tohji K (2003) Mn-Zn ferrite with magnetization for temperature sensitivity magnetic fluid. J Appl Phys 93(10):8450
5. Parekh K, Upadhyay RV, Mehta RV (2000) Electron spin resonance study of temperature sensitive magnetic fluid. J Appl Phys 88(5):2799
6. Upadhyay T, Upadhyay RV, Mehta RV (1997) Characterization of a temperature sensitive magnetic fluid. Phys Rev B 55(9):5585
7. Dey S, Ghose J (2003) Synthesis, characterization and magnetic studies on nanocrystalline $Co_{0.2}Zn_{0.8}Fe_2o_4$. Mater Res Bull 38:1653
8. Creanga D, Calugaru Gh (2005) Physical investigations of a ferro fluid based on hydrocarbons. J Magn Magn Mater 289:81
9. Dong C (1999) PowderX: Windows -95 based program for powder X-ray diffraction data processing. J Appl Crystallogr 32:838
10. Ahn Y, Choi EJ, Kim EH (2003) Superparamagnetic relaxation in cobalt ferrite nanoparticles synthesized from hydroxide carbonate precursors. Rev Adv Mater Sci 5:477
11. Ma M, Zhang Y, Yu W, Shen H-Y, Zhang H-Q, Gu N (2003) Preparation and characterization of magnetic nano particles coated by aminosilane. J Colloid and Surface A: Physicochem Eng Aspects 212:219
12. Ahmed SR, Kofinas P (2005) Magnetic properties and morphologyof block co-polymer-cobalt oxide nanoparticles. J Magn Magn Mater 288:219
13. Ishikawa T, Nakazaki H, Yasukawa A, Kandori K, Seto M (1999) Influences of Co2 +, Cu2 + and Cr3 + ions on the formation of magnetite. Corros Sci 41:1665
14. Wu N, Fu L, Su M, Aslam M, Wong KC, Dravid VP (2004) Interaction of fatty acid monolayers with cobalt nanoparticles. Nano Letters 4:383

Resonant Frequency Analysis of Nanowire-Based Sensor and Its Applications Toward Ethanol Sensing

D. Parthiban, M. Alagappan and A. Kandaswamy

Abstract In this work, a nanowire-based sensor is designed and its applications toward ethanol sensing is reported. Here, Zinc oxide has been taken as the suitable material for designing nanowire because of its ease of enhanced molecular adsorption while interacting with the target gas molecules. The resonant frequency shift after attaching the ethanol gas molecule is analyzed mathematically and compared with the COMSOL simulation results. The Simulation results showed that the theoretical model leads to a small percentage error of less than 3 %. Hence, this sensor can be used to detect the ethanol molecule in human breath and it provides painless breath diagnosis for alcoholic patients.

Keywords Zincoxide • Nanowire • Ethanol molecule and COMSOL

1 Introduction

Nowadays, nanosensors have attracted tremendous interest among all of us because of its ease of miniaturization, faster response and enhanced power consumption. Recently, metal oxides play an important role due to their small dimensions, low cost, and high compatibility. Some of the metal oxides used for gas sensing applications are SnO_2, GeO_2, ZnO, GaN, WO_3, indium tin oxide, and TiO_2 [1]. Zinc oxide (ZnO) is an interesting chemically and thermally stable n-type semiconductor with a large exciton binding energy of 60 meV and large

---

D. Parthiban (✉) · M. Alagappan · A. Kandaswamy
PSG College of Technology, Coimbatore 641004, India
e-mail: parthibankvd@gmail.com

M. Alagappan
e-mail: alagappanpsg@gmail.com

S. Sathiyamoorthy et al. (eds.), *Emerging Trends in Science, Engineering and Technology*, Lecture Notes in Mechanical Engineering, DOI: 10.1007/978-81-322-1007-8_56, © Springer India 2012

bandgap energy of 3.37 eV at room temperature. Due to these properties, ZnO is widely used in nanoelectronics, optoelectronics, and detectors which are sensitive to toxic and combustible gases [1, 2].On reducing the size of the material to nanoscale, the material exhibits high surface to volume ratio and quantum effects starts playing a role [3, 4]. The reason for choosing nanostructures is to make high precision sensors for precise measurements and to get a better sensitivity. ZnO is an ideal candidate for making electromechanical coupled devices with ease of forming different nanostructures and also it is biodegradable and possibly biocompatible material. In present years, one-dimensional nanostructures like nanowires and nanorods have attracted great attention in the field of gas sensors when compared with the bulk and thin film gas sensors [5].The gas sensing mechanism of ethanol sensor involves adsorption of ethanol molecules on the surface of the semiconducting material leading to change in resonant frequency. By analyzing the change in resonant frequency, the existence of the attached ethanol gas molecule can be easily detected.

2 Design Principles and Analysis

In this work, ZnO nanowire has been choosen for sensor application. Here, one end of the nanowire is anchored to a support and the other end is left free. The resonant frequency depends on the spring constant of the nanowire and the mass distribution along the nanowire. If another atom or molecule is attached to the ZnO nanowire, the mass distribution along the beam will be changed, and hence the resonant frequency of the nanowire will be shifted. By measuring the resonant frequency shift of the ZnO nanowire resonator, the mass of the attached atom/molecule can be precisely measured. Hence, extremely high resolution can be achieved with the use of this mass sensor. A theoretical model is developed to describe the vibration of the nanowire cantilever structure [6]. The resonant frequency change of the cantilever due to attached mass is analyzed. The resonant frequency of the structure depends on the spring constant of the nanowire and the mass distribution along the wire. Hence, this sensor can be used as an important biomarker for disease diagnosis. As shown in the Fig. 1, the nanowire is activated to vibrate freely along Y direction.

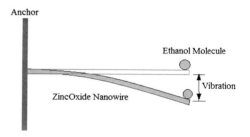

Fig. 1 Model of Zinc Oxide nanowire resonator with attached mass at its end

Assuming small-angle deflection, the vibration of the nanowire can be described by the following Eq. (1) [6]

$$EI\frac{d^2y}{dt^2} + \rho A\frac{d^2y}{dt^2} = 0 \tag{1}$$

Where E is Young's modulus of the nanowire, I is the inertial moment of the nanowire, ρ is the density of the nanowire, A is the cross-section area of the nanowire at location, and x and y are the deflection of nanowire at location x. The nanowire cantilever can be treated as a simplified spring-mass model. Hence, the resonant frequency of the cantilever can be calculated by using Eq. (2).

$$f = \frac{1}{2\pi}\sqrt{\frac{k}{m_{eq}}} \tag{2}$$

Where k is the effective spring constant of the cantilever, and m_{eq} is the equivalent vibration mass of the nanowire with the attached mass at its end tip. From this equation, the resonant frequency of the nanowire is closely related to the spring constant of the nanowire and the mass distribution along the nanowire. A certain mass attached to its end tip of the nanowire will introduce a certain resonant frequency shift to the nanowire resonator structure. The spring constant k and equivalent mass m_{eq} of cantilever with attached mass at its end tip is given by Eq. (3)

$$k = \frac{3EI}{L^3}, m_{eq} = \frac{33}{140}\rho AL + m \tag{3}$$

When no mass is attached, the original resonant frequency of the nanowire resonator is calculated with the help of Eq. (4)

$$f_o = \frac{1}{2\pi}\sqrt{\frac{140EI}{11\rho AL^4}} \tag{4}$$

The resonant frequency of nanowire with attached ethanol mass is given by Eq. (5)

$$f = f_o\frac{1}{\sqrt{1 + \frac{140m}{33\rho AL}}} \tag{5}$$

Thus, the resonant frequency shift due to attached ethanol mass is calculated by Eq. (6)

$$\Delta f = f_o - f = \left(1 - \frac{1}{\sqrt{1 + \frac{140m}{33\rho AL}}}\right) \tag{6}$$

The percentage of resonant frequency shift compared to original resonant frequency is given by Eq. (7)

Table 1 Material properties of ZnO nanowire

Material properties	Values
Youngs modulus E	$0.210*10^{15}$ ng/nm$-$s^2
Poissons ratio	0.33
Acceleration g	$9.8*10^9$ nm/s^2
Density ρ	$5.676*10^{-12}$
Radius r	0.55 nm
Length L	8 nm

$$\frac{\Delta f}{f_o} = 1 - \frac{1}{\sqrt{1 + \frac{140m}{33\rho AL}}} \qquad (7)$$

Thus, by measuring percentage resonant frequency shift ($\Delta f/f_0$), the attached mass can be calculated by Eq. (8)

$$m = \frac{33\rho AL}{140}\left(\frac{1}{\left(1-\frac{\Delta f}{f_o}\right)^2} - 1\right) \qquad (8)$$

That is, by measuring the relative resonant frequency shift ($\Delta f/f_0$), we know the value of the attached ethanol mass. This is the working principle of using nanowire as a gas sensor [1]. The material properties of ZnO nanowire for predicting the resonant frequency shift are given in the Table 1.

3 Design Implementation

Figure 2a shows a 3D visualisation of ZnO nanowire-based gas sensor for sensing ethanol gas. The nanowire with the dimensions of 0.5 nm radius and 8 nm in length. The mass of the molecule is represented in the form of solid cylinder with 0.2 nm as the radius and 0.3455 nm as the height.

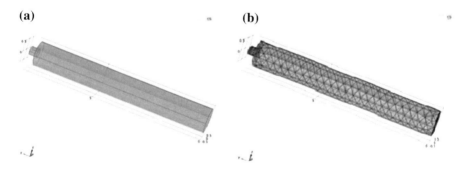

Fig. 2 a Geometry of Zinc oxide Nanowire with attached mass at its end and **b** Mesh Model of an Ethanol Sensor

This model is solved using solid mechanics physics by considering eigen frequency analysis. Free tetrahedral meshing is chosen to solve the above model and the above Fig. 2b shows the solved mesh model an ethanol sensor

4 Performance Analysis

The frequency of the nanowire without attached mass is $1.48393188e^{10}$ Hz and for the attached mass of $7.64729*10^{-14}$ ng which corressponds to one ethanol molecule the frequency change of $1.478436004e^{10}$ Hz is observed. Theoritical values are listed in the Table 2 after adding an additional ethanol molecule to the nanowire.

Using the values listed in the Table 2, the theoretical results show the percentage frequency shift of 0.3738 % for one ethanol molecule. The performance of the theoretical frequency shift result is verified by comparing with COMSOL simulation results.

After analyzing with COMSOL, the frequency of the nanowire without attached mass is $1.538097e^{10}$ Hz and for the attached mass of $7.64729e^{-14}$ ng which corressponds to one ethanol molecule, the frequency change of $1.532504e^{10}$ Hz is observed as shown in the Fig. 3a and b respectively.

Table 2 Resonant frequency versus attached mass

Attached mass m (ng)	Resonant frequency (Hz)
$7.64729*10^{-14}$	$1.478436e^{10}$
$1.52945*10^{-13}$	$1.472950e^{10}$
$2.29418*10^{-13}$	$1.467525e^{10}$
$3.05891*10^{-13}$	$1.462160e^{10}$

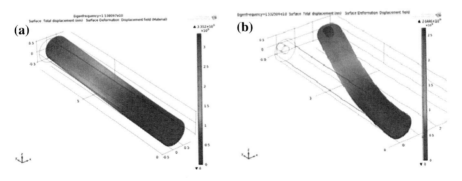

Fig. 3 a Simulation result of no mass attached and b Simuation result of attached mass

5 Results and Discussion

The values listed in the Table 3 show the percentage frequency shift of 0.3636 %. The COMSOL simulation results show that the theoretical model leads to a small percentage error of 2.805 %. Similarly, the results and theoretical prediction of other three cases are verified as well, and the results are listed in Table 4. This shows that the results are more accurate.

The relationship between the attached ethanol mass and the percentage of resonant frequency change is plotted, as shown in Fig. 4a and b respectively for theoretical calculation and practical verification. From the graphs, it is clear that the linearity is established.

Table 3 COMSOL simulation result of resonant frequency versus various attached mass

Attached mass m (ng)	Resonant frequency (Hz)
$7.64729*10^{-14}$	$1.532504e^{10}$
$1.52945*10^{-13}$	$1.526881e^{10}$
$2.29418*10^{-13}$	$1.521317e^{10}$
$3.05891*10^{-13}$	$1.515809e^{10}$

Table 4 Theoretical frequency shift versus comsol frequency shift

Case	Attached mass m in (ng)	Theoritical shift in %	COMSOL shift in %	Relative error in %
1.	$7.64e^{-14}$	0.3738	0.3636	2.805
2.	$1.52e^{-13}$	0.7435	0.7272	1.960
3.	$2.29e^{-13}$	1.1090	1.0900	1.174
4.	$3.05e^{-13}$	1.4710	1.4490	1.510

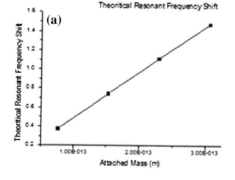 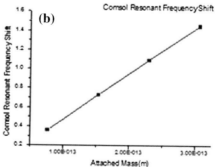

Fig. 4 a Relationship between theoretical resonant frequency shift versus attached mass and b. Relationship between COMSOL resonant frequency shift versus attached mass

6 Conclusion

A 3D multiphysics modeling of ZnO nanowire-based gas sensor is designed and the resonant frequency change of the resonator due to attached mass is analyzed. The effectiveness of the theoretical model is verified with COMSOL simulation results. The COMSOL simulation results show that the percentage frequency shift $\Delta f/f_o$ based on theoretical model leads to a small percentage error of less than 3 %. Thus, the proposed ethanol sensor can achieve extremely high sensitivity in molecular level.

References

1. Hseuh TJ (2007) Zno nanowire-based Co sensors prepared at various temperatures. J Electrochem 154(12):J393–J396
2. Lu CY, Chang SP, Chang SJ (2009) ZnO Nanowire based oxygen sensor. Department of Electrical Technology, Advanced optoelectronic technology center and centre for micro/nano science and technology, National Cheng Kung University, Tainan 70101
3. Bajpai R, Tigli O, Zaghloul M (2009) Modeling and simulation of nanowire based cantilever structure. The George Washington University, USA
4. Bajpai R, Zaghloul M (2009) Modeling and simulation of a ZnO nanowire bridge and development of an electrical equivalent circuit in liquid. The George Washington University, USA
5. Gupta SP, Aditee J, Manmeet K (2010) Development of gas sensors using ZnO nanostructures. J Chem Sci 122(1):57–62
6. Xia M, Xiong X (2011) Mathematical model of a proposed carbon nanotube sensor for ultra sensitive acetone sensing. University of Bridgeport, Bridgeport

Spectroscopic Analysis on Sm^{3+} Doped Fluoroborate Glasses

S. Arunkumar and K. Marimuthu

Abstract Optical behaviors of the Sm^{3+} doped $(50 - x) B_2O_3 + 25ZnO + 25NaF + xSm_2O_3$, (where x = 0.05, 0.1, 1, 2 and 3 wt%) glasses have been studied and reported. The glasses were prepared by following melt-quenching technique and the optical absorption spectra in the UV-vis-NIR region between 350 and 1800 nm and the luminescence spectra between 550 and 750 nm, have been measured for the prepared glasses at room temperature. Optical band gap, band tail parameter, and Urbach energy have been calculated from the fundamental absorption edges for direct and indirect transitions. The band gap value is found to decrease with the increasing concentration of the Sm^{3+} ions. The band gap values are found to be 3.075–2.011 eV for direct transitions and 3.408–2.661 eV for indirect transitions. The origin of the Urbach energy has been identified and discussed. The Urbach energy is found to increase considerably with the increase in Sm^{3+} ions and is found to have values from 0.2268 to 0.8854 eV. A strong luminescence in the reddish-orange spectral region has been observed for the prepared Sm^{3+} glasses.

Keywords Glasses • Bonding parameters • Luminescence • Urbach's energy

1 Introduction

Rare-earth (RE) doped B_2O_3 glasses are used for the development of electronic and optoelectronic devices, lasers, sensor devices, optical fibers, amplifiers, and hole burning high-density memories because of their excellent properties like super hardness, high insulation, linear, and nonlinear optical behavior [1]. There has been a considerable attention in the study on optical, structural, and dielectric behavior of borate-based glasses [2–4]. The excitations and emissions are due to the transition among 4f electronic states of trivalent RE ions, which are highly sensitive to the symmetry, structure of the local environment, and phonon energy of the host matrix [5].

S. Arunkumar · K. Marimuthu (✉)
Department of Physics, Gandhigram Rural University, Gandhigram 624 302, India
e-mail: mari_ram2000@yahoo.com

S. Sathiyamoorthy et al. (eds.), *Emerging Trends in Science, Engineering and Technology*, Lecture Notes in Mechanical Engineering, DOI: 10.1007/978-81-322-1007-8_57, © Springer India 2012

The Sm^{3+} ion exhibits a strong luminescence in the reddish-orange spectral region. Through the optical analysis, the effect of host matrix on the local environment of the given RE cation with its first nearest neighbor could be elucidated using Judd-Ofelt theory [6, 7]. Jayasankar et al. [8] reported the optical properties of Sm^{3+} ions in lithium borate and lithium fluoroboarte glasses, and Sanmuga Sundari et al. [9] studied and reported the structural and optical behavior of Sm^{3+} doped sodium borate and fluoroborate glasses. Ali [10] studied the optical band gap of the Sm^{3+} doped bismuth borate glasses for various inter band transitions using Davis and Mott theory [11, 12]. The present work reports, spectroscopic and band gap studies on Sm^{3+} doped alkali fluoroborate glasses. Through the ultraviolet absorption edges, the optical band gap (E_{opt}), band tailing parameter (B), and Urbach tail energy (ΔE) have been calculated and reported for direct and indirect transitions.

2 Experimental

The Sm^{3+} doped alkali fluoroborate glasses were prepared following conventional melt-quenching technique. The high purity (Sigma Alrich) analytical grade chemicals (99.99 % purity) H_3BO_3, ZnO, NaF, and Sm_2O_3 were used in the present work. The relevant weight fractions of the chemical composition $(50 - x)$ $B_2O_3 + 25ZnO + 25NaF + xSm_2O_3$, (where x = 0.05, 0.1, 1, 2 and 3 wt%) were weighed, thoroughly mixed, ground in an agate mortar and taken into a porcelain crucible and melted in a furnace by keeping it at a temperature of 1050 °C for 45 min. The crucibles were stirred for homogeneous mixing. The melt is poured on to a preheated brass plate and the glasses were annealed for 7 h at 450 °C to remove the thermal strain and to avoid the formation of air bubbles. These glasses were polished to plane faces for optical measurements. Optical absorption spectra of the Sm^{3+} doped alkali fluoroborate glasses were measured using CARY 500 UV-vis-NIR spectrometer in the wavelength range 350–1,800 nm with a spectral resolution of ±0.1 nm. The luminescence measurements were made using Perkin-Elmer LS55 spectrometer in the wavelength range 400–750 nm with a resolution of ±0.1 nm. The refractive index (n_d) of the prepared glasses was measured using Abbe refractometer taking mono bromonaphthalene as a contact liquid at sodium wavelength (589.3 nm). All these measurements were carried out at room temperature.

3 Results and Discussion

3.1 Absorption Spectra

The room temperature optical absorption spectra of the Sm^{3+} doped alkali fluoroborate glasses recorded in the UV-vis-NIR region is shown in Fig. 1. The band assignments of the Sm^{3+}: xBZNS glasses were made referring earlier reports [1–4]

Spectroscopic Analysis on Sm³⁺ Doped Fluoroborate Glasses

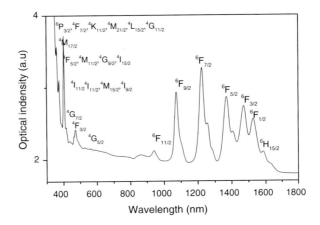

Fig. 1 Absorption spectra of the Sm⁺: 1BZNS glass

and the transitions in the absorption spectra originate from the ground state $^6H_{5/2}$. The observed absorption band transitions are presented in Table 1. The excited state transitions which originate from the ground state is due to the transition between multiplets of the 4f⁵ samarium configuration. In the present study, the absorption transitions occur almost at same position which are similar to other reported Sm³⁺ glasses [5]. There is no variation in band positions with varying glass compositions and the change in energy band positions has been observed from absorption spectra. According to Jorgensen and Judd [6, 7], the position and intensity of certain electric dipole transitions of the RE ions are found to be very sensitive to the

Table 1 Observed band positions ($\bar{\beta}$ cm⁻¹) and bonding parameters (and δ) of Sm³⁺: BZNS glasses

Energy level	0.05BZNS	0.1BZNS	1BZNS	2BZNS	3BZNS	Aquo [17]
$^6F_{1/2}$	6266	6278	6282	6268	6277	–
$^6H_{15/2}$	6556	6548	6572	6572	6542	–
$^6F_{3/2}$	6782	6788	6794	6808	6815	6630
$^6F_{5/2}$	7286	7304	7314	7308	7328	7100
$^6F_{7/2}$	8166	8182	8184	8178	8159	8000
$^6F_{9/2}$	9306	9296	9288	9312	9310	9200
$^6F_{11/2}$	10556	10576	10580	10660	10560	10500
$^4G_{5/2}$	17872	17892	17896	17874	17852	17900
$^4F_{3/2}$	18942	19024	19072	18986	18979	18900
$^4G_{7/2}$	20082	20072	20106	20074	20054	20050
$^4M_{15/2}$	21030	21042	21052	21072	21072	20800
$^4I_{13/2}$	21624	21636	21642	21618	21642	21600
$^4F_{5/2}$	22306	22352	22362	22364	22333	22200
$^4M_{17/2}$	23142	23170	23162	23124	23144	–
$^6P_{3/2}$	24452	24552	24624	24562	24569	24950
$\bar{\beta}$	1.008	1.009	1.007	1.007	1.009	–
δ	−0.794	−0.891	−0.695	−0.695	−0.891	–

environment of the RE ion and such transitions are termed as hyper sensitive transitions and in the present Sm^{3+}: glasses $^6H_{5/2} \rightarrow {}^6P_{3/2}$ and $^6H_{5/2} \rightarrow {}^6F_{7/2}$ are the hypersensitive transitions and the same is confirmed in the present work [8]. The sample which starts to absorb completely at a characteristic wave length (λ_{edge}) in the ultraviolet region is defined as fundamental absorption edge and in the present work the value of λ_{edge}, is found to increase with increasing Sm_2O_3 content. It is observed from the Table 2 that the λ_{edge} is found to be blue shift for lower concentration and it makes red shift with increasing Sm^{3+} ion concentration.

3.2 Nephelauxetic Ratio and Bonding Parameters

The nephelauxetic ratio is defined as $\beta = v_c/v_a$. Where v_c and v_a are the energies of the corresponding transitions in the complex and aquo-ion, respectively, and N is the number of energy levels that are used to compute β value. Depending on the environment δ may be positive or negative indicating covalent or ionic bonding [15]. The values of $\bar{\beta}$ and δ are calculated and presented in Table 1. The δ value in present glass is found to be negative which indicates that the bonding is of ionic in nature. Among the prepared Sm^{3+} doped glasses Sm^{3+}: 2BZNS glass exhibit high ionic nature among the prepared glasses. It is observed from the Table 1 that Sm^{3+} ions exhibit stronger ionic bonding in fluroborate glasses. Based on the relative magnitudes of δ (cm-1) values, among the prepared glasses, it is found that the ionic character decreases in the order 3BZNS $<$ 2BZNS $<$ 1BZNS $<$ 0.1BZNS $<$ 0.05BZNS. It can be concluded that the glass 0.05BZNS possess higher covalency because of lower ionic.

3.3 Luminescence Spectra

The luminescence spectra of the Sm^{3+}: xBZNS glasses were recorded at room temperature between 550 and 750 nm on excitation with 488 nm. Since all the spectra are alike, a representative Sm^{3+}: 0.05BZNS glass spectra is shown in

Table 2 the fundamental absorption edge (λ_{edge}), Optical energy gap (E_{opt}), Band tailing parameter (B), and Urbach energy (ΔE) corresponding to n = 1/2, 2, 3 for the prepared Sm^{3+}: xBZNS glasses

Sample code	λ_{edge} (nm)	n = 1/2		n = 2		n = 3		ΔE (eV)
		E_{opt} (eV)	B (cm^{-2}eV)	E_{opt} (eV)	B (cmeV)$^{-1/2}$	E_{opt} (eV)	B (cmeV)$^{-2/3}$	
0.05BZNS	327	3.076	8.161	3.408	640.2	2.873	2.598	0.227
0.1 BZNS	336	2.781	6.248	3.236	492.4	2.687	1.552	0.307
1 BZNS	340	2.369	5.012	2.967	386.7	2.339	1.655	0.423
2 BZNS	345	2.238	4.416	2.836	20.13	2.137	0.587	0.762
3 BZNS	357	2.011	3.553	3.385	502.2	1.936	1.378	0.885

Fig. 2. The Sm^{3+}: xBZNS glasses exhibit four emission bands corresponding to $^4G_{5/2} \to {}^6H_J$ (J = 5/2, 7/2, 9/2 and 11/2) transitions similar to other reported Sm^{3+} glasses [2–5]. Among the emission transitions $^4G_{5/2} \to {}^6H_{7/2}$ band exhibit strong intense emission centered around 602 nm and its full-width at half maximum is 17.6 nm. Thulasiramudu et al. [13] studied Sm^{3+} doped ZnO-PbO-B$_2$O$_3$ glasses and reported that the transition $^4G_{5/2} \to {}^6H_{7/2}$ with $\Delta J = \pm 1$ is a magnetic dipole (MD) allowed but it is an electric dipole (ED) dominated. The $^4G_{5/2} \to {}^6H_{5/2}$ and $^4G_{5/2} \to {}^6H_{9/2}$ transitions are MD and ED allowed transitions, respectively [7–9, 11]. The intensity ratio of the hypersensitive transition to nonhypersensitive transition is useful to explain the symmetry of the local environment of the 4f^{3+} ions. The observed $^4G_{5/2} \to {}^6H_{9/2}$ transition has greater intensity than $^4G_{5/2} \to {}^6H_{5/2}$ transition, which specify the asymmetry nature of the host matrix [13, 14]. The value of red to orange intensity ratio is found to be maximum (1.338) for Sm^{3+}: 0.1BZNS glass and the R/O ratio of the prepared glasses are found to be 1.186, 1.338, 1.265, 1.263 and 1.181 for 0.05BZNS, 0.1BZNS, 1BZNS, 2BZNS and 3BZNS, respectively. The value of R/O ratio is found to decrease with the increase in Sm$_2$O$_3$ concentration beyond 0.1 wt%.

3.4 Optical Band Gap and Urbach's Energy Analysis

The absorption coefficient near and below the fundamental optical absorption edges of the absorption spectra were determined using Eq. (1).

$$\alpha(\nu) h\nu = B \left(h\nu - E_{\text{opt}} \right)^n \quad (1)$$

The optical absorption at the fundamental edges has been determined using Eq. (1) for various transitions by fitting n = 1/2, 1/3, 2, and 3 corresponding to direct allowed, direct forbidden, indirect allowed, and indirect forbidden

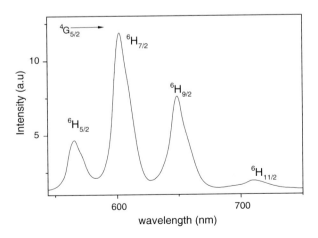

Fig. 2 Luminescence spectra of the Sm$^+$: 1BZNS glass

transitions, respectively. Many researchers were studied and reported the direct and indirect band gap studies on glasses using Tauc's, Mott and Davis theory [11, 12]. Sindhu et al. [15] and Rani et al. [16] suggested and reported that the indirect forbidden transition (n = 3) is also possible in glass materials.

Figure 3 shows the Tauc's plot which represents the variation of $(\alpha h\nu)^2$ with photon energy $h\nu$, by substituting the value $n = 1/2$ in Eq. (1) corresponding to direct allowed transitions. The band gap values were obtained through extrapolation to zero absorption $((\alpha h\nu)^2 = 0)$ in the linear region of the curves of the prepared glasses. The band gap values are found to decrease with increasing Sm_2O_3 content. The trend may be attributed to the increase in the degree of disorder which causes more defects or localized states in the band gap. The band tail parameter B is obtained through the slope of the linear region of the curves. The observations have been carried out for the indirect allowed and forbidden transitions through the variation of $(\alpha h\nu)^{1/3}$ versus $h\nu$ and $(\alpha h\nu)^{1/2}$ versus $h\nu$ using Eq. (1). The optical band gap values were obtained by extrapolating to zero absorption in the linear part of the Tauc's plot at $(\alpha h\nu)^{1/2} = 0$ and $(\alpha h\nu)^{1/3} = 0$ for indirect allowed and indirect forbidden transitions. The band gap values corresponding to each transition of the prepared glasses are presented in Table 2. The band tail parameter values are also estimated and presented in Table 2. Eraiah et al. [12] reported the optical band gap energy values for Sm^{3+} doped Zinc-phosphate glasses and is found to decrease with increase in Sm_2O_3 content and the values are found to be in the range of 4.20–2.89 eV. The Urbach's plot corresponding to ln (α) versus $h\nu$ have been shown in Fig. 6 for all the prepared glasses. It is observed that the absorption coefficient near the fundamental absorption edge has an exponential behavior and obeys the empirical Urbach rule (4). The band tails associated with the valence band and the conduction band arises due to the potential fluctuations of internal fields in the material, extend into the band gap and exhibit exponential behavior. Materials with larger Urbach energy used to have greater tendency to convert weak bonds into defects. The ΔE values evaluated

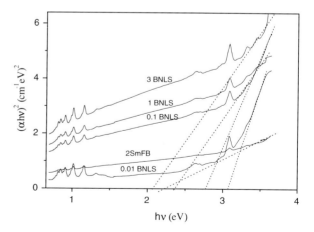

Fig. 3 Tauc's plot for the Sm^{3+}: xBZNS alkali fluroborate glasses for n = 1/2

from the slope of the straight line of Urbach's plot lies between 0.2288 and 0.8854 eV for the prepared glasses.

Thulasiramudu et al. [13] reported that the band gap values for direct and indirect transitions are 2.96, 2.87 eV, respectively, for Sm^{3+} doped borate glasses. Ali [10] reported that the band gap values are found to decrease with increase in Sm_2O_3 content. This causes because of the structural changes in the glasses, including the formation of greater number of nonbridging oxygens, which decrease the E_{opt} values due to the addition of the Sm_2O_3. The region where the absorption coefficient $(\alpha(\nu)) < 10^2$ cm^{-1} involves low-energy absorption and is observed to be due to the optical transitions between the localized states. The Urbach's energy and the band gap of the Sm^{3+}: xBZNS glasses are depicted as a function of the Sm_2O_3 content in Fig. 8. The Urbach's energy increases with the increase in Sm_2O_3 content, whereas the band gap values are found to decrease. The Sm^{3+}: 0.05BZNS glass posses lower Urbach's energy and is concluded that the possibility of the long range order arises locally from the number of defects. Among the prepared glasses Sm^{3+}: 0.05BZNS glass exhibits lower Urbach's energy (Fig. 4).

4 Conclusion

Through the optical absorption spectra of the Sm^{3+} doped alkali fluroborate glasses, it is concluded that the observed band position occurs almost at the same position irrespective of the change in the chemical composition and the intensity of the observed spectra increases linearly with the increase in Sm_2O_3 content. Among the observed transitions $^6P_{3/2}$ and $^6F_{7/2}$ bands are observed to be hypersensitive. The intense reddish-orange emission observed at 602 nm corresponding

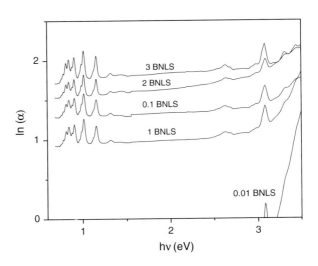

Fig. 4 Urbach's plot for the Sm^{3+}: xSmFB alkali fluroborate glasses

to the $^4G_{5/2} \rightarrow {}^6H_{7/2}$ transition can be used for laser applications in the visible region. The optical absorption coefficient ($\alpha(\nu)$) and various other optical parameters such as optical energy gap (E_{opt}), band tailing parameter (B), Urbach energy (ΔE) were evaluated and discussed. The fundamental absorption edges exhibit red shift (lower energy) with the increase in Sm_2O_3 content. The optical energy gap values are found to decrease with the addition of Sm_2O_3 content, whereas the Urbach's energy is found to increase.

Acknowledgments One of the authors (S.Arunkumar) is grateful to University Grants Commission (UGC) for the award of the Rajiv Gandhi National Fellowship (RGNF) to pursue Ph.D degree.

References

1. Ali AA (2009) J Lumin 129: 1314
2. Babu P, Jayasankar CK (2000) Opt Mater **15**: 65
3. Carnall WT, Fields PR, Rajnak K (1968) J Chem Phys 49: 4412
4. Davis EA, Mott NF (1970) Philos Mag 229: 903
5. Eraiah B, Bhat SG (2007) J Phys Chem Solids 68: 581
6. Jayasankar CK, Babu P (2000) J Alloys Compd 307: 82
7. Judd BR (1962) Phys Rev 127: 750
8. Karunakaran RT, Marimuthu K, Surendra Babu S, Arumugam S (2010) J Lumin 130: 1067
9. Lavin V, Babu P, Jayasankar CK, Martin IR, Rodriguez VD (2001) J Chem Phys 115: 10935
10. Maheshvara K, Linganna K, Marimuthu K (2011) J Lumin 2746–2753
11. Mott NF, Davis EA (1979) Electronic processes in non-crystalline mater, 2nd edn. Clarendon Press, Oxford
12. Ofelt GS (1962) J Chem Phys 37: 511
13. Rani S, Sanghi S, Agarwal A, Seth VP (2009) Spectrochim Acta Part A 74: 673
14. Rayappan A, Selvaraju K, Marimuthu K (2011) Physica B 406: 548–555
15. Sanmuga Sundari S, Marimuthu K, Sivaraman M, surendra Babu S (2010) J Lumin 130: 1313
16. Sindhu S, Sanghi S, Agarwal A, Seth VP, Kishore N (2005) Mater Chem Phy 90: 83
17. Thulasiramudu A, Buddhudu S (2007) Spectrochim Acta Part A 67: 802

Synthesis and Spectroscopic Characterization of Pure and L-Arginine Doped KDP Crystals

K. Indira and T. Chitravel

Abstract Optically good quality single crystals of pure and L-arginine-doped KDP crystals were grown by a slow evaporation method. The grown crystals were subjected to optical and dielectric studies. The unit cell parameters were calculated by X-ray diffraction. The UV-Vis spectrum shows the transmitting ability of the crystals in the entire visible region and transmittance percentage is increased for the doped KDP crystals. The presence of the functional groups has been identified by FT-IR spectrum. Hence, L-arginine doped KDP crystals are found to be more beneficial from an application point of view as compared to pure KDP crystals. From Microhardness test it has been observed that the addition of organic additives improves the mechanical strength of the crystal.

Keywords Doping • Slow evaporation • X-ray diffraction • UV-Vis • FT-IR • Microhardness

1 Introduction

Materials with large optical nonlinearity are needed to realize applications in opto-electronics, telecommunication industries, laser technology, and optical storage devices. Potassium dihydrogen phosphate (KDP) is an excellent inorganic nonlinear optical (NLO) material and has a considerable interest among several research workers because of its wide frequency, high efficiency of frequency conversion, and high damage threshold against high power laser. With the aim of improving the SHG efficiency of KDP, researchers have attempted to modify KDP crystals either by doping different type of impurities or by changing the growth conditions [1–9].

K. Indira (✉)
Department of Physics, Pandian Saraswathi Yadav Engineering College, Sivagangai, India

T. Chitravel
Department of Physics, Anna University, Madurai Region, India

S. Sathiyamoorthy et al. (eds.), *Emerging Trends in Science, Engineering and Technology*, Lecture Notes in Mechanical Engineering, DOI: 10.1007/978-81-322-1007-8_58, © Springer India 2012

Most of amino acids possess NLO property; therefore, it is of interest to dope them in KDP crystals. The effects of amino acid on the NLO efficiency of KDP crystals were already published [1–3]. L-arginine is a potential material to produce crystals for nonlinear optical applications [3]. In the present study, L-Arginine doped KDP crystals were grown by slow aqueous solvent evaporation technique [4, 5]. The optical and dielectric behavior of the both pure and L-arginine doped KDP crystals has been studied and discussed in detail.

2 Experiment

Commercially available KDP was used for the growth. Without any further purification, KDP was dissolved in double distilled water. After obtaining the saturation, the solution was stirred well for about 3 h, filtered and kept separately for the slow evaporation. Similar procedure was followed for the addition of 0.5 % L-arginine to the saturated solution of KDP. Within 12–15 days, transparent crystal of both pure and L-arginine doped KDP crystals was yielded.

3 Characterization

3.1 X-Ray Diffraction Studies

Powder X-ray diffraction studies of pure and L-arginine doped KDP crystals were carried out using XPERT PRO diffractometer with CuKα ($\lambda = 1.5418$ Å) radiation. The samples were scanned for 2θ values from $10°$ to $80°$ at a rate of $2°$/min. Figures 1 and 2 show the powder XRD pattern of the pure and doped KDP crystals. The powder patterns were indexed and the lattice parameter values for the pure and doped KDP crystals were calculated. It is observed that both the pure and doped crystals crystallize in Tetragonal structure. The lattice parameters of the samples are presented in Table 1. There are slight variations in the lattice parameters and cell volume of the pure and doped crystals. These variations are due to the incorporation of L-arginine in the KDP crystal lattice.

3.2 FT-IR Studies

The FT-IR spectrum of pure KDP and L-arginine doped KDP crystals has been recorded within the wave number range from 500 to 4,000 cm^{-1}. Pellets with the constituents of each sample with KBr have been prepared and used in the experiment and the spectral result has been shown in Figs. 3 and 4. In the FT-IR spectra

Synthesis and Spectroscopic Characterization 629

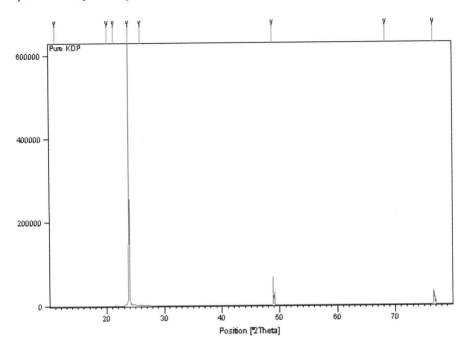

Fig. 1 XRD pattern for pure KDP

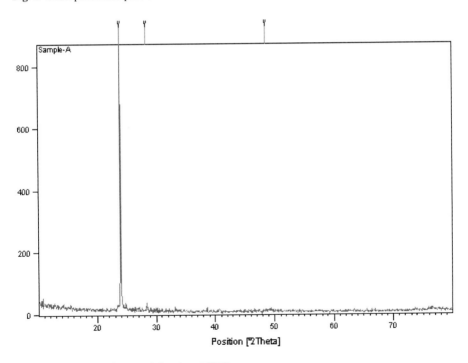

Fig. 2 XRD pattern for L-arginine doped KDP

Table 1 The unit cell parameter values

Sample	Lattice parameter		Volume (V)	Crystal structure
	$a = b$	c		
Pure KDP	7.457	6.796	387.42	Tetragonal $(\alpha = \beta = \gamma = 90°)$
L-arginine doped KDP	7.454	6.796	384.65	Tetragonal $(\alpha = \beta = \gamma = 90°)$

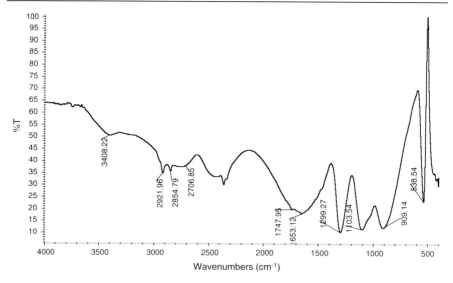

Fig. 3 FT-IR spectrum for pure KDP

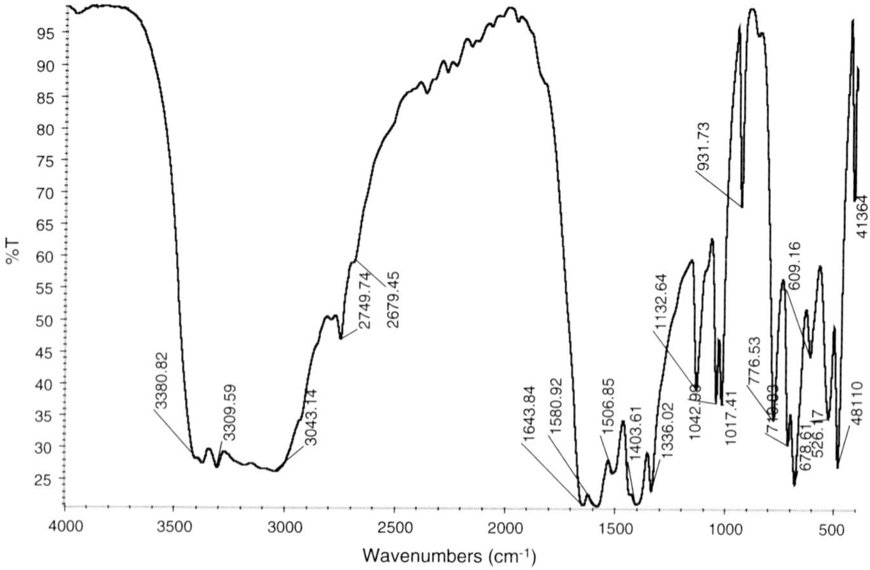

Fig. 4 FT-IR spectrum for L-arginine doped KDP

Synthesis and Spectroscopic Characterization

of pure KDP crystal, the observed absorption peaks corresponding to the P–OH stretching, P–O–H bending, C $=$ O stretching, and CH$_2$ bending. Based on the chemical structure of KDP and L-arginine the frequency assignments have been made to establish the functional groups present in the grown crystal. The broad band which ends at 3,380.82 cm^{-1} is due to P–OH stretching of H$_2$PO$_4$ [10]. The peak at 3,446.81 and 1,589.37 cm^{-1} is due to asymmetric stretching of NH$_2$ and bending of NH$_2$. The band at 948.34 and 8,760.31 cm^{-1} is due to stretching and symmetric stretching of C–C. The band at 1,458.00 cm^{-1} is due to CH$_2$ bending.

In the FT-IR spectra of L-arginine doped KDP crystals, the same peaks have been observed with some additional peaks. These additional peaks correspond to the functional groups of L-arginine, which confirms the doping of L-arginine in the KDP crystals in Table 2. There is a slight shift in the peak positions because of the hydrogen bonding.

3.3 Microhardness Studies

The microhardness of the material was measured by using the Vickers diamond pyramidal indentor attached to a REICHERT POLYVAR 2 MET microscope. The microhardness measurements were made for the applied loads varying from 25 to 100 g. The hardness of the crystal can be calculated using the relation.

$$Hv = 1.8544\, P \big/ d^2$$

where Hv is the Vickers hardness number, P is the indenter load in gm and d is the diagonal length of the impression in mm. The hardness of the Pure KDP crystal and L-Arginine doped KDP crystal was shown in the measurement Table 3.

It seems that the hardness value of the doped KDP crystal is higher than the hardness of the pure KDP crystal. The addition of L-arginine increases the hardness of the crystal. This is because of the incorporation of the Aminos into

Table 2 FT-IR peak assignments

S. no	KDP	Dop.KDP	Assignments
1.	3,408.22, 2,921.96, 2,854.79, 2,706.85	3,380.82, 3,309.59, 3,043.14	P–OH stretching of H$_2$PO$_4$, O–H stretching of COOH, N–H stretching of NH$_3$ and C–H stretching of CH$_2$ and C–H
2.	1,747.95	1,643.84	C $=$ O stretching, P–O–H bending and –C $=$ NH$_4$ stretching
3.	1,299.27	1,336.02	CH$_2$ bending and P $=$ O stretching of KDP
4.	909.14	1,017.41, 931.73	P–OH and C–H stretching
5.	538.54	526.17	P–OH deformation

Table 3 Hardness measurement

S. no	Compound	Load		
		25 gm	50 gm	100 gm
1.	KDP	20.04	31.67	51.89
2.	L-arginine doped KDP	66.69	98.71	140.64

superficial crystal lattice and removing defect centers which reduce the weak lattice stresses on the surface [10].

3.4 UV-Visible Spectral Studies

The optical properties of a material are important, as they provide information on the electronic band structure, localized state and types of optical transitions. Pure and L-arginine doped KDP crystals plates with a thickness of 2 mm without antireflection coating were cut and used for optical measurement. The UV-Visible transmission spectrum was recorded using Perkin Elmer Model-Lambda 35

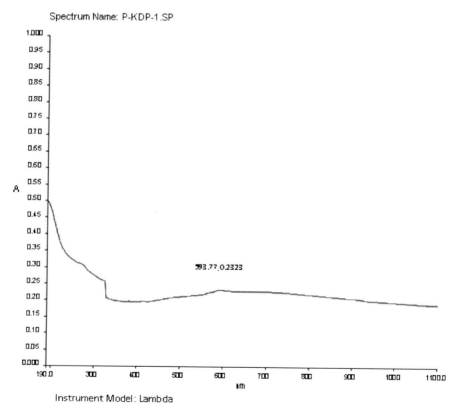

Fig. 5 UV-Visble absorption spectra for pure KDP

Synthesis and Spectroscopic Characterization

spectrometer in the range 190–1,100 nm. From the graph, it is evident that the pure KDP crystal has a transparency window from 320 nm but in case of L-arginine doped KDP has transparency window from 235 nm onwards suggesting the suitability for SHG of the 1,064 nm radiation. The Pure KDP crystal has 65 % transmission in the entire visible region, but the L-arginine doped KDP crystal has 100 % transmission in the entire visible region. The large transmittance window in the visible region enables very good optical transmission of second harmonic frequencies of Nd:YAG lasers.

Figures 5 and 6 show the UV-Visible spectrum of both pure and doped KDP crystals. For optical applications, the crystal should be highly transparent in the considerable region of wavelength [11]. The good transmission of the crystal in the entire visible region suggests its suitability for second harmonic generation devices [12]. The UV-Visible spectral analysis shows that both the crystals are transparent in the entire visible region. There is less absorption near the wave length of 235 nm, which is slightly shifted to higher wavelength side from pure KDP and it may be assigned to electronic excitation in the L-arginine doped KDP crystal. The absence of absorption and excellent transmission in entire visible region makes this crystal a good candidate for optoelectronic application.

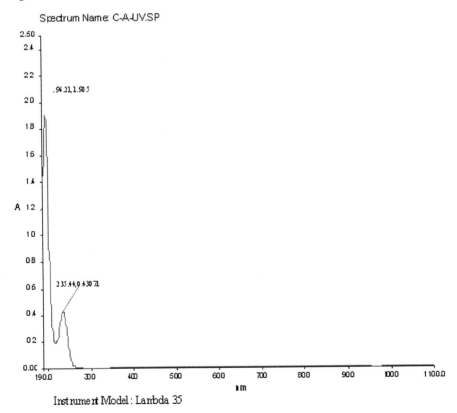

Fig. 6 UV-Visble absorption spectra for L-arginine doped KDP

4 Conclusion

Optical quality, colorless, and transparent single crystals of Pure and 0.5 % L-Arginine doped KDP crystals were grown by slow evaporation method. The X-ray diffraction studies were carried out, and the lattice parameters were calculated. The FT-IR spectral studies confirm the presence of all functional groups in the grown crystal. The microhardness values of doped KDP crystals are found to be increased by the presence of the dopant L-arginine. This result indicates that the grown crystals are useful for device application. The UV-Visible spectra analysis reveals that the transmittance efficiency improves considerably for the addition of L-arginine. The absence of absorption and excellent transmission in entire visible region makes this crystal a good candidate for optoelectronic application. Hence, L-arginine doped KDP crystals are found to be more beneficial from an application point of view as compared to pure KDP crystals.

References

1. Parikh KD, Dave DJ, Parekh BB, Joshi MJ (2007) Thermal, FT-IR and SHG efficiency studies of L-arginine doped KDP crystals. Bull Mater Sci 30(2):105–112 View at Publisher. View at Google Scholar. View at Scopus
2. Parikh KD, Dave DJ, Parekh BB, Joshi MJ (2010) Growth and characterization of L-alanine doped KDP crystals. Cryst Res Technol 45(6):603–610 View at Publisher. View at Google Scholar. View at Scopus
3. Vijayan N, Rajasekaran S, Bhagavannarayana G (2006) Growth and characterization of nonlinear optical amino acid single crystal: L-alanine. Cryst Growth and Des 6(11):2441–2445 View at Publisher. View at Google Scholar. View at Scopus
4. Ozga K, Krishnakumar V, Kityk IV, Jasik-Ślezak J (2008) L-lysine mono hydrochloride dihydrate as novel elasto and electro optical materials. Mater Lett 62(30):4597–4600 View at Publisher. View at Google Scholar. View at Scopus
5. Krishnakumar V, Nagalakshmi R, Manohar S et al (2010) Elasto optical spectra of novel L-Lysine mono hydrochloride dehydrate single crystals. Int J Mod Phys B 24(5):629–645 View at Scopus
6. Dhanaraj PV, Rajesh NP, Ramasamy P, Jeyaprakasan M, Mahadevan CK, Bhagavannarayana G (2009) Enhancement of stability of growth, structural and NLO properties of KDP crystals due to additive along with seed rotation. Cryst Res Technol 44(1):54–60 View at Publisher. View at Google Scholar. View at Scopus
7. Kumar PP, Manivannan V, Sagayaraj P, Madhavan J (2009) Growth and characterization of pure and doped NLO L-arginine acetate single crystals. Bull Mater Sci 32(4):431–435 View at Publisher. View at Google Scholar. View at Scopus
8. Kumar BS, Babu KR (2008) Effect of L-arginine, L-histidine and glycine on the growth of KDP single crystals and their characterization. Indian J Pure Applied Phys 46(2):123–126 View at Scopus
9. Rahman A, Podder J (2010) Effect of EDTA on the growth kinetics and structural and optical properties of KDP crystal. International Journal of Optics, vol 2010. Article ID 978763, 5 pages. View at Publisher. View at Google Scholar. View at Scopus
10. Delci Z, Shyamala D, Karuna S, Senthil A, Thayumanavan A (2012) Enhancement of optical, thermal and hardness in KDP crystals by boron doping. Int Chem Tech Res 4(2):816–826 View at Publisher. View at Google Scholar. View at Scopus

Synthesis and Spectroscopic Characterization

11. Krishnakumar V, Nagalakshmi R (2005) Crystal growth and vibrational spectroscopic studies of the semi organic non-linear optical crystal-bis (thiourea) zinc chloride. Spectrochim Acta A 61(3):499–507 View at Publisher. View at Google Scholar. View at PubMed
12. Venkataramanan V, Maheswaran S, Sherwood JN, Bhat HL (1997) Crystal growth and physical characterization of the semi organic bis (thiourea) cadmium chloride. J Cryst Growth 179(3–4):605–610

Structural and Dielectric Studies on Dy^{3+} Doped Alkaliborate Glasses

S. Arunkumar and K. Marimuthu

Abstract Alkali borate glasses doped with trivalent dysprosium were prepared by following melt quenching technique with the chemical composition $(50-x)B_2O_3 + 20Li_2O + 20LiF + 10BaO + xDy_2O_3$ by varying the concentration of the rare-earth dopant (where $x = 0.01, 0.1, 1, 2$ and 3 mol %). The structural analyses have been carried out using FTIR spectra. The fundamental stretching vibrations of various borate network (BO_3 and BO_4 groups) were indentified and reported. The dielectric constant (ε') and dielectric loss (ε'') of the Dy^{3+} ions doped alkali borate glasses have been studied as a function of frequency and temperature. AC conductivity measurements of the prepared glasses have been studied as a function of temperature. The ε' and ε'' values as a function of temperature and frequency are found to be maximum for $x = 1$ mol % alkali borate glass. The AC conductivity is found to obey the universal power law and is found to explain the electrical conduction in the prepared glasses. The value of dielectric parameters ε' and ε'' as a function of frequency and temperature are found to increase upto $x = 1$ mol % Dy^{3+} ions doped glasses and after that it is found to decrease.

Keywords Melt quenching • FTIR • Dielectric constant • AC conductivity

1 Introduction

The glasses with good ionic conduction have always a high concentration of monovalent ions that are used as charge carrier. The frequency-dependent conductivity and dielectric behavior found to give important details on the ionic or electric transport phenomena in disordered solids [1]. In order to study of ionic and electrical conductivity of glasses was revealed that all glasses need not be a insulator, but can be fast ionic conductor, super ionic conductor, high ionic conductor, and semiconducting glasses under certain circumstances and these ionic conductivity

---

S. Arunkumar · K. Marimuthu (✉)
Department of Physics, Gandhigram Rural University, Dindigul 624 302, India
e-mail: mari_ram2000@yahoo.com

S. Sathiyamoorthy et al. (eds.), *Emerging Trends in Science,*
Engineering and Technology, Lecture Notes in Mechanical Engineering,
DOI: 10.1007/978-81-322-1007-8_59, © Springer India 2012

materials have technological importance due to their potential application in fuel cells, electrochromic displays, solid-state batteries semiconductors, and electrochemical devices [2]. Borate glasses are commonly insulating in nature and with the addition of transition metal oxide (TMO), transition metal ions (TMI) and rare-earth ions (RE^{3+}) makes these as semiconducting in nature [3]. In case of oxide glasses B_2O_3 is a basic glass former because of their higher bond strength, lower cation size, smaller heat of fusion, and trivalency of boron [4]. The ionic conductivity of glasses also depends upon the nature, content of modifier oxide, and glass former composition. In the present study, the temperature and frequency dependent dielectrics and conductivity of glass series $(50-x)B_2O_3 + 20Li_2O + 20LiF + 10BaO + xDy_2O_3$ ($x = 0.01, 0.1, 1, 2$ and 3 wt %) were discussed. The variation of dielectric constant (ε'), dielectric loss (ε''), and AC conductivity (σ_{ac}) of these glasses with wide range of frequency and temperature was studied and reported.

2 Experimental

In the present work, the high purity (Sigma Alrich) chemicals H_3BO_3, $LiCO_3$, LiF, La_2O_3, and Dy_2O_3 were used for glass formation and all samples were prepared by following melt-quenching techniques. The sample composition and code representation of these glasses are 0.01BLBD, 0.1BLBD, 1BLBD, 2BLBD, and 3BLBD with respect to Dy^{3+} ion concentration namely 0.01, 0.1, 1, 2, and 3 mol %, respectively. The mixture of the relevant chemicals in appropriate portion has been taken in porcelain crucible and chemicals were melted in electrical furnace at 950 °C for 45 min. The chemicals in the crucible were stirred often for homogeneous mixing of the chemical components. The glasses were annealed for 9 h at 350 °C to remove the strain and improve the mechanical strength. These glasses were slowly quenched to room temperature and polished to obtain uniform thickness and plane faces for conducting measurements.

3 Results and Discussion

3.1 FTIR Studies

The FTIR spectra of Dy^{3+} doped 1BLBD alkali borate glass is shown in Fig. 1. The bands observed around 3,400–3,500 cm^{-1} are attributed to the hydroxyl (OH) groups in the prepared glasses [5]. The bands indentified around 2,853 and 2,922 cm^{-1} are due to the presence of hydrogen bond. The bands appear around 1,400–1,410 cm^{-1} are assigned to the $B-O^-$ stretching vibrations of BO_3 triangles [6]. The band observed at 1,343–1,362 cm^{-1} is attributed to the asymmetric stretching modes of BO_3 borate triangles and the sharing of other borate units such as penta, tri, and

Fig. 1 FTIR spectra of Dy^{3+} doped 1BLBD glass

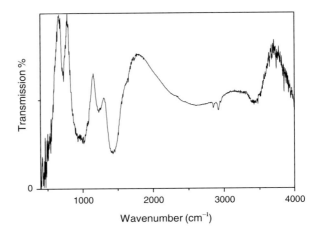

divalent groups. The bands around 1,235–1,245 cm^{-1} are assigned to the asymmetric stretching vibrations of B–O bonds from BO_3 units [7]. The weak bands observed at 934, 1,020, and 1,070 cm^{-1} are attributed to the B–O stretching vibrations of BO_4 units in tri-borate, tetra borate, and pyroborate groups. The band observed at 982 cm^{-1} is attributed to the attributed to the B–O bonds stretching vibrations of borate units in which boron atoms are coordinated with three oxygen atoms (BO_3 units) [8]. The band centered at 870 cm^{-1} which is the characteristic for stretching vibrations of B–O bonds in BO_4 tetrahedra from diborate groups. The band around 742 cm^{-1} is mainly attributed to the B–O bonds stretching vibrations of tetrahedral BO_4 units.

3.2 Frequency and Temperature Dependence Dielectric Studies

The real part of the dielectric (ε') at different frequencies and temperatures derived from the measured capacitance (C_p) and geometrical area (A) and thickness(d) of the sample using following formula as, $\varepsilon' = C_p d / \varepsilon_0 A$.

The ε' of materials is determined by electronic, ionic, and space charge polarization [9–11]. The dielectric loss factor is the phase difference due to loss of energy within the sample at a particular frequency [11]. The variation of dielectric constant with frequency of the glass sample 1BLBD glass at different temperature is given in Fig. 2 and from the given graph value of ε' for the sample is below 15. It is evident from this figure that ε' increases with temperature this value of ε' slightly agree with the value of ε' reported for other borate glasses. In present glasses dielectric constant varies from 4.0923 to 12.345. The value of dielectric constant for all other glasses is given in Table 1. The frequency dependent dielectric loss (ε'') is given in Fig. 3 and from this we can conclude that the value of dielectric loss decreases with increasing frequency [12]. The dielectric loss is found to follow a power with frequency that is

Fig. 2 The frequency dependence $\varepsilon'(\omega)$ of 1BLBD glass for varies temperature

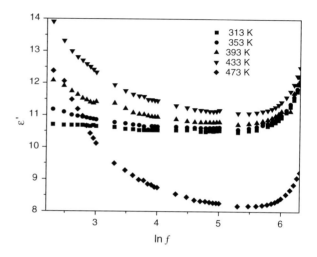

Table 1 The value of dielectric constant (ε'), Dielectric loss (ε''), the power of angular frequency (m) at temperature 34 and the AC conductivity ($\sigma_{ac} \times 10^{-11}$ Ω cm^{-1}), DC conductivity ($\sigma_{dc} \times 10^{-11}$ Ω cm^{-1}), power law exponent factor (s), and activation energy (W) of all glasses at frequency 50 kHz

Sample	ε'	ε''	m	s	$\sigma_{ac} \times 10^{-11}$ (Ω cm^{-1})	$\sigma_{dc} \times 10^{-11}$ (Ω cm^{-1})	W (eV)
0.01BLBD	10.356	0.0399	−0.239	0.6221	16.071	4.32	0.8923
0.1BLBD	10.982	0.0451	−0.159	0.4902	12.091	5.47	1.164
1BLBD	11.344	0.0459	−0.184	0.5115	7.4371	3.33	1.775
2BLBD	11.078	0.0433	−0.780	0.6421	9.4325	5.12	1.349
3BLBD	9.804	0.0348	−0.307	0.5901	11.092	3.91	0.9923

Fig. 3 The frequency dependence $\varepsilon''(\omega)$ of 1BLBD glass for varies temperature

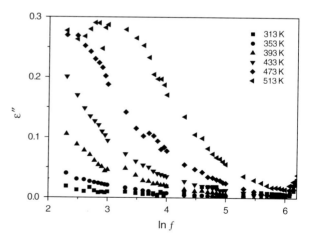

$\varepsilon'' = A\omega^m$. The value of m was calculated from slope of these lines and it is found that values of m at different temperature are negative and less than unity [13].

The temperature-dependent dielectric constant and dielectric loss are given in Figs. 4 and 5. The dielectric constant (ε') and dielectric loss (ε'') of the sample increase with increase in temperature and the variation of ε' with temperature was more. A similar behavior was exhibited by all other glasses. The slow variation of dielectric constant with temperature is the usual trend in ionic conducting glasses. The increasing values of ε' and ε'' with increase in temperature are normally associated with the decrease in bond energies [11]. While increase in temperature, the intermolecular forces and thermal agitation get increased, and hence they strongly change the orientational vibrations due to the fact that the dielectric constant values become larger at lower frequency and at high temperature. It could be suggested that at the lower temperature, the contribution of the electronic and ionic components to the total polarizability is small.

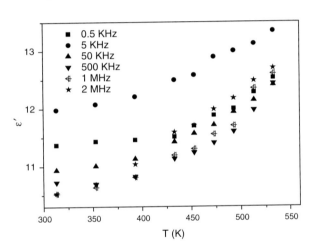

Fig. 4 The temperature dependent $\varepsilon'(\omega)$ for varies frequencies

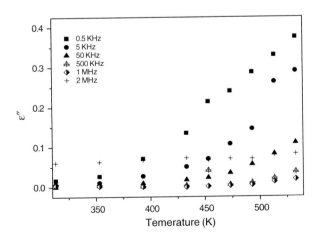

Fig. 5 The temperature dependent $\varepsilon''(\omega)$ for varies frequencies

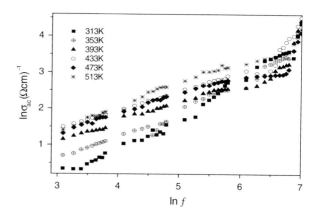

Fig. 6 The frequency dependent AC conductivity of 1BLBD glass

3.3 Frequency and Temperature Dependence AC Conductivity

Figure 6 shows the frequency dependence of AC conductivity (σ_{ac}) at different temperature for the glass 1BLBD. It is observed that the conductivity dispersion, which is strongly dependent on frequency and temperature. The σ_{ac} values are increased with increase in frequency and as well as temperature. This could be suggested that the mechanism responsible for the AC conductivity is hopping conduction. The temperature-dependent AC conductivity of 1BLBD glass at different frequency is given in Fig. 7 and from this shows that conductivity increases with increase in temperature and it reveals that the activation energy is temperature dependent which is the characteristic of small polaron hopping conduction mechanism in present glasses [12]. The logarithmic conductivity in temperature range (313–533 K) exhibiting a linear dependence on reciprocal temperature [13, 14]. The activation

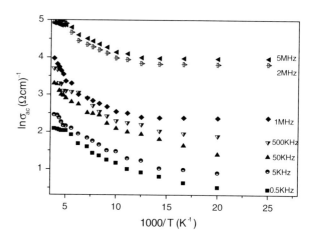

Fig. 7 The temperature dependence conductivity of 1BLBD glass

of electrical conduction decreases and electrical conductivity increases. The value of AC conductivity (σ_{ac}), DC conductivity (σ_{dc}), power law exponent factor, and activation energy (W) are given in Table 1. It is observed that the values of σ_{ac} are strongly dependence as a function of temperature at the higher temperature, whereas it is independent at the lower temperature [15, 16].

4 Conclusion

The Dy^{3+} doped alkali borate glasses were prepared and the structural and dielectric properties were discussed. The FTIR spectra reveal the presence of the fundamental stretching vibrations of the various borate networks in the prepared glasses. The slow variation of dielectric constant with temperature is the usual trend in ionic conducting glasses. The electrical conductivity of present glasses revealed a frequency dependence and $\sigma_{ac}(\omega)$ increasing with increasing frequency conductivity increase with increase in temperature and it reveals that the activation energy is temperature dependent which is the characteristic of small polaron hopping conduction mechanism in present glasses. The activation of electrical conduction decreases and electrical conductivity increases.

Acknowledgments One of the authors (S. Arunkumar) is grateful to University Grants Commission (UGC) for the award of the Rajiv Gandhi National Fellowship (RGNF) to pursue Ph.D degree.

References

1. Prashant Kumar M, Sankarappa T (2009) J Non-Cryst Solids 355:295–300
2. Tarsikka PS, Singh B (2009) Asia J Chem 21(10):162–166
3. Krishna Murthy M, Murthy KSN, Veeraiah N (2000) Bull Mater Sci 23(4):285–293
4. Karunakaran RT, Marimuthu K, Surendra Babu S, Arumugam S (2010) J Lumin 130:1067–1072
5. Arul Rayappan I, Marimuthu K, Surendra Babu S, Sivaraman M (2010) J Lumin 130:2407–2412
6. Chakrabarti R, Das M, Karmakar B, Annapurana K (2007) J Non-Cryst Solids 353:1422–1426
7. Alemi AA, Sedghi H, Mirmohseni AR, Golsanamlu V (2006) Bull Mater Sci 29(1):55–58
8. Lucacel RC (2007) J Non-Cryst Solids 353:2020–2024
9. Gedam RS, Deshpande VK (2009) Bull Mater Sci 32(1):83–87
10. Nagaraja N, Sankarappa T, Prashant Kumar M (2008) J Non-Cryst Solids 354:1503–1508
11. Mohmoud KH, Abdel Rahim FM, Atef K, Sadeek YB (2011) Curr Appl Phys 11:55–60
12. Prashant Kumar M, Sankarappa T, Santhosh Kumar J (2008) J Alloys Compd 464:393–398
13. Sankarappa T, Prashant Kumar M, Devidas GB, Nagaraja N, Ramakrishnareddy R (2008) J Mol Struct 889:308–315
14. Inder P, Ashish A, Sujata S, Anshu S, Neetu A (2009) J Alloys Compd 472:40–45
15. Veeranna Gowda VC, Anavekar RV (2005) Solid State Ion 176:1393–1401
16. El-Desoky MM, Al-Shahrani A (2006) Physica B 383(2):163

Part XI
Chemistry

Crystal Growth and Characterization of Biologically Essential Drug Materials

K. Bhavani, K. Sankaranarayanan and S. Jerome Das

Abstract Biologically essential drug material of Paracetamol was crystallized with the nonessential amino acid in equimolar ratio by slow solvent evaporation method. Good quality of transparent drug crystals was grown within 3 weeks and its stability was analyzed by thermal studies. The unit cell parameters of the grown crystals were determined by single X-ray diffraction studies. FTIR, optical, and dielectric studies were also carried out to the grown biomolecular crystal which is very important in the drug development phase.

Keywords Paracetamol • Drug crystals • Stability • S-XRD • FTIR • Optical • Dielectric

1 Introduction

Active pharmaceutical ingredient (API) of Paracetamol has been extensively used as antipyretic and analgesic drugs [1, 2]. Paracetamol is a nonsteroidal, anti-inflammatory drug (NSAID) and it is also an important intermediate in manufacturing of azo-dyes and photographic chemicals [3] and it reduces nephrotoxicity [4]. Most of the drugs are delivered to the patients in crystalline form [5]. Crystallization of an API is not only an important art but also provides knowledge about physical properties of crystalline nature such as crystal form, shape,

---

K. Bhavani (✉)
Department of Physics, SRM University, Vadapalani, Chennai 600026, India
e-mail: bhavani.k2006@yahoo.co.in

K. Sankaranarayanan
Department of Physics, Alagappa University, Karaikudi, India

S. Jerome Das
Department of Physics, Loyola College, Chennai, India

S. Sathiyamoorthy et al. (eds.), *Emerging Trends in Science, Engineering and Technology*, Lecture Notes in Mechanical Engineering, DOI: 10.1007/978-81-322-1007-8_60, © Springer India 2012

size and system, and so on. Owing to the medicinal and biological importance of Paracetamol, it was crystallized and this research work system can be employed at various stages of drug development phase.

2 Experimental Section

2.1 Solubility

In pharmaceutical substance solubility data are an essential parameter to achieve required concentration of drug in systemic for pharmacological response. The drug solubility in saturated solution is a static property that relates more closely to the bioavailability rate [6]. Active ingredients are poorly soluble in water [7], but they dissolve easily in ethanol. The solubility of Paracetamol in ethanol was done by polystat thermostatic bath using water as a circulation fluid.

The solution was prepared and maintained at 33 °C with continuous stirring to ensure homogenous temperature and concentration over the entire volume of the solution. On reaching saturation, the content of the solution analyzed gravimetrically. This process was repeated for every 5 °C in water from 33 to 43 °C. The solubility curve of Paracetamol in ethanol was shown in Fig. 1.

2.2 Crystal Growth from Drug Material

In the present study, we investigate the active ingredient crystal growth with non-essential α-amino Aspartic acid. It can be synthesized from central metabolic pathway intermediates in humans. Aspartic acid is found in luncheon meats, sausage meat, sugarcane, and molasses from sugar beets [8].

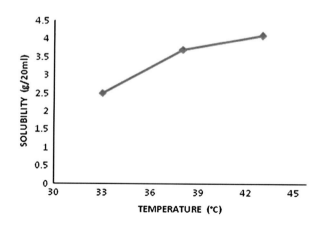

Fig. 1 Solubility of paracetamol in ethanol

Fig. 2 Paracetamol crystals

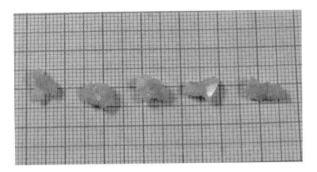

A saturated solution of Paracetamol with Aspartic acid in equimolar ratio was prepared separately using ethanol as a solvent. Now this solute–solvent mixture was transferred to crystal growth vessels and the crystallization was allowed to taken place by slow evaporation technique under room temperature. Irregular shape of white color was harvested within 3 weeks of growth period is shown in Fig. 2.

The final product drug crystal has the dimensions of 5 × 3 × 2 mm^3 and it was characterized by various analytical techniques to know the physicochemical properties, since it is an essential parameter for the design and formulation of the oral dosage [9].

3 Experimental Results and Discussion

3.1 Single X-Ray Studies

Single crystal X-ray diffraction analysis of the grown crystal has been carried out to identify the crystal structure and to get the lattice parameters using Bruker-Kappa APEX2 diffractometer with MoKα ($\lambda = 0.71073$ Å) radiation. The calculated cell parameters are $a = 7.13$ Å, $b = 9.41$ Å, $c = 11.73$ Å, $\beta = 97.42°$, $V = 780$ Å^3 and space group $P_{21/c}$. It was observed from the single XRD measurement that the grown Paracetamol crystals belong to monoclinic system. In the pharmaceutical industry, monoclinic system is thermodynamically stable at room temperature in all atmospheres and is one of the essential required physicochemical properties in this field [10].

3.2 FTIR Spectral Analysis

Fourier transform infrared (FTIR) spectroscopy accurately detects crystallinity ranging from 1 to 99 % in pure material [11]. The recorded FTIR spectrum of Paracetamol is shown in Fig. 3.

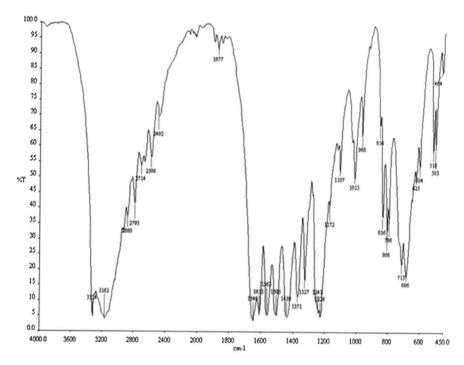

Fig. 3 FTIR spectrum of paracetamol crystal

Table 1 gives the observed vibration wave number and their tentative assignments of Paracetamol crystal.

Table 1 The observed vibration wave number and their assignment of Paracetamol crystal

Paracetamol	
Observed wave number cm^{-1}	Assignments
3,325	N–H hydroxyl
3,162	O–H hydroxyl
2,880	C–H hydroxyl
2,793	C = O and O–H carboxylic acid
2,714	C = O and O–H stretching
1,649	C = O amide
1,610	C–C aromatic function
1,563	C–N amide
1,505	C–C aromatic function
1,439	C–C aromatic function
1,371	CH$_3$ rocking
1,327	O–H stretching
1,226	C–N amide
1,015	C–C–C aromatic function

Fig. 4 UV-Vis spectrum of paracetamol crystal

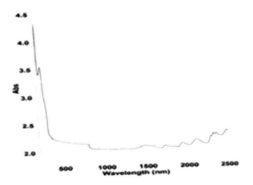

From the FTIR analysis, it shows active ingredients and amino acid which were presented in the grown crystal [12].

3.3 UV-Vis Spectral Analysis

The UV-Vis spectrum was recorded using Varian Cary 5E UV-Vis spectrometer in the range between 500 and 2,500 nm. Figure 4 shows the UV-Vis absorption spectrum of the grown biomedicine crystal. From the obtained UV results illustrate clearly that the grown crystal was highly transparent between 800 and 1,500 nm ranges.

The maximum absorbance of Paracetamol was found to be 250 and 300 nm, it shows the low concentration of additives which was presented in the crystals which can provide a means of controlling of water contents, crystal energy and order, dissolution rate, and bioavailability. Such parameters are very useful in quality control testing of drugs and their products [13].

3.4 Thermal Analysis

Thermo gravimetric and differential thermal analyses (TG/DTA) were carried out simultaneously for the grown drug crystals. A powder sample of API crystal 9.87 mg was used for the analysis in the nitrogen atmosphere at the heating rate of 20 K/min. Figure 5 shows the TG/DTA curves of the Paracetamol crystals.

From the thermal analysis, results reveal that no loss of weight observed around 100 °C showing the absence of any absorbed water molecules in the sample, but the crystals were starts to lose weight at 250 °C and the weight loss ends at 350 °C, due to the decomposition of the compound. In the DTA, curve of Paracetamol crystal shows two significant exothermic peaks at 169.6 and 335.4 °C. The first peak was the melting point of Paracetamol and the second one

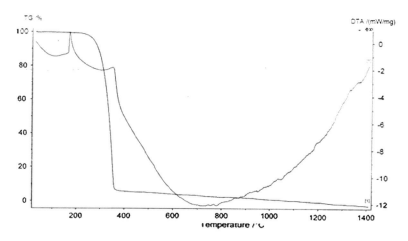

Fig. 5 TG/DTA analysis of paracetamol crystal

may assign to solvent molecules evaporation. The preferred solid form of the compounds supports the previous literatures [14, 15].

3.5 Dielectric Studies

Good quality of Paracetamol crystals was selected for dielectric measurements using HIOKI 3532-50 LCR HITESTER. The selected sample was cut by a diamond saw and polished using paraffin oil. Figure 6 shows the variation of dielectric constant with frequency.

The dielectric constant has high values in the lower frequency region and then it decreases with the applied frequency. The high value of dielectric constant at low

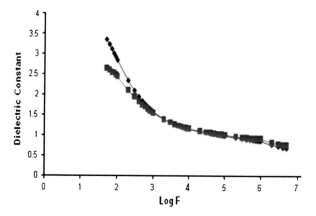

Fig. 6 Dielectric constant of paracetamol

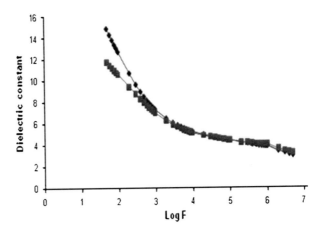

Fig. 7 Dielectric loss of paracetamol

frequencies may be due to the presence of all the four polarizations namely space charge, orientation, electronic, and ionic polarization and its low value at higher frequencies may be due to the loss of significance of these polarizations gradually [16]. The dielectric loss of Paracetamol crystals was also studied as a function of frequency as shown in the Fig. 7. These curves suggest that dielectric loss was strongly depends on the frequency of the applied field similar to that of dielectric constant.

4 Conclusions

Medicinal cum biological importance of drug material crystals was successfully grown from aqueous solution at room temperature. Single X-ray diffraction results reveal that the grown drug crystal belongs to monoclinic system. The FTIR spectrum confirms Paracetamol and amino acid were present in the sample. From the UV spectrum results the low concentration of additives was present in the crystal. The thermal analysis reveals that the biological crystal has the maximum chemical stability at 250 °C was a very important factor for drug dosage, manufacturing, and packaging. Dielectric studies indicate that the dielectric constant and loss depends on the frequency of the applied field. This research attempt could be useful for drug discovery, delivery, development, and synthesis process.

References

1. Sweetman SCM (2002) The complete drug reference, 33rd edn. Pharmaceutical Press, London
2. Hardmam JG, Limbird LE (1996) Goodman and Gilman's the pharmacological basis of therapeutics, 9th edn. Mc Graw-Hill Companies, New York, USA

3. Granberg RA, Rasmuson AC (1999) Solubility of paracetamol in pure solvents. J Chem Eng Data 44(6):1391–1395
4. Cazin M, Cazin YC, Lhermitte M, Paluszerak D, Binochi A (1987) Acta Ther 13:224
5. Li T, Li B, Tomassone MS (2006) Chemical Engineering Science 61:5159–5169
6. Rathore KS et al (2007–08) A review on hydrops, compounds for solubility enhancement of poorly water soluble substances. The pharma review
7. Banerjee R, Bhatt PM, Ravindra NV, Desiraju GR (2005) Saccharin salts of active pharmaceutical ingredients, their crystal structures, and increased water solubilities. Cryst Growth Des 5(6):2299–2309
8. The Merck Index, 862 Aspartic acid (11th edn.) 1989, p 132. ISBN 0-911910-28-X
9. Chaturvedi AK, Amita V (2012) Solubility enhancement of poorly water soluble drugs by solid dispersion. Int Pharm Sci Res 3:26–34
10. Marti E, Kaiserberger E, Kaiser G (eds) (2000) Thermoanlytical crystallization of pharmaceuticals, Netzsch annual 2000, Netzsch–Geraetebu Gmbh 12 Selb, p 29
11. Broman E, Khoo C, Taylor LS (2001) A comparison of alternations ploymer excipient and processing methods for moving solid dispersion of poorly water soluble drug. Int J Pharm 222(1):139–151
12. Ciurczak EW, Drennen JK III (2001) Pharmaceutical and medical applications of near-infrared spectroscopy. Marcell Dekker, NewYork, pp 73–105
13. Murtaza G et al (2011) Development of a UV-spectrophotometric method for the simultaneous determination of aspirin and paracetamol in tablets. Sci Res Essays 6(2):417–421
14. Byrn SR, Pfeiffer RR, Stowell JG (1999) Solid state chemistry of drugs. SSCI, West Lafayette
15. Brittain H (ed) (1999) Polymorphism in pharmaceutical solids, vol 95. Marcel Dekker, New York
16. Smyth CP (1965) Dielectric behavior and structure. McGraw-Hill, New York, p 132

Indigo Dye Removal by Using Coconut Shell Adsorbent and Performance Evaluation by Artificial Neural Network

S. K. Deshmukh

Abstract The present investigation was carried out to study the removal of indigo dye using a nonconventional adsorbent coconut shell. The adsorption isotherm is obtained in batch reactor. It is found that dye adsorption capacity (pollutant removal efficiency) of a steady system depends on adsorbent material, pH of solution, adsorbent dose, particle size, initial concentration, and type of activation. It is observed that the process uptake gives insufficiency information to completely understand the adsorption mechanisms, when database is generated for similar characteristics trends in adsorption. In consistency in the characteristic of several adsorbents that is being reported. The attempt made in proposing the empirical model could not fit the experimental data well, which may be due to nonlinear relationship and incomplete understanding between input and output variables. Artificial neural network has the abilities to relate the input and output variables without having any knowledge on physics of the system provided an accurate and large amount of data on system variable to train the networks is available. In this adsorption studies, we are modeling an adsorption system by using ANN for dyes removal from wastewater.

Keywords Nonconventional adsorbent • Neural network • Indigo dye • Isotherm

1 Introduction

Adsorption is one of the established unit operation used for the treatment of contaminated water, i.e., raw water and wastewater. Dyes used in chemical industries are difficult to remove by conventional wastewater treatment methods since they stable to light, heat, oxidizing agents, and are resistance to aerobic digestion. The presence of dye in water is highly visible and affects water transparency, resulting in reduction of light penetration, gas solubility in water and reduces

---

S. K. Deshmukh (✉)
Faculty Chemical Engineering Department, Jawaharlal Darda Institute of Engineering
and Technology, Yavatmal 445001, Maharashtra, India

S. Sathiyamoorthy et al. (eds.), *Emerging Trends in Science,*
Engineering and Technology, Lecture Notes in Mechanical Engineering,
DOI: 10.1007/978-81-322-1007-8_61, © Springer India 2012

photosynthetic activity. An adsorption process is often used at the end of a treatment sequence for pollution control due to the high degree of purification that can be achieved to avoid hazardous effect for carcinogenic and mutagenic public health problem and aquatics eco-system problem.

The aim of our study was to develop a atificial neural network-based (ANN) adsorption model for the dye removal from industrial wastewater by using low cost active carbon adsorbent from biodegradable material such as coconut shell, as an alternative to the classical molecular sieve adsorbents, and to determine the main parameters which play a important role in the removal of synthetic dyes from industrial wastewater. This low cost adsorbent found to be an effective binder of dyes from its aqueous solution, which is capable of removing 92 % of the dyes present in industrial wastewater.

In this adsorption studies, we are modeling an adsorption system by using ANN for dyes removed from wastewater. We have used a double layer model to fit these experimental adsorption isotherms as it was preceded in the case of dye adsorption. The double layer model is established using a statistical physics treatment. This treatment gives us the expressions of the different model parameters, and thus will allow us to get physical interpretations of the behavior of their values. The advantage to apply this ANN model is to give a physical meaning to the parameters involved in the model and then give new interpretations of the adsorption system. An ANN model involves algorithms under which information is accumulated in programmed objects that are capable of learning through much iteration using simulated or real data. This form of artificial intelligence can handle problems, in which relationships are less known or very complex.

2 Preparation of Bioadsorbent

Two things are mostly essential for the choice of bioadsorbent, they are cost and availability, particularly when the adsorbent used is not changeable or it cannot be recharged. Bioadsorbents used in this respective work are abundantly available in India as well as rest part of the world and these are also cheapest one.

2.1 Coconut Shell

The raw material, i.e., coconut shell is collected and was used as an adsorbent. It was observed that active adsorption sites were occupied by lignin. The coconut shell was first washed thoroughly with distilled water to remove the dirt and other foreign matter and dried at 40 °C for a period of 4–5 h. Then again dry the coconut shell at 200–220 °C for about 2 days in closed furnace. After complete carbonization of coconut shell they are washed with distilled water to remove the ash adhere over the surface of shells. Then again dried in sunlight for 2 days. Then ground for size of adsorbent particle up to 3 mm. There are two activation processes for increasing the active site on the surface of adsorbent, in activation process, pore on the surface is evacuated, and hence increases the surface area for adsorption of dye. There are two methods for the activation of bioadsorbent.

2.2 Acid Treatment

Acid treatment is given by using HNO3 (nitric acid). The 50 gm of corn cob powder sample is treated with 500 ml of 1 N acid in 1,000 ml of conical flask, and then mixture was gently heated on burner up to its boiling stage constantly for 15 min. Washing of treated bioadsorbent is carried out using distilled water, washing is done to maintain the pH in between 6 and 6.5. Washing must be done to remove the traces and lignin along with of the color present. Further, the bioadsorbent is dried in oven at 40°C for one day. This material is used for rest of adsorption studies.

2.3 Steam Treatment

The effective activation occurs by supplying steam to the bioadsorbent at a temperature of 900–1100°C. Hence, the activation capacity of steam is much better than acid. But if the steam temperature is below 900°C the steam activation process is slow. Adsorption of dyes on surface of adsorbent is more in case of acid activated coconut shell as compared to steam activated coconut shell, because nitric acid is to activate the particle of adsorbent more better than simply by steam by producing active sites on surface.

3 Materials and Methods

Adsorption studies were performed using batch technique to obtain rate and equilibrium data. Experiments were carried by shaking 5 gm of coconut shell adsorbent dose with 50 ml of aqueous solution containing known concentration of dye. All experiments were conducted using double distilled water and analytical grade chemicals. Also sample-containing dye was maintain at desired pH by adding 0.1 N HCl or 0.1 N NaOH. All the experiments were performed at room temperature 29°C. The analysis of dye concentration is one by using tree point software used in Raymond Ltd. (denim section), yavatmal for testing Indigo dye concentration in a dye bath is based on principle of spectroscopy and pH was measured by pH meter(Elico Pvt. Ltd, Hyderabad). Electrical Shaker (Remi model, Mumbai) was used for agitating the samples.

4 Model Development Procedure for Adsorption

Understanding the dynamics of batch adsorption systems for modeling is a demanding task due to the strong nonlinearities in the equilibrium isotherms, interference effects of competition of solutes for adsorbent sites, mass transfer resistances between fluid phase and solid phase, and fluid-dynamic dispersion phenomena. The mathematical modeling in the case of adsorption of solids present in the mobile phase involves writing down balance equations for each species in the form of mathematical equations. These equations are complemented by appropriate isotherm, kinetic, or hysteresis relations, which usually involve cross-dependence of the isotherms between species. The main disadvantage of this mathematical modeling methodology is that it results in a complicated mathematical formulation that makes it difficult to be utilized in system design.

In this study, a computer-based model by applying ANN is studied for isothermal adsorption process for dye removal with porous and low-cost adsorbent. An ANN model has been used to predict the fraction of the organic component, which is adsorbed by the active carbon adsorbent. The effect of material (Type of adsorbent), pH, adsorbent dose, particle size, type of activation, and initial concentration of dye in wastewater in this adsorption process is studied. Neural networks can predict any continuous relationship between these inputs and the target, i.e., final concentration for calculating pollutant removal efficiency. Similar to linear or nonlinear regression, artificial neural networks develop a gain term that allows prediction of target variables for a given set of input variables. An ANN model developed for pollutant removal efficiency is described in this work. The procedure of applying neural networks to develop a model for adsorption of dye removal of industrial wastewater has the following steps.

1. Problem formulation
2. ANN training by ANN software.

4.1 Problem Formulation

For batch studies of adsorption, the PRE is considered as function of PRE = f (material, type of adsorbent, pollutant, pH, adsorbent dose, initial conc. of adsorbate, particle size, and type of activation used)

$$\text{Pollutant removal efficiency} = \frac{(C_i - C_0)}{C_i} \times 100$$

where,

C_i Initial steady-state dye concentration of sample
C_o Final steady-state dye concentration of sample.

Other pollutants can be omitted from the above equation, since we are considering only one major pollutant that is dye. A database of 173 data points was collected. The six variables listed in the right hand side of the above equation are considered as input variables while PRE is the output variable.

We have coded for various adsorbents considered here as,

Coconut shell-01

For the sake of identifying the type of activation process, we have coded also these different types of activation as,

Acid activation-10

Steam activation-20

After coding the all-non-numerical data we have proceeded for ANN training by using ANN software.

4.2 Back Propagation Algorithm

1. Initialize Weights
2. Read set of Input/Output data
3. Forward Pass Computation
   Consider

i-Output Layer
j-Hidden Layer
k-Input Layer

Output data of the hidden layer are computed using the equation as follows:

$$h = \sum_{K=i}^{p} W_{ik} X_k \tag{1}$$

$$V_j = \frac{1}{1 + e^{-bh_j}} \tag{2}$$

The actual output of the network, i.e., between hidden layer and output neuron, is computed by

$$g_i = \sum_{j=1}^{m} W_{ij} V_j \tag{3}$$

$$y_i = g_i \tag{4}$$

The neural networks are entirely data driven. The implication is that data must be available to train the network and issue of "how much data" is an important factor. In practice, a good model can be

4. Computation for Backward Pass

$$\text{compute error, } e = y^d - y \tag{5}$$

$$\text{compute } \delta_i = y(1 - y)(y^d - y) \tag{6}$$

5. Update the weights connecting hidden layer to output layers using the following rule

$$w_{ij}(t + 1) = w_{ij}(t) + \eta \delta_i v_j \tag{7}$$

6. Update the weights connecting input layer to hidden layers using the following rule

$$w_{jk}(t + 1) = w_{jk}(t) + \eta \delta_i \delta_j x_k \tag{8}$$

7. Make Square of the error
8. Repeat steps 2–7 for all the given datasets
9. Compute the square mean error using

$$\text{error} = \frac{\sum (\text{error})^2}{N X^2} \tag{9}$$

If square mean error > specified error = 0.001 for the present case, then repeat the steps 2 to 9 else, break. From the generated data (173 sets), 20 datasets were randomly selected as testing sets (approximately 12 % each of total). The rest, 153 datasets were used for ANN training. After every dataset training, ANN weights were adjusted. In the beginning of training, testing error decreased with the training process. Training was continued until testing error did not decrease. After training, both training and test sets were used to verify ANN performance.

4.3 Neural Network Topology

ANN of following architecture is used for prediction.
No of neurons:

Input layer	First hidden layer		Second hidden layer	Output layer	
6	5		5	1	
Data points training	Test	Learning rate	Momentum	Best root mean square	
				Training	Test
153	20	0.3	1,532	0.072831	0.106985

4.4 Validation of Neural Network Model

The primary risk in developing a model is that of over training, a situation in which the neural network starts to reproduce the noise specific to a particular sample in the training data, which may cause it to lose its ability to predict accurately. There are certain techniques that are used to avoid this; the most popular of this is *network validation*. The testing of the network predictivity is done by reserving some of the data, which are excluded from the training datasets. The network is used to predict the outputs for these reserved data records, and the calculated outputs are compared with the observed values. If they are found to be sufficiently close, the network is considered to be sufficiently predictive and the network is said to be validated.

After successful training of neural network, the values of output variable as a function of various input parameters are predicted. The graphs are plotted between predicted and actual values of PRE. The comparison between predicted and experimental values of PRE for training and test data file indicates that the trained artificial neural network can be successfully used as a model for estimation of PRE for this case.

4.5 Prediction Using the Developed ANN Model

In this case, we have given random changes in two or more input variables. The prediction for this type of data points is not possible by an ordinary simple model since these models are flexible to changes in any one or utmost two input variables only. For changes in more than two input variables at a time, a simple model is not tolerable, since adsorption phenomena have a very complex nonlinear relationship. In this case, our ANN model is exploited and is best suited for this case of dye removal from wastewater. By using this model, we got predicted output values close to experimental values.

The table showing the random changes in input variables with predicted outputs is shown as follows.

1st input	2nd input	3rd input	4th input	5th input	6th input	Actual output	Predicted output
2	3.5	2	115	0.71	10	32.17	30.78156
2	6.3	3	75	0.71	10	42.67	40.37279
2	4.9	3	115	0.6	10	31.05	35.12498
2	7.7	2.5	75	1.4	10	34.67	33.75433
2	10.4	2.5	115	1.4	10	33.48	33.60871
Correlation coefficient=							0.845755
Root mean square error=							2.221056
Average % error=							5.16982

5 Results and Discussion

The comparison of actual data obtained from experimental runs and predicted values by using ANN is shown by plotting graphs as below:

1. Effect of adsorbent dose on pollutant removal efficiency (Graphs 1, 2, 3, 4):

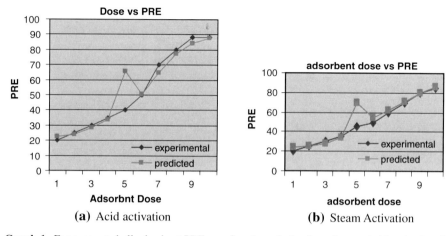

Graph 1 For coconut shell adsorbent PRE as a function of adsorbent dose. **a** Acid activation. **b** Steam activation

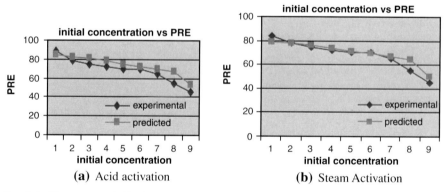

Graph 2 For coconut shell adsorbent PRE as a function of initial concentration. **a** Acid activation. **b** Steam activation

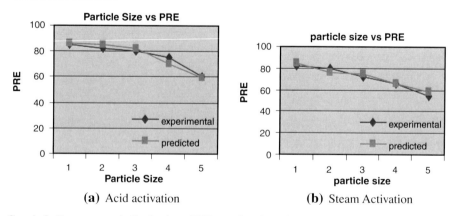

Graph 3 For coconut shell adsorbent PRE as a function of particle size. **a** Acid activation. **b** Steam activation

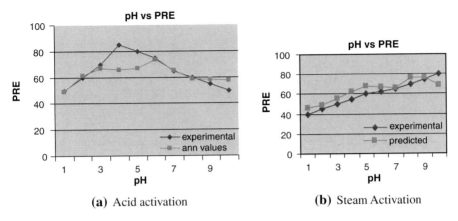

Graph 4 For coconut shell adsorbent PRE as a function of pH. a Acid activation. b Steam activation

6 Conclusion

The study presents the capability of ANN model in capturing intricate relationship between various process parameters affecting the PRE of an adsorption system. The prediction accuracy of the ANN model developed for estimation of pollutant removal efficiency is found to be 94.83 %. ANN has extremely correlated PRE as a function of all the input variables such as type of adsorbent, pH, adsorbent dose, initial conc. of adsorbate, particle size, and type of activation used. The modeling power of ANN promises their use in complex industrial process application with out much difficulty. This work demonstrates that the present ANN model can effectively be used for the prediction of PRE for adsorption of dye from wastewater.

The mean square errors of the overall set of data are better in neural network. However, more number of datasets for training the network is more likely to enhance the overall predictability of the network. Also, ANN is good in interpolation tool and these interpolative predictions by ANN are very close to the experimental data. Thus, it can be concluded that neural network can be treated as a potent means for PRE prediction in a fast and reliable ways, compared to the conventional mathematical modeling.

Structural and Electrical Conductivity of ZnO Nano Bi-pyramids-like Structure

Subbaiyan Sugapriya, Rangarajalu Sriram and Sriram Lakshmi

Abstract In this paper, we report the structural and electrical properties of ZnO nano bi-pyramid-like structure synthesized by a one-step process of hydrothermal redox route. X-ray diffraction (XRD) result shows that the prepared ZnO exhibits hexagonal phase and no secondary phase was observed. The crystallite size has been calculated by Scherrer's equation and was found to be in the range 30–50 nm. SEM images reveal that ZnO nano bi-pyramid-like structure has been formed which has a length of ~4.5 μm and base of ~4 μm. The I–V characteristics have been carried out to study the electrical behavior of the prepared ZnO nano bi-pyramid-like structure.

Keywords Hydrothermal • ZnO nano bi-pyramid • I–V measurement • Electrical conductivity • SEM

1 Introduction

Nanoscale semiconductor materials have attracted great interests of researchers because of their importance not only in fundamental research areas but also in practical applications. Among many nanoscale semiconductor materials, ZnO nanostructures have been studied intensively and extensively over the last decade not only for their remarkable chemical and physical properties, but also for their

---

S. Sugapriya (✉) · R. Sriram · S. Lakshmi
Department of Chemistry, Coimbatore Institute of Technology, Coimbatore,
Tamil Nadu 641014, India
e-mail: sugapriya0314@gmail.com

R. Sriram
e-mail: drrsriram@rediffmail.com

S. Lakshmi
e-mail: sshyamsriram@gmail.com

S. Sathiyamoorthy et al. (eds.), *Emerging Trends in Science,*
Engineering and Technology, Lecture Notes in Mechanical Engineering,
DOI: 10.1007/978-81-322-1007-8_62, © Springer India 2012

current and future diverse technological applications. ZnO is a typical inorganic semiconducting and piezoelectric material; this material has a direct wide band gap of 3.37 eV and a large exciton binding energy of 60 meV at room temperature [1]. It has enormous applications in electronic, optoelectronic, electrochemical, and electromechanical devices [2, 3], ultraviolet (UV) lasers [4], light-emitting diodes [5], field emission devices [6], high performance nanosensors [7], solar cells [8, 9], piezoelectric nanogenerators [10], and nanopiezotronics [11]. In order to grow one-dimensional (1D) ZnO nanostructures, various techniques have been developed for the synthesis of the ZnO nanostructures like wet chemical methods [12], physical vapor deposition [13], pulsed laser deposition [14], sputtering [15], flux methods [16], and electrospinning [17].

2 Preparation

ZnO nano bipyramids have been prepared using the required precursors by chemical method. An aqueous solution of 1 M Zinc acetate dihydrate $\{Zn(CH_3(COO))_2.2 H_2O\}$ was dissolved in water and stirred for about 20 min at room temperature. Sodium hydroxide (NaOH) (0.5 M) was added drop wise to the above-mentioned solution. The color of the solution changed into milky-white color, indicating the formation of ZnO nano particles in the solution. The solution was stirred for 4 h at room temperature. After 4 h, the solution was transferred into a stainless steel autoclave with a Teflon liner, which was then filled with distilled water to 70 % of its capacity. The autoclave was sealed and maintained at 180 °C for 24 h, then allowed cool to room temperature. The supernatants were removed and the deposited precipitate was centrifuged and washed with water and ethanol several times. The samples were then suspended in ethanol and allowed to age for 2 h without stirring. After centrifugation, the samples then dried in oven at 70 °C for 2 h. Then, the as prepared ZnO was annealed at 450 °C for 1 hour.

The structural properties of the sample have been studied using PANalytical X-ray diffractometer. The morphology of the sample has been studied using scanning electron microscope (JEOL Model JSM -6360). Compositional analysis of the samples has been carried out using energy dispersive analysis of X-rays (JEOL Model JSM -6360). The characteristic has been studied using four-probe method.

3 Result and Discussion

Figure 1 shows the X-ray diffraction (XRD) patterns of the ZnO nano structure. The diffraction peaks at 2θ (degrees) of 31.63°, 34.61°, 36.32°, 47.66°, 56.94°, 62.97°, 66.57°, 68.12°, 69.48°, 72.11° and 72.26° are respectively indexed as the (100), (002), (101), (102), (110), (103), (200), (112), (201), (004), and (202) planes of ZnO.

Structural and Electrical Conductivity of ZnO Nano Bi-pyramids-like Structure

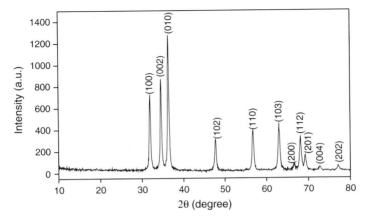

Fig. 1 X-ray diffraction pattern of ZnO nano bi-pyramid-like structure samples

All the diffraction peaks in the 2θ range measured correspond to the hexagonal structure of ZnO with lattice constants $a = 3.253 Å$ and $c = 5.214 Å$ and are in good agreement with those on the standard data card (JCPDS card No. 36–1451). The sharpness of the diffraction peaks suggests that the product is well crystallized. The crystallite size of ZnO is calculated using Scherrer's equation

$$D = \frac{K\lambda}{\beta \cos \theta}$$

where, D is the grain size, K is a constant taken to be 0.94, λ is the wavelength of the X-ray radiation, β is the full width at half maximum and θ is the angle of diffraction. The crystallite size has been calculated and is found to be in the range 30–50 nm.

Figure 2a,b,c shows the SEM image of the ZnO sample. The image shows that the sample has bi-pyramid-like structure. The sizes of the bi-pyramid-like structures are about ~5.5 μm on an average. The length and breadth of the bi-pyramid-like structure are about ~4.5 and ~4 μm, respectively.

Fig. 2 a, b, c SEM images of as prepared ZnO nano bi-pyramid-like structure

Energy dispersive X-ray analysis (EDS) of ZnO nano bi-pyramid-like structures is shown in Fig. 3. The chemical constituents present in the ZnO sample are of Zn-49.65 and O−50.35 %. In the EDS, Zn and O are the elements detected, indicating that the sample is highly pure.

For I–V measurements in bulk, pellets of 13 mm diameter and thickness ≈1 mm were prepared under a load of 5 tons. These pellets were used in four-probe method. The I–V characteristics of the samples were studied at the room temperature as well as at various temperatures (80, 125, 200, 273, and 320 K) using the four-probe method. The temperature dependence of resistivity was measured at constant current by varying the temperature continuously. The silver paste was used for ohmic contact between the sample and the copper probes. DC voltage across the electrodes was measured by varying the current. I–V plots are shown in Fig. 4a show the temperature dependence of I–V characteristic of ZnO nano bi-pyramid-like structure as a representative case.

I–V characteristics of the samples are measured in presence of argon gas at low as well as at high temperatures. Liquid nitrogen is used for lowering the temperature. Argon gas is necessary to eliminate the moisture content otherwise the moisture present in air will change the electrical properties of the sample (especially at low temperatures). Figure 4a depicts the characteristics of pure ZnO at the constant current, where the voltage decreases as we go on lowering the temperature. The conductivity values of the sample are plotted against to the temperature (K^{-1}), as shown in the Fig. 4b, c. This figure depicts that the conductivity increases with temperature. The conductivity of the ZnO increases with increasing temperature values are given in the Table 1.

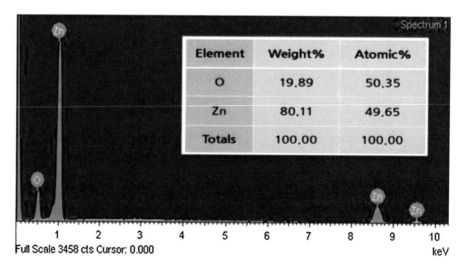

Fig. 3 EDS spectra of as prepared ZnO nano bi-pyramid-like structure sample

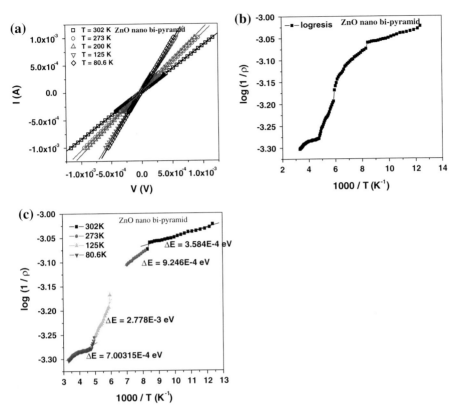

Fig. 4 a I–V characteristics. b, c Conductivity Vs Temperature of as prepared ZnO nano bi-pyramid-like structure sample

Table 1 Temperature with conductivity

T(K)	σ (Ω^{-1})	R (Ω)
80.6	0.85095	1.1751
125	1.06281	0.9429
200	1.14752	0.8714
273	1.72682	0.5791
302	1.82652	0.5475

(When a semiconductor is extremely cold, almost all electrons are held tightly by individual atoms so hard to move through the material. When a semiconductor is heated, the heat energy knocks loose some of the electrons thus it can move through the material easily. Thus, higher temperature means greater conductivity).

4 Conclusion

The structural and electrical properties of ZnO nano bi-pyramid-like structure was synthesized by a one-step process of hydrothermal redox route. The size of the bi-pyramid-like structure is about ~5.5 μm on an average. The length and breadth of the bi-pyramid-like structure are about ~4.5 and ~4 μm, respectively. The conductivity of the ZnO nano bi-pyramid-like structure was estimated from the I–V characteristic from lower temperature to higher temperature. The conductivity of the sample increases with increasing temperature. The increase in the conductivity can be attributed to the increase in the electrical conductivity of the prepared sample.

References

1. Ozgur U, Alivov YI, Liu C, Teke A, Reshchikov MA, Dogan S, Avrutin V, Cho SJ, Morkoc H (2005) A comprehensive review of ZnO materials and devices. J Appl Phys 98:041301
2. Wang ZL (2009) ZnO nanowire and nanobelt platform for nanotechnology. Mat Sci Eng R 64:33–71
3. Wang ZL (2009) Ten years' venturing in ZnO nanostructures: from discovery to scientific understanding and to technology applications. Chinese Sci Bull 54:4021–4034
4. Govender K, Boyle DS, O'Brien P, Binks D, West D, Coleman D (2002) Room-temperature lasing observed from ZnO nanocolumns grown by aqueous solution deposition. Adv Mater 14:1221–1224
5. Park WI, Yi GC (2004) Electroluminescence in n-ZnO nanorod arrays vertically grown on p-GaN. Adv Mater 16:87–90
6. Wang WZ, Zeng BQ, Yang J, Poudel B, Huang JY, Naughton MJ, Ren ZF (2006) Aligned ultralong ZnO nanobelts and their enhanced field emission. Adv Mater 18:3275–3278
7. Zhou J, Gu YD, Hu YF, Mai WJ, Yeh PH, Bao G, Sood AK, Polla DL, Wang ZL (2009) Gigantic enhancement in response and reset time of ZnO UV nanosensor by utilizing Schottky contact and surface functionalization. Appl Phys Lett 94:191103
8. Weintraub B, Wei YG, Wang ZL (2009) Optical fiber/nanowire hybrid structures for efficient three-dimensional dye-sensitized solar cells. Angew Chem Int Ed 48:8981–8985
9. Wei YG, Xu C, Xu S, Li C, Wu WZ, Wang ZL (2010) Planar waveguide-nanowire integrated three-dimensional dye sensitized solar cells. Nano Lett 10:2092–2096
10. Yang RS, Qin Y, Dai LM, Wang ZL (2009) Power generation with laterally packaged piezo-electric fine wires. Nat Nanotechnol 4:34–39
11. Wang ZL (2008) Towards self-powered nanosystems: from nanogenerators to nanopiezotronics. Adv Funct Mater 18:3553–3567
12. Vayssieres L, Keis K, Lindquist SE, Hagfeldt A (2001) Purpose-built anisotropic metal oxide material: 3D highly oriented microrod array of ZnO. J Phys Chem B 105:3350–3352
13. Yao BD, Chan YF, Wang N (2002) Formation of ZnO nanostructures by a simple way of thermal evaporation. Appl Phys Lett 81:757–759
14. Hong JI, Bae J, Wang ZL, Snyder RL (2009) Room temperature, texture-controlled growth of ZnO thin films and their application for growing aligned ZnO nanowire arrays. Nanotechnology 20:085609
15. Chiou WT, Wu WY, Ting JM (2003) Growth of single crystal ZnO nanowires using sputter deposition. Diam Relat Mater 12:1841–1844
16. Xu CK, Xu GD, Liu YK, Wang GH (2002) A simple and novel route for the preparation of ZnO nanorods. Solid State Commun 122:175–179
17. Lin D, Wu H, Pan W (2007) Photoswitches and memories assembled by electrospinning aluminum-doped zinc oxide single nanowires. Adv Mater 19:3968–3972

Part XII
Mathematics

An Ordering Policy for Deteriorating Items with Quadratic Demand, Permissible Delay, and Partial Backlogging

K. F. Mary Latha and R. Uthayakumar

Abstract This paper develops an economic order quantity (EOQ) model for deteriorating items with time-dependent quadratic demand and permissible delay in payments. Shortages are allowed and are partially backlogged. The objective is to develop an optimal policy that minimizes the total cost.

Keywords Inventory · Quadratic demand · Permissible delay · Shortages · Partial backlogging

1 Introduction

In today's globally challenging environment, customer satisfaction bounds to play a crucial role to enhance trade. Settlement of accounts is one of the factors which influence it. In traditional inventory models, either in deterministic or probabilistic, it is often assumed that payment will be made to the supplier for the goods, immediately after receiving them. But in reality this is not so. In order to promote demand and market share or decrease inventories of certain items, a supplier provides credit period to the customers. During that period, the customer earns interest on the payment received for the goods sold and thus enhances revenue. Therefore, customers prefer to delay payment until the deadline given by the supplier. Goyal [1] was the first to develop an inventory model with permissible delay in payments. A more general economic order quantity (EOQ) model with permissible delay in payments, price-discount effect, and different types of demand rate were developed by Sana and Chaudhury [2]. Recently, a permissible delay model for noninstantaneous deteriorating items with price and time-dependent demand and partial backlogging has been developed by Reza and Isa [3].

In formulating inventory models, deterioration of items must be given due to consideration. During the storage period, items like foodstuff, pharmaceuticals,

---

K. F. Mary Latha (✉)
Jayaraj Annapackiam College for Women, Periyakulam, Theni, Tamil Nadu, India
e-mail: kfm.latha@gmail.com

R. Uthayakumar
The Gandhigram Rural Institute-Deemed University, Gandhigram, Dindigul, Tamil Nadu, India

---

S. Sathiyamoorthy et al. (eds.), *Emerging Trends in Science,*
Engineering and Technology, Lecture Notes in Mechanical Engineering,
DOI: 10.1007/978-81-322-1007-8_63, © Springer India 2012

chemicals, etc., deteriorate significantly with time which in turn decreases the amount or value of these products. Ghare and Schrader [4] developed a model for an exponentially decaying inventory. Researchers [4–6] relaxed the assumption of constant deterioration rate and developed models by considering the varying rate of deterioration. Deteriorating inventory models with trended demand were developed by researchers [7–15]. Inventory models with exponentially time-varying demand patterns were developed by Jalan and Chaudhuri [16]. Sometimes stock may be inadequate to fulfill the customer's demand. In stock out period, if the customers prefer to wait for the items they can be backlogged; if they move to other suppliers the demand is lost. Several researchers have developed models under the assumption that stock out items are partially backordered. Many EOQ models were developed in the literature considering constant demand focusing on the control of inventories. Of late, many researchers [17–23] etc., have concentrated on realistic models with time-varying demand. The review of the literature shows that the researchers have focused only on linear and exponential time-dependent demand. A linearly time-varying demand indicates an unrealistic uniform change in demand rate of the item per unit time. Also, the exponentially time-varying demand indicates an improbable rapid change in demand rate. Researchers [24, 25] have developed EOQ models with quadratic time-dependent demand. Recently Khanra et al. [26] developed an EOQ model for deteriorating items, considering quadratic demand with permissible delay in payments without shortages.

This paper focuses on an EOQ model for deteriorating items considering time-dependent quadratic demand. Shortages are allowed and are partially backlogged. Among the various time-varying demands in EOQ models, the more realistic demand approach is to consider a quadratic time-dependent demand rate, because it represents both rise and fall in demand. The demand rate is of the form $D(t) = a + bt + ct^2$. Here, $c = 0$ represents a linear demand rate and $b = c = 0$ represents the constant demand rate. In addition, permissible delay in payment is taken into consideration. An algorithm is presented to derive the optimal replenishment policy when the total cost is minimized. Numerical examples are provided to illustrate the optimization procedure.

2 Assumptions and Notations

2.1 Assumptions

1. The demand rate for the item is represented by a quadratic and continuous function of time.
2. Time horizon is infinite.
3. The lead time is zero and the replenishment rate is infinite i.e., replenishment is instantaneous.
4. Shortages are allowed to occur. Only a fraction δ $(0 \leq \delta \leq 1)$ of it is backlogged and the remaining fraction $(1 - \delta)$ is lost.

An Ordering Policy for Deteriorating Items with Quadratic Demand 675

5. The constant rate of deterioration is known and only applied to on hand inventory.

2.2 Notations

$D(t)$ The time-dependent demand rate is $D(t) = a + bt + ct^2, a > 0, b \neq 0, c \neq 0$.

Here a is the initial rate of demand, b is the rate with which the demand rate increases. The rate of change in the demand rate itself changes at a rate c.

A Cost per replenishment order.
p Per unit purchase cost of the item.
h_p Inventory holding cost (excluding interest charges) per rupee of unit purchase cost per unit time.
s Per unit shortage cost per unit time.
π Per unit opportunity cost due to lost sales.
θ Constant deterioration of an item.
I_p Interest charges per rupee investment in stock per year.
I_e Interest earned per rupee in a year.
M Permissible period of delay in settling the accounts with the supplier.
T Time interval in year between two consecutive orders.
t_1 Time at which the inventory level becomes zero.

3 Model Formulation

Consider an inventory system in which I_m units arrive the system at the beginning of each cycle. During the interval $[0, t_1]$, the inventory depletes due to demand and deterioration and it becomes zero at time t_1. The model is depicted in Fig. 1.

The instantaneous inventory level at any time t during the cycle time T can be represented by the following differential equations with $I(0) = I_m$, $I(t_1) = 0$.

$$\frac{dI(t)}{dt} + \theta I(t) = -(a + bt + ct^2); \quad 0 \leq t \leq t_1 \tag{1}$$

$$\frac{dI(t)}{dt} = -a\delta; \quad t_1 < t \leq T \tag{2}$$

Using the boundary conditions the solutions of the above differential equations are

$$\begin{aligned} I(t) &= \frac{1}{\theta^3} \left[\{\theta^2 \left(a + bt_1 + ct_1^2\right) - \theta \left(2ct_1 + b\right) + 2c\} e^{\theta(t_1 - t)} \right. \\ &\quad \left. - \{\theta^2 \left(a + bt + ct^2\right) - \theta \left(2ct + b\right) + 2c\} \right] \end{aligned} \tag{3}$$

$$I(t) = -a\delta (t - t_1) \tag{4}$$

Fig. 1 Graphical representation of inventory system

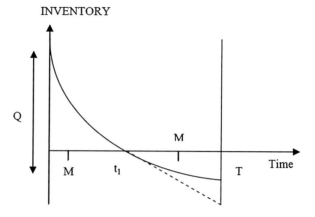

The maximum inventory level is given by $I_m = I(0)$

$$I_m = \frac{1}{\theta^3}\left[\left\{\theta^2\left(a + bt_1 + ct_1^2\right) - \theta\left(2ct_1 + b\right) + 2c\right\}\left(e^{\theta t_1} - 1\right) - \left\{a\theta^2 - b\theta + 2c\right\}\right] \quad (5)$$

The maximum amount of shortage to be backlogged is

$$I_b = a\delta(T - t_1) \quad (6)$$

Based on the assumptions and description of the model, the total annual relevant cost is the sum of the following elements.

1. **Deterioration Cost:** Number of deteriorated items in $[0, t_1)$ is

$$I(0) - \int_0^{t_1} D(t)dt = \frac{1}{\theta^3}\left[\left\{\theta^2\left(a + bt_1 + ct_1^2\right) - \theta\left(2ct_1 + b\right) + 2c\right\}e^{\theta t_1} - \left(a\theta^2 - b\theta + 2c\right)\right]$$
$$- \left(at_1 + \frac{bt_1^2}{2} + \frac{ct_1^3}{3}\right)$$

Deterioration cost for the cycle $[0, T]$

$$= p\left(\frac{1}{\theta^3}\left[\left\{\theta^2\left(a + bt_1 + ct_1^2\right) - \theta\left(2ct_1 + b\right) + 2c\right\}e^{\theta t_1} - \left(a\theta^2 - b\theta + 2c\right)\right]\right.$$
$$\left. - \left(at_1 + \frac{bt_1^2}{2} + \frac{ct_1^3}{3}\right)\right) \quad (7)$$

2. **Holding Cost:** Inventory occurs during period t_1, therefore the holding cost in the interval $[0, t_1)$ is $C_h = h \int_0^{t_1} I(t)dt$ where $h = ph_p$

$$= \frac{h}{\theta^4} \left[\{\theta^2 (a + bt_1 + ct_1^2) - \theta (2ct_1 + b) + 2c\} \left[e^{\theta t_1} - 1 \right] \right.$$
$$\left. - \left\{ \theta^2 \left(a + \frac{bt_1}{2} + \frac{ct_1^2}{3} \right) - \theta (ct_1 + b) + 2c \right\} \theta t_1 \right] \tag{8}$$

3. **Shortage Cost:** During the shortage period there are two cases need to be considered. They are depicted in Fig. 1.
   Shortage cost over the period $[t_1, T)$ is

$$C_s = s \int_{t_1}^{T} -I(t)dt = \frac{sa\delta}{2} (T - t_1)^2 \tag{9}$$

4. **Ordering Cost:** Since replenishment is done at the start of the cycle,

$$\text{ordering cost is } C_r = A \tag{10}$$

5. **Opportunity Cost:** Opportunity cost due to lost sales during the replenishment cycle is

$$C_0 = \pi \int_{t_1}^{T} a(1 - \delta)dt = \pi a (1 - \delta) (T - t_1) \tag{11}$$

6. **Interest Payable: Case 1:** $M \leq t_1$ Since the credit period is shorter than or equal to the replenishment cycle time, the product still in stock is assumed to be financed with an annual rate I_p and the interest payable is

$$\text{IP}_1 = pI_p \int_{M}^{t_1} I(t)dt$$
$$= \frac{pI_p}{\theta^4} \left[\{\theta^2 (a + bt_1 + ct_1^2) - \theta (2ct_1 + b) + 2c\} \left[e^{\theta(t_1 - M)} - 1 \right] \right.$$
$$\left. -\theta (t_1 - M) \left\{ \theta^2 \left(a + \frac{b (t_1 + M)}{2} \right) + \frac{c (t_1^2 + t_1 M + M^2)}{3} \right\} \right.$$
$$\left. -\theta (c (t_1 + M) + b) + 2c \right]$$

Case 2: $M > t_1$ In this case, the buyer pays no interest for the items.

7. **Interest Earned: Case 1:** $M \leq t_1$ As the length of the period with positive inventory stock of the item is larger than the credit period, the buyer can use the sale revenue to earn interest with an annual rate I_e in $[0, t_1)$.
   Therefore the interest earned is

$$\text{IE}_1 = pI_e \int_{0}^{t_1} (t - t_1)D(t)dt = \frac{pI_e t_1^2}{12} \left(6a + 2bt_1 + ct_1^2 \right) \tag{12}$$

Case 2: $M > t_1$ In this case, the buyer earns the interest on the sales revenue during the period $[0, M]$.

$$IE_2 = pI_e \left[\int_0^{t_1} (t - t_1)D(t)dt - (M - t_1) \int_0^{t_1} (t - t_1)D(t)dt \right]$$

$$= \frac{pI_e}{12} \left[6at_1^2 + 2bt_1^3 + ct_1^4 + 2t_1(M - t_1)(6a + 3bt_1 + 2ct_1^2) \right]$$

Therefore, the total variable cost per unit time is

$$TC(t_1, T) = TC_1(t_1, T), \quad \text{if } M \le t_1 \text{ where}$$
$$= TC_2(t_1, T), \quad \text{if } M > t_1 \tag{13}$$

$$TC_1(t_1, T) = \frac{1}{T} \left[A + \frac{sa\delta}{2}(T - t_1)^2 + \pi a(1 - \delta)(T - t_1) - \frac{pI_e t_1^2}{12}(6a + 2bt_1 + ct_1^2) \right.$$

$$+ \frac{h}{\theta^4} \left[\{\theta^2(a + bt_1 + ct_1^2) - \theta(2ct_1 + b) + 2c\}[e^{\theta t_1} - 1] - \left\{ \theta^2 \left(a + \frac{bt_1}{2} + \frac{ct_1^2}{3} \right) - \theta(ct_1 + b) + 2c \right\} \theta t_1 \right]$$

$$+ p \left(\frac{1}{\theta^3} \left[\{\theta^2(a + bt_1 + ct_1^2) - \theta(2ct_1 + b) + 2c\}e^{\theta t_1} - (a\theta^2 - b\theta + 2c) \right] - \left(at_1 + \frac{bt_1^2}{2} + \frac{ct_1^3}{3} \right) \right)$$

$$+ \frac{pI_p}{\theta^4} \left[\{\theta^2(a + bt_1 + ct_1^2) - \theta(2ct_1 + b) + 2c\} \left[e^{\theta(t_1 - M)} - 1 \right] \right.$$

$$\left. \left. - \theta(t_1 - M) \left\{ \theta^2 \left(a + \frac{b(t_1 + M)}{2} \right) + \frac{c(t_1^2 + t_1 M + M^2)}{3} \right\} - \theta(c(t_1 + M) + b) + 2c \right] \right]. \tag{14}$$

$$TC_2(t_1, T) = \frac{1}{T} \left[A + \frac{sa\delta}{2}(T - t_1)^2 + \pi a(1 - \delta)(T - t_1) \right.$$

$$+ \frac{h}{\theta^4} \left[\{\theta^2(a + bt_1 + ct_1^2) - \theta(2ct_1 + b) + 2c\}[e^{\theta t_1} - 1] - \left\{ \theta^2 \left(a + \frac{bt_1}{2} + \frac{ct_1^2}{3} \right) - \theta(ct_1 + b) + 2c \right\} \theta t_1 \right]$$

$$- \frac{pI_e}{12} \left[6at_1^2 + 2bt_1^3 + ct_1^4 + 2t_1(M - t_1)(6a + 3bt_1 + 2ct_1^2) \right]$$

$$+ p \left(\frac{1}{\theta^3} \left[\{\theta^2(a + bt_1 + ct_1^2) - \theta(2ct_1 + b) + 2c\}e^{\theta t_1} - (a\theta^2 - b\theta + 2c) \right] - \left(at_1 + \frac{bt_1^2}{2} + \frac{ct_1^3}{3} \right) \right). \tag{15}$$

Case 1: $M \le t_1$. Now for minimizing the total average cost per unit time, the optimal values of T and t_1 can be obtained by solving the following simultaneous equations:

$$\frac{\partial TC_1(t_1, T)}{\partial T} = 0; \quad \text{and} \quad \frac{\partial TC_1(t_1, T)}{\partial t_1} = 0 \tag{16}$$

provided they satisfy the sufficient conditions $\frac{\partial^2 TC_1(t_1, T)}{\partial T^2} > 0$; $\frac{\partial^2 TC_1(t_1, T)}{\partial t_1^2} > 0$;

$$\frac{\partial^2 TC_2(t_1, T)}{\partial T^2} \frac{\partial^2 TC_2(t_1, T)}{\partial t_1^2} - \left(\frac{\partial^2 TC_2(t_1, T)}{\partial T \partial t_1} \right)^2 > 0$$

An Ordering Policy for Deteriorating Items with Quadratic Demand   679

Case 2: $M > t_1$. Now for minimizing the total average cost per unit time, the optimal values of T and t_1 can be obtained by solving the following simultaneous equations:

$$\frac{\partial TC_2(t_1, T)}{\partial T} = 0 \text{ and } \frac{\partial TC_2(t_1, T)}{\partial t_1} = 0 \qquad (17)$$

provided they satisfy the sufficient conditions $\frac{\partial^2 TC_1(t_1,T)}{\partial T^2} > 0$; $\frac{\partial^2 TC_1(t_1,T)}{\partial t_1^2} > 0$; and

$$\frac{\partial^2 TC_2(t_1, T)}{\partial T^2} \frac{\partial^2 TC_2(t_1, T)}{\partial t_1^2} - \left(\frac{\partial^2 TC_2(t_1, T)}{\partial T \partial t_1}\right)^2 > 0$$

Algorithm:

Step 1: Determine T_1^* and t_{11}^* from Eq. (16). If $M \leq t_{11}^*$ evaluate $TC_1(t_{11}^*, T_1^*)$ using Eq. (14).

Step 2: Determine T_2^* and t_{12}^* from Eq. (17). If $M > t_{12}^*$ evaluate $TC_2(t_{12}^*, T_2^*)$ using Eq. (15).

Step 3: If the condition $M \leq t_{11}^*$ and $M > t_{12}^*$ is satisfied go to step 4, otherwise go to step 5.

Step 4: Compare $TC_1(t_{11}^*, T_1^*)$ and $TC_2(t_{12}^*, T_2^*)$ and find the minimum cost.

Step 5: If $M \leq t_{11}^*$ is satisfied but $M < t_{12}$, then $TC_1(t_{11}^*, T_1^*)$ is the minimum cost, else if $t_{11}^* < M$ but $M > t_{12}^*$ then $TC_2(t_{12}^*, T_2^*)$ is the minimum cost. Using the optimal solution procedure described above, we can find the optimal order quantity to be

$$Q = \frac{1}{\theta^3}\left[\{\theta^2\left(a + bt_1 + ct_1^2\right) - \theta\left(2ct_1 + b\right) + 2c\}e^{\theta t_1} - (a\theta^2 - b\theta + 2c)\right]$$
$$+ a\delta\left(T - t_1\right)$$

$$(18)$$

4 Numerical Example

Example 1

Let us consider an inventory system with the following data:
$A = 100$, $h = 15$, $s = 30$, $p = 80$, $\pi = 25$, $a = 10$, $b = 5$, $c = 1$, $\theta = 0.20$, $M = 0.0411$, $\delta = 0.56$, $I_p = 0.15$, $I_e = 0.13$ in appropriate units.
Using the solution procedure described above, and using MATLAB 7.0, the minimum average cost is $TC(t_1^*, T^*) = 7{,}774.20$ and the optimal values of the cycle length and the shortage period are $T^* = 0.2131$ and $t_1^* = 0.0658$, respectively.

Example 2

Let us consider an inventory system with the following data:
$A = 100$, $h = 15$, $s = 30$, $p = 80$, $\pi = 25$, $a = 10$, $b = 5$, $c = 1$, $\theta = 0.20$, $M = 0.0411$, $\delta = 0.56$, $I_p = 0.15$, $I_e = 0.13$ in appropriate units.

Using the solution procedure described above, and using MATLAB 7.0, the minimum average cost is $\mathrm{TC}(t_1{}^*, T^*) = 8{,}920.40$ and the optimal values of the cycle length and the shortage period are $T^* = 0.2797$ and $t_1{}^* = 0.2607$, respectively.

5 Conclusion

In this paper, we have developed a time-dependent quadratic inventory model for deteriorating items with permissible delay in payments. Shortages are allowed and partially backlogged. An analytic formulation of the model and an optimal solution procedure to find the optimal replenishment policy was also presented. A rapidly increasing demand can be represented by an exponential function of time. An exponential rate of change in demand is high and the fluctuation or variation of any commodity in the real market cannot be so high. Hence, a quadratic demand seems to be a better representation of time-varying demands. Thus a possible future research issue is to consider two-level trade credit with variable demand.

Acknowledgments The research work has been supported by University Grants Commission (UGC–SAP), New Delhi, India.

References

1. Goyal SK (1985) EOQ under conditions of permissible delay in payments. J Oper Res Soc 36:335–338
2. Sana SS, Chaudhuri KS (2008) A deterministic EOQ model with delay in payments and price discount offers. Eur J Oper Res 184:509–533
3. Reza M, Isa NKA (2012) Joint control of inventory and its pricing for non-instantaneously deteriorating items under permissible delay in payments and partial backlogging. Math Comput Model 55:1722–1733
4. Ghare PM, Schrader GH (1963) A model for an exponentially decaying inventory. J Ind Eng 14:238–243
5. Covert RP, Philip GC (1973) An EOQ model for items with Weibull distribution deterioration. AIIE Trans 5:323–326
6. Philip GC (1974) A generalised EOQ model for items with Weibull distribution. AIIE Trans 6:159–162
7. Bahari-Kashani H (1989) Replenishment schedule for deteriorating items with time proportional demand. J Oper Res Soc 40:75–81
8. Chung KJ, Ting PS (1993) A heuristic for replenishment of deteriorating items with a linear trend in demand. J Oper Res Soc 44(12):1235–1241

9. Dave U, Patel LK (1981) (T, Si) policy inventory model for deteriorating items with time proportional demand. J Oper Res Soc 32:137–142
10. Giri BC, Goswami A, Chaudhuri KS (1996) An EOQ model for deteriorating items with time varying demand and costs. J Oper Res Soc 47:1398–1405
11. Goswami A, Chaudhuri KS (1991) An EOQ model for deteriorating items with a linear trend in demand. J Oper Res Soc 42(12):1105–1110
12. Hariga M (1995) An EOQ model for deteriorating items with shortages and time-varying demand. J Oper Res Soc 46:398–404
13. Jalan AK, Giri RR, Chaudhuri KS (1996) EOQ model for items with Weibull distribution deterioration, shortages and trended demand. Int J Syst Sci 27(9):851–855
14. Jalan AK, Chaudhuri KS (1999) Structural properties of an inventory system with deterioration and trended demand. Int J Syst Sci 30(6):627–633
15. Lin C, Tan B, Lee WC (2000) An EOQ model for deteriorating items with shortages. Int J Syst Sci 31(3):391–400
16. Jalan AK, Chaudhuri KS (1999) An EOQ model for deteriorating items in a declining market with SFI policy. Korean J Comput Appl Math 6(2):437–449
17. Donaldson WA (1977) Inventory replenishment policy for a linear trend in demand—an analytical solution. Oper Res Q 28:663–670
18. Mitra A, Fox JF, Jessejr RR (1984) A note on deteriorating order quantities with a linear trend in demand. J Oper Res Soc 3:5141–5144
19. Ritchie E (1980) Practical inventory replenishment policies for a linear trend in demand followed by a period of steady demand. J Oper Res Soc 31:605–613
20. Ritchie E (1984) The EOQ for linear increasing demand: a simple optimal solution. J Oper Res Soc 35:949–952
21. Ritchie E (1985) Stock replenishment quantities for unbounded linear increasing demand: an interesting consequence of the optimal policy. J Oper Res Soc 36:737–739
22. Silver EA, Meal HC (1969) A simple modification of the EOQ for the case of a varying demand rate. Prod Inventory Manage 10:52–65
23. Silver EA (1979) A simple inventory replenishment decision rule for a linear trend in demand. J Oper Res Soc 30:71–75
24. Ghosh SK, Chaudhuri KS (2006) An EOQ model with a quadratic demand, time-proportional deterioration and shortages in all cycles. Int J Syst Sci 37(10):663–672
25. Khanra S, Chaudhuri KS (2003) A note on an order level inventory model for a deteriorating item with time dependent quadratic demand. Comput Oper Res 30:1901–1916
26. Khanra S, Ghosh SK, Chaudhuri KS (2011) An EOQ model for a deteriorating item with time dependent quadratic demand under permissible delay in payment. Appl Math Comput 218:1–9

Complex Dynamics of BRD Sets

Bhagwati Prasad and Kuldip Katiyar

Abstract The intent of the paper is to study the dynamics of the Mandelbrot like Baker-Rippon-Devaney (BRD) sets for complex exponential family under Mann iterates.

Keywords Mandelbrot set • BRD set • Complex exponential function • Mann iteration

1 Introduction

The Julia and Mandelbrot sets are of vital importance in the study of the complex dynamics of functions. A lot of work has been done on the structures of these sets of the complex analytic functions such as polynomial, rational and exponential functions. The importance of the transcendental functions lies in the fact that the Mandelbrot and Julia sets for such functions have an alternative characterisation suitable for easier computations. Such functions are studied by authors of this paper in [1, 2]. Misiurewicz [3] was the first to explore the mathematical aspects of the complex exponential maps of the type $z_{n+1} = e^{z_n}$, Baker and Rippon [4] studied the complex exponential family $e^{\lambda z}$ and Devaney [5–7] extensively studied the family λe^z of maps. Romera et al. [8] noticed that the two exponential families $z_{n+1} = e^{\lambda z_n}$ and $z_{n+1} = \lambda e^{z_n}$ have the same Mandelbrot-like sets. Thereafter, they call these sets as Baker-Rippon-Devaney (BRD) sets. They studied these BRD and Julia sets for the complex families λe^z, $e^{z^2+\lambda}$ and $e^{z/\lambda}$ from graphical point of view and obtained interesting results. The sequences $\{z_{n+1}\}$ defined above are generated iteratively using Picard iteration scheme. Many authors have studied these families

---

B. Prasad (✉) · K. Katiyar
Department of Mathematics, Jaypee Institute of Information Technology,
A-10, Sector-62, Noida 201307, UP, INDIA
e-mail: b_prasad10@yahoo.com

K. Katiyar
e-mail: kuldipkatiyar.jiitn@gmail.com

S. Sathiyamoorthy et al. (eds.), *Emerging Trends in Science, Engineering and Technology*, Lecture Notes in Mechanical Engineering, DOI: 10.1007/978-81-322-1007-8_64, © Springer India 2012

by using different iteration procedures. Prasad and Katiyar [2] used the Mann iteration scheme in their bifurcation analysis related results for the complex exponential family λe^z. Further, they used Ishikawa iteration scheme to study the fractal patterns of contractive maps (see [9]). In this paper, our aim is to study the structures of Mandelbrot like sets for the complex exponential family λe^z and $e^{z/\lambda}$ by using the Mann iterative scheme.

2 Preliminaries

In this section, we present the basic definitions and concepts required for our study.

Definition 2.1. The Mandelbrot-like set of a family of complex maps $z_{n+1} = f_\lambda(z_n)$ for the initial value z_0 (usually the one corresponding to the critical point of the family of maps) is defined as the set of $\lambda \in C$ (the set of complex numbers) for which the n-th iteration of the function $f_\lambda^n(z_0)$ does not tend to ∞ as n tends to ∞

$$M = \{\lambda \in C \,|\, f_\lambda^n(z_0) \text{K}\infty \text{ when } n \to \infty\}$$

We denote the complex exponential family of the map $z_{n+1} = f_\lambda(z_n)$ by $E_\lambda(z)$.

Definition 2.2. Let X be a non-empty set and $f: X \to X$. The orbit of a point z in X is defined as a sequence

$$\{f^n(z) : n = 0, 1, 2, \ldots\}$$

If we generate the orbits by using Peano-Picard iteration, the generated orbit, represented by

$$PO(f, z_0) := \{z_n : z_n = f(z_{n-1}), n = 1, 2, \ldots\}$$

is called the Picard orbit. This iteration requires one number as input to return a new number as output and popularly called as one-step feedback machine.

The Mann iteration [11] is defined in the following manner:

$$MO(f, z_0, \alpha_n) := \{z_n : z_n = \alpha_n f(z_{n-1}) + (1 - \alpha_n) z_{n-1}\} \tag{1}$$

for $n = 1, 2, 3, \ldots$, where $0 < \alpha_n \leq 1$ and $\{\alpha_n\}$ is convergent away from 0.

The sequence $\{z_n\}$ constructed as above is two-step feedback system.

In this paper, we shall study the Mann orbit for $\alpha_n = \alpha$. It is remarked that (1) with $\alpha_n = 1$ is the Picard iteration.

2.1 Escape Criterion for Exponential Family

We follow the escape criterion of Peitgen and Saupe [12] to generate graphical images of BRD sets for the families λe^z and $e^{z/\lambda}$. For the exponential family $E_\lambda(z)$, we observe that the orbits tend to infinity in the direction of the positive real axis. That is, $\lim_{n\to\infty} \left| E_\lambda^n(z) \right| = \infty$ whenever $\lim_{n\to\infty} \mathrm{Re}(E_\lambda^n(z))$ tends to infinity.

The number of iteration has to be finite due to the computational limitation. We use 40 iterations for our study of the complex dynamics of the above families. Therefore, if the real part of $E_\lambda^n(z)$ exceeds 50, then we say that the orbit of z escapes (also see, [2, 4–8, 13–15]).

2.2 Algorithm to Generate Graphical Images

We follow the following steps to generate the computer graphical images of BRD sets of complex exponential families:

1. Take complex plane λ and compute the orbit of z_0 up to N iterations using Mann iteration corresponding to different values of λ.
2. If the orbit of z_0 enters the region $\mathrm{Re}(z) \geq 50$ at iteration $j \leq N$ then colour λ with a colour corresponding to j.
3. If the orbit never enters this half plane, then colour z black and declare that $\lambda \notin J(E_\lambda)$.

The colouring scheme of the graphics presented in the figures depends upon the rate of escape to infinity. We follow the colour schemes and algorithm of Pietgen and Saupe [cf. 9] for our study. A point λ is coloured black if the orbit of z_0 is not escaped within the first N iterates, red is used to denote points which escape to infinity fastest. Shades of orange, yellow and green are used to colour points which escape less quickly and shades of blue and violet represent the points which escape, but only after a significant number of iterations.

This colouring scheme is well depicted in the graphical patterns given in Figs. 1 and 2.

3 BRD Sets with Mann Orbits

Let $z_n = x_n + iy_n$ and $\lambda = \lambda_1 + i\lambda_2$. Then for complex exponential family λe^z, the real and imaginary part of $z_{n+1} = f(z_n) = \lambda e^{z_n}$ are given by

$$
\begin{aligned}
\mathrm{Re}z_{n+1} &= \alpha \left\{ e^{x_n} \left(\lambda_1 \cos y_n - \lambda_2 \sin y_n \right) \right\} + (1 - \alpha) x_n, \\
\mathrm{Im}z_{n+1} &= \alpha \left\{ e^{x_n} \left(\lambda_1 \sin y_n + \lambda_2 \cos y_n \right) \right\} + (1 - \alpha) y_n
\end{aligned}
\tag{2}
$$

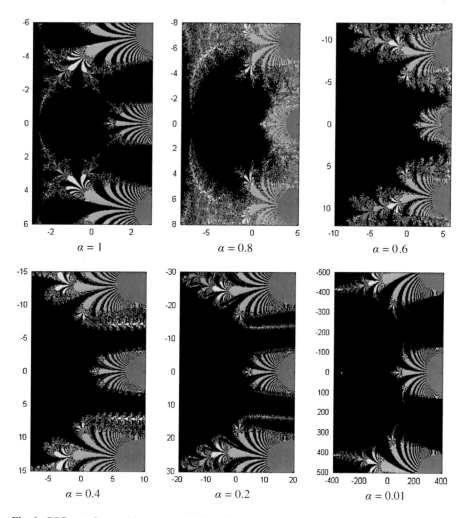

Fig. 1 BRD sets for complex exponential family λe^z

and for complex exponential family $e^{z/\lambda}$, the real and imaginary part of $z_{n+1} = f(z_n) = e^{z_n/\lambda}$ are given by

$$\begin{aligned}
\text{Re} z_{n+1} &= \alpha \left\{ e^{\frac{\lambda_1 x_n + \lambda_2 y_n}{\lambda_1^2 + \lambda_2^2}} \cos\left(\frac{\lambda_1 y_n - \lambda_2 x_n}{\lambda_1^2 + \lambda_2^2}\right) \right\} + (1-\alpha) x_n, \\
\text{Im} z_{n+1} &= \alpha \left\{ e^{\frac{\lambda_1 x_n + \lambda_2 y_n}{\lambda_1^2 + \lambda_2^2}} \sin\left(\frac{\lambda_1 y_n - \lambda_2 x_n}{\lambda_1^2 + \lambda_2^2}\right) \right\} + (1-\alpha) y_n
\end{aligned} \quad (3)$$

Equations (2) and (3) give us the formule to compute the main body of the loop in the program. Using the algorithm 2.2, we generate the BRD sets for the

Complex Dynamics of BRD Sets

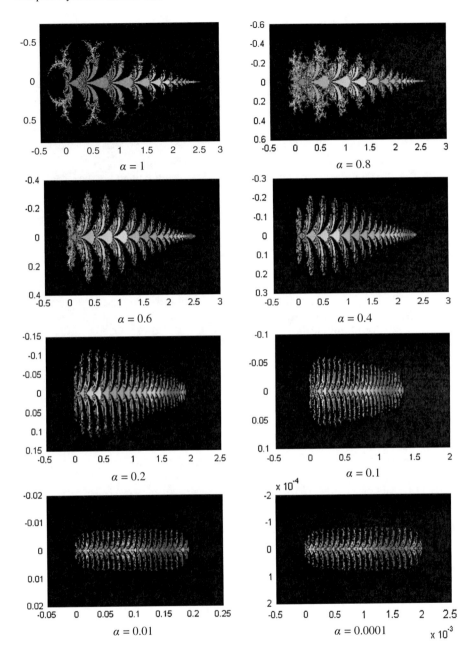

Fig. 2 BRD sets for complex exponential family $e^{z/\lambda}$

Mann iterates (1) of the exponential functions for the specific choices of parameter α by fixing $N = 40$ and $z_0 = 0$ in figures.

4 Conclusions

Some appreciable changes are observed in the patterns of the BRD sets for the complex exponential families when Mann iterative scheme is used. For the family λe^z, as the value of the parameter α decreases from 1 towards zero, the size of the fingers (in the sense of Devaney [7]) also increases. It is also noticed that for this family the fractal growth of BRD sets increases for $1 > \alpha > 0.5$ and decreases for $0.5 > \alpha > 0$ for the same initial z. For the family $e^{z/\lambda}$, when we decrease α from 1 towards 0.01, the shape and size of the BRD sets change and show shrinking pattern. But beyond $\alpha = 0.01$ towards zero, its shape remains unchanged and only shrinking pattern is observed.

References

1. Prasad B, Katiyar K (2010) Julia sets for a transcendental function. In: Proceedings of the IEEE international conference on computer engineering and technology, Jodhpur, India, pp E59-E61
2. Prasad B, Katiyar K (2011) Fractals via Ishikawa iteration. In: Balasubramaniam P (ed) ICLICC 2011. CCIS, vol 140. Springer, Heidelberg, pp 197–203
3. Misiurewicz M (1981) On iterates of e^z. Ergod Theor Dyn Syst 1:103–106
4. Baker N, Rippon PJ (1984) Iteration of exponential functions. Ann Acad Sci Fenn A1(9):49–77
5. Devaney RL (1984) Julia set and bifurcation diagrams for exponential maps. Am Math Soc 11:167–171
6. Devaney RL (1992) A first course in chaotic dynamical systems: theory and experiment. Addison-Wesley, Reading
7. Devaney RL, Henk Broer FT, Hasselblatt B (eds) (2010) Complex exponential dynamics. Elsevier Science vol 3, pp 125–223
8. Romera M, Pastor G, Alvarez G, Montoya F (2000) Growth in complex exponential dynamics. Comput Graph 24(1):115–131
9. Prasad B, Katiyar K (2012) Dynamics of Julia sets for complex exponential functions. In: Balasubramaniam P, Uthayakumar R (eds) ICMMSC 2012. CCIS, vol. 283(19). Springer, Heidelberg, pp 185–192
10. Barnsley MF (1993) Fractals everywhere, 2nd edn. Revised with the assistance of and a foreword by Hawley Rising, III. Academic Press Professional, Boston
11. Mann WR (1953) Mean value methods in iteration. Proc Am Math Soc 4(3):506–510
12. Peitgen HO, Saupe D (1988) The science of fractal images. Springer, New York
13. Baranski B (2007) Trees and hairs for some hyperbolic entire maps of finite order. Math Z 257(1):33–59
14. Schleicher D, Zimmer J (2003) Escaping points of exponential maps. J Lond Math Soc 67(2):380–400
15. Xingyuan W, Qijiang S (2006) Growth in complex exponential dynamics. Appl Math Comput 81(2):816–825

Coordinating Supply Chain Inventories with Shortages

R. Uthayakumar and M. Rameswari

Abstract In this paper, we present the benefit of coordination in supply chain with quantity discount strategy and backorders. First, the buyer's economic order quantity with shortages is discussed by taking into account the effects of fill rate (backordering). Then an integrated production-inventory lot sizing model under cooperation mode is presented, by introducing a quantity discount policy. The centralized decision-making model is formulated to examine the effectiveness of the proposed quantity discount model with backordering. A numerical example illustrates validity of the coordination mechanism.

Keywords Coordination • Inventory • Supply chain • Fixed lifetime product • Quantity discount • Backordering

1 Introduction

Inventory models often have the common assumption of infinite lifetime of items. Clearly, in many practical situations this assumption is very unrealistic. Mostly, foodstuff, photographic film, and pharmaceutical products have their expiry date, that is, they have a fixed known lifetime. Any units which remain unused by their expiry date are considered outdated, and must be removed from inventory to be discarded with disposal cost or sold at a discounted price. Due to the limited product lifetime, an ineffective inventory management at each stage in the supply chain from production to consumers can lead to high system costs.

R. Uthayakumar
Department of Mathematics, Gandhigram Rural Institute-Deemed University,
Gandhigram, Dindigul, Tamil Nadu, India
e-mail: uthayagri@gzmail.com

M. Rameswari (✉)
Department of Mathematics, SSM Institute of Engineering and Technology,
Dindigul, Tamil Nadu, India
e-mail: sivarameswari1977@gmail.com

S. Sathiyamoorthy et al. (eds.), *Emerging Trends in Science,*
Engineering and Technology, Lecture Notes in Mechanical Engineering,
DOI: 10.1007/978-81-322-1007-8_65, © Springer India 2012

689

2 Literature Review

Viswanathan and Piplani [1] adopted the common replenishment epochs (CRE) strategy and suggested that the supplier offers a discount that is the maximum of the discount required by all buyers to participate. Gurnani [2] examined the effect of quantity discounts in different ordering structures in a single-vendor heterogeneous buyers system. Sarmah et al. [3] provided an intensive survey on the coordination mechanisms between the vendor and the buyers. Stadtler [4] considered a quantity discount and suppliers selection problem. Mendoza and Ventura [5] incorporated quantity discounts and transportation costs in an inventory model. Duan et al. [6] modeled a single-vendor, single-buyer supply chain for item with fixed lifetime was considered without stockouts; neither backorders nor lost sales. The remainder of this paper is organized as follows: Sect. 3 describes the notations and assumptions used throughout this study. In Sect. 4, the decentralized model for EOQ model with and without coordination, and centralized model are formulated. In Sect. 5, numerical examples are presented to illustrate the proposed model. Finally, we draw the conclusions and give suggestions for future research.

3 Assumptions and Notations

3.1 Assumptions

The following assumptions are made throughout this paper:

1. Demand is known and constant.
2. Shortages are allowed and completely backordered.
3. Lead time is zero.

3.2 Notations

For developing the proposed models, the following notations are used throughout this paper:

D Annual demand
A_v Setup cost per production run for the vendor
A_b Buyer ordering cost
h_v Holding cost for the vendor per item per unit time
h_b Holding cost for the buyer per item per unit time
C_b The cost to keep a unit backordered for a year
P_b The delivered unite price paid by the buyer

Coordinating Supply Chain Inventories with Shortages

L_t — Lifetime of product
Q_0 — Buyer's EOQ
f — Fill rate
m — The vendor's order multiple in the absence of any coordination
K — The buyer's order multiple under coordination, KQ_0 buyer's new order quantity
$d(K)$ — Discount per unit to the buyer if he orders KQ_0 every time

4 Mathematical Formulation for Decentralized System Without Coordination

4.1 EOQ Model with Backorders

We assume that the buyers' inventory policies can be described by the widely used EOQ model with backorders. In the EOQ model with backorders, we use the fill rate, f and express the proportion of immediate filling demand in the inventory cycle. The proportion $(1 - f)$ that is not filled from inventory is backordered. The buyer's total cost and order quantity without coordination is expressed as

$$Q_0 = \sqrt{\frac{2A_b D}{h_b f^2 + C_b(1 - f)^2}}$$

with the total cost $TC_b = \sqrt{2A_b D \left[h_b f^2 + C_b(1 - f)^2\right]}$

The vendor's order size should be some integer multiple of Q_0 denoted by mQ_0. Therefore, the total annual cost for the vendor is given by

$$
\begin{aligned}
TC_v &= \frac{DA_v}{mQ_0} + \frac{(m - 1)Q_0 h_v}{2} \\
&= \frac{A_v}{m}\sqrt{\frac{D\left[h_b f^2 + C_b(1 - f)^2\right]}{2A_b}} + (m - 1)h_v\sqrt{\frac{DA_b}{\left[h_b f^2 + C_b(1 - f)^2\right]}}
\end{aligned}
$$

In the absence of any coordination, the vendor's total cost can be formulated as follows:

$$
\min TC_v(m) \\
\text{subject to } \begin{cases} mt_0 \le L_t \\ m \ge 1 \end{cases} \tag{1}
$$

where $mt_0 \le L_t$ is to ensure that items are not overdue before they are sold up by the buyer.

Theorem 1

Let m^* be the optimal value of (1), if $L_t^2 \geq \frac{2A_b}{D[h_b f^2 + C_b(1-f)^2]_0}$, then

$$m^* = \min\left\{ \sqrt{\frac{A_v\left[h_b f^2 + C_b(1-f)^2\right]}{A_b h_v} + \frac{1}{4}} - \frac{1}{2}, \left\lceil \frac{L_t}{\sqrt{\frac{2A_b}{D[h_b f^2 + C_b(1-f)^2]}}} \right\rceil \right\}$$

(2)

where $\lceil a \rceil$ is the least integer greater than or equal to a, $L_t^2 \geq \frac{2A_b}{Dh_b}$ is to ensure that $m^* \geq 1$.

Proof

$TC_v(m)$ is strictly convex in m. Let m_0^* be the optimum of $\min_{m \geq 1} TC_v(m)$ then

$$m_0 = \max\left\{\min\left\{m \mid TC_v(m) \leq TC_v(m+1)\right\}, 1\right\}$$

$$= \max\left\{\min\left\{m \mid m(m+1) \leq \frac{2DA_v}{Q_0^2 h_v}\right\}, 1\right\}$$

$$= \left\lceil \sqrt{\frac{A_v\left[h_b f^2 + C_b(1-f)^2\right]}{A_b h_v} + \frac{1}{4}} - \frac{1}{2} \right\rceil \geq 1$$

By using $t_0 = \sqrt{\frac{2A_b}{D[h_b f^2 + C_b(1-f)^2]}}$ and substitute in (1),

we get $m\sqrt{\frac{2A_b}{D[h_b f^2 + C_b(1-f)^2]}} \leq L_t$.

Set $m1^* = \left\lceil \frac{L_t}{\sqrt{\frac{2A_b}{D[h_b f^2 + C_b(1-f)^2]}}} \right\rceil$. Since $L_t^2 \geq \frac{2A_b}{D[h_b f^2 + C_b(1-f)^2]}$, $m_1^* \geq 1$ holds.

In view of $TC_v(m)$ is a convex function, if $m_0^* \leq m_1^*$, $m^* = m_0^*$, else $m^* = m_1^*$. So if $L_t^2 \geq \frac{2A_b}{D[h_b f^2 + C_b(1-f)^2]}$, $m^* = \min\left\{m_0^* > m_1^*\right\}$. The proof of Theorem 1 is complete $\square$.

4.2 Mathematical Formulation for Decentralized with Coordination

Under the quantity discount coordination strategy, the vendor requests the buyer to alter his current order size by a factor K ($K > 0$), and compensate the buyer by a quantity discount at a discount factor $d(K)$. The

Coordinating Supply Chain Inventories with Shortages 693

vendor's order quantity is nKQ_0, where n is a positive integer and KQ_0 is the buyer's new order quantity. The total cost $\overline{TC_v(n)}$ of the vendor is given by $\overline{TC_v(n)} = \frac{DA_v}{nKQ_0} + \frac{(n-1)KQ_0h_v}{2} + P_bDd(K)$. The problem with coordination can be formulated as follows:

$$\min \overline{TC_v(n)}$$

$$\text{subject to} \begin{cases} nKt_0 \leq L_t \\ \frac{DA_v}{nKQ_0} + \frac{Q_0}{2}\left[h_b f^2 + C_b(1-f)^2\right] \\ \quad -\sqrt{2DA_b\left[h_b f^2 + C_b(1-f)^2\right]} + P_bDd(K) \\ n \geq 1 \end{cases} \tag{3}$$

The first constraint $nKt_0 \leq L_t$ is to ensure the items are not overdue before they are sold up by the buyer. The second constraint is the retailer's participation constraint, i.e., the buyer's cost under coordination cannot exceed that in the absence of any coordination.

Theorem 2

Let m^* and n^* be the optimum of (2) and (3), respectively, then the following inequality holds: $\overline{TC_v(n*)} \leq TC_v(m)$.

Proof

The term $P_bDd(K)$ in the right hand side of the second constraint of (3) is just the compensation to the buyer by the vendor, which is component of the vendor's costs. By the second constraint, $P_bDd(K)$ takes the smallest value only when the second constraint is an equation, so if $\overline{TC_v(n)}$ is minimized, the second constraint must be an equation.

$$\frac{DA_v}{nKQ_0} + \frac{Q_0}{2}\left[h_b f^2 + C_b(1-f)^2\right] - \sqrt{2DA_b\left[h_b f^2 + C_b(1-f)^2\right]} = P_bDd(K)$$

$$d(K) = \frac{\frac{DA_v}{nKQ_0} + \frac{Q_0}{2}[h_b f^2 + C_b(1-f)^2] - \sqrt{2DA_b[h_b f^2 + C_b(1-f)^2]}}{P_bD}$$

$$\tag{4}$$

$d(1) = 0$. Hence (3) is equivalent to (2) if $K = 1$, i.e. (1) is a special case of (3), so (4) holds. The proof is complete $\square$.

Since $d(K)$ is convex in K, $\overline{TC_v(n)}$ is obviously convex in K.
Let K^* be the minimum of $\overline{TC_v(n)}$, by a simple calculation.

$$K^*(n) = \sqrt{\frac{2D\left[\frac{A_v}{n} + A_b\right]}{Q_0^2\left[(n-1)h_v + \left[h_b f^2 + C_b(1-f)^2\right]\right]}}$$

Substituting $K^*(n)$ and t_0 into $nKt_0 \leq L_t$, we have

$$\frac{L_t^2 Q_0^2}{4A_b} + \left[h_b f^2 + C_b(1-f)^2\right]\left[(n-1)h_v + \left[h_b f^2 + C_b(1-f)^2\right]\right]$$
$$- A_v n + A_b n^2 \geq 0$$

Let $G(n) = -A_b n^2 + \left(\frac{L_t^2 Dh_v}{2} - A_v\right)n + \frac{L_t^2 D\left[h_b f^2 + C_b(1-f)^2 - h_v\right]}{2\left[h_b f^2 + C_b(1-f)^2\right]}$

Then the first constraint of (3) is equivalent to $G(n) \geq 0$.

$$\min \ \overline{TC_Y(n)} = \sqrt{2D\left[\frac{A_v}{n}(n-1)h_v + \frac{A_v}{n}\left[h_b f^2 + C_b(1-f)^2\right] + A_b\left[(n-1)h_v + \left[h_b f^2 + C_b(1-f)^2\right]\right]\right]}$$
$$- \sqrt{2DA_b\left[h_b f^2 + C_b(1-f)^2\right]}$$

$$\text{subject to} \begin{cases} G(n) \geq 0 \\ n \geq 1 \end{cases}$$

(5)

Since $\sqrt{x}$ is a strictly increasing function for $x \geq 0$.

$$\min \overline{TC_Y(n)} = D\left[\frac{A_v}{n}(n-1)h_v + \frac{A_v}{n}\left[h_b f^2 + C_b(1-f)^2\right] + A_b\left[(n-1)h_v + \left[h_b f^2 + C_b(1-f)^2\right]\right]\right]$$
$$- \sqrt{2DA_b\left[h_b f^2 + C_b(1-f)^2\right]}$$

$$\text{subject to} < error - \begin{cases} G(n) \geq 0 \\ n \geq 1 \end{cases}$$

It is obvious that (5) is a nonlinear programming. To solve (5), we must discuss the properties of $\overline{TC_Y(n)}$ and $G(n)$. Since $\overline{TC_Y(n)}'' = \frac{2DA_v}{n^3}\left[h_b f^2 + C_b(1-f)^2 - h_v\right].\overline{TC_Y(n)}$ is convex when $(h_b + c_b)f^2 + C_b - 2fC_b > h_v$ and concave otherwise. Since $G''(n) = -2A_v < 0$, $G(n)$ is strictly concave.

4.3 Model Formulation for Centralized System

In this section, the centralized system optimization problem with a single decision maker will be analyzed. Under the proposed quantity discount contract, the vendor's total cost can be reduced while the buyer's cost remains the same. If the vendor offers to share its cost saving with the buyer, the buyer can also be better off. If there is a common decision maker for both the buyer and vendor, the objective is to minimize the total cost of the system.

$$TC_c(n) = \frac{DA_v}{nQ} + \frac{(n-1)Qh_v}{2} + \frac{DA_b}{Q} + \frac{Q}{2}\left[h_b f^2 + C_b(1-f)^2\right]$$

$$subject\ to \begin{cases} \frac{nQ}{D} \leq L_t \\ n \geq 1 \end{cases}$$

(6)

Coordinating Supply Chain Inventories with Shortages

Theorem 3

The proposed quantity discount strategy can achieve system coordination.

Proof

Let Q^* be the buyer's optimal order quantity, then Q^* satisfies

$$Q* = \sqrt{\frac{2D\left[\frac{A_v}{n} + A_b\right]}{(n-1)h_v + h_b f^2 + C_b(1-f)^2}} \tag{7}$$

$$\min TC_c(n) = \sqrt{2D\left[(A_v - A_b)h_v + \frac{A_v\left[h_b f^2 + C_b(1-f)^2 - h_v\right]}{n} + nA_b + A_b\left[h_b f^2 + C_b(1-f)^2\right]\right]}$$

$$subject\ to \begin{cases} -A_b n^2 + \left(\frac{L_t^2 Dh_v}{2} - A_v\right)n + \frac{L_t^2 D[h_b f^2 + C_b(1-f)^2 - h_v]}{2} \\ n \geq 1 \end{cases}$$

$$\tag{8}$$

Since $\sqrt{x}$ is strictly increasing for $x \geq 0$. Equation (8) is equivalent to the following problem:

$$\min TC_c(n) = D\left[(A_v - A_b)h_v + \frac{A_v\left[h_b f^2 + C_b(1-f)^2 - h_v\right]}{n} + nA_b + A_b\left[h_b f^2 + C_b(1-f)^2\right]\right]$$

$$subject\ to \begin{cases} -A_b n^2 + \left(\frac{L_t^2 Dh_v}{2} - A_v\right)n + \frac{L_t^2 D[h_b f^2 + C_b(1-f)^2 - h_v]}{2} \\ n \geq 1 \end{cases}$$

$$TC_c(n) = \overline{TC_v(n)} + \sqrt{2DA_b\left[h_b f^2 + C_b(1-f)^2\right]}$$

where $\sqrt{2DA_b\left[h_b f^2 + C_b(1-f)^2\right]}$ is the buyer's actual cost under coordination. Hence, the buyer's optimal order quantity under coordination is equal to that under centralized system. The proof of Theorem 3 is complete $\square$.

5 Numerical Example

To illustrate the results obtained in this paper, the proposed model is applied to efficiently solve the following numerical example. Consider an inventory system with the following characteristics:

$D = 15{,}000$ units/year; $f = 0.9$ year; $L_t = 0.25$ year; $P_b =$ Rs. 20 per unit order; $C_b =$ Rs. 15 unit; $A_v =$ Rs. 25 per order; $A_b =$ Rs. 35 per order; $h_v =$ Rs. 3 per unit per year; $h_b =$ Rs. 10/unit/unit time. The optimal values are $K^* = 0.0304$;

$Q^* = 565.6854$ units; $d(K^*) = 0.0286$; TCDWO (Total Cost for Decentralized Without Coordination) = Rs. 4,614; TCDW (Total Cost for Decentralized With Coordination) = Rs. 4,677; TC_c (Total Cost for centralized system) = Rs. 5,686.

6 Conclusion

In this study, we have investigated a quantity discount coordination strategy for a single-vendor single-buyer with backorders. In addition, to validate the efficiency of the proposed quantity discount strategy, the system optimization problem under centralized decision making is discussed analytically. We prove that the decentralized quantity discount strategy can achieve system optimization and win-win outcome.

Acknowledgments Authors thank the anonymous referees for their useful comments and suggestions for improving this paper. This research work is supported by UGC-SAP (Special Assistance Programme), Department of Mathematics, Gandhigram Rural Institute (Deemed University), Gandhigram, Tamil Nadu, India.

References

1. Viswanathan S, Piplani R (2001) Coordinating supply chain inventories through common replenishment epoch. Eur J Oper Res 129(2):277–286
2. Gurnani H (2001) A study of quantity discount pricing models with different ordering structures: order coordination, order consolidation, and multi-tier ordering hierarchy. Int J Prod Econ 72(3):203–225
3. Sarmah SP, Acharya D, Goyal SK (2006) Buyer vendor coordination models in supply chain management. Eur J Oper Res 175(1):1–15
4. Stadtler H (2007) A general quantity discount and supplier selection mixed integer programming model. OR Spectrum 29(4):723–744
5. Mendoza A, Ventura JA (2008) Incorporating quantity discounts to the EOQ model with transportation costs. Int J Prod Econ 113(2):754–765
6. Duan Y, Luo J, Huo J (2010) Buyer-vendor inventory coordination with quantity discount incentive for fixed lifetime product. Int J Prod Econ 128(1):351–357

Corrosion Inhibition of Mild Steel in Hydrochloric Acid Medium by 1-Methyl-3-Ethyl-2, 6-Diphenyl Piperidin-4-One Oxime

K. Tharini, K. Raja and A. N. Senthilkumar

Abstract The corrosion inhibitory effect of 1-methyl-3-ethyl-2, 6-diphenyl piperidin-4-one oxime (PO) against mild steel (MS) in 1 M hydrochloric acid medium has been investigated by weight loss study, electrochemical methods, SEM, and theoretical studies. The weight loss studies were conducted at four different temperatures such as 30, 40, 50, and 60 °C for various concentrations (0, 25, 50, 100, 150, 200, 250, 300, 400 and 500 ppm) over 2 h duration. The study showed that inhibition efficiency (IE) increases with increase of PO concentration and decreases with increase of temperature. It was found that inhibition was due to adsorption of PO on the MS surface obeying Temkin's adsorption isotherm. The calculated values of free energy of adsorption (ΔG_{ads}) support physisorption mechanism. Electrochemical data for corrosion processes such as corrosion potential (E_{corr}), corrosion current (i_{corr}), and Tafel slopes (b_a and b_c) were determined using Tafel plot, which showed that increase in concentration of PO decreases corrosion current and behaves as mixed mode inhibitor. AC impedance measurement as determined by Nyquist plot revealed that charge transfer resistance increases with increase of concentration, whereas double layer capacitance decreases with increase of concentration complimenting each other. SEM studies revealed the surface protecting ability of PO in HCl medium. Quantum chemical studies illustrated that the corrosion IE of PO is due to the electrons of phenyl ring.

Keywords 1-methyl-3-ethyl-2 • 6-diphenyl piperidin-4-one oxime • Corrosion inhibition • Polarization • Impedance • SEM and quantum chemical studies

---

K. Tharini
PG and Research Department of Chemistry, Government Arts College,
Trichy 620022, India

K. Raja
Department of Chemistry, Raja Saraboji Government College, Thanjavur 613005, India

A. N. Senthilkumar (✉)
PG and Research Department of Chemistry, Alagappa Government Arts College,
Karaikudi 630003, India
e-mail: ansent@gmail.com

S. Sathiyamoorthy et al. (eds.), *Emerging Trends in Science,*
Engineering and Technology, Lecture Notes in Mechanical Engineering,
DOI: 10.1007/978-81-322-1007-8_66, © Springer India 2012

1 Introduction

Hydrochloric acid is employed for acid pickling, industrial cleaning, acid descaling, and oil well acidizing operations in industries. Mild Steel (MS) is an industrial structural material that undergoes aggressive corrosion in HCl medium. One of the important methods of protection of metals against corrosion is by employing inhibitors. Many organic compounds are recognized as effective inhibitors, for the corrosion of MS in HCl acid medium because organic molecules containing hetero atoms such as N, O, and S form coordinate bond with metal or alloys owing to their free electron pairs and act as good inhibitors. Compounds with π electrons also exhibit inhibitive character due to interaction between metal and alloy with π orbitals [1–6]. In the present study, the inhibition property of 1-methyl-3-ethyl-2, 6-diphenyl piperidin-4-one oxime (PO) on the acid corrosion of MS is investigated.

2 Experimental

2.1 Sample Preparation

MS specimens of following composition C–0.13 %, P–0.032 %, Si–0.014 %, S–0.025 %, Mn–0.48 %, and Fe remainder were mechanically cut into specification of $4 \times 1 \times 0.2$ cm, cleaned and scrubbed with emery paper to expose clean shining surface and degreased with acetone.

2.2 Preparation of PO

The precursor ketones viz., 1-methyl-3-ethyl-2, 6-diphenyl piperidin-4-one were prepared by the method of Baliah et al. [7]. Then this ketone was treated with filtrate formed from hydroxyl amine hydrochloride and sodium acetate in ethanol and refluxed for 4 h and finally poured into water. The product viz., 1-methyl-3-ethyl-2, 6-diphenyl piperidin-4-one oxime was re-crystallized from ethanol and duly characterized using ^{1}H and ^{13}C NMR spectra and their structures are depicted in Fig. 1.

2.3 Weight Loss Method

MS coupons were immersed in pure 1 M HCl containing various concentrations of PO for 2 h time interval at temperatures 30–60 °C. The percentage inhibition efficiency (IE) and rate of corrosion (CR) were calculated using Eqs. 1, 2, and 3.

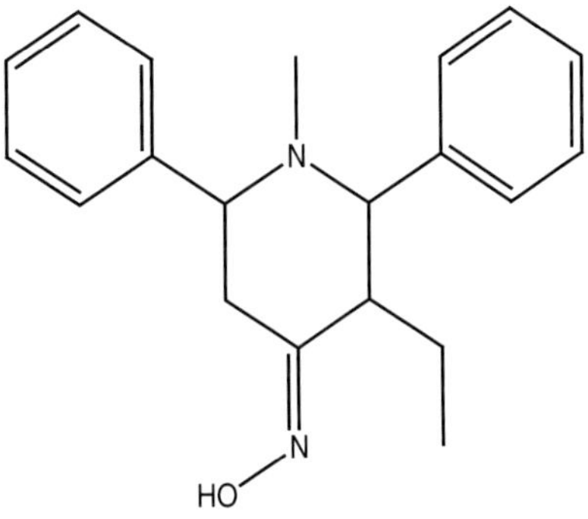

Fig. 1 Structure of 1-methyl-3-ethyl 2, 6-diphenyl piperidin-4-one oxime (PO)

$$\theta = \frac{W_0 - Wi}{W_0} \tag{1}$$

$$IE = \theta \times 100 \tag{2}$$

$$CR = \frac{\text{Weight loss in mg}}{\text{Surface area in cm}^2 \times \text{immersion period in h}} \tag{3}$$

where

W_o	Weight loss without PO
W_i	Weight loss with different concentrations of PO

2.4 Polarization and Impedance Studies

Potentiodynamically, the Tafel polarization curves were recorded using computerized CHi 604 c model. In this setup, Pt electrode, calomel electrode, and MS specimens were used as auxiliary, reference, and working electrodes, respectively, which were immersed in the presence and absence of PO. Impedance studies were carried out in the frequency range of 10 kHz–10 mHz for MS in 1 M HCl with and without different concentrations of PO. IE was calculated using R_{ct} as follows:

$$IE = 1 - \frac{R_{cto}}{R_{cti}} \times 100 \tag{4}$$

where

R_{cto}	Charge transfer resistance in absence of PO
R_{cti}	Charge transfer resistance in presence of various concentrations of PO

2.5 SEM Analysis

SEM micrographs were taken using computerized electron microscope (Philips XL series). All the micrographs were taken under the resolution of 750x.

2.6 Quantum Chemical Studies

The quantum chemical study was done using Dewar's LCAO–SCF–MO semiempirical method, AM1 in the commercially available computer package program in an Intel Pentium duo core processor computer.

3 Results and Discussion

Table 1 shows calculated IE and CR values for PO inhibition for MS dissolution in acid medium. From the table, it is clear that IE increases with increase of PO concentration. Increase of IE with increase of PO concentration revealed that process is under adsorption control. CR increases with increase of temperature indicating physisorption [8]. Several adsorption isotherms were assessed and the Temkin's adsorption isotherm was found to be the best. From the slopes and intercepts of plot, free energy of adsorption was calculated (<40 kJ/mol) which showed that adsorption of PO on MS surface obeys physisorption mechanism [9].

Tafel polarization curves of MS in 1 M HCl solution with different concentrations of PO are shown in Fig. 2. Increase in concentration of PO causes slight

Table 1 Weight loss measurements for the MS dissolution immersed in 1 M HCl and in different concentrations of PO

Concen-tration (ppm)	Temperature (K)							
	303		313		323		333	
	IE	CR	IE	CR	IE	CR	IE	CR
0	–	3.09	–	7.24	–	9.57	–	12.80
25	39.2	1.88	37.9	4.50	35.6	6.16	31.4	8.78
50	42.6	1.77	39.7	4.37	37.7	5.96	31.9	8.72
100	47.8	1.61	42.3	4.18	40.9	5.66	33.4	8.53
150	51.4	1.50	47.3	3.82	45.7	5.20	37.6	7.99
200	55.8	1.37	50.9	3.55	50.8	4.71	40.3	7.65
250	60.5	1.22	56.4	3.15	53.4	4.46	41.8	7.46
300	77.8	1.03	58.2	3.02	54.8	4.33	43.4	7.25
400	80.8	0.84	63.1	2.67	57.9	4.03	47.2	6.76
500	82.2	0.64	67.9	2.32	60.3	3.80	54.3	5.85

Corrosion Inhibition of Mild Steel in Hydrochloric Acid

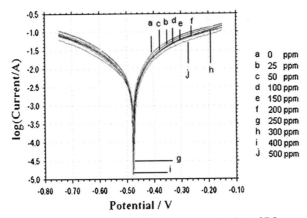

Fig. 2 Tafel curves in presence and absence of various concentration of PO

Table 2 Electrochemical parameters for the corrosion process of MS immersed in 1 M HCl and in various concentrations of PO

Concentration (ppm)	E_{corr} mV vs SCE	I_{corr} $\mu A/cm^2$	b_a mV/dec	b_c mV/dec	C_{dl} $\mu F/cm^2$	R_{ct} Ohm cm^2	IE Rct
0	−479	3612	220	280	212	20	–
25	−476	2511	160	210	157	27	25.9
50	−472	1995	120	180	132	32	37.5
100	−474	1584	100	110	103	41	51.2
150	−475	1258	180	220	92	46	56.5
200	−475	1000	200	260	88	48	58.3
250	−469	794	130	220	69	61	67.2
300	−471	630	170	230	61	70	71.4
400	−479	501	120	180	59	72	72.2
500	−475	398	70	90	54	78	74.3

shifting of corrosion potential on both the directions with variation in Tafel slopes indicating mixed mode inhibiting action of PO. i_{corr} values decreased with increase of PO concentration (Table 2) which indicates the corrosion controlling property of PO. Nyquist plot in the presence and absence of various concentrations of PO in 1 M HCl is shown in Fig. 3. The dispersion obtained in Nyquist plot was due to the dispersive capacitive loop and the inhomogeneities on the electrode surface [10]. Rct increases with increase of PO concentration whereas C_{dl} decreases with increase of PO concentration indicating the protection efficiency of PO.

SEM micrographs which were taken under 750x resolution is shown in Figs. 4, 5, and 6 for brightly polished MS surface, MS specimen exposed to 1 M HCl and MS exposed to PO in 1 M HCl, respectively. Figure 5 revealed the oxide inclusion on the surface of MS which were damaged in presence of 1 M HCl as seen from Fig. 6. Figure 7 explained surface protection efficiency exhibited by PO.

Fig. 3 Nyquist plot for pure 1 M HCl and different concentration of PO

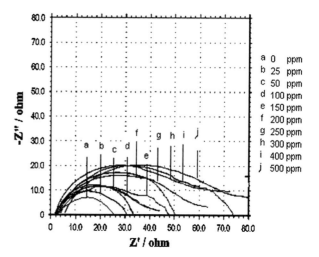

Fig. 4 Micrograph of brightly polished MS surface

Quantum studies done for PO with its E_{HOMO} are shown in Fig. 7. E_{HOMO} of PO is -9.121939 eV and that of its E_{LUMO} is 0.362866 eV. The high E_{HOMO} values of PO indicate the good inhibiting nature of PO. Moreover, energy diagram of HOMO revealed that the protection of PO is due to π electrons present in the benzene ring of the molecule and lone pair of electrons present on N atom attached to the ring. The adsorption of PO on MS can take place via following four ways [11]

1. Electrostatic attraction between charged PO and MS surface.
2. Interaction of uncharged electron pairs of PO on MS surface.
3. Interaction of π electrons of PO with MS.
4. Combination of 1–4.

Corrosion Inhibition of Mild Steel in Hydrochloric Acid   703

Fig. 5 SEM analysis of MS surface exposed to 1M HCl

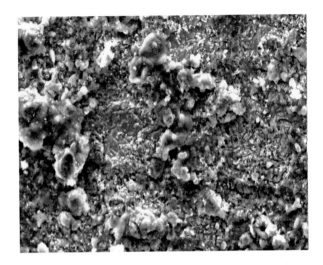

Fig. 6 SEM picture of MS surface exposed to 1 M HCl in presence of PO

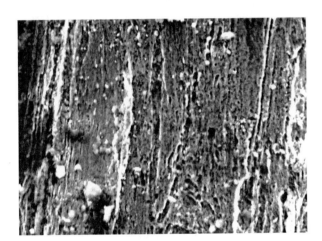

Fig. 7 HOMO surface of PO

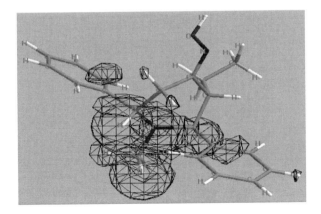

The third way would be the most probable mode for the protection efficiency of PO on MS surface as evidenced from computational studies.

4 Conclusions

The following conclusions were made from the study. PO decreases the corrosion rate of MS in acid medium in a dose dependent manner. Adsorption of PO on MS Surface follows Temkin's adsorption isotherm. E_{corr}, b_a, and b_c values have not been shifted to particular direction indicating PO as mixed mode indicator. Decrease of I_{corr} with increase of PO shows PO as a good inhibitor. R_{ct} increase with increase of PO shows PO as a surface protector. SEM revealed surface film forming ability of PO. Quantum chemical studies showed clearly the adsorption of PO on MS surface is by the π electrons and lone pairs of electron present in aniline moiety of PO.

Acknowledgments Authorities of AGAC, Karaikudi, GAC, Trichy and RSGC, Thanjavur are thanked for their help and encouragement.

References

1. Senthilkumar AN, Tharini K, Sethuraman MG (2012) Steric effect of alkyl substituted piperidin-4-one oximes for corrosion control of mild steel in H_2SO_4 medium. Acta Phys Chim Sin 28(2):399–406
2. Senthilkumar AN, Tharini K, Sethuraman MG (2009) Corrosion inhibitory effect of few piperidin-4-one oximes on mild steel in hydrochloric acid medium. Surf Rev Lett 16(1):141–147
3. Mernari B, Elattari H, Traisnel M, Bentiss F, Lagrenee M (1998) Inhibiting effects of 3,5-bis(n-pyridyl)-4-amino-1,2,4-triazoles on the corrosion for mild steel in 1 M HCl medium. Corros Sci 40(2–3):91–399
4. Stoyanova AE, Sokolova EI, Raicheva SN (1997) The inhibition of mild steel corrosion in 1 M HCl in the presence of linear and cyclic thiocarbamides—Effect of concentration and temperature of the corrosion medium on their protective action. Corros Sci 39(9):1595–1604
5. Cheng XL, Ma HY, Chen SH, Yu R, Chen X, Yao ZM (1999) Corrosion of stainless steels in acid solutions with organic sulfur-containing compounds. Corros Sci 41(2):321–323
6. Elmorsi MA (1998) Using acrylamide and oligo (oxyethylene) methacrylate in corrosion protection of steel. Corros Sci 41(2):305–320
7. Baliah V, Noller CR (1948) The preparation of some piperidine derivatives by the mannich reaction. J Am Chem Soc 70(11):3853–3855
8. Ebenso EE, Okafor PC, Ekpe Uf (2003) Studies on the inhibition of aluminium corrosion by 2-acetylpheno thiazine in chloro acetic acid. Anti Corros Methods and Mater 50(6):414–421
9. Ashassi-Sorkhabi H, Majidi MR, Seyyedi K (2004) Investigation of inhibition effect of some amino acids against steel corrosion in HCl solution. Appl Surf Sci 225(1–4):176–185
10. Yurt A, Balaban A, Ustun Kandemir S, Bereket G, Erk B (2004) Investigation on some schiff bases as HCl corrosion inhibitors for carbon steel. Mater Chem Phys 85(2–3):420–426
11. Shorky H, Yuasa M, Sekine I, Issa RM, El-Baradie HY, Gomma GK (1998) Corrosion inhibition of mild steel by schiff base compounds in various aqueous solutions. Corros Sci 40(12):2173–2186

New Mutation Embedded Generalized Binary PSO

Yograj Singh and Pinkey Chauhan

Abstract Particle Swarm Optimization (PSO) emerged as a potential global optimizer among other population based heuristics for solving continuous as well as discrete valued problems. This paper proposes a new modified binary PSO, which employs a generalized sigmoid function as a binary number generator with an additional benefit of controlling its only parameter to exploit the combined effect of sigmoid as well as linear function. Further, a logical mutation operator is also introduced to prevent stagnation of particles when algorithm does not show any improvement in objective function value for certain number of iterations. The proposed variant is termed as "Generalized Binary PSO with Mutation (GBPSOM)". The local and global version of proposed variant are tested against a set of well-known benchmark functions and results are compared with corresponding versions of standard Binary PSO. Numerical results indicate the efficiency and reliability of proposed variant over standard version.

Keywords Binary PSO • Generalized sigmoid function • Parameters • Mutation

1 Introduction

Swarm intelligence deals with the synchronized and collective behaviors of simple foragers interacting locally with each other and their environment which results in geometrically coordinated path patterns. These models are inspired by the social behavior of insects and other animals. Particle Swarm Optimization (PSO) is categorized as a swarm intelligence paradigm which simulates the swarming behavior of bird flocks and fish schools. The idea was first acknowledged by Kennedy and

---

Y. Singh (✉)
Department of Mathematics, University of Delhi, Delhi, India
e-mail: yograjchauhan26@gmail.com

P. Chauhan
Department of Mathematics, Indian Institute of Technology Roorkee, Roorkee, India
e-mail: pinkeychauhan030@gmail.com

S. Sathiyamoorthy et al. (eds.), *Emerging Trends in Science,*
Engineering and Technology, Lecture Notes in Mechanical Engineering,
DOI: 10.1007/978-81-322-1007-8_67, © Springer India 2012

Eberhart in 1995 [1]. An easily implementable code and less complex computational environment make it preferable over other population-based heuristics. Primarily, it was introduced for solving continuous optimization problems, but later on a discrete version [2] was proposed by Kennedy and Eberhart to handle problems generated in discrete search space. In PSO, particles are considered as simple flying agents roaming around a multidimensional search space. Each flying agent (i.e. particle) refers to a potential solution and the group of particles refers to a set of solutions called 'swarm'. Each particle is characterized by its velocity and position along with its personal and social experience. The movement of each particle is regulated by the best performer of entire swarm or of its neighborhood. The continuous and binary version adopts the same searching process up to termination of algorithm, except the change of position update equation which includes a different strategy to generate required binary or discrete numbers. In standard binary PSO, binary patterns are generated by employing a sigmoid function having a range within [0,1] with an input from basic velocity update equations. Despite having several attractive features and a potential global optimizer, PSO alike several other population based search algorithms have certain drawbacks. The main problem arises in case of multimodel or complex combinatorial problems when PSO is likely to be trapped in bad regions or local optima, while approaching a near optimal solution. A slow convergence rate and increased computational complexity are other problems associated with it.

Several amendments have been proposed in various studies to improve PSO's performance and get over its weaknesses. A binary PSO variant with a newly defined velocity vector and having a better interpretation of continuous problems into discrete problems was proposed in [3], which claims better performance over old standard versions. Some studies have implemented ideas, such as mutation [4] and genotype-phenotype concept [5], borrowed form other heuristics like evolutionary algorithm for improving the performance of BPSO. The idea of using logical functions like Boolean functions for generating binary numbers in PSO has been proposed in [6], which is further applied to solve antenna design problems. A binary PSO variant, based on the concept of angle modulation has been proposed in [7], which reduces the complexity of binary problems by mapping high-dimensional problems to a four-dimensional problem defined in continuous space. Since the notion of chaos is gaining popularity in all fields, therefore some studies [8, 9] have also embedded chaos theory in binary PSO for performance enhancement. In Ref. [10], binary PSO is integrated with lambda-iteration method for solving unit commitment problem. In Ref. [9], correlation-based feature selection (CFS) and the taguchi chaotic binary particle swarm optimization (TCBPSO) were intertwined to develop a hybrid method for handling gene selection and classification problems. Many other application-oriented versions of binary PSO have been proposed in various studies [11–14], resulting an improved problem specific version of BPSO.

This paper proposes a new modified binary PSO, where basic sigmoid function (Logistic Function) in BPSO is replaced by generalized logistic function, which is used as binary number generator. In the present study, the basic logistic function is generalized by adding a new parameter in order to create the combined effect of sigmoid as well as linear function. The basic sigmoid function as employed in standard BPSO has good exploration capability in initial stages but may not be able to perform exploitation efficiently in later stages due to a disadvantage of decreasing diversity as the search progresses. The linear function has a fair chance of producing a diversified population even in later stages. Herein, we tried to combine the positive sides of both the function into a single one by controlling this new parameter of generalized logistic function. Further, a logical mutation operator is also introduced for avoiding stagnation of particles, when algorithm does not show any improvement for certain number of iterations.

The rest of this chapter is organized as follows: Sect. 2 presents a brief introduction of traditional PSO. The newly proposed binary PSO has been described in Sect. 3. The computational studies are carried out in Sect. 4. Discussion of results is performed under Sect. 5, and chapter ends with concluding remarks in Sect. 6.

2 Standard PSO

Here, first a brief introduction to continuous PSO version is presented and afterwards, the discussion is extended to explain the standard binary PSO.

2.1 Standard Continuous PSO

The mathematical equations that guide particle's flight through a "multidimensional search space" are velocity and position update equations. The updating process combines information collected from three components; previous memory, personal experience, and experience gained through social interaction of particles in a specified neighborhood. The basic PSO update rules are as:

$$v_{id}(t+1) = w * v_{id}(t) + c_1 r_1 \left(p_{\text{best}}(t) - x_{id}(t)\right) + c_2 r_2 \left(p_{\text{gbest}}(t) - x_{id}(t)\right) \tag{1}$$

$$x_{id}(t=1) = v_{id}(t) + x_{id}(t) \tag{2}$$

where notation $d = 1, 2,\dots$ D stands for the dimension of search space and $i = 1, 2,\dots$ S represents the size of the swarm. Parameters "$c1$" and "$c2$" are constants, called cognitive and social scaling parameters, respectively, which quantify the contribution of personal and social experiences. The parameters "$r1$" and "$r2$" maintain diversity in the population and are uniformly distributed over the range [0,1]. $x_{id}(t)$, $v_{id}(t)$ are defined, respectively, as the position and velocity of dth dimension of ith particle at tth iteration. $p_{\text{best}}(t)$ is the best position of ith particle in dth dimension and is evaluated based on particle's personal experience. $p_{\text{gbest}}(t)$ is the best position found by whole swarm so far. w is inertia weight, large values of w facilitate exploration, with increased diversity, while a small value promotes local exploitation. Also, there is a parameter called V_{max}, which bounds the velocities of particles and prevents them from leaving the feasible search space and jumping into infeasible space.

2.2 Standard Binary PSO

The basic binary PSO differs from the above-described continuous version at the position updating stage only, while rest of the searching process is same for both the variants. In binary PSO, the position updating rule converts into a binary number generator. The whole swarm updates its directions using basic velocity update equation as in continuous PSO, while the particle's position is decided by a probability-based function named sigmoid function. In standard binary PSO, sigmoid function is employed to determine the probability for generating binary numbers, 0 and 1, or map all the positions in the range [0, 1]. The velocity update equation is fed as input to the sigmoid function so that particles could be benefited with its previous positions and previous memory. The sigmoid function is given by;

$$\text{sigm}(V_{id}) = \frac{1}{1 + \exp(-V_{id})} \tag{3}$$

where, V_{id} is the velocity of ith particle in d-dimension. The position update equation turns into a probabilistic equation and is expressed as;

$$x_{id}(t+1) = \begin{cases} 1 & \text{if } U(0, 1) < \text{sigm}(V_{id}(t)) \\ 0 & \text{otherwise} \end{cases} \tag{4}$$

where $U(0, 1)$ is a quasirandom number selected from a uniform distribution in the range [0, 1]. It is evident from Eq. (4) that as V_{id} increases then $\text{sigm}(V_{id}(t)) \rightarrow 1$ i.e., the chance of generating 1 bit is increased in solution patterns. This implies that most of generated particles will have the same bit pattern which causes diversity loss, a big problem, which affects the performance of BPSO to greater extent. Therefore, it is suggested to set $V_{id} \in [-6, 6]$.

3 Proposed Binary PSO: Modified Binary PSO (GBPSOM)

In standard binary PSO, the probability of flipping a bit from 0 to 1 or vice versa is decided by basic logistic function, which maps the velocity vector in the range [0, 1]. The Logistic function follows S-shape trajectory with nonmonotonic property. Further, the concave shape of sigmoid function causes low exploration for some bigger V_{id} values. Thus for a more diversified search, we have generalized the basic sigmoid function by introducing a new parameter and then exploit the properties of sigmoid as well as linear function into a single one by controlling this new parameter. The generalization helps in better exploration even in case of large velocity values, and on the other hand maintains diversity even in the later iterations. Additionally, a logical mutation operator is also proposed in this study to avoid stagnation of particles if no improvement is observed in objective function value for certain number of iterations. The mathematical relation for generalized logistic function is given by;

$$\text{Gen\_sigm}(X) = \frac{1}{1 + \exp(-b * X)} \quad (5)$$

where, the parameter b decides the steepness of function. As the parameter b varies, the shape of generalized logistic function swings between sigmoid and linear shape. The behavior of generalized logistic function for different values of parameter b is illustrated in Fig. 1. It is observed that as b changes its value from the set {0.2, 0.3, 0.4, 0.5, 0.7, 1.0}, the graph changes its shape from linear to S-shape, respectively. This behavior predicts that the manipulation of parameter b can change the probability of generating binary bits at every iteration and in this way the benefits of a linear and sigmoid function can be intertwined. In the present study, parameter b is linearly decreased iterationwise in the range (1.0, 0.2) to provide better exploration in early iterations and maintaining diversity in later stages as well.

After being determined Gen_sigm(X) using (5) with specified parameter settings, the particle positions are generated as :

$$x_{id}(t+1) = \begin{cases} 1 & \text{if } U(0,1) < \text{Gen\_sigm}(V_{id}(t)) \\ 0 & \text{otherwise} \end{cases} \quad (6)$$

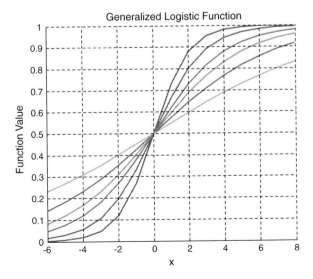

Fig. 1 Behavior of generalized logistic function for different values of parameter b

Further, mutation process is introduced to prevent the stagnation of particles. The proposed mutation operator is inspired by the logics of Boolean function which is basically used to handle binary inputs. Here, if for t number of iterations, the objective function value remains same then global best particle of the swarm will be mutated using proposed mutation operator. Here, only 10 % of total dimensions are randomly selected for mutation. A selected particle is mutated using Boolean mutation operator as follows:

$$X_{\text{Mut}}(d) = ! X_{\text{Gbest}}(d) \tag{7}$$

Hence, the proposed variant maintains the exploration and exploitation tradeoffs by controlling parameter b and also provides better chances for escaping out of local optima using a boolean mutation operator when particle starts to stagnate in poor valleys. The working of proposed GBPSOM is shown by an algorithm as follows:

Algorithm for GBPSOM

BEGIN
Create and Initialize a D-dimensional swarm, S
For t = 1 to the maximum number of iterations
For i = 1 to S,
For d = 1 to D,
Apply the particle velocity using Eq. (1)
Update particle position using Eqs. (5) and (6)
End-for-d;
Compute fitness of updated position;
End-for-i;
Update personal best, Pi, and global best information, P_g;
If(Pg(t + 1) == Pg(t) for 10 % of Max_iter)
Apply binary mutation using Eq. (7);
Terminate if P_g meets problem requirements;
End-for-t;

---

4 Experimental Evaluation: Testing on Benchmark Problems

The efficiency of local and global versions of GBPSOM and BPSO is validated on a suite of five scalable benchmark functions given in Table 1. The problem size for benchmark problems is taken as 10. The four variants, BPSO with ring topology (BPSO-L), BPSO with star topology (BPSO-G), GBPSOM with ring topology (GBPSOM-L), and GBPSOM with star topology (GBPSOM-G) are considered to select the best-performing algorithm which can be recommended for solving real world problems successfully.

New Mutation Embedded Generalized Binary PSO

Table 1 Benchmark problems

Sr. no	Function	Formula	Range	Dimension	Minima
1	Sphere	$f(x) = \sum_{i=1}^{n} x_i^2$	$[-5.12, 5.12]$	10	0
2	Griewank	$f(x) = \sum_{i=1}^{n} \frac{x_i^2}{4000} - \prod_{i=1}^{n} \cos\left(\frac{x_i}{\sqrt{i}}\right) + 1$	$[-600, 600]$	10	0
3	Rosenbrock	$f(x) = \sum_{i=1}^{n-1} \left(100*(x_{i+1} - x_i^2)^2 + (x_i - 1)^2\right)$	$[-30, 30]$	10	0
4	Rastrigin	$f(x) = 10n + \sum_{i=1}^{n} \left(x_i^2 - 10\cos(2\pi x_i)\right)$	$[-5.12, 5.12]$	10	0
5	Ellipsoidal	$f(x) = \sum_{i=1}^{n} (x_i - i)^2$	$[-30, 30]$	10	0

4.1 Parameter Setting

For a fair comparison, GBPSOM-L, GBPSOM-G, BPSO-L, and BPSO-G run with the same parameter setting and on same computing environment. Although parameter values vary from problem to problem but for the sake of comparison the following parameter values are chosen after extensive computations.

- Swarm size(S) = 5 * number of variables.
- Acceleration coefficients: $c_1 = 2.8$, $c_2 = 1.3$.
- Max velocity $(V_{max}) = 6$.
- Number of runs = 50.
- Maximum number of iteration = 500.

4.2 Performance Evaluation Criteria

The efficiency of all considered variants is tested on 5 well-known benchmark functions existing in the literature. All the algorithms are implemented in C++ and the experiments are performed on a DELL System with 1.5 GHz speed and 4 GB of RAM under WINXP platform. For all benchmarks with known global optima, the termination criterion for all considered PSO variants is set to be as maximum number of iterations or when the known optimum is within 100 % of accuracy, whichever occurs earlier. A run in which the algorithm finds a solution satisfying $f_{min} - f_{opt} \leq 0.0$, where f_{min} is the best solution found when the algorithm terminates and f_{opt} is the known global minimum of the problem, is considered to be successful. For each method and problem, the following are recorded:

- Success rate (SR) = $\frac{\text{Number of successful runs}}{\text{Total number of runs}} \times 100$
- Average number of function evaluations (AFE).

- Average error (AE) $= \dfrac{\sum_{n} (f_{min} - f_{opt})}{n}$

  Where, n is the total number of runs.

- Average computational time over n runs.
- Standard deviation of error over n runs.
- Minimum error over n runs.

  Where, n is number of runs, taken as 50.

5 Results and Discussion

The computational studies, which verified the efficiency of BPSO-L, BPSO-G, GBPSOM-L, and GBPSOM-G are carried out in this section. A comparative study is performed based on different performance aspects such as Average Error, Standard Deviation, Minimum Error, Success Rate, Average Function Evaluation, and Average Computational Time. Table 2 summarizes the results of continuous set of considered benchmark problems. The convergence analysis of BPSO-L, BPSO-G, GBPSOM-L, and GBPSOM-G for Rosenbrock function is shown in Figs. 2 and 3, respectively, which predicts that both

Table 2 Comparison results for benchmark problems

Fun	Method	AE	ME	SD	SR(%)	AFE	ACT(Sec)
F1	**BPSO-L**	0	0	0	100	41567	2.0104
	GBPSOM -L	0	0	0	100	39463	1.789033
	BPSO-G	0	0	0	100	23566.6	1.557267
	GBPSOM -G	0	0	0	100	22716	1.431667
F2	**BPSO-L**	0	0	0	100	55100	2.682267
	GBPSOM -L	0	0	0	100	54833.33	2.393733
	BPSO-G	0.045149	0	0.047923	26	195133.333	16.41403
	GBPSOM -G	0.04072	0	0.035195	30	188350	15.57343
F3	**BPSO-L**	9.13	0	26.0547	70	174716.666	8.9828
	GBPSOM -L	8.82	0	30.482126	84	153983.3	6.715333
	BPSO-G	167.26	0	478.702	25	201216.6	16.7401
	GBPSOM -G	157.3	0	350.53	42	167800	13.65103
F4	**BPSO-L**	0	0	0	100	40716.66	2.1229
	GBPSOM -L	0	0	0	100	38783.33	1.7776
	BPSO-G	3.856628	0	9.832512	86	64750	5.357267
	GBPSOM -G	2.892	0	8.677	92	51583.33	4.067167
F5	**BPSO-L**	0	0	0	100	120166.66	6.322367
	GBPSOM -L	0	0	0	100	108300	4.758833
	BPSO-G	2.633333	0	2.750555	13	224500	19.0578
	GBPSOM -G	3.7	0	5.355	20	218633.3	18.33957

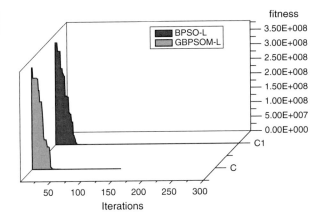

Fig. 2 Convergence graph of Rosenbrock function for local versions

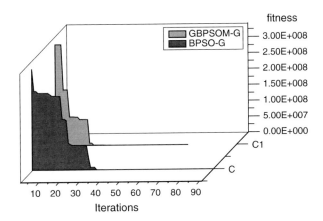

Fig. 3 Convergence graph of Rosenbrock function for global versions

GBPSOM-L and GBPSOM-G attain optimal solution in lesser number of iterations as compared to the corresponding versions of BPSO-L and BPSO-G. A comparative analysis of BPSO and GBPSOM could be seen at a glance using boxplots as in Fig. 4 on all performance aspects. The best-performing algorithm on each performance criteria is marked with star. The order of performance for considered set of benchmark problems with respect to different performance aspects is as:

$$\text{GBPSOM-L} > \text{BPSO-L and GBPSOM-G} > \text{BPSO-G}$$

The proposed variants GBPSOM-L and GBPSOM-G perform better than the corresponding BPSO variants. While GBPSOM-L clearly outperform all other variants on almost all performance criteria which indicate the capability and efficiency of proposed variant.

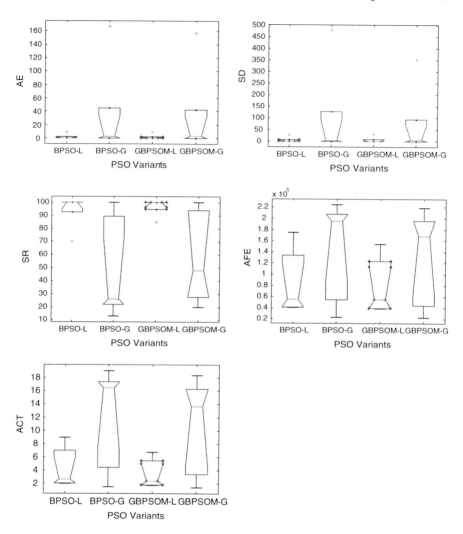

Fig. 4 Boxplots for different performance aspects for considered PSO variants

6 Conclusion

This paper proposes a modified scheme of employing generalized logistic function instead of basic sigmoid function as the binary number generator in binary PSO. Generalized logistic function has an additional benefit of combining the positive sides of both linear and sigmoid function by controlling its parameter. Further, a Boolean mutation operator is also embedded for preventing stagnation phenomena. The corresponding variant is termed as GBPSOM. The efficiency of Local and global versions of proposed GBPSOM is validated on a set of scalable benchmark functions and results are compared with corresponding versions of standard BPSO. The Comparative analysis shows that the local and global versions of proposed GBPSOM perform much better over the corresponding variants of BPSO. The computational analysis manifests the competency of GBPSOM. Thus, GBPSOM could be considered as a good alternative for handling binary as well as real-valued problems.

References

1. Kennedy J, Eberhart RC (1995) Particle swarm optimization. In: Proceedings of IEEE international joint conference on neural networks, pp 1942–1948
2. Kennedy J, Eberhart RC (1997) A discrete binary version of the particle swarm algorithm. In: Proceedings of IEEE 1997 conference systems man cybernetics, Piscataway, pp 4104–4108
3. Khanesar MA, Teshnehlab M, Shoorehdeli MA (2007) A novel binary particle swarm optimization. In: Proceedings of 15th mediterranean conference on control and automation, Athens-Greece (27–29 July 2007)
4. Lee S, Park H, Jeon M (2007) Binary particle swarm optimization with bit change mutation. IEICE Trans Fundam Electron Commun Comput Sci E-90A(10): 2253–2256
5. Lee S, Soak S, Oh S, Pedrycz W, Jeon M (2008) Modified binary particle swarm optimization. Prog. Nat. Sci 18:1161–1166 Elsevier
6. Marandi A, Afshinmanesh F, Shahabadi M, Bahrami F (2006) Boolean particle swarm optimization and its application to the design of a dual-band dual-polarized planar antenna. In: Proceedings of IEEE congress on evolutionary computation,CEC-2006, pp 3212–3218
7. Pampara G, Franken N, Engelbrecht AP (2005) Combining particle swarm optimization with angle modulation to solve binary problems. Evolut Comput IEEE Congress 1:89–96
8. Chuang L-Y, Yang C-H, Li JC (2011) Chaotic maps based on binary particle swarm optimization for feature selection. Appl Soft Comput 11:239–248
9. Chuang L-Y, Yang C-S, Wu K-C, Yang C-H (2011) Gene selection and classification using taguchi chaotic binary particle swarm optimization. Expert Syst Appl 38:13367–13377
10. Yuan X, Nie H, Su A, Wang L, Yuan Y (2009) An improved binary particle swarm optimization for unit commitment problem. Expert Syst Appl 36:8049–8055
11. Chuang L-Y, Chang H-W, Tu C-J, Yang CH (2008) Improved binary pso for feature selection using gene expression data. Comput Biol Chem 32:29–38
12. Chuang L-Y, Tsai S-W, Yang C-H (2011) Improved binary particle swarm optimization using catfish effect for feature selection. Expert Syst Appl 38, 12699–12707
13. Khalil TM, Youssef HKM, Aziz MMA (2006) A binary particle swarm optimization for optimal placement and sizing of capacitor banks in radial distribution feederes with distorted substation voltages. In: Proceedings of AIML 2006 international conference, Sharm El Sheikh, Egypt(June 13-15, 2006), pp 137–143
14. Luh G-C, Lin C-Y, Lin Y-S (2011) A binary particle swarm optimization for continuum structural topology optimization. Appl Soft Comput 11:2833–2844 Elsevier

On Fuzzy Fractal Transforms

R. Uthayakumar and M. Rajkumar

Abstract Approximation of an image, a union of spatially—contracted and gray-scale modified copies of subsets of itself is known as fractal image coding. Generally, images are considered as function u(x) in L^2 or L^∞, and fractal coding method is developed. In this article, a complete metric space $(\mathfrak{F}, d_\infty)$ of images on fuzzy fractal space $\tilde{X}$ is constructed. A method of fractal image coding on fuzzy fractal space is introduced. A fuzzy fractal transform T is defined over the space $(\mathfrak{F}, d_\infty)$, under suitable conditions, T: $\mathfrak{F} \to \mathfrak{F}$ is contractive, implying the existence of a unique fixed point.

Keywords Fractal transform • Fuzzy fractal space • Fuzzy iterated function systems • Fuzzy fractal transforms • Iterated function systems

1 Introduction

The basic principle of fractal image compression (FIC) was introduced by Barnsley in 1988. It is also known as fractal image encoding, because compressed image is represented by contractive transforms and mathematical functions which are necessary for reconstruction of original image. Contractive transform ensures that the distance between any two points on transformed image will be less than the distance of same points on the original image. These transforms are composed of the union of a number of affine mappings on the entire image, known as iterated function system (IFS) [1, 2]. Barsnley has derived a special form of the *Contractive Mapping Transform* (CMT) applied to IFS's called the *Collage Theorem* [3]. The usual approach of FIC is based on the collage theorem, which provides distance between the image to be encoded and the fixed point of a transform, in terms of the distance between the transformed image and the image itself. This distance is known as *collage error* and it should be as small as possible.

R. Uthayakumar · M. Rajkumar
Department of Mathematics, Gandhigram Rural Institute, Deemed University, Gandhigram, 624302 Dindigul, Tamil Nadu, India
e-mail: uthayagri@gmail.com

R. Uthayakumar · M. Rajkumar (✉)
Department of Mathematics, B.S.Abdur Rahman University, Vandalur, 600048 Chennai, Tamil Nadu, India
e-mail: m.rajkmar@gmail.com

S. Sathiyamoorthy et al. (eds.), *Emerging Trends in Science, Engineering and Technology*, Lecture Notes in Mechanical Engineering, DOI: 10.1007/978-81-322-1007-8_68, © Springer India 2012

Concisely, FIC seeks to approximate function of an image as a union of its modified copies. The net result is the target is approximated by the attractive fixed point of a contractive *fractal transform* operator that performs the shrinking and gray-level modifying operations on image functions. The approximation may be generated by the iteration procedure, where is a suitable starting "seed", for example, a blank screen.

Fuzzy sets are very good representatives which represent vagueness in everyday life. After Zadeh introduced it in 1965, many authors extended various parts of sub disciplines in mathematics, science and engineering to include fuzzy cases. Fuzzy set theory finds in image processing as a growing application domain. This may be explained not only by its ability to model the inherent imprecision of images together with expert knowledge, but also by the large and powerful toolbox it offers for dealing with spatial information under imprecision.

In this article, we attempt to define a fractal transform operator on the fuzzy fractal space which was introduced by Saidi [4, 5]. Saidi defined iterated function system for fuzzy fractal space. Using these IFS and by constructing new fuzzy function space we define a fuzzy fractal transform operator. The structure of the article is as follows: In Sect. 2, we state the fundamental concepts of the fractal transform using IFSM on the function spaces. In Sect. 3, we describe the development of iterated function system over fuzzy sets. In Sect. 4, we construct the image function space for defining fractal transform. In Sect. 5, we defined a fractal transform on the constructed function space and stated some results.

2 Iterated Function Systems and Fractal Transforms

In this section we give the outline of the fractal transforms. For more detailed reading refer [1–3, 6–9].

Let us begin by defining Iterated function systems. Let (X, d) denote a compact metric space, typically $[0, 1]^n$. Let $w = \{w_1, w_2, w_3, \ldots, w_N\}$ be a set of one-to-one contraction maps $w_i : X \to X$, with the contractive factors $c_i \in [0, 1)$, known as an N map IFS [2]. Define $c = \max_{1 \le i \le N} c_i$. The suitable complete metric space for IFS is the Hausdorff space $(H(X), h)$ formed by the nonempty compact subsets of X. Connected with the IFS maps w_i, define a set-valued mapping $\widehat{w}$ on the complete metric space $((H(X), h)$ [10], the action of which is defined to be $\widehat{w}(S) = \bigcup_{i=1}^n w_i(S)$, $S \in H(X)$, where $w_i(S) = \{w_i(x), x \in S\}$ is the image of S under w_i, $i = 1, 2, \ldots, N$. The contractivity of the mapping $\widehat{w}$ is follows.

Theorem 1 $\widehat{w}$ is a contraction mapping on $(H(X), h)$: [2] i.e.

$$h\left(\widehat{w}(A), \widehat{w}(B)\right) \le ch(A, B), A, B \in H(X)$$

On Fuzzy Fractal Transforms

Corollary 1

There exists a unique set $A \in H(X)$, such that $W(A) = A$, the so-called attractor of the IFS W [2]. Moreover, for any $S \in H(X), h(W^n(S), A) \to 0$ as $n \to \infty$.

After defining IFS, we are now ready to define fractal transform operator. We consider a general function space $\mathcal{F}(X)$ supported on X. The essential components of a fractal transform operator on $\mathcal{F}(X)$ are as follows.

1. A set of N one-to-one contraction maps $w_i : X \to X$ with the condition that $\bigcup_{i=1}^{N} w_i(X) = X$.
2. A set of Lipchitz grayscale maps $\varphi_i : R \to R$.i.e. for each φ_i there exist a $k_i \geq 0$ such that $|\varphi(t_1) - \varphi(t_2)| \leq k_i |t_1 - t_2|$ for all $t_1, t_2 \in R$

These two maps together said to comprise an "Iterated Function System with Greyscale Maps" (IFSM), denoted as (w, φ) [6, 7]. For each $x \in X$, this IFSM produces one or more fractal components defined as,

$$g_i(x) = \begin{cases} \varphi_i \left(u \left(w_i^{-1}(x) \right) \right), & \text{if } x \in w_i(X) \\ 0, & \text{otherwise} \end{cases}$$

Now define the fractal transform operator $T : \mathcal{F}(X) \to \mathcal{F}(X)$ as, $T(u(x)) = \sum_{i=1}^{N} g_i(x)$. The operator T combines all the fractal components. The following result establishes the condition for the contraction of the fractal transform operator.

Theorem 2

Let (w, φ) be an IFSM as defined above, with spatial contractions w_i and Lipschitz greyscale maps φ_i [7]. Then for $p \geq 1$ and $u, v \in L^p(X)$,

$$\|Tu - Tv\|_P \leq \sum_{i=1}^{N} c_i^{1/p} k_i \|u - v\|_p$$

Corollary 2 If $c = \sum_{i=1}^{N} c_i^{1/p} k_i < 1$, [7] then T is contractive in $L^p(X)$ with fixed point $\bar{u} \in L^p(X)$. The fixed point equation, $\bar{u}(x) = (T\bar{u})(x) = \sum_{i=1}^{N} \varphi_i \left(\bar{u} \left(w_i^{-1} \right) \right)(x)$, indicates that $\bar{u}$ is "self—similar", i.e. that it can be written as a sum of spatially—contracted and grayscale-modified copies of itself.

In short, generalized fractal transforms works as follows:

1. It produces a set of N spatially—contracted copies of u.
2. It modifies the values of these copies by means of a suitable range mapping.
3. It recombines the modified copies of u using an appropriate operator in order to get the element $v \in L^p(X), v = Tu$. The inverse problem can be posed by the Banach contraction mapping principle which is as follows.

Theorem 3 Let (Y, d_Y) be a complete metric space [6, 8]. Also, let $T : Y \to Y$ be a contraction mapping with contraction factor $c \in [0, 1)$,

i.e. for all $u, v \in Y, d_Y(Tu, Tv) \le c d_Y(u, v)$,
Then there exists a unique $\bar{u} \in Y$ such that $\bar{u} = T\bar{u}$. Moreover, for any $v \in Y, d_Y(T^n v, \bar{u}) \to 0$ as $n \to \infty$.

Theorem 4 (Continuity of fixed points [6, 8])

Let (Y, d_Y) be a complete metric space and $T : Y \to Y$ be a contraction mapping with contraction factor $c \in [0, 1)$. Then for any $u \in Y$,

$$d_Y(u, \bar{u}) \le \frac{1}{1-c} d_Y(u, Tu)$$

where u is the fixed point of T.

3 Fuzzy Fractal Space and Fuzzy Iterated Function Systems

In this section we state results derived by Saidi [4, 5] on IFS for fuzzy sets. Let (X, d) be a compact metric space. A fuzzy singleton x_t is defined as $x_t = \{(x, \mu_{x_t}) : x \in X, t \in [0, 1]\}$. It is also defined as $x_t(u) = \begin{bmatrix} 0, & x \ne u \\ t, & x = u \end{bmatrix}$.

We denote the set of all fuzzy singletons x_t by $\tilde{X}$ and which represent the fuzzy image base space. Define a distance function $d^* : \tilde{X} \times \tilde{X} \to R^+$ as $d^*(x_t, y_s) = \max\{d(x, y), |t - s|\}, \forall x_t, y_s \in \tilde{X}$.

Then d^* is a metric on $\tilde{X}$. Setup a space, $F(X) \subseteq P(\tilde{X})$ where $F(X) = \{A \in P(\tilde{X}) \setminus A : X \to I, A(x) = t\}$, and $P(\tilde{X})$ is the power set of $X \times I$, called as fuzzy space. The space $(F(X), d^*)$ is a complete metric space. Then define a mapping $w^* : \tilde{X} \to \tilde{X}$ as $w^*(x, t) = (w(x), rt), r \in [0, 1)$.

On Fuzzy Fractal Transforms 721

Theorem 5

w^* is a contraction mapping on the space $(F(X), d^*)$ [5].

Definition 1

The space $H(F(X)) = \{A \in F(X) / A \neq \emptyset, A \text{compact}\} \subseteq F(X)$, is called fuzzy Hausdorff space or fuzzy fractal space [5]. A distance function D^* on $H(F(X))$ will be defined by putting,

$$D^* = \max \left\{ \begin{array}{l} \sup_{x \in X} \inf_{y \in Y} \max \{d(x, y), |A(x) - A(y)|\}, \\ \sup_{y \in Y} \inf_{x \in X} \max \{d(x, y), |A(x) - A(y)|\} \end{array} \right\}$$

D^* can be proven to be a metric on $H(F(X))$. Also $H(F(X))$ complete metric space. Let us attempt to define contraction mappings on $H(F(X))$.

Let $w^* : H(F(X)) \to H((F(X))$ be defined as
$w^*(A) = \{w^*(x, t) \mid (x, t) \in A\}$, $A = \{(x, A(x)) \mid x \in X\}$, and

$w^*(A) = \{w^*(x), rt) \mid (x, t) \in A\} = \{(w(x), rA(x)) \mid x \in X\} \in F(X)$.

Then w^* is a contraction mapping on the space $(H(F(X)), D^*)$. From this, we write the fuzzy iterated function system as $(F(X), w_1^*, w_2^*, \ldots, w_N^*)$ where $F(X)$ is a complete metric space and each $w_r^* : F(X) \to F(X)$, is a contractive function with corresponding contractivity factor $s_r, r = 1, 2, 3, \ldots, N$.

Define $\qquad W^* : H(F(X)) \to H(F(X)) \qquad$ as

$W^*(B) = \bigcup_{i=1}^{N} w_i^*(B)$ for any $B \in H(F(X))$. $B = \{(x, B(x))/x \in X\}$.

Theorem 6

W^* is a contraction mapping provided w^* is a contraction [5].

4 Complete Metric Space for Fuzzy Fractal Transform

We turn our discussion to our main results. In this section, we develop a complete metric space of image functions for defining fractal transform.

Definition 2 Let u is a multifunction from $\tilde{X}$ to $F(X)$. The image $u(x_t)$ of a fuzzy singleton x_t is a fuzzy set in $F(X)$. $u(x_t, y)$ stands for the grade of membership of y in $u(x_t)$. The set $u(x_t)$ in $F(X)$ is defined by,

$$u(x_t)(y) = \sup_{x_t \in \tilde{X}} (u(x_t, y) \wedge t), \ y \in X.$$

If $u\ (x_t)$ is a closed, compact or convex we say that u is closed, compact or convex valued, respectively. In the following, we will suppose that u is compact for each $x_t \in \tilde{X}$. Define, $\mathfrak{F}\ (F\ (X)) = \left\{ u/u : \tilde{X} \to H(F\ (X)), \text{ i.e.}, u \text{ is compact}\right\}$ with the distance function defined as,

$$d_\infty\ (u, v) = \sup_{x_t \in \tilde{X}} D^*\ (u\ (x_t), v\ (x_t))$$

Theorem 7

The space $(\mathfrak{F}, d_\infty)$ is a complete metric space.

Proof From the definition of d_∞, it is clear that $d_\infty\ (u, v) \geq 0, \forall u, v.$

1.
$$d_\infty\ (u, v) = 0 \Leftrightarrow \sup_{x_t \in \tilde{X}} D^*(u\ (x_t), v\ (x_t)) = 0$$
$$\Leftrightarrow D^*(u\ (x_t), v\ (x_t)) = 0$$
$$\Leftrightarrow u\ (x_t) = v\ (x_t)$$
$$\Leftrightarrow u = v$$

2.
$$d_\infty\ (u, v) = \sup_{x_t \in \tilde{X}} D^*(u\ (x_t), v\ (x_t))$$
$$= \sup_{x_t \in \tilde{X}} D^*(v\ (x_t), u\ (x_t))$$
$$= d_\infty\ (v, u)$$

3.
$$d_\infty\ (u, v) = \sup_{x_t \in \tilde{X}} D^*(u\ (x_t), v\ (x_t))$$
$$\leq \sup_{x_t \in \tilde{X}} \left[D^*(u\ (x_t), w\ (x_t)) + D^*(w\ (x_t), v\ (x_t)) \right]$$
$$\leq \sup_{x_t \in \tilde{X}} D^*(u\ (x_t), w\ (x_t)) + \sup_{x_t \in \tilde{X}} D^*(w\ (x_t), v\ (x_t))$$
$$= d_\infty\ (u, w) + d_\infty\ (w, v)$$
$$\Rightarrow d_\infty\ (u, v) \leq d_\infty\ (u, w) + d_\infty\ (w, v), \ \forall u, v, w \in \mathfrak{F}$$

$\therefore (\mathfrak{F}, d_\infty)$ is a metric space. Now we prove the completeness of $(\mathfrak{F}, d_\infty)$. Let u_n be a Cauchy sequence of elements of $(\mathfrak{F}, d_\infty)$. So, $\forall \in> 0$, there exists $n_0\ (\in) > 0$ such that for all $n.\, m \geq n_0\ (\in)$. We have

On Fuzzy Fractal Transforms

$$d_\infty (u_n, u_m) \leq \in$$
$$\Rightarrow D^* (u_n (x_t), u_m (x_t)) \leq \in, \text{ for all } n. m \geq n_0 (\in)$$

$\Rightarrow u_n (x_t)$ is a Cauchy sequence in $H(F(X))$. Since $H(F(X))$ is a complete metric space, there exists $\bar{A} \in H(F(X))$ such that $D^* (u_n (x_t), \bar{A}) \to 0$ when $n \to \infty$. So for, all $x_t \in \tilde{X}$ and for all $n. m \geq n_0 (\in)$, we have

$$D^* (u_n (x_t), u_m (x_t)) \leq \in,$$

On making $m \to \infty$, we have,

$$D^* (u_n (x_t), \bar{A}) \leq \in,$$
i.e. $$\sup\nolimits_{x_t \in \tilde{X}} D^* (u_n (x_t), \bar{A}) \leq \in,$$
i.e. $$d_\infty (u_n, A) \leq \in$$
$\therefore (\mathfrak{F}, d_\infty)$ is a complete metric space $\qquad\qquad\qquad\qquad \Box.$

5 Fuzzy Fractal Transform

In this section, we construct a fuzzy fractal transform operator T on the space $(\mathfrak{F}, d_\infty)$. We now list the ingredients for a fuzzy fractal transform operator.

1. A set of N one-to-one contraction maps $w_i^* : \tilde{X} \to \tilde{X}$, with the condition that $\bigcup_{i=1}^{N} w_i^* \left(\tilde{X} \right) = \tilde{X}$.

2. A set of N grayscale maps $\varphi_i : H(F(X)) \to H(F(X))$ assumed to be Lipchitz. (i.e.) for each i, there exists a $\alpha_i \geq 0$ such that,

$$|\varphi_i (A) - \varphi_i (B)| \leq \alpha_i |A - B|$$

(i.e.) $D^* (\varphi_i (A), \varphi_i (B)) \leq \alpha_i D^* (A, B), \ \forall A, B \in H(F (X))$

Now, define the N fractal components of the fractal transform operator $T : \mathfrak{F} \to \mathfrak{F}$ defined by the above are as follows.

$$g_i (x_t) = \begin{cases} \varphi_i \left(u \left(w_i^{*-1} (x_t) \right) \right), & x_t \in w_i^* \left(\tilde{X} \right) \\ 0, & x_t \notin w_i^* \left(\tilde{X} \right) \end{cases}$$

And given an $u \in \mathfrak{F}$, $(T u) (x_t) = \sum_{i=1}^{N} g_i (x_t) = \sum_{i=1}^{N} \varphi_i \left(u \left(w_i^{*-1} (x_t) \right) \right)$.

Thus, the ith fractal component $g_i (x_t)$, scales the gray-level value of u in the prcimage $w_i^{*-1} (x_t)$ if it exists. We state some conditions for the contraction of the fuzzy fractal transform operator

Theorem 8 For $u, v \in \mathfrak{F}$, $d_\infty (Tu, Tv) \leq \left(\sup_{x_t \in \tilde{X}} \sum_{i=1}^N \alpha_i \right) d_\infty (u, v)$.

Proof

$$d_\infty (Tu, Tv) = d_\infty \left(\sum_{i=1}^N \varphi_i \left(u \left(w_i^{*-1}(x_t) \right) \right), \sum_{i=1}^N \varphi_i \left(v \left(w_i^{*-1}(x_t) \right) \right) \right)$$

$$= \sup_{x_t \in \tilde{X}} D^* \left(\sum_{i=1}^N \varphi_i \left(u \left(w_i^{*-1}(x_t) \right) \right), \sum_{i=1}^N \varphi_i \left(v \left(w_i^{*-1}(x_t) \right) \right) \right)$$

$$\leq \sup_{x_t \in \tilde{X}} \sum_i D^* \left(\varphi_i \left(u \left(w_i^{*-1}(x_t) \right) \right), \varphi_i \left(v \left(w_i^{*-1}(x_t) \right) \right) \right)$$

$$\leq \sup_{x_t \in \tilde{X}} \sum_i \alpha_i D^* \left(u \left(w_i^{*-1}(x_t) \right), v \left(w_i^{*-1}(x_t) \right) \right)$$

$$\leq \left[\sup_{x_t \in \tilde{X}} \sum_i \alpha_i \right] d_\infty (u, v)$$

$\square$

Corollary 3 It is clear that If $\sup_{x_t \in X} \sum_i \alpha_i < 1$, then T is contraction on $\mathfrak{F}$. Consequently, there exists a fixed point such that $\bar{u} = T\bar{u}$. The inverse problem can be stated as follows.

Theorem 9 Given an $u \in \mathfrak{F}$ suppose that there exists a contractive operator T such that $d_\infty (u, T(u)) < \in$. If u^* is the fixed point of T and $c = \sup_{x_t \tilde{X}} \sum_i \alpha_i$, then $d_\infty (u, u^*) \leq \frac{\in}{1-c}$.

Proof

$$d_\infty \left(u, u^* \right) \leq d_\infty (u, T(u)) + d_\infty \left(T(u), u^* \right)$$
$$= d_\infty (u, T(u)) + d_\infty \left(T(u), T \left(u^* \right) \right)$$
$$\leq d_\infty (u, T(u)) + c d_\infty \left(u, u^* \right).$$

Therefore, $d_\infty (u, u^*) \leq \frac{\in}{1-c}$..

References

1. La Torre D, Vrscay ER (2011) Generalized fractal transforms and self-similarity: recent results and applications, Rev Art Image Anal Stereol 30:63–76
2. Hutchinson J (1981) Fractals and self-similarity. Indiana Univ J Math 30:713–747

3. Barnsley MF, Demko S (1985) Iterated function systems and the global construction of fractals. Proc. Roy. Soc. London Ser A 399:243–275
4. Al-Saidi N (2002) On multi fuzzy metric space. Ph.D. thesis, Al-Nahreen University, Baghdad
5. AL-Sa'idi NMG, Said MRM, Ahmed AM (2009) IFS on the multi—fuzzy fractal space. In: World Academy of Science, Engineering and Technology, p 53
6. Alexander SK (2005) Multiscale methods in image modeling and image processing. PhD thesis. Department of Applied Mathematics, University of Waterloo
7. Forte B, Vrscay ER (1995) Theory of generalized fractal transforms, NATO ASI on fractal image encoding and analysis, Norway, 8–17 July 1995
8. Forte B, Vrscay ER (1994) Solving the inverse problem for function/image approximation using iterated function systems, 1. theoretical basis, fractals, vol 2(3), World Scientific Publishing Company, pp 325–334
9. Centore P, Vrscay ER (1994) Continuity of fixed points for attractors and invariant measures for iterated function systems. Can Math Bull 37:315–329
10. Barnsley MF (1993) Fractals everywhere, 2nd edn. Academic Press Professional, Inc., San Diego

Part XIII
Management Studies

An Application of Graph Theory Towards Portfolio Judgement

Tuhin Mukherjee and Arnab Kumar Ghoshal

Abstract This paper introduces a new outlook in portfolio management with the help of graph theory, a widely used discrete mathematical tool in engineering branches. For a given portfolio, this paper describes a methodology to compute a balance index. Empirical study of this paper finds that high balance index portfolio is recommended for better return. With such a new approach of risk analysis, investors will also be able to take their portfolio restructuring decision subject to the availability of investment capital. This new outlook opens number of challenging research area in both portfolio management and graph theory.

Keywords Graph • Complete graph • Portfolio • Balance index • Structural balance

1 Introduction

A portfolio refers to a collection of securities held by an investor. Selection of securities within a portfolio obviously depends on risk aversion of the investor. Thus, a good mix of securities implies different meaning to different people. Risk analysts usually consider return appreciation of securities pairwise within a portfolio. Types of securities included in a portfolio, may not necessarily be those which are positively correlated to each other [1]. The rationale of such selection is to protect against adverse performance by limiting the risk. For example, to hedge against adverse downside risk of stock market, investors can simultaneously hold a put option to offset potential losses. This paper will address such issues using a graph theoretical framework. To the best of the knowledge, authors claim that

---

T. Mukherjee
Department of Business Administration, University of Kalyani, Nadia, India
e-mail: tu_2002@rediffmail.com

A. K. Ghoshal (✉)
A.K.C.S.I.T, University of Calcutta, Calcutta, India
e-mail: mr.arnabghoshal@gmail.com

S. Sathiyamoorthy et al. (eds.), *Emerging Trends in Science,*
Engineering and Technology, Lecture Notes in Mechanical Engineering,
DOI: 10.1007/978-81-322-1007-8_69, © Springer India 2012

sufficient literature is presently not available to link graph theory and portfolio management.

The next section describes graph theoretical concept followed by graph modelling of portfolio. Subsequently proposed methodology is illustrated followed by empirical evidence in Indian context. The paper terminates with vital conclusions along with scope future expansion.

2 Concept of Graph Theory

A graph G is mathematically defined as tuple (V, E, φ) where V is a non-empty set whose elements are called vertices of G, E is a set whose elements are called edges of the graphs and φ is a function from E to $V \times V$ and called incidence function of the graph [2].

Geometrically, the graph $G = (\{v_1, v_2, v_3\}, \{e_1, e_2\}, \{(e_1, v_1, v_2), (e_2, v_1, v_2)\}$ can be represented as in Fig. 1.

Extending the notion of a graph, a signed graph is simply a graph where each edge corresponds to either positive or negative sign. For example, Fig. 2 is a signed graph.

Signed graph is relevant in describing the concept of structural balance [3]. According to the principle extolled by balance theory, a structure is balanced or stable iff all the cycles (closed walk between vertices and edges) of the signed graph are positive. Otherwise, it will be unbalanced. A cycle is said to be positive iff it has even number of negative edges in the cycle. Otherwise, it will be negative.

Fig. 1 A simple graph

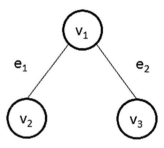

Fig. 2 A signed graph

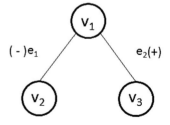

According to balance theory, any real-life scenario with unbalanced structure has a tendency always to undergo changes, (i.e. correction in bias) in order to restore its state of balance. Several approaches for determining the degree of balance associated with signed graph has been reported [4]. Among them, the triangular degree of balance index is perhaps the most elegant. The index is calculated by taking the ratio of the number of positive triangles (i.e. 3 vertices cycle) to the total number of triangles in the signed graph.

This paper requires the concept of complete graph also. G is said to be complete iff it includes all possible edges between its vertices. A complete graph with n vertices is denoted by K_n as in Fig. 3.

If we denote negative edges by dotted line then some possibilities of complete graphs with 3, 4, and 5 vertices are shown in Figs. 4, 5, and 6, respectively.

With these concepts of graph theory, now we are in a position to move in the next section describing link between graph theory and portfolio management.

3 Graph Modelling of Portfolio

A portfolio of n securities can be represented as a complete graph whose vertices are securities and edges are sign of correlation between vertices. Hence, a negative edge implies purpose of hedging. By analysing such portfolio graph, this paper recommends following benefits.

- The nature of portfolio graph can be used to judge the motive of investor. If protection against market shocks is the intention then graph should be balanced with at least one negative edge. On the other hand, a balanced graph with all

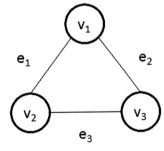

Fig. 3 A complete graph with three vertices (k_3)

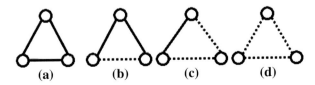

Fig. 4 Possible K_3

Fig. 5 Possible K_4

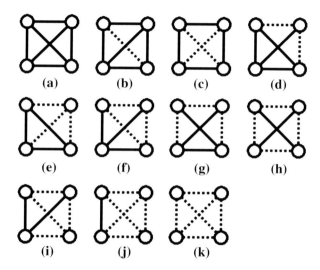

Fig. 6 Possible K_5

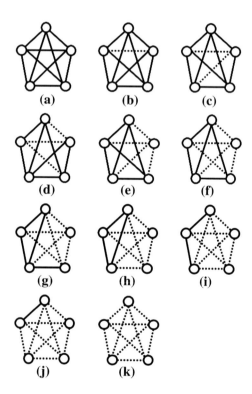

positive edges offers tendency to move either upside or downside. Hence, it will be speculative in nature.
- A given portfolio graph is able to suggest possible course of actions for investors. If the portfolio graph is not balanced then suggestive restructuring can

enhance the state of balance. On the other hand, for a given balanced graph, we can suggest structural transformations to move towards either hedging or speculation subject to risk.

4 Illustration and Proposed Methodology

Let us consider some possibilities of the portfolio graphs of 3, 4, and 5 vertices, respectively, as shown earlier in Figs. 4, 5, and 6. Only the balanced graphs are marked asterisk. Now, we will discuss two possible situations. Case-I depicts the portfolio is balanced initially and case-II considers when it is unbalanced initially.

Case-I: Evidently 5(a) is balanced and has all positive edges and hence unprotected against downside risk. In order to hedge, the paper finds two possible courses of actions (recommended for an investor):

- Replacement of one or more vertices within Portfolio graph.
- Removal or Incorporation of one or more vertices within portfolio graph.

Evidently 5(e) and 5(h) both graphs are balanced. A transformation from 5(a) to 5(h) will involve replacing more than one securities while 5(e) will need only one replacement.

A balanced K_n can be transformed to either balanced K_{n+m} or balanced K_{n-m} for some suitable natural number m. For example hedging can be done by incorporating stock option or future. It will imply transformation like 5(a) to 6(e).

Case-II: In this case, also we have two courses of actions as like as first case. To illustrate restructuring of unbalanced portfolio, we consider 6(a) which is unbalanced which can be transformed to balanced 6(e) without changing number of vertices. Even if a stable structure like 6(e) is not achievable, in order to reduce level of risk, investors may seek to restructure to one that has a higher degree of balance such as 6(h) or 6(g). Similarly to case-I, we can also restructure unbalanced portfolio by removal or incorporation of securities.

5 Empirical Analysis

This section finds the evidence that goodness of a portfolio can be determined by its degree of balance. To arrive at such empirical findings, we have considered four portfolios, each of which consists of 10 randomly selected stocks from National Stock Exchange of India, such that no two stocks within the same portfolio, belong to same sector. The list of stocks is given in sorted order in Table 1.

The sample is drawn with replacement among portfolios but without replacement within a portfolio. Naturally, the same stock may participate in different portfolios and restrict our choice in such a way that no two portfolios have more than 5 stocks (50 % of 10 stocks) in common. The merit of such selection is not to

Table 1 Four portfolios (ten stocks each)

No	Stock name (industry affiliation)-portfolio1	Stock name (industry affiliation)-portfolio2	Stock name (industry affiliation)-portfolio3	Stock name (industry affiliation)-portfolio4
1	Axis Bank Ltd (finance)	ABB LTD (capital goods)	ACC LTD (Housing related)	Axis Bank Ltd (finance)
2	BPCL (refineries)	Adani power Ltd (power)	Axis Bank Ltd (finance)	Bajaj Hindustan Ltd. (agriculture)
3	CESC Ltd (power)	Berger paints India Ltd (chemical and petrochemical)	Bilcare Ltd (health care)	Berger paints India Ltd (Chemical and Petrochemical)
4	CMC Ltd (software)	CIPLA (pharmaceutical)	CIPLA (pharmaceutical)	Bharat Forge Ltd (transport equipments)
5	DLF(housing related)	Dish TV India Ltd (media and publishing)	DLF (housing related)	DLF (housing related)
6	EIH Ltd (tourism)	EIH Ltd (tourism)	Escorts Ltd (transport equipments)	EIH Ltd (tourism)
7	FDC Ltd (healthcare)	Emami Ltd (FMCG)	Finolex industries Ltd (chemical and petrochemical)	Entertainment network(India) Ltd (media and publishing)
8	Grasim industries Ltd (textile)	Grasim industries Ltd (textile)	Grasim industries Ltd (textile)	Essar oil Ltd (refineries)
9	Havel's India Ltd (capital goods)	HCL Technologies Ltd (software)	GTL Infrastructure Ltd (telecom)	Havel's India Ltd (capital goods)
10	ITC Ltd (FMCG)	Hindustan Unilever Ltd (diversified)	Hindustan Unilever Ltd (diversified)	Hindalco industries Ltd (aluminium)

Source of Industry Affiliation NSE India

be biased with predefined sectorial correlation among stocks (as nicely pointed out by [5]. Depending on their average prices (i.e. (closing prices + opening prices)/2) from 1 April 2008 to 31 March 2012 (four consecutive financial years), we have completed pairwise correlation $\rho = \frac{\text{Cov}(x,y)}{\sigma x \, \sigma y}$ using statistical package SPSS and use the sign of ρ in our empirical study. Now for each of these portfolios, we constructed their portfolio graphs and recorded their balance index shown in Table 2.

It is evident from the finance literature that only return cannot be the proper tool for measuring performance of any portfolio. One must consider the risk undertaken by each portfolio also. For measuring effectiveness of our portfolios, we have used Sharpe, Treyner, and Jenson which are the proven tools in financial management.

With these measures, we computed our four portfolios to rank them and results are presented in Table 2. From those results, it is clear that portfolio graph with highest degree of balance is proved to be ranked first and similarly remaining

An Application of Graph Theory Towards Portfolio Judgement 735

Table 2 Rank of four portfolios (ten stocks each) versus balanced index of their graph

Portfolio	Balance index	Sharpe	Treyner	Jenson	Rank
1	0.31	−0.0089	−0.1370	−0.0640	4
2	0.72	0.0041	0.2056	0.2957	1
3	0.56	0.0015	0.0527	0.1492	3
4	0.62	0.0023	0.1663	0.2141	2

Result Computed

portfolios are ranked according as decreasing balance of index. Although the author agrees personally that to keep the degree of balance as the sole criteria to invest may not work fully but to significant extent this paper proved it viable for randomly selected portfolios from Indian stock market. It opens a new area of research towards stock market using a tool like graph theory which has almost never been used previously.

6 Conclusion

Main objective of this paper is to illustrate a methodology describing how complete signed balanced graph representation of portfolio can be used for risk analysis. It also provides a surprisingly elegant method to describe state of balance for a given portfolio graph. This paper recommends a portfolio graph which is complete signed and balanced for effective hedging but market dynamics can anytime change these characteristics of an existing portfolio. Proactive measures to restructure the portfolio graph may, therefore, be a necessary challenge in the finance literature (future scope). In the empirical part, this paper attempts to study the goodness of a portfolio in terms of its balance index. It has been found that more the balance index, more suitable the portfolio is. Hence, this paper recommends the investors to go for that portfolio which has more degree of balance. Although our illustrations are based on simple portfolio but the idea is generally applicable for portfolio involving larger cluster of assets. As the number of assets increases, the geometrical representation of portfolio graph should provide attractive alternatives for exposing structurally weak portfolios. Hence, finding optimal transformation (of portfolio graph structure) subject to the available investment capital constraint is another challenge in future. Such optimality of a restructuring can be perceived in terms of number of vertices (securities) involved, changing amount of investment capital, and so on.

References

1. Hull JC (1993) Options features and other Derivative Securities, (2nd edn). Prentice Hall, New Jersey
2. Donald CW (2001) Signed graphs for portfolio analysis in risk management. J Manage Math 13:201–210
3. Harary F (1961) A structural analysis of the situation in the middle east in 1956. J Conflict Resolut 5:167–178
4. Harary F (1961) A graph-theoretic approach to analysis. Research note
5. Basu S (1977) The investment performance of common stocks in relation to their price earning ratios. J Finance 2:663–682

A Study on CRM Influence in Small and Medium Retail B2B Industries in India

T. Narayana Reddy and G. Silpa

Abstract Retail Industry has a constant race to increase profits, keep the current customers, and gain or poach new ones, competing for customers on a globalized market like never before. Many sets of tools aimed at aiding the interaction between supplier and customer are the customer relationship methodologies. CRM is aimed at building strong and long-term relationships with customers. CRM aims to help organizations to build individual customer relationships in such a way that both the firm and the customer get the most out of the exchange, with long-term benefits. Retailers have bought into the CRM concept, but they are not fully implementing their CRM systems. This study discusses how retailers can gather customer data and how they can analyze these data to gain useful customer insights. We provide an overview of the methods predicting customer responses and behavior over time, and also discuss the existing knowledge on the application of marketing actions in a CRM context.

Keywords Customer relationship management (CRM) • Retail industry • Customers • Profit

1 Introduction

Concept of Customer relationship management (CRM) is not new, the old mom and pop stores are a good example of this, they knew exactly about each customer and their preferences and also their financial capabilities. However when firms grew, information about individual customers became lost among the masses.

To deal with this problem, the term relationship management (RM) started with an idea of work more with direct customer relationships around 1980s. Relationship

---

T. Narayana Reddy
JNTU, Anantapur, Andhra Pradesh, India
e-mail: tnreddyjntua@gmail.com

G. Silpa (✉)
YITS, Tirupathi, Andhra Pradesh, India
e-mail: Silpasg@gmail.com

S. Sathiyamoorthy et al. (eds.), *Emerging Trends in Science, Engineering and Technology*, Lecture Notes in Mechanical Engineering, DOI: 10.1007/978-81-322-1007-8_70, © Springer India 2012

Management is a group of methodologies and terms that describes how corporations should strive for long-term relations, work with quality goods strive for good customer service [1]. Slowly Interest on RM was rekindled and disappeared because it was proving difficult to achieve good short-term results and maintaining database also proved very expensive during 1990s. Several companies have stopped the relationship management, but with the introduction of more advanced information technologies, it came back with a new name Customer Relationship Management. With the introduction of advanced IT systems entire databases of customer information can be made available at all points in the organization, enabling every member of the organization to have a complete view of each customer. CRM is an integration of technologies and business processes used to satisfy the needs of a customer during any given interaction. It involves acquisitions analysis and use of knowledge about customers in order to sell more goods or services and to do it more efficiently [2]. A company's product can be compared quickly to another, and many companies are offering very similar products or services to each other. With this in mind the service, quality, and relationship experience becomes one of the greatest competitive aspects for business survival. Companies are also realizing that they can more easily lock in customers by understanding their needs and competing with exceeded expectations, something which CRM systems can help organize.

The realization of the benefits of CRM are also noted in the market of related software products; in 2008 the CRM market reached 8.9 billion USD, and by 2014 it is expected to reach 16.3 billion USD.

2 Brief Overview of CRM Influence in Small and Medium Retail Industries

2.1 CRM in Retailing

CRM is the practice of analyzing and utilizing marketing databases and leveraging communication technologies to determine corporate practices and methods that will maximize the lifetime value of each individual customer, the collective set of interactions between the retailer and firm communicates the brand concept to shoppers, thereby associating it with store. The value of CRM is depicted in (Fig. 1).

CRM enables the firm to take this process further by identifying smaller groups of customers with homogeneous needs, which are sometimes called customer segments or subsegments [3].

2.2 Objectives of the Study

- To describe the objectives and benefits of CRM
- To know the CRM Strategies
- To measure the CRM performance

2.3 Methodology

This study will go through variable stages of research and describes the area of research to create an understanding of the area. This study will make generalizations with collected data. This is the basis for descriptive research and also exploratory in its methods. It also tries to explain the different reasons or underlying causes for the observed events, which is synonymous with explanatory research.

2.4 Research Approach

This study is starting in theory and moving to data, collecting qualitative information through structured interviews. The information gathered through interviews cannot be derived from numbers, as the study is performed through conceptualization of the theory. Qualitative research is well suited for this type of research that deals with the try to understand events regarding management and decision making, which require a close look at details. The drawback being that the results often need to be verified by using quantitative methods.

2.5 Sample Selection

It is impractical to collect data from the entire population so it is important to reduce the sample size, it can be collected according to time constraints. In order to get a representative subset of the population nonprobability sampling is used. This is the preferred method, the interviewee will be asked point to point for someone who responds to the interview questions in a more detailed manner.

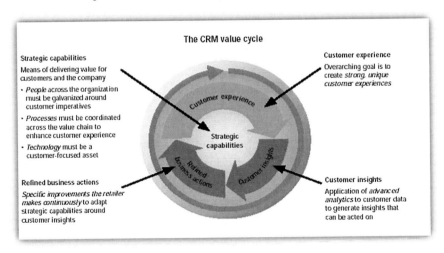

Fig. 1 The value cycle of CRM

2.6 Data Analysis

The analysis of the collected data consists of examining, categorizing, and structuring of collected data to test the proposition of the study. The analysis will be based on the framework developed from theory; this will be used to test the respondent's answers and check if they are agreeing with theory.

3 Description of Objectives and Benefits of CRM

The research shows that there are a number of reasons why companies choose to work with CRM in B2B, objectives related to both the customers and the business itself. The objectives that viewed as important are organization data, customer satisfaction, retention, and understanding customers better. The belief is that if the firms could meet requests and respond quicker as a result of being better structured, this could have a positive effect on the customer satisfaction and therefore their loyalty and the retention as well [4]. It was however pointed out that such effects would probably be minimal as price is the major factor in B2B purchasing decision. The real gains are instead viewed as the firm being more efficient in dealing with customers, resulting in time savings.

Objectives in the area of CRM process being able to help them to meet customer needs better to increase sales by any significant amount. Although responses indicate that the firms do not view meeting needs is an important factor in CRM system. Firms explained that they must customize their offering to every customer or they would be out of business. The meeting of needs is really an integral part of the B2B interaction.

CRM in B2B is to organize information around customer. By having complete view of each customer and making it accessible for all parts of the firm, CRM allows the organization to hold full information so they cannot depend on employees, when an employee leaves organization the information about customers are easier to maintain and all the information regarding them stays with the company.

Using CRM, customer data can be easily be reviewed and this will increase efficiency of a person who has not previously worked with a customer also can be able to know what has happened (Tables 1, 2).

The study shows that there are number of CRM strategies that are viewed to be important in B2B process. The areas which have been found important are, customer touch points, establishing customer value, product customization, interdepartmental communication, and also starting small and expanding the use of CRM overtime.

It was found that the customer touchpoints are an important aspect of the B2B interactions. Each of these touchpoint should be able to quickly access all the relevant information, so that the customer can be served quickly and also the new information can be entered into the system as it arrives. Email is the primary and most important type of communication in B2B firms. Many of the customer touchpoints are outside, the sales department cash in hand calculations are easier and

A Study on CRM Influence in Small and Medium Retail B2B Industries in India 741

Table 1 Comparison of objectives

Objectives	CRM Implemented Firms			CRM not implemented Firms		
Grading used to response	Agree	Neutral	Disagree	Agree	Neutral	Disagree
Increase loyalty	52	48	–	11.4	56.8	31.8
Increase satisfaction	81.4	18.6	–	9.7	53.2	37.1
Increase retention	72.6	37.4	–	11.3	61.2	27.5
Collect information	89.2	10.8	–	3.4	55.3	41.3
Understanding customers	81	19	–	12.8	63.8	23.4
Meet needs	45	51.2	3.8	5.2	61.9	32.9
Increase sales	43.7	48.2	8.1	2.8	67.9	29.3
CRM Main goal is to organize	92.8	7.2	–	27	59	14

Source: Primary data

Table 2 Comparison of strategies

Strategies	CRM Implemented firms			CRM not implemented firms		
Grading used to response	Agree	Neutral	Disagree	Agree	Neutral	Disagree
Culture touch points	57.2	36.9	5.9	11.4	36.8	51.8
Customer value and influence	59.8	38.4	1.76	5.7	47.2	47.1
Interdepartmental communication	61.2	37.4	1.4	9.3	61.2	29.5
Product customization	82.7	14.8	2.5	1.4	67.3	31.3
Analyzing reasons of customer behavior	63.8	29	7.2	9.8	56.8	33.4
Classifying and differentiating customers		40.8	59.2	3.2	53.9	42.9
Interaction points	71.9	28.1		2.8	68.9	28.3
Customer life cycle	78.4	21.6		–	47.9	52.1

Source: Primary data

better to use than values based on estimated values such as loyalty and satisfaction. Price is the main factor in keeping customers. Although how company treats the customers is important.

The study found that the most important aspect of CRM is that it helps the firm to organize the customer contacts in such a way that the customers feel better taken care of and the employees spend less time in finding relevant customer information. The study did not find reliable methods of measuring this factor, other than the opinions of the users.

The study shows that there was no interest in measuring the overall effect CRM has on the shareholder value. There was also very little interest in measuring the overall loyalty and value of customers or forecasting using these values. This is probably an effect of the strong relations that are present in the study (Table 3).

Table 3 Measurement of CRM performance

Measurement	CRM Implemented firms			CRM not implemented firms		
Grading used to respondents	Agree	Neutral	Disagree	Agree	Neutral	Disagree
Total firm benefit	72	28	–	9.3	52.6	38.1
Loyalty, retention, value	82.3	17.7	–	9.7	51.3	37.1
Process evaluation	69	31	–	11.3	59.2	27.5
System evaluation	84.2	15.8	–	3.4	51.1	41.3
Human capital	78.1	21.9	–	12.8	56.7	23.4
Strategic alignments	55.4	41.2	4.4	5.2	68.2	32.9
Cultural effects	51.7	48.3		2.8	61.9	29.3

Source: Primary data

4 Finding and Suggestions

4.1 Findings

CRM helps the firm to organize customer contacts in a smarter manner. It gives a good impression to the customers, and employees spend less time for coordinating and finding relevant customer information. The study showed that customization plays a major role in meeting the needs of the B2B customers.

B2B firms measure the value of customers as an ongoing process. The firms did not like terms, such as loyalty and expected retention rate, because they viewed them as uncertain measurements which could not be linked directly to any real income.

Satisfying price is the greatest factor in all sales, which basically means that if a competitor can lower prices.

4.2 Suggestions

Identify all the places where the firm interacts with customers, these touchpoints need to be considered when developing a CRM process. Repeated training to the employees can go a long way in reaching this goal. Find the key factors in customers mind, is it just price or is there some weight in that systems, these can boost loyalty by help providing good customer service. CRM is crucial for the success of the project that it is well defined with a limited scope of objectives. Organization of information is the strongest reason for implementing CRM but there is no objective way to measure the effect. This provides best way to measure the performance of CRM through collection of opinions from the users of the system.

The market insecurity and the nonpredictability of the financial crises and their dimensions stimulate the development of at least two stipulations: the first is related to sharing the risk among the representative agents of the exchange. The second is related to the desire to share information about the present and expected market development among the market participants, which considerably increases the market effectiveness. Therefore, the market instability initiates information symmetry among the market agents, which results in improvement of the market environment and exchange stabilization.

5 Conclusion

In this paper, we discussed the role of CRM in retailing. The enormous amount of customer data in retailing environments and the integration of channels, which now allow observation of online search behavior, will create new research challenges. For retail managers, our overview provides useful insights on how to execute CRM in their daily practice. The availability of a vast amount of data developed useful methods for targeting the right customers with the right offer at the right time and for predicting future behavior and customer value. Furthermore, research has produced findings on how specific marketing actions affect customer performance. This knowledge can be used to improve marketing decision making in the increasingly multichannel retail environment.

References

1. Nevin JR (1995) Relationship marketing and distribution channels: exploring fundamental issues. J Acad Mark Sci Fall 23(4): 327–334
2. Mascarenhas OA, Kesevan R, Bernachhi M (2006) Lasting customer loyalty: a total customer experience approach. J Consum Mark 23(7):397–405
3. Parvatiyar A, Sheth JN. (2001) Conceptual freamework of customer relationship management
4. McMullan R, Gilmore A (2005) The conceptual development of customer loyalty, measurement: a proposed scale. J Target Meas Anal Mark 11(3):230–243

A Study on the Effectiveness of Campus Recruitment and Selection Process in IT Industries

Geeta Kesavaraj and Manjula Pattnaik

Abstract The aim of this study is to find out the effectiveness of campus recruitment and selection at IT industries in Chennai, i.e., TCS. Campus recruitment provides a platform for the organizations to meet the aspirants and pick up intelligent, committed youth from various colleges and education institutes who have the requisite enthusiasm and zeal to prove themselves. In order to find the right candidates, organizations employ various recruitment and selection procedures. With the growth in IT industries, the need for talented and self-motivated young people who can work tirelessly has grown. Thus, organizations do not confine themselves to only the selected institutes but spread their net far and wide. This study was undertaken at one of the leading IT company in Chennai to know about the effectiveness of the various activities involved in campus recruitment and to see how much the candidate is aware about recruitment. It also includes the candidates' perception on various selection methods and the overall level of satisfaction among the new recruits in the organization. This project is based on the primary data which was collected as per the sample size. The following are the statistical tools, which are used for analysis and interpretation viz., Karl Pearson's Rank Correlation, One-Way ANOVA, Cohen's Kappa, Simple Weighted Average Method, and Chi Square test. The study concluded with findings and suggestions that would help to know the effectiveness of campus recruitment and selection.

Keywords Campus recruitment • Recruitment practices • Interviews • Education programs

---

G. Kesavaraj · M. Pattnaik (✉)
Sree Sastha Institute of Engineering and Technology, Chennai, India
e-mail: drmanjula23@gmail.com

G. Kesavaraj
e-mail: geetamurli@gmail.com

S. Sathiyamoorthy et al. (eds.), *Emerging Trends in Science,*
Engineering and Technology, Lecture Notes in Mechanical Engineering,
DOI: 10.1007/978-81-322-1007-8_71, © Springer India 2012

1 Introduction

Graduate recruitment or campus recruitment refers to the process whereby employers undertake an organized program of attracting and hiring students who are about to graduate from schools, colleges, and universities. Graduate recruitment programs are widespread in most of the developed world. Employers commonly attend campuses to promote employment vacancies and career opportunities to students who are considering their options following graduation. Selection methods used by employers include interviews, aptitudes tests, role plays, written assessments, group discussions, and presentations. Many schools, colleges, and universities provide their students with independent advice via a career advisory service which is staffed by professional career advisor. The career advisory service often organizes a career fair or job fair where a large number of employers visit the campus at once giving students the opportunity to meet a range of potential employers. Employers involved in graduate recruitment programs often form themselves into professional bodies or associations to share best practice or to collaborate in setting a recruitment code of practice. Career advisors also form themselves into professional bodies or associations to ensure that current best practice is shared across members and passes onto students [1–7].

2 Review of the Literature

Bratton and Gold (2007) differentiate the two terms while establishing a clear link between them in the following way: 'Recruitment is the process of generating a pool of capable people to apply for employment to an organization. Selection is the process by which managers and others use specific instruments to choose from a pool of applicants a person or persons more likely to succeed in the job(s), given management goals and legal requirements.' According to the National Association of Colleges and Employers (NACE) employers should target only campuses that "produce both the type of students that best fit into its corporate culture and the number of students it needs to achieve its hiring goals." Although there is no set formula determining which campuses to visit, campuses can be rated by several objective factors [8–13]. According to Julie Cunningham, president of The Cunningham Group, these include:

1. Curriculum/ranking—Is the school accredited: Is the curriculum relevant to the needs of your organization?
2. Location—Will the distance to campus justify the time and money it takes to recruit there? Will the distance create relocation and retention issues?
3. Demographics—Does the overall enrollment and percentages of woman and minority candidates meet the company's recruiting needs?
4. Graduation dates—When will candidates be available for work?
5. Career services/faculty/student organizations—Are the services the company needs available through the career center? Is the faculty accessible and interested

in career opportunities for their students? Can the company collaborate with the school's student organizations?

6. Competitive environment—Are the student's expectations in line with what the organization can offer?
7. Potential recruiters/team leaders—Does the company have enough alumni to create a recruiting team?
8. Internal opinion of the school—What is the general opinion of the school within the company? Would the school be accepted as part of the campus recruiting program?

Ten On-Target Recruitment Tips (WetFeet Inc.)

1. Focus on schools that most closely fit company needs, not just prestigious schools.
2. Arrive early. By April of senior year, most top candidates are taken.
3. Network with student clubs and organizations.
4. Nail the campus info session.
5. Define what you offer, including downsides, for good job fit.
6. Send enthusiastic and informed recruiters.
7. Take time for personal touches such as e-mail, calls, and thank you.
8. Follow up on promises.
9. Make people feel wanted.
10. Provide a great work experience especially in internships. People will hear about it.

According to the survey of students by Wet Feet Inc., a San Francisco-based recruitment research and consulting firm, the best campus recruiters were firms that sincerely communicated their interest and sponsored such events as one-on-one coffee chats and a "day on the job," simulating some aspect of the company's actual work environment.

3 Statement of the Problem

Campus recruitment provides a platform for the organizations to meet the aspirants and pick up intelligent, committed youth from various colleges and education institutes who have the requisite enthusiasm and zeal to prove themselves. In order to find the right candidates, organizations employ various recruitment and selection procedures. With the growth in IT industries, the need for talented and self-motivated young people who can work tirelessly has grown. The findings indicate that the majority of companies rely on traditional recruitment and personnel selection techniques over the use of online assessment instruments. Thus, organizations do not confine themselves to only the selected institutes but spread their net far and wide. It can be concluded that this company is doing very well in the area of different fields if it is compared with the other competitors [14–17].

This study was undertaken in IT industries to know about the effectiveness of campus recruitment and selection, and also to identify the effectiveness of the various activities involved in campus recruitment. It also explains that how much the candidate is aware about recruitment process. It also includes the candidates' perception on various selection methods and the overall level of satisfaction among the new recruits.

4 Scope of the Study

In today's world, there is a war to hire the right kind of talent. Capable and hardworking manpower is the best asset that any company can have. The scope of my study is to determine the effectiveness of the various activities involved in campus recruitment and to see how much the candidate is aware about recruitment process. It also includes the candidates' perception on the various selection methods which are adopted by the company. This study will also help us to find out the satisfaction level of the new recruits in the organization.

5 Objectives of the Study

5.1 Primary Objective

To study the effectiveness of campus recruitment and selection at IT industries in Chennai.

5.2 Secondary Objectives

1. To study the effectiveness of the activities involved in campus recruitment.
2. To know how much the candidate is aware about the recruitment process.
3. To study the candidates' perception on the various selection methods.
4. To identify the overall level of satisfaction among the new recruitments.

6 Research Design to Meet the Objectives

Location	Chennai
Sampling unit	IT employees
Sampling size	150
Sampling method	Convenience sampling
Instrument for information	Structured questionnaire

Table 1 The following are the response of the respondent on different factors experienced during recruitment and selection process

Sl. No.	Particulars	Happiness	Environment	Time management	Recruitment cycle	Overall experience	Communication flow	Overall facilities	Behavior
1	Highly dissatisfied	0	0	0	0	20	0	0	40
2	Dissatisfied	0	17	0	33	27	0	0	53
3	Neutral	17	40	40	17	23	17	17	7
4	Satisfied	50	43	23	50	30	60	55	0
5	Highly satisfied	33	0	37	0	0	23	28	0

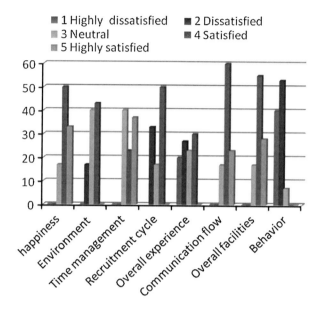

Following are the responses of respondents on different factors experienced during recruitment and selection process.

7 Statistical Testing For Research Hypothesis

7.1 Karl Pearson's Correlation

Aim: To analyze the relation between behavior of the interviewers and level of happiness in interview session.

Null Hypothesis (H$_0$):

There is a positive relation between behavior of the interviewers and level of happiness in interview session (Table 2, and 3).

Table 2 Analysis of relation between behavior of the interviewers and level of happiness in interview session

X	X^2	Y	Y^2	XY
60	3,600	55	3,025	3,300
80	6,400	65	4,225	5,200
10	100	30	900	300
0	0	0	0	0
0	0	0	0	0
$\Sigma X = 150$	$\Sigma X^2 = 10100$	$\Sigma Y = 150$	$\Sigma Y^2 = 8150$	$\Sigma XY = 8800$

A Study on the Effectiveness of Campus Recruitment

Table 3 Calculation

Factors	Behavior of the interviewers (X)	Level of Happiness in the interview session (Y)
Excellent	60	55
Very good	80	65
Good	10	30
Average	0	0
Poor	0	0

Table 4 Table shows satisfaction by the growth at workplace from opportunities

Factors	Opportunities provided in workplace	Factors	Satisfaction in work place
Not at all	0	Yes	100
Somewhat	25	No	50
Neutral	60		
Happy	65		
Very happy	0		
Total	150		150

Alternate Hypothesis (H_1):

There is negative relation between behavior of the interviewers and level of happiness in the interview session.

$$r = \frac{N \sum XY - \sum X \sum Y}{\sqrt{N \sum X^2 - (\sum X)^2} \sqrt{N \sum Y^2 - (\sum Y)^2}}$$

$$= 21500/22545$$

$$r = 0.9536$$

Inference:

Since the r is positive, there is positive relation between behavior of interviewers and level of happiness in interview session.

7.2 One-Way Anova

Aim: To identify the significant difference between the opportunities that is provided at workplace to learn and grow and satisfaction with the working place.

Null Hypothesis (H_0)

There is no significant difference between the opportunities that are provided at work place to learn and grow and satisfaction with the working place.

Alternative Hypothesis (H_1)

There is significant relationship between the opportunities that are provided at work place to learn and grow and satisfaction with the working place (Table 4).

Grand Total = G = ΣYij = 150 + 150 = 300
Correction Factor = G^2/N = $[(300)^2]/7$ = 12,857
Total Sum of Squares
SST = $\Sigma i + \Sigma j \, y(ij)^2 - G^2/N$
$= [(25)^2 + (60)^2 + (65)^2 + (100)^2 + (50)^2] - 12,857$
$= 8093$
Between Samples Sum of Squares:
R_1 = SSB = [150^2/5 + 150^2/2] −12,857 = 2,893
Within Samples Sum of Squares:
R_2 = SSW = SST−R1 = 8,093–2,893 = 5,200
Anova Table

Source of variation	Degrees of freedom	SOS	Mean square	F ratio
Between ropes	K−1 = 2−1 = 1	R1 = 2,893	2,893	F = (2,983/1,040) = 2.781
Error	N−K = 7−2 = 5	R2 = 5,200	1,040	
Total	N−1 = 7−1 = 6			

F(1,5) = 6.61 (table value)
F < F (1,5)
Thus, H_0 is accepted
Inference:
Since there is no significant difference between the opportunities that are pro-
vided at work place to learn and grow and satisfaction with the working place.

7.3 Simple Weighted Average Method

Tables 5, 6
Weighted Frequency = No. of Respondents × Rank Allotted

Table 5 Table to identify the reason to leave the organization

Sl. No.	Factors	No. of respondents
1	Marriage	40
2	Team fitment	20
3	Career opportunity	75
4	Health problems	15
5	Others	0

Table 6 Calculation: table showing reasons to leave the organization

Sl. No.	Factors	No. of respondents	Rank	Weighted frequency	W.F/100
1	Marriage	40	1	40	0.4
2	Team fitment	20	2	40	0.4
3	Career opportunity	75	3	225	2.25
4	Health problems	15	4	60	0.6
5	Others	0	0	0	0

A Study on the Effectiveness of Campus Recruitment

Inference:

The above table inferred that career opportunity will be the main reason to leave the organization by the employees.

7.4 Cohen's Kappa

Aim: To find out the significant relationship between communication of the requirements of the organization and staff responses in a timely manner.

Null Hypothesis (H_0):

There is a positive relationship between the communication of the requirements of the organization and staff responses in timely manner.

Alternate Hypothesis (H_1):

There is negative relationship between the communication of the requirements of the organization and staff responses in timely manner.

$$k = \frac{Pr(a) - Pr(e)}{1 - Pr(e)}$$

(Tables 7, 8, 9)

Expected Frequency $= 0.5676$

$$k = \frac{Pr(a) - Pr(e)}{1 - Pr(e)}$$

$$k = \frac{0.60 - 0.5676}{1 - 0.5676}$$

$$k = 0.0749$$

Table 7 Communication of the requirements of the organization

Communication of the requirements of the organization

Particulars	Yes	No	Total
Yes	60	40	100
No	50	0	50
Total	110	40	150

Table 8 Staff responses in timely manner

Staff responses in timely manner

Particulars	Yes	No	Total
Yes	0.40	0.26	0.66
No	0.33	0	0.33
Total	0.73	0.26	0.99

Table 9 Overall status of the response

Particulars	Yes	No
Yes	0.4818	–
No	–	0.0858

Inference:

Since the calculated value is positive, we accept the null hypothesis and hence there is positive relationship between the communication of the requirements of the organization and staff responses in timely manner.

7.5 Chi Square Test

Aim: To find out the relationship between overall facilities offered at working place and working place satisfaction.

Null Hypothesis (H_0)

There is no significant difference between overall facilities offered at working place and working place satisfaction.

Alternative Hypothesis (H_1)

There is significant difference between overall facilities offered at working place and working place satisfaction.

Expected frequency = Row total × Column total/Grand total (Table 10, 11)

Level of significance = 0.05

$$\text{Degree of freedom} = (r - 1)(c - 1)$$
$$= (2 - 1)(5 - 1)$$
$$= 1 \times 4 = 4 \text{ at } 5\% \text{ level}$$

Calculated value = 7.633

Tabulated value = 9.488

$$Z = Z_{cal} < Z_{tab}$$
$$Z = 7.633 < 9.488$$

Hence, the Null Hypothesis (H0) is accepted.

Table 10 Working place satisfaction

Overall facilities	Working place satisfaction					
	Highly dissatisfied	Dissatisfied	Neutral	Satisfied	Highly satisfied	Total
Yes	0	0	20	50	30	100
No	0	0	05	20	25	50
Total	0	0	25	70	55	150

Table 11 Calculation

O	E	(O–E)	$(O–E)^2$	$(O–E)^2/E$
20	16.66	3.34	11.15	0.669
50	46.66	3.34	11.15	0.669
30	36.66	−6.66	44.35	1.209
05	8.33	−3.33	11.08	1.330
20	23.33	−3.33	11.08	1.330
25	18.33	6.67	44.48	2.426
Total				7.633

Inference:

Hence, H0 is accepted. Thus, there is no significant difference between overall facilities offered at working place and working place satisfaction

8 Findings

1. In all the IT industries, 100 % of the employees are joined through campus recruitment.
2. Majority of the representatives (40 %) were from HR department who were involved in the process of recruitment and selection and 23 % representatives were from administration department and 20 and 17 % representatives were from marketing and sales department.
3. All the employees state that the initial contact of the representatives was polite and in professional manner.
4. Majority of the employees were satisfied with the flow of communication throughout the recruitment and selection process.
5. Most of the employees (60 %) state that their overall experience in recruitment and selection process was very good.
6. 100 % of the employees have stated that they were not called by the representatives for organizational visit in recruitment and selection process.
7. Almost 57 % of the employees stated that preplacement talk was held for 45 min and 43 % of the employees stated that it was held for 1 h.
8. About 83 % of the employees were happy with the recruitment cycle.
9. Majority of the employees (70 %) stated that the organization communicated their requirements properly to employees.
10. 63 % of the employees stated that a response from staff was timely.
11. All the employees stated that written test in the selection process was divided into three parts.
12. Only 70 % of the employees are satisfied with the time management in written test.
13. 100 % of the employees are highly satisfied with the environment of the interview.
14. Most of the employees (53 %) feel that the behavior of the interviewers was very good.
15. Majority of the employees (80 %) stated that their overall experience in the interview session was very good.
16. Almost 83 % of the employees are satisfied with the overall facilities provided in the workplace and with the opportunities that are provided at work to learn and grow.
17. Every employee in the organization is satisfied with the overall behavior of their superior and 100 % of the employees are satisfied with their workplace.
18. Most of the employees (50 %) of the employees would like to leave the organization for career opportunities.

9 Conclusion

This study helped to understand the effectiveness of campus recruitment and selection process of various IT industries in Chennai. To make college recruiting effective, the recruiting organization must first determine how many and which colleges should be targeted. It may prove cost-effective to do intensive recruiting in a few, carefully selected institutions, establishing a presence and building the organization's reputation among students and faculty. Timely and frequent dissemination of the literature, the offer of internships and awards of prizes for academic or social prowess help to advertise the organization as a preferred place of employment. Subsequent invitations to the organization's offices, made to students identified as potential employees, may serve to solidify the firm's image.

To be successful at campus placement, IT industries need to strategically position and market themselves to the students not just during the recruitment season but all through the session. Students relate more with employer brands that sponsor college events, arrange for learning sessions with students, organize events to involve the student fraternity in the corporate. Students have started to strategically place their resumes with different companies. For quality candidates, organizations are not just competing with competitors from the same sector but also from different industries. In coming times, it is expected that more innovative contracts and business models would come up that would not only strengthen the institute–industry relationship in the long run but also aid in placing the right candidate into the right jobs.

References

1. Kothari CR Business research methodology, 1st edn. New Age Publications
2. Roberts G (2005) Recruitment and selection (developing practice), 2nd edn. Chartered Institute of Personnel and Development, 1 June 2005
3. Nankervis A, Compton R, Morrissey B (2009) Effective Recruitment and Selection Practices, 5th edn. Mc Phersons Printing Group, Australia
4. Cooper DR, Schindler PS (1988) Business research methods, 6th edn. McGraw Hills Publisher Book Co, Singapore
5. Luthans F (2005) Organizational behavior, 9th edn. McGraw Hills Publisher Co, New York
6. Aswathapa K (2002) Human resource management, 6th edn. McGraw Hills Publisher
7. Tirupati AN, Human values, 3rd edn. New Age International Publisher
8. www.citehr.com-Human Resource selection-recruitment and selection
9. www.managementparadise.com- Human resource management- recruitment and selection
10. minmaxmin.blogspot.com/human resource management
11. www.tcs.com
12. www.inc.com/guides/7-ways-to-improve-employee-satisfaction
13. www.fanshawec.ca/kpi/graduate-and-employer-satisfaction
14. Ployhart RE (2006) Staffing in the 21st century; new challenges and strategic opportunities. J Manage, 32: 868–897

15. Kumar N, Garg P (2010) Impact of online-recruitment on recruitment performance. Asian J Manage Res
16. Piotrowswi C, Armstrong T (2006) Current recruitment and selection practices: a national survey fortune 1000 firms. North Am J Psychol 18(3): 489–496
17. Recruitment and selection process A study of Hindustan Coco Cola Beverage Pvt. Ltd. Ganguly, Jammu, India. Int J Innovation Manage Technol 1(4) Oct 2010

Financial Transactions Over Wired and Wireless Networks: Technical Perspective

J. Gnana Jayanthi, J. Felicita, S. Albert Rabara, A. Arun Gnanaraj and P. Manimozhi

Abstract While electronic commerce (E-commerce) which is over wired communication continues to impact the global business environment profoundly, technologies and applications are beginning to focus more on transactions over the Wireless Web referred to as Mobile Commerce (M-commerce). With this trend, a new set of issues arises. The purpose of this paper is to focus more on the emerging technologies which attract the M-Business and to examine some of these challenges and issues in M-Commerce so that researchers, developers and managers have to focus their research efforts in these emerging domains. This paper also highlights the way to promote business, by efficiently handling financial transactions over wired/wireless communications.

Keywords Mobile commerce · M-commerce · M-business · Financial crisis · Wireless technologies for transactions

---

J. Gnana Jayanthi (✉)
Department of Computer Applications, J.J. College of Engineering and Technology, Tiruchirappalli, Tamil Nadu, India
e-mail: jgnanamtcy@yahoo.com

J. Felicita
Department of Management Studies, St. Joseph's College, St. Joseph's Institute of Management, Tiruchirappalli, Tamil Nadu, India
e-mail: pio03davis@hotmail.com

S. Albert Rabara
Department of Computer Science, St. Joseph's College, Tiruchirappalli, Tamil Nadu, India
e-mail: a_rabara@yahoo.com

A. Arun Gnanaraj
Axiom Telecom, Dubai, UAE
e-mail: arun.art06@gmail.com

P. Manimozhi
Department of Computer Applications, J.J. College of Engineering and Technology, Tiruchirappalli, Tamil Nadu, India

S. Sathiyamoorthy et al. (eds.), *Emerging Trends in Science, Engineering and Technology*, Lecture Notes in Mechanical Engineering, DOI: 10.1007/978-81-322-1007-8_72, © Springer India 2012

1 Introduction

Mobile and wireless communications are the technologies with user-centric based and have highly penetrated in the global. In recent days, different types of systems are being applied for different application areas. Further developments are expected to evolve from existing and emerging systems. In future, most of the systems and applications will be mainly designed from a user-centric perspective [1]. The rapid development in the Internet and its related technologies changes the ways people live. People are more attracted towards wireless and mobile devices such as cell phones, laptops, notepads, Personal Digital Assistants (PDAs), palmtops. They like to buy theatre tickets while waiting to board a plane, or monitor financial markets and scan e-mail between meetings, play games, surf Internet wherever they may be. These facilities are provided by the advent of electronic commerce (E-Commerce) [2].

While E-Commerce continues to impact the global business environment profoundly, technologies, and applications are beginning to focus more on mobile computing and the wireless web. Mobile technology is the most pervasive communications technology in the world [3]. The mobile Internet is accessible from anywhere and at any time and this is the advantage that the carriers are trying to exploit in a variety of services [4]. This advanced mobile and Internet technologies coined together, enable people to be connected any time, any place, without being tied to a wired infrastructure. This feature allows mobile users to have mobile business transactions termed as Mobile Commerce [5], allowing Business-to-Employees (B2E), Business-to-Business (B2B), and Business-to-Consumer (B2C) applications [6]. These applications move processes, information, products, and services closer to clients, partners and consumers. Meanwhile, "Anywhere/anytime" access and its potential for B2E, B2B, and B2C via wireless technology accounts for M-Commerce's tremendous demand.

The Chap. 2 briefs the technologies for the mobile business emerged in two aspects (1) Network and (2) Service. The advantages of M-Commerce are elaborated in Chap. 3. The Chap. 4 discusses the constraints in M-Commerce development. The Chap. 5 outlines the measures to promote the Electronic/ Mobile (E/M) Commerce Development. Finally the paper is concluded in Chap. 6.

2 Advanced Technologies for E/M-Commerce

The analysis of wireless protocols and their interaction with existing Internet infrastructures are based on two specific areas: protocol efficiency and Internet-wireless communication security (including authentication, access control, and authorization functions). These two often play important role in protocol viability. The analysis focuses on the Network Technologies like GSM, HSCSD, GPRS, EDGE, and

Financial Transactions Over Wired and Wireless Networks: Technical Perspective 761

UMTS/3G and with the Service Applications like WAP, iMode, and LEAP specifications. J2ME, EZWeb, and J-SkyWeb documentations have a limited use.

2.1 Network Technologies

In general all the mobile protocols are very similar to each other, being client–server based, enabling a continuously increasing amount of services to be provided for both E/M-commerce users. Although the protocols are very similar to each other but still few of the protocols are introducing some challenges to their adoption. This is because it is more difficult to get a certain critical mass of subscribers to use a universal technology to enable frictionless service providing. The future will show which of the following protocols is going to deliver the strongest commercial value at any point in time and will be supported by the largest number of attractive applications.

Global System for Mobile Communication (GSM) [5] operates in the 900 MHz and the 1800 MHz frequency band and is the prevailing mobile standard in Europe and most of the Asia–Pacific region. GSM is being used by most of the people. GSM network was commercially launched and in recent years it is the world's leading and fastest growing mobile standard. GSM standard provides the critical mass to make it economically feasible to develop a large variety of innovative applications and services.

High Speed Circuit Switched Data (HSCSD) is a circuit switched protocol based on GSM. HSCSD transmits data up to 4 times the speed of the typical theoretical wireless transmission rate of 14.4 kbps. The key problem in the emergence of this market is that there are currently only few manufacturers.

General Packet Radio Service (GPRS) is a packet switched wireless protocol offering access to data networks and permits burst transmission speeds up to 115 kbps when it is completely rolled out. The real advantage of GPRS is that it provides a connection between the mobile terminal and the network. However, GPRS will require new terminals that support the higher data rates and these seem to be the bottleneck to the early adaption of the technology [5].

Enhanced Data Rates for Global Evolution (EDGE) is a higher bandwidth version of GPRS permitting transmission speeds up to 384 kbps. While a number of mobile operators are considering implementing EDGE as an interim data technology between GPRS and UMTS, the success of EDGE depends very much on the timely availability of the products and applications. Many operators were also advancing plans to deploy EDGE technology to increase the speed and capacity of mobile services offered in their current GSM frequency allocations.

Universal Mobile Telecommunications System (UMTS) represents an evolution in terms of services and data speeds from today's "second generation" mobile networks [6]. Many operators worldwide are also giving their customers with faster data services named "2.5G" systems based on GPRS technology—a natural evolutionary stepping-stone towards UMTS.

2.2 Service Technologies

Wireless Application Protocol (WAP) is a patented protocol widely marketed in the US and an "official" wireless standard in the European Union. WAP was first published in April of 1998 by WAP Forum. WAP's design accommodates all types of wireless networks. WAP utilizes a set of WAP developed transmission protocols to transfer content from Internet to users' devices. These underlying protocols include WCMP, WDP, WTLS, WTP, and WML. The current WAP-based services charge users by the time duration of their data transfer, the prices being closely correlated with the phone-service charges on their devices. All web-content accessible through the web is developed in a standard HTML. When a user makes an access request to a Web site through a WAP-enabled wireless device, the Web site content is translated by the user's Wireless Service Provider (WSP) from HTML to WML, and then sent to the user. The connection from WSP to a content provider is an Internet link with SSL encryption enabled as required. The wireless transmission of content radio packets is encrypted using Wireless Transport Layer Security Protocol (WTLS).

J2ME technology is developed using JAVA in the wireless communication field. Java's security model is matured and provides sound and secure Java-based wireless communication. On the other hand, Java's performance ratings have not been all satisfactory, especially on low-powered machines with limited processing capacities, such as the current generation of wireless devices.

I-mode gained a wide acceptance in Japan and is now expanding to Europe. I-mode uses compact HTML for delivery of content and packet switching to sustain continuous connection at a data transfer speed of 9.6 kbps. Content charges are billed by content providers extra to the subscription fees. I-mode terminals do not support JavaScript. The content providers create I-mode sites in compact HTML format and then upload the content to a Web server. No encryption exists on the I-mode wireless transmission, leaving the wireless transport untrusted.

3 Application of E/M-Commerce in Financial Transactions

3.1 E/M-Commerce can Enhance Sales and Buying Opportunities

With the opportunity of M-business, an enterprise can search and browse suppliers, find new suppliers and trading partners, to take the initiative, online publication of procurement information and increase sales opportunities for the seller. M-business also increases the number of buyers. M-Commerce, through the network, opens the market, free from time and space constraints, makes use of the Internet market penetration region, a channel for the profits to share with consumers Producers understand the actual needs of users, to gain more orders especially.

3.2 Savings of Transaction Cost by E/M-Commerce

Compared with traditional E-commerce, mobile business enables more speed, less links, no pressure on commodity stocks, a well-run E/M-shopping mall. Even inventory can be done without the pressure of any stock, which will save the cost of a lot of money. With E/M-commerce platform, businesses can make use of Trade link direct bargaining and transmission terms of the deal, increased information exchange between enterprises of the speed and accuracy, reducing transaction costs on both sides.

3.3 E/M-Commerce can Increase the Transactions Efficiency

The economic crisis prompted many enterprises to actively try to have better cost-effective Internet mobile business to reduce the operating cost of sales and to facilitate the customers to do business or any transactions while they move. M-commerce applications enhance transaction speed; reduce the flow of time items; links the producers and consumers around the world through the network, reducing the space on the transaction from both sides. The transactions are realized in the online collection of information from the signing of the transaction to pay the entire contract process, physical goods from the customer by the latest logistics enterprises speed directly to the customers.

4 Constraints in E/M-Commerce Development in Business Transactions

4.1 Payment Issues

Adoption of E/M-business transactions lacks direct, face-to-face contact and communication. The credit card payment system introduces some risks such as the acts of bad faith on the transaction, doubted situation and a serious impediment [8].

4.2 Hacking Passwords

Hacking passwords is another area to be focused. Computer hackers decipher the banking, business procedures and passwords. If an enterprise has been deciphering trade secrets, contracts have been tampered with, huge sums of money lost in the flow leading to serious losses. This makes traders think that the insecurity of online transactions which affects E/M transactions enthusiasm [7].

4.3 Legal Issues

Present M/E-Commerce architectures do not offer any laws, rules, and regulations or any legal polices. The construction of logistics infrastructure is weak and cannot fully meet the requirements of logistics operation. The enterprises do not have any standard legal polices and this reduces the customers interests which in turn results in imbalance between supply and demand [9].

5 Measures to Promote the Development of E/M-Commerce in Business Transactions

Credit card management information system is to be still researched to strengthen the insecurity and privacy of passwords. Certification Center is to be established in order to ensure the security of E/M-commerce transactions. Continually the protocols and operating systems is to be improved to guard against security vulnerabilities, to prevent the system has been eavesdropping, tampering, denial of service attacks, acts deny unauthorized access and dissemination of the virus. The use of encryption, digital signatures, access control, data integrity, authentication, information flow to fill routing control, is to be enhanced the network security technology content. E/M-commerce transactions are to be improved. To accelerate the construction of credit system of E/M-commerce, new ideas are to be incorporated. Publicity and training efforts are required. Innovative procedures are to be found out for the way of disclosures. New methodologies are to be enhanced for the enterprises and individuals to the concept of credit. The Government has to undertake and frame polices for business through M/E-commerce credit monitoring credit system to explore the relevance of e-commerce legislation, actively carried out on E/M-commerce businesses.

6 Conclusion

Financial crisis triggered a global economic recession many times. This paper clearly briefed the technical concepts and practical issues related with E/M-businesses. The highlight of the paper is that this paper has opened up many of the issues behind the Electronic and Mobile Commerce architecture including imperfect credit system, ineffective protection of on-line transaction, lack of logistics, and distribution system. Some suggestions are also pointed out in this paper like Government undertaking of Policy Issues to enhance E/M-commerce architecture and to promote more E/M-business transactions.

References

1. W Mohr (2008) Vision for 2020?. Springer J Wireless Pers Commun 44(1):27–49. ISSN: 0929-6212 (Print) 1572-834X (Online)
2. Little MC, Shrivastava SK (1998) Integrating the object transaction service with the web. In: Proceedings of the second international IEEE conference on enterprise distributed object computing workshop, EDOC '98, La Jolla, CA, pp 194–205. ISBN: 0-7803-5158-4
3. Baladron C, Aguiar J, Carro B, Sanchez-Esguevillas A (2008) Integrating user-generated content and pervasive communications. IEEE Pervasive Comput 7(4):58–61. ISSN: 1536-1268
4. Pierre S (2001) Mobile computing and ubiquitous networking: concepts, technologies and challenges. Sci Dir Telematics Inform 18(2–3):109–131
5. Tiwari R, Buse S, Herstatt C (2008) From electronic to mobile commerce opportunities through technology convergence for business services. CACCI J 1:38–45
6. Carlsson C, Walden P, Veijalainen J (2003) Mobile commerce: decision technologies for management track. In: IEEE Proceedings of the 36th Hawaii international conference on system sciences (HICSS'03), IEEE computer society. ISBN: 0-7695-1874-5
7. Liu YC, Sun X, Mao X (2011) Security problems and countermeasures with commercial banking computer networks. In: Proceedings of seventh international conference on computational intelligence and security, pp 821–825
8. Wei W, Li J, Cao L, Ou Y, Chen J (2012) Effective detection of sophisticated online banking fraud on extremely imbalanced data. Springer J World Wide Web Aust, ISSN 1386-145X
9. AlSudiari MAT, Vasista TGK (2012) Cloud computing and privacy regulations: an exploratory study on issues and implications. Int J Adv Comput (ACIJ) 3(2):159–162

Job Satisfaction as Predictor of Organizational Commitment

P. Na. Kanchana

Abstract The aim of the study is to determine the relationship between the Job Satisfaction and Organizational Commitment amongst employees in Tamil Nadu News Print and Papers Limited (TNPL), Kagidapuram, Karur District. The samples were selected by proportionate stratified sampling method. The data were collected by distributing structured 142 items questionnaires to 500 employees in TNPL. The result of the analyses showed that there was significant relationship between Job satisfaction and Organizational commitment and Job satisfaction as predictor of organizational commitment. The analysis proved that there is a significant positive relationship between job satisfaction and organizational commitment.

Keywords Affective commitment · Continuance commitment · Normative commitment · Organizational commitment · Job satisfaction

1 Introduction

With the change of the industrial structure in recent years, the output value of the service industry has become more than 70 % of the GDP in most countries. Thus, the service industry plays a significant role in national economic development. There is a need to understand a link between individuals and organizations in terms of conceptual framework of organizational commitment. Organizational commitment is widely described in the management and behavioral sciences literature as a key factor in the relationship between individuals and organizations. Specific employee attitudes relating to job satisfaction and organizational commitment are of major interest to the field of organizational behavior and the practice of human resources management. This article discusses how the job satisfaction acts an important predictor of organizational commitment.

P. Na. Kanchana (✉)
J.J.College of Engineering and Technology, Ammapettai, Tiruchirappalli, India
e-mail: pankanch@yahoo.co.in

S. Sathiyamoorthy et al. (eds.), *Emerging Trends in Science, Engineering and Technology*, Lecture Notes in Mechanical Engineering, DOI: 10.1007/978-81-322-1007-8_73, © Springer India 2012

Organizational commitment is a feeling of dedication to one's employing organization, willingness to work hard for that employer, and the intent to remain with that organization [1]. This theory proposes that there are three components to organizational commitment: affective, continuance, and normative commitment [2]. Affective commitment is an emotional attachment to the organization. Continuance commitment is characterize by a more rational analysis of the costs of staying versus leaving the organization. Normative commitment is a sense of moral obligation to stay with the organization.

The term "job satisfaction" was proposed by Hoppock [3] who suggested that job satisfaction means employees' emotions and attitude toward their jobs, and is their subjective reaction toward their jobs. Tai et al. [4] observed that Job satisfaction and Organizational Commitment are highly correlated. Markovits et al. [5] suggested that affective organizational commitment was found to be most influential with respect to levels of intrinsic and extrinsic job satisfaction.

2 Brief Overview of Organizational Commitment

Sarminah Samad [6], study examined the relationship between organizational commitment and job performance. Consequently the study determined the effect of job satisfaction on the relationship between organizational commitment and job performance. The results revealed that there was a positive relationship between organizational commitment and job performance. The hierarchical analysis found that job satisfaction (both the hygiene and motivator factors) played moderating role on the relationship between organizational commitment and job performance. Findings and implications for managerial practices are discussed.

Asad Afzal Humayoun [7], the purpose of this study is to investigate the relationship between employee retention, job satisfaction, perceived supervisory support and compensation by considering the organizational commitment as mediating variable in pharmaceutical industry in Pakistan.

3 Research Methodology

3.1 Objectives

This study is being carried out to identify the relationship between Job satisfaction and Organizational commitment. The aim of the study is to determine the Job satisfaction as predictor of organizational commitment amongst employees in Tamil Nadu News Print and Papers Limited, Kagidapuram, Karur District.

3.2 Hypotheses

H_1 There exist a significant difference about components of organizational commitment viz., affective commitment, normative commitment and continuance commitment, of employees at TNPL across the various levels of Job satisfaction.

H_2 There exist a significant difference in components of organizational commitment viz., affective commitment, normative commitment and continuance commitment of employees before and after joining in TNPL.

H_3 There exist a significant difference in job satisfaction before and after joining in TNPL Table 1.

4 Results and Discussion

The majority of the respondent taking part in this study were men 304 (93.1 %) and 26 (7.9 %) were women. Workmen category were 215 (65.2 %), Staff 21 (6.4 %) and Executives 94 (28.5 %). Whereas, the majority of the respondents aged between 36 and 45 years 117 (35.5 %) and most of them i.e. 140 (42.4 %) earn between `10,001 and `20,000 monthly. For educational level, most of the respondents 74 (22.4 %) have Bachelor degrees.

The Table 2 on F-test shows the component affective commitment at three different levels (Low, Medium, High) has significant difference across the various levels of job satisfaction at TNPL (F-value $= 2.035, p < 0.05$).

The Table 3 on F-test shows the component continuance commitment at three different levels (Low, Medium, High) has significant difference across the various levels of job satisfaction at TNPL (F-value $= 11.002, p < 0.01$).

The Table 4 on F-test shows the component normative commitment at three different levels (Low, Medium, High) has significant difference across the various levels of job satisfaction at TNPL (F-value $= 9.499, p < 0.01$).

The Table 5 on F-test shows the organizational commitment at three different levels (Low, Medium, High) has significant difference across the various levels of job satisfaction at TNPL (F-value $= 13.536, p < 0.01$).

Table 1 Research methodology

Particulars	Description
Types of research used	Descriptive research
Research approach	Survey
Research instrument	Questionnaire
Sample size	330 respondents (TNPL employees)
Sampling technique	Proportionate stratified random sampling

Table 2 One way ANOVA on job satisfaction against affective commitment

Variable	Level	Mean and S.D.	Source	Sum of squares	Df	Mean square	F
Job satis-faction	Low ($N = 87$)	2.26 (0.769)	Between groups	2.122	2	1.061	2.035*
	Medium ($N = 122$)	2.36 (0.669)					
	High ($N = 121$)	2.14 (0.734)	Within groups	170.496	327	0.521	
	Total ($N = 330$)	–	–	172.618	329	–	

*$p < 0.05$

Table 3 One way ANOVA on job satisfaction against continuance commitment

Variable	Level	Mean and S.D.	Source	Sum of squares	Df	Mean square	F
Job satis-faction	Low ($N = 87$)	2.26 (0.769)	Between groups	10.59	2	5.29	11.02**
	Medium ($N = 122$)	2.36 (0.669)					
	High ($N = 121$)	2.14 (0.734)	Within groups	157.32	327	0.48	
	Total ($N = 330$)	–	–	167.91	329	–	

**$p < 0.01$

Table 4 One way ANOVA on job satisfaction against normative commitment

Variable	Level	Mean and S.D.	Source	Sum of squares	Df	Mean square	F
Job satis-faction	Low ($N = 87$)	2.26 (0.769)	Between groups	11.15	2	5.57	9.49**
	Medium ($N = 122$)	2.36 (0.669)					
	High ($N = 121$)	2.14 (0.734)	Within groups	191.85	327	0.59	
	Total ($N = 330$)	–	–	202.99	329	–	

**$p < 0.01$

The two tail significance for the change in variables of affective commitment before and after joining TNPL indicates that $p < 0.01$ and, therefore, is significant at 1 % level. The membership of joining in TNPL improves the affective commitment ($t = -7.515, p < 0.01$).

The two tail significance for the change in variables of continuance commitment before and after joining TNPL indicates that $p < 0.01$ and, therefore, is

Job Satisfaction as Predictor of Organizational Commitment

Table 5 One way ANOVA on job satisfaction against organizational commitment

Variable	Level	Mean and S.D.	Source	Sum of squares	Df	Mean square	F
Job satisfaction	Low ($N = 87$)	1.68 (0.758)	Between groups	13.22	2	6.61	15.534**
	Medium ($N = 122$)	2.05 (0.661)					
	High ($N = 121$)	2.18 (0.691)	Within groups	159.64	327	0.49	
	Total ($N = 330$)	–	–	172.85	329	–	

** $p < 0.01$

significant at 1 % level. The membership of joining in TNPL improves the continuance commitment ($t = -7.042, p < 0.01$).

The two tail significance for the change in variables of normative commitment before and after joining TNPL indicates that $p < 0.01$ and, therefore, is significant at 1 % level. The membership of joining in TNPL improves the normative commitment ($t = -7.756, p < 0.01$).

The two tail significance for the change in variables of job satisfaction before and after joining TNPL indicates that $p < 0.01$ and, therefore, is significant at 1 % level. The membership of joining in TNPL improves the job satisfaction ($t = -5.294; p < 0.01$).

KMO measure of sampling adequacy is an index to examine the appropriateness of factor analysis. It is seen that Kaiser-Meyer-Olkin measure of sampling adequacy index is 0.851 and hence the factor analysis is appropriate for the given data set. Bartlett's test of Sphericity Chi square statistics is 2155.751, that shows the fourteen statements are correlated and hence as inferred in KMO, factor analysis is appropriate for the given data set.

It is evident that, the two factors acted together and accounted for 54.130 % of the total variance. Hence, the variables have been reduced from nineteen to four underlying factors.

Looking at Table 6, the variables namely Ventilation facility, Salary, Welfare facility, Achievement recognition, Incentive scheme, Suggestions have loadings of 0.502, 0.538, 0.679, 0.598, 0.711, and 0.702 on factor-1 indicating that it is a combination of these six variables and it was named as Empowered work Environment. The variables like Proud to work, challenging, Relationship with colleagues, Raw materials, and Management tools, Team work have loadings of 0.754, 0.662, 0.660, 0.681, and 0.566 on factor-2, indicating that it is a combination of these five variables which was named as Job context. Next for factor-3, it is evident that Count on good work, Fair system of rewards, Medical facilities, Work policies have loadings of 0.682, 0.603, 0.649, and 0.525 on factor-3, indicates that it is a combination of these four variables and it was named as Job content. The variables like Working Hours, Impartial treatment by supervisors, Relevant matters

772

Table 6 Factor analysis-rotated factor loadings

Factor	Variables	Factor loading
I	Ventilation facility	0.502
Empowered work environment	Salary	0.538
	Welfare facility	0.679
	Achievement recognition	0.598
	Incentive scheme	0.711
	Suggestions	0.702
II	Proud to work	0.754
Job context	Challenging	0.662
	Relationship with colleagues	0660
	Raw materials and management tools	0.681
	Team work	0.566
III	Count on good work	0.682
Job Content	Fair system of rewards	0.603
	Medical facilities	0.649
	Work policies	0.525
IV	Working hours	0.722
Work ethics	Impartial treatment by supervisors	0.728
	Relevant matters by superiors	0.526
	Standard of living	0.652

by superiors, Standard of living have loading of 0.722, 0.728, 0.526, and 0.652 on factor-4 which indicates that it is a combination of these four variables which was named as work ethics.

From Table 7, all the components of organizational commitment had a strong and significant positive relation to job satisfaction. The independent variable job satisfaction had significant positive relationship with affective commitment ($r = 0.105$, $p < 0.05$), continuance commitment ($r = 0.251$, $p < 0.01$) and

Table 7 Inter-correlation matrix between job satisfaction and the components of organizational commitment

Factors	Job satisfaction	Affective commitment	Continuous commitment	Normative commitment	Organizational commitment
Job Satisfaction	1				
Affective commitment	0.105*	1			
Continuance commitment	0.251**	0.134*	1		
Normative commitment	0.234**	0.081	0.162**	1	
Organizational commitment	0.264**	0.277**	0.303**	0.203**	1

$** p < 0.01$; $* p < 0.05$

normative commitment ($r = 0.234$, $p < 0.01$). Also job satisfaction had significant positive relationship with overall organizational commitment ($r = 0.264$, $p < 0.01$).

5 Conclusion

The results obtained in this study showed that job satisfaction had a significant positive correlation with organizational commitment and the components of organizational commitment. The results of all the studies mentioned above shows that organizational commitment is important to organizations and the higher level of organizational commitment will be attained when the employees have higher level of Job Satisfaction. Also, the study clearly indicates that job satisfaction is one of the predictors of organizational commitment.

The results of the study should be interpreted with caution as a probability sample Stratified random sample was utilized in the study. Therefore, the results obtained from the research may be specific to the sample that was selected for the investigation, and cannot be generalized with confidence to other public sector entities.

Acknowledgments The Author acknowledge Tamil Nadu News Print and Papers Limited(TNPL), Kagidapuram, Karur District for providing to carry out this work.

References

1. Meyer JP, Allen NJ (1988) Links between work experience and organizational commitment during the first year of employment: a longitudinal analysis. J Occup Psychol 61:195–209
2. Meyer JP, Allen NJ (1991) A three-component conceptualisation of organizational commitment. Hum Resour Manage Rev 11(1):61–89
3. Hoppock R (1935) Job satisfaction, New York: Harper and brothers P-47
4. Tai TW, Bame SI, Robinson CD (1998) Review of nursing turnover research, 1977–1996, Social science medicine 47(12):1905–1924
5. Yannis M, Davis Ann J, Van DR (2007) Organizational commitment profiles and job satisfaction among greek private and public sector employees. Int J Cross Cult Manage 7(1)
6. Samad S (2011) The effects of job satisfaction on organizational commitment and job performance relationship: a case of managers in malaysia's manufacturing companies. Eur J Soc Sci 18(4)
7. Anis A, Rehman KU, Rehman IU, Asif Khan M, Humayoun AA (2011) Impact of organizational commitment on job satisfaction and employee retention in pharmaceutical industry. Afr J Bus Manage 5(17):7316–7324

Knowledge Management Key to Competitive Advantage

V. Maria Tresita Paul and G. Prithiviraj

Abstract

Knowledge is like light. Weightless and intangible, it can easily travel the world, enlightening the lives of people everywhere. Yet billions of people still live in the darkness of poverty (World Bank 1999).

The above quote sheds light on the significance of knowledge and the need to manage this vast knowledge for optimum utilization. In the last decade, knowledge management has emerged as a very successful organization practice and has been extensively treated in a large body of academic work. In today's knowledge-based society and business, "knowledge" is seen as the main source of competitive advantage. It is often a fuzzy concept, with no direct referent in the real world. In order to conceptualize this, people use the synonym knowledge asset and view it as a strategy for attaining competitive advantage, that can be managed, evaluated, invested in, and that becomes one of the main sources of value creation within an organization. Such is the importance accorded to knowledge management in the today's scenario that it has become a popular discipline even though the field is only 10 years old. From research perspective, knowledge management is a vast area that covers all aspects of knowledge creation, storage, sharing, development, and uses and maintenance. This chapter attempts to discuss about knowledge management as a theory which has evolved as a wholesome practice, various methods used in business for knowledge sharing and the various issues associated to knowledge management's widening significance.

Keywords Knowledge • Knowledge management • Competitive advantage • Knowledge asset

---

V. Maria Tresita Paul (✉) · G. Prithiviraj
Tirupur and Shri Krishna Institute of Management Science, PARK'S College,
Coimbatore, India
e-mail: maria.tresi@gmail.com

G. Prithiviraj
e-mail: pritvi5588@gmail.com

S. Sathiyamoorthy et al. (eds.), *Emerging Trends in Science,*
Engineering and Technology, Lecture Notes in Mechanical Engineering,
DOI: 10.1007/978-81-322-1007-8_74, © Springer India 2012

1 Introduction

Knowledge management—a set of management activities, aimed at designing and influencing processes of knowledge creation and integration including processes of sharing knowledge—has emerged as one of the most influential new organizational practices. Numerous companies have experimented with KM initiatives in order to improve their performance. The society is now facing a shift from an information era to one of knowledge era. We are now living in a knowledge-based society, where individual and organizational knowledge, as well as brainpower, have replaced physical assets as critical resources in the corporate world [1]. The shift made both managers and management scholars reconsider the sources of competitive advantage. Therefore, knowledge and the ability to create and manage it, represents the main source of sustainable competitive advantage within the business environment.

Knowledge management thus seems to be one of those areas, where managerial practice and the academic literature develop simultaneously and perhaps even co-evolve. In many organizations today, knowledge is dispersed and fragmented; organizations are in a state of flux, as they strive to adjust to their changing business environment. Knowledge exists in various forms—embedded in products and processes, lurking in databases, but above all, in people's heads. The exploitation of its potential requires fusion between an organization's technical systems, processes, and people (the socio-technical system) and between theory and practice (the learning system). Hence, there arises a need for effective and efficient knowledge management. Firms view knowledge and knowledge management as part of their strategic orientation to gain competitive advantage over other firms. Thus, the researchers attempt to discuss about knowledge management as a theory which has evolved as a wholesome practice and the various issues associated to knowledge management's widening significance.

2 Data, Information, Knowledge

Data have no meaning or significance in themselves. Example, individual customer responses about an item fed in a computer spreadsheet. Information is data which have meaning because of a relational connection. In other words, information is data which have been processed to be useful. Information aims to provide answers to the questions 'Who?', What?', 'Where?' and 'When?' It is worth noting that although information is intended to be useful, it is not necessarily so. Merely aggregating data and identifying relationships between variables do not guarantee utility. Knowledge is information to which a process has been applied, which may eventually become expertise [2]. It is 'the collation of information for a particular purpose, intended to be useful'. Knowledge aims to answer the question 'How?' Developing new knowledge from that which already exists to answer the question 'Why?' may be defined as understanding. Knowledge is a justified personal belief that increases an individual's capacity to take effective action.

3 Knowledge Management

Knowledge management is scoped out broadly as, any process or practice of creating, acquiring, capturing, sharing, and using knowledge, wherever it resides to enhance learning and performance in organization. Knowledge management has been defined in many different ways. However, the most common description of knowledge management is as a business practice, which emphasizes the creation, dispersion, and use of knowledge [3]. The following definition summarizes knowledge, management as, "the explicit and systematic management of vital knowledge and its associated processes of creating, gathering, organizing, diffusion, use and exploitation, in pursuit of organizational objectives". The underlined words are important:

- *Explicit*— unless something is made explicit it frequently does not get properly managed.
- *Systematic*— this helps to create consistency of methods and the diffusion of good practice.
- *Vital*— every conversation and every new document in an organization adds to the organization's knowledge pool.
- *Processes*— as well as being an important dimension of management and business processes, knowledge processes are important in their own right.

The main processes of knowledge management are knowledge sharing (of existing knowledge), knowledge creation, and knowledge conversion (innovation). The purpose of knowledge management is to enable the organization to gain access to the knowledge held within the individuals of the firm.

4 Knowledge Asset

Knowledge is built up from data and information as well as prior knowledge.

Any prior knowledge is viewed as an asset in business organizations; and it is referred as "knowledge base" or "knowledge asset". The concept of knowledge assets is not something new; it has been long used by the sixteenth-century alchemists that were undertaking precise measures to protect the secrets of their craft. What is new in the late twentieth century is that knowledge assets are coming to constitute the very basis of postindustrial economies [4]. In 1983, Polanyi introduced distinction with regard to knowledge, that there is *explicit knowledge*; can be expressed in words and numbers and can be easily communicated and shared in the form of hard data, scientific formulae, codified procedures, or universal principles, having as main characteristic the ease of codification; and there is *tacit knowledge*, that is highly personal and hard to formalize. Subjective insights, intuitions, and hunches fall into this category of knowledge, knowledge embedded in experience, values, beliefs that are very difficult to articulate and to codify.

In 2000, the Japanese authors Nonaka, Toyama, Konno defined knowledge assets as firm-specific resources that are indispensable to creating value for the firm. Knowledge assets are regarded as having moderating character in the knowledge creation process. Their classification of knowledge asset has four dimensions which are discussed in Table 1.

5 Managing Knowledge Assets

One of the main problems with knowledge is that, if trapped inside the minds of key employees, in filling drawers, databases, it is of little to no value to the company or the project undertaken. It has to be supplied to the right people at the right time in order to be of full value. As knowledge is boundary less, the management has to redefine the project on the basis of the knowledge it owns. Without a proper administration in a project, it is very difficult to know exactly what that the team knows. In order to overcome this situation, the management has to constantly read the situation, to determine what kinds of knowledge assets are available for them. This needs the creation of an environment where knowledge assets can flow freely from the people who own them to the people who are in need. In order to create the environment the leaders have to supply the necessary conditions: autonomy, creative chaos, redundancy, requisite variety, trust. *Autonomy* is considered to increase the chances of finding valuable information and motivating the members to create new knowledge. Autonomous individuals set task boundaries for themselves in the pursuit of the goal set by the organization, customer, etc. In project environments, autonomy is a prerequisite condition. An autonomous team can perform many functions, amplifying and sublimating individual perspectives to higher levels [5].

Table 1 Categories of knowledge assets. *Source* Nonaka, Toyama, Konno (2000)

Experiential knowledge assets	*Conceptual knowledge assets*
Tacit knowledge shared through common experiences	Explicit knowledge articulated through images, symbols, and language
• Skills and know-how of individuals	• Product concepts
• Care, love, trust, and security	• Design
• Energy, passion, and tension	• Brand equity
Routine knowledge assets	*Systemic knowledge assets*
Tacit knowledge routinized and embedded in actions and practices	Systemized and packaged explicit knowledge
• Know-how in daily operations	•Documents, specifications, manuals
• Organizational routines	• Database
• Organizational culture	• Patents and licenses

6 Knowledge Management Processes

As discussed earlier, the main processes of knowledge management are *knowledge sharing* (of existing knowledge), *knowledge creation*, and *knowledge conversion* (innovation). Within organizations, much of the emphasis of early knowledge management programs was on knowledge sharing "knowing what we know". More recently, there has been growing interest in the knowledge processes that underlie innovation [6]. The activities of innovation processes in knowledge management are:

- Create. New ideas are created.
- Codify. Here, a prototype design or a process description is developed.
- Embed. At this stage, the knowledge is encapsulated in manufacturing processes and organizational procedures.
- Diffuse. Products are distributed in the marketplace or processes are implemented throughout the organization. Their application then generates ideas for improvements, and so the process repeats.

The main activities of knowledge sharing in knowledge management processes are:

- Collect. Existing knowledge is gathered either on a routine basis or as needed. Often its existence is formally recorded in a knowledge inventory or knowledge map.
- Organize/store. The knowledge is classified and stored, often using an organization or industry-specific thesaurus or classification schema. This makes subsequent retrieval easier.
- Share/disseminate. Information may be sent routinely to those people who are known to be interested in it—this is information 'push'. Meetings and events act as vehicles to share tacit knowledge.
- Access. Information is made easily accessible from a database, for example over an intranet. Users access it as they need it—this is information 'pull'.
- Use/exploit. The knowledge is used as part of a work process. It is refined and developed. Through use, additional knowledge is created and the cycle repeats itself.

A useful form of knowledge that can result from the proper knowledge management process is Meta knowledge—knowledge about knowledge.

7 Competitive Advantage

Knowledge management is a way of doing business just like marketing or Six Sigma is. However, the understanding of concept of knowledge management by different individuals within same organization may be incoherent [7]. The quest to innovate through research and development is essential for firms to remain ahead of competitors. Indeed, many firms view the acquisition of new knowledge as a way to gain and maintain competitive advantage. However, few firms fully realize the benefits from highly valued knowledge. Knowledge that is isolated in one department or in a specific segment of the value chain is not utilized to its full extent. This leads to wastage or stagnation of unused productive knowledge. In today's world, knowledge and the capability to create and utilize knowledge are considered to be the most important source of a firm's sustainable competitive advantage. Any knowledge management strategy that is designed to improve business performance must address three components:

- The work processes or activities that create and leverage organizational knowledge;
- A technology infrastructure to support knowledge capture, transfer, and use; and
- Bringing in change in the organization to improve operational efficiencies, enhance organizational learning, intensify innovation, or speed up response to the market that are essential to effective knowledge use.

In this perspective, an organization's competitive edge is determined by the continuous generation and synthesis of collective, organizational knowledge. The development of knowledge alone however is not sufficient; an organization must also command 'the ability to effectively apply the existing knowledge to create new knowledge', taking action based on its knowledge-based assets. Thus, knowledge management focuses explicitly on the design of organizational processes in such a way that the benefits of organizational knowledge can be maximized in terms of its innovation potential and competitive advantage. Some of the advantages are listed below:

- Fostering knowledge-supportive culture—Characteristics of the general culture include a safe environment, focused on delivering quality work without delay— i.e., "getting the right thing done as soon and with as little fuss as possible!"
- Focused knowledge management practice to align with enterprise direction— Practitioners of knowledge management identify the intended business direction of the enterprise to ascertain that the associated knowledge-related factors receive appropriate attention and are well maintained.
- Practice accelerated learning—Pursue a broad range of knowledge transfer activities to ascertain that valuable knowledge is captured, organized, and structured, deployed widely, and used and leveraged.
- Pursuing, 'Four Success Factors'—These factors focus on providing employees with: provide shared understanding of the following personal traits,

Knowledge and resources	Permission
Opportunities	Motivation

- Create supportive infrastructure capabilities—Implement of new or adapt existing capabilities to provide needed and effective supports for KM.
- Provide effective governance for the knowledge management practice—Monitor, evaluate, and guide their activities and their plans, results, and opportunities.

8 New Trends and Widening Significance

Knowledge management in large organizations is now well established, with new techniques and new technology tools continually adding to the corporate armory for exploiting knowledge management. The result of more effective knowledge management has led to reduced operating costs, faster time-to-market for new products, better customer service, reduced risk, and many other reported bottom line benefits. However, over the course of the last 5 years three general shifts of new focus stand out:

- Sharing existing knowledge: this was the thrust of many early knowledge initiatives and is reflected in the knowledge sharing process.
- Creating and converting new knowledge: this is the innovation thrust.
- A growing external focus: this has led to an upsurge in interest in customer relationship management systems and interest in knowledge markets.

As the innovation agenda grows in prominence, and e-business will become more established, we can expect that organizations will continue to exploit their knowledge assets in novel ways, and create knowledge–intensive businesses.

9 Conclusion

The shift toward a knowledge-based society has made knowledge as the center stage of obtaining the sustainable competitive advantage within every aspect of businesses. Knowledge-based capabilities are considered to be the most strategically important ones to create and sustain competitive advantage [8]. Nowadays, knowledge is perceives as being of great value, of great importance, as an asset for the future, asset that can be invested in, it is additive, it allows for returns and most importantly can and must be managed and measured, thus, reinventing the concept of knowledge assets. Despite the multiple perspectives on knowledge assets, the management scholars all agree on the fact that knowledge assets are the basis for value creation within any organization, project. Knowledge management activities are adding value to organizations by enhancing innovation and innovativeness. The processes of knowledge management are knowledge sharing of existing knowledge, knowledge creation, and knowledge conversion through innovation, when properly monitored adds up vitally to organizational value. Thus, the concept of Knowledge management which is still evolving and has vast scope in future needs to be better understood and practiced by the business firms to gain a competitive upper hand.

References

1. Drucker PF (1993) Post-capitalism society. Harper Business, New York
2. Licbowitz J (2000) Building organisational intelligence: a knowledge management primer. CRC Press, Boca Raton

3. Alavi M, Leidner D (2001) Review: knowledge management and knowledge management systems: conceptual foundations and research issues. MIS Q 25(1):107–136
4. Boisot M (1998) Knowledge assets. Securing competitive advantage in the information economy. Oxford University Press, Oxford
5. Grant RM (1996) Toward a knowledge-based theory of the firm. Strate Manage J Spec Issue Knowl Firm 17:109–122
6. Skyrme DJ (2002) Knowledge management: approaches and policies, special issue: knowl manage
7. Cader Y (2007) Knowledge management and knowledge based marketing. J Bus Chem 4(2)
8. De Nisi A, Hitt M, Jackson S (2003) The knowledge-based approach to sustainable competitive advantage. In: Jackson H, DeNisi (eds) Managing knowledge for sustained competitive advantage. Jossey-Bass, San Francisco, pp 3–33

Part XIV
Digital Libraries: Information Management for Global Access

Knowledge Management and Academic Libraries

K. Mahalakshmi and S. Ally Sornam

Abstract Knowledge management is not one single discipline. It is an integration of numerous activities and fields of study. The field of knowledge management helps in improving the effectiveness of academic libraries and for their parent institutions. This chapter introduces knowledge management principles and its application in academic libraries and librarians as knowledge managers.

Keywords Knowledge management • Academic libraries • Application • Knowledge managers

1 Introduction

Both knowledge and knowledge management (KM) are difficult to define. Academics in the field of KM typically define knowledge as a derivative of information, which is derived from data. Knowledge is information or data, organized in a way that is useful to the organization. The central idea is that KM efforts work to create, codify, and share knowledge valuable to the organization.

Another idea is that KM shifts the focus from process to practice. Over the past 20 years, academic libraries have generated increasing amounts of information about their operations. Libraries do not consider organizational knowledge as a resource in its own right as they do personnel, collections, or facilities. The basic idea influencing KM is that knowledge is a strategic asset that must be managed. It should be managed as an asset or resource. KM is essentially about tacit knowledge (TK). It is aimed at making TK explicit and then sharing that for reusing across an organization. The following can be expressed as KM lifecycle:

Knowledge generation $\rightarrow$ Knowledge codification $\rightarrow$ Knowledge transfer.

K. Mahalakshmi (✉)
Faculty of Engineering, Avinashilingam University, Coimbatore, Tamilnadu, India
e-mail: nilamaha@gmail.com

S. Ally Sornam
PGDRLIS, Bishop Heber College, Tiruchy, Tamilnadu, India
e-mail: ally_jelen@yahoo.co.in

S. Sathiyamoorthy et al. (eds.), *Emerging Trends in Science, Engineering and Technology*, Lecture Notes in Mechanical Engineering, DOI: 10.1007/978-81-322-1007-8_75, © Springer India 2012

KM initiatives are expressions of part of this process. Some are concerned with the first part: knowledge creation, innovation, or organizational learning. Others are concerned with capturing Tacit Knowledge for codification. This means recording videos or feeding data into a database. Knowledge transfer is sharing knowledge. This means a database of information with access methods. It can mean.

Fostering networks of people for sharing knowledge or creating knowledge maps showing who has what expertise. There are different tools for the KM. It can be divided into two parts: Information technologies tools and web-based (IT) tools. Web, in coming years, will present a set of new tools for managing knowledge by providing an extremely rich common language for representing knowledge.

2 Definition-KM

Knowledge management is "a process, which deals with knowledge creation, acquisition, packaging and application or reuse of knowledge." It basically consists of the following four steps:

- Knowledge Collection,
- Organization,
- Data protection and presentation,
- Dissemination of Knowledge Information.

Knowledge management is the way to keep knowledge growing through sharing and such sharing is best done either in material or human terms.

Karl Sveiby [1] defined KM as, "the art of creating value from an organization's intangible assets." Davenport and Prusak [2] defined KM as "KM is concerned with the exploitation and development of the knowledge assets of an organization with a view to furthering the knowledge objectives." Despres Charles and Chauvel Daniele [3] defined KM as, the purpose of knowledge management is to enhance organizational performance by explicitly designing and implementing tools, processes, systems, structures, and cultures to improve the creation, sharing, and use of different types of knowledge that are critical for decision-making.

3 Types of Knowledge

Knowledge is classified into two types.

- Explicit knowledge
- Tacit knowledge

Explicit knowledge: It is formal and easy to communicate to others. It is the knowledge of rationality. That is, policies, rules, specifications, and formulae. It is also known as declarative knowledge.

Tacit knowledge: It is complex form of knowledge. It has two dimensions namely technical and cognitive. This is personal knowledge, which is in human mind and difficult to formalize and also difficult to communicate.

4 Evolution of KM

As early as 1965, Peter Drucker [4] already pointed out that "knowledge" would replace land, labor, capital, machines, etc., to become the chief source of production. His foresight did not get much attention back then. It was not until 1991 when Ikujiro Nonaka [5] raised the concept of "tacit" knowledge and "explicit" knowledge as well as the theory of "spiral of knowledge" in the *Harvard Business Review* that the time of "knowledge-based competition" finally came. In his latest book, Building Organizational Intelligence: a Knowledge Management Primer, Jay Liebowitz [6] stated: "In today's movement toward knowledge management, organizations are trying to best leverage their knowledge internally in the organization and externally to their customers and stakeholders. They are trying to capitalize on their organizational intelligence to maintain their competitive edge." The thrust of knowledge management is to create a process of valuing the organization's intangible assets in order to best leverage knowledge internally and externally. Knowledge management, therefore, deals with creating, securing, capturing, coordinating, combining, retrieving, and distributing knowledge. The idea is to create a knowledge sharing environment whereby sharing knowledge is power as opposed to the old saying that, knowledge is power.

5 KM Operations in Academic Libraries

Knowledge management in libraries refers to the management of the production, diffusion, and transfer of knowledge as well as of the network system constructed by related institutions and organizations. It includes three aspects:

- Theoretical innovation management of knowledge,
- Technical innovation management,
- Organizational innovation management.

Theoretical innovations management is to enrich and enlarge the theoretical and practical research fields of library and information science through pursuing the latest development trends in library science world over. Technical innovation management is to manage the network system constructed by institutions and organizations that relate to the full course of technical innovation. In their evolution from conventional libraries to electronic/digital libraries, libraries should make technical breakthrough and progress and build up technical facilities to support knowledge management. Organizational innovation management is to create a set of effective organizational management systems adaptable to the requirements in the electronic library era, to support and strengthen the knowledge management activities, by optimizing the functional departments and operation procedures of libraries (Fig. 1).

Knowledge dissemination is of equal importance as compared to knowledge innovation. Knowledge creators do not have much time to look for knowledge users. Although there are a multitude of knowledge users, it is very difficult to acquire knowledge that already exists in the minds of knowledge creators as restricted by various objective and subjective conditions. Therefore, libraries may play the part of knowledge tosser, use diverse media and channels to disseminate various new knowledges. The Internet, with its mass information approach and extensive contents, will provide people with the main approach to searching knowledge and acquiring information.

6 KM in Libraries

An effective knowledge management program is a long-term project and requires significant commitment from the organization. How to manage knowledge will become an important subject facing libraries in new future. Knowledge management in libraries should be focused on effective research and development of knowledge, creation of knowledge bases, exchange and sharing of knowledge between library staffs (including its users), training of library staff, speeding up explicit processing of the implicit knowledge and sharing up explicit processing of the implicit knowledge, and realizing of its sharing. With the help of the knowledge management processes, libraries convert data and information stored in various sources into knowledge and deliver only relevant knowledge to users. Knowledge management within libraries involves organizing and providing access to intangible resources that help librarians and administrators carry out their tasks more effectively and efficiently. Knowledge management in libraries is the combination of different processes such as acquisition of knowledge from different sources (print, electronic and human) and classification, storing, indexing, and

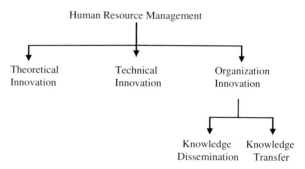

Fig. 1 Knowledge innovation management

dissemination of that knowledge using people, process, and technology in such a way by which library could fulfill the mission of the parent organization in term of users' satisfaction. It is to promote relationship in and between libraries, between library and user, to strengthen knowledge internetworking and to quicken knowledge flow. The libraries are moving from collection management to knowledge management and digital technologies offering new information services and products. The application of information technology (IT) enlarges the scope of knowledge acquisition, raises knowledge acquisition speed and reduces knowledge acquisition cost. IT is indispensable in the application and exchange of knowledge and other fields. It functions as a source and tool for knowledge innovation.

7 Librarians as Knowledge Managers

Librarians as successful knowledge manager should have the following qualities:

- Librarians should have theoretical and practical experiences in designing and implementing information system.
- Knowledge about library's information sources or assets, products, and services.
- Understanding the needs of the users including faculties and researchers and sharing the information.
- Knowledge about the emerging library trends and technologies and their application in libraries.
- Knowledge about the digital library concepts, current and new advances in the scholarly communication systems, electronic publishing, and creating knowledge databases.
- Having knowledge about library's own competencies and capabilities.

8 Conclusion

Knowledge management requires a holistic and a multidisciplinary approach to management processes and an understanding of the dimensions of knowledge work. It is an evolution of good management practice sensibly and purposely applied. Knowledge management occupies a very outstanding position in the creation of the knowledge innovation system. Libraries in a knowledge-based society should develop their own knowledge management systems. An efficient knowledge management system is one that will enable libraries to store information sources manually or electronically and facilitate the process of retrieving, sharing, tracking, and distributing these information sources efficiently with their users. Knowledge management and the sharing of knowledge can help libraries with the improvement of the quality of their services as well as the creation and maintenance of a learning culture.

References

1. Karl S (2003) The facts about knowledge. In: Suliman AH (ed) Knowledge management: cultivating knowledge professionals. Chandos Publishing, Oxford
2. Davenport TH, Prusak L (1998) Information ecology: mastering the information and knowledge environment. Oxford Univ Press, New York
3. Charles D, Daniele C (2001) In knowledge horizons: the present and the promise of knowledge management, 2nd ed. Butterworth-Heinemann, Boston
4. Drucker P (1999) Managing oneself. Harvard Bus Rev 77:65–74
5. Nonaka I, Takeuchi H (1995) The knowledge-creating company: how japanese companies create the dynamics of innovation. Oxford Univ Press, New York
6. Liebowitz J (2000) Building organizational intelligence: a knowledge management primer. CRC Press, Boca Raton, p 1

Author Index

A
Abhinav, T., 305
Agarwal, P., 47
Akresh, M. S., 517
Alagappan, M., 611
Albert Rabara, S., 759
Ally Sornam, S., 785
Alwarsamy, A., 305
Annamalai, K., 55
Anthony Xavior, M., 241, 343, 391
Aravind Kumar, D., 421
Arora, P., 81
Arun Gnanaraj, A., 759
Arun, M., 421
Arun Prakash, G., 297
Arun Premnath, A., 305
Arunkumar, S., 619, 637

B
Bak, K. M., 231
Barath, P., 217
Baskar, A., 391
Behera, D. K., 373, 383
Bhavani, K., 647
Bhuvaneswari, M., 603
Boopathi Raja, V., 3

C
Chandrasekaran, K., 261
Chauhan, P., 705
Chitravel, T., 627

D
Darshan Kumar, J., 421

D
Deepanraj, B., 285
Deepthi, S., 421
Deshmukh, S. K., 655
Dhanalakshmi, G., 505
Dhinakarraj, C. K., 285
Durairaj, R. B., 351
Durga Prasad, P. V., 33

E
Elamaran, T., 93, 111
Evangeline, A., 297

F
Felicita, J., 759

G
Ghoshal, A. K., 729
Gnana Jayanthi, J., 759
Goyal, V., 81
Gupta, A. V. S. S. K. S., 33

H
Harish, G., 325
Hari Vignesh, J., 273

I
Indira, K., 627
Ingle, S., 123

J
Janarthanan, B., 579

S. Sathiyamoorthy et al. (eds.), *Emerging Trends in Science, Engineering and Technology*, Lecture Notes in Mechanical Engineering, DOI: 10.1007/978-81-322-1007-8, © Springer India 2012

J

Jeevarenuka, K., 487
Jerome Das, S., 647
Joseph Stalin, M., 217
Joshua Amarnath, D., 551

K

Kalaichelvan, K., 231
Kalaivanan, K., 185
Kanagaraj, G., 207
Kanchana, P. Na., 767
Kandaswamy, A., 611
Kannan, T. T. M., 157
Karthikeyan, S., 561
Katiyar, K., 683
Kawamura, J., 571
Kesavan, R., 429
Kesavaraj, G., 745
Krishna, P. V., 197
Kulasekharan, N., 477
Kumaresh, S. K., 421

L

Lakshmi, S., 665
Lim, W. C. E., 207

M

Mahalakshmi, K., 785
Maheshvaran, 595
Manikandan, 459
Manimozhi, P., 759
Maria Tresita Paul, V., 775
Marimuthu, K., 595, 619, 637
Marimuthu, P., 157
Marimuthu, P., 261
Mary Latha, K. F., 673
Mathana Krishnan, S., 217
Mohan, B., 409
Mohankumar, S., 497
Mukherjee, T., 729

N

Nagarajan, D., 497
Nagarhalli, M., 123
Nandedkar, V., 123
Narayana Reddy, T., 737
Natarajan, E., 447
Nayak, P., 473
Nirmala, J., 505

P

Padmanaban, K. P., 19, 63, 185
Padmini, R., 197
Palanikumar, K., 325
Panneer, R., 361
Parakh, B., 197
Parammasivam, K. M., 165
Parthiban, D., 611
Pathak, P. K., 541
Pattnaik, M., 745
Paulraj, G., 297
Pillai, N., 459
Ponnambalam, S. G., 207
Prabaharan, S. R. S., 571
Pradeep Kumar, A. R., 55
Prasad, B., 683
Premkartikkumar, S. R., 55
Prerna, C., 541
Prithiviraj, G., 775
Pugazhenthi, R., 343

R

Raja, K., 261
Raja, K., 697
Rajaraman, A., 505
Rajeswari, N., 571
Rajkumar, D., 251
Rajkumar, M., 717
Rajmohan, T., 325
Ramachandran, T., 19
Rameswari, M., 689
Ranjithkumar, P., 251
Rao, S., 523

S

Sabareeswaran, M., 185
Saha, K. N., 473
Said, A. E., 517
Sankaranarayanan, G., 285
Sankaranarayanan, K., 647
Sankaran Pillai, G., 487
Santhosh, M., 63
Saravana Kumar, A., 273
Saravana Prabhu, R., 273
Saravanakumar, D., 409
Sarkar, A., 383
Satheeshkumar, G., 487
Sathish, S., 551
Sathiyanarayanan, C., 251
Sathya, R. I., 561

Author Index

Selvakumar, K., 165
Selvasekarapandian, S., 571
Sendhilnathan, S., 603
Sengar, K. K. S., 399
Senthilkumar, A. N., 697
Senthil Kumar, N., 285
Senthilvelan, T., 333
Shahul Hameed, P., 487
Shankar Ganesh, N., 93, 111
Shankar, R., 47, 103
Shanker, J., 351
Shanmugam, V., 3
Shanmugasundaram, K., 579
Shukla, A. C., 399
Shyamaprasad, R., 273
Silpa, G., 737
Singh, D., 541
Singh, Y., 705
Sivakumar, K., 447
Sivakumar, R., 145
Sivaramakrishnan, V., 145, 361
Sivasankar, M., 351
Srikant, R. R., 197
Srinivas, T., 47, 103
Sriram, G., 333
Sriram, R., 665
Srivastava, A. K., 541

Sugapriya, S., 665
Sundaram, Y. S., 497
Sunder Selwyn, T., 429
Sureshkumar, 459

T
Tharini, K., 697

U
Umashankaran, M., 273
Uthayakumar, R., 673, 689, 717

V
Vasundara, M., 185
Vetrivel, M., 333
Vinayagamoorthy, J., 19
Vinayagamoorthy, R., 241, 317
Vinoth Kumar, P., 351
Vinoth Raj, K., 93, 111, 131

X
Xavior, M. A., 317

Printed by Publishers' Graphics LLC